From My Vast Repertoire ...

Guido Altarelli's Legacy

From My Vast Repertoire ...

Guido Altarelli's Legacy

Editors

Stefano Forte
Università di Milano, Italy

Aharon Levy
Tel Aviv University, Israel

Giovanni Ridolfi
Università di Genova, Italy

World Scientific

NEW JERSEY • LONDON • SINGAPORE • BEIJING • SHANGHAI • HONG KONG • TAIPEI • CHENNAI • TOKYO

Published by

World Scientific Publishing Co. Pte. Ltd.
5 Toh Tuck Link, Singapore 596224
USA office: 27 Warren Street, Suite 401-402, Hackensack, NJ 07601
UK office: 57 Shelton Street, Covent Garden, London WC2H 9HE

British Library Cataloguing-in-Publication Data
A catalogue record for this book is available from the British Library.

FROM MY VAST REPERTOIRE ...
Guido Altarelli's Legacy

ISBN 978-981-3238-04-6

For any available supplementary material, please visit
https://www.worldscientific.com/worldscibooks/10.1142/10934#t=suppl

Typeset by Stallion Press
Email: enquiries@stallionpress.com

Printed in Singapore

Preface

At the 1985 Cargèse school, Guido Altarelli was invited to lecture on physics beyond the Standard Model [Altarelli (1987)]. One of us (SF), a first year graduate student, was in the audience, about to meet Guido for the first time. Guido had already acquired semi-mythical status. His first lecture, delivered with Guido's trademark flair, only heightened expectation for the sequel, which was supposed to come on the afternoon of day two. However, on the subsequent day, to everbody's surprise, at the first morning lecture Guido was already standing at the lectern. "Professor Wess" said Guido "is indisposed and cannot lecture today. So the organisers have asked me to stand in for him. After careful consideration, I decided to pull out from my vast repertoire of slides a couple of lectures on the status of QCD [Altarelli (1985)]". His face then opened in the broad grin that we got to know well in the years to come.

A few years after that Cargèse school, two of us (SF and GR) started a long and close collaboration with Guido Altarelli. It was a fantastic experience, and the beginning of a deep friendship. Guido's teaching, both in the strict sense of physics subjects we have learned from him, and in the broader sense of the attitude towards research, is still very present in our minds and in our conversations. Having collaborated with Guido Altarelli, and discussed physics with him for many years, is something we are very proud of.

For one of us (AL), Guido is inextricably tied to the deep inelastic scattering conference series: Guido was considered by all as the 'father' of DIS and was the first one I, as one of the founders of the DIS conference series, approached and invited to serve on the International Advisory Committee. After Guido's passing, a prize was founded to honour his memory, the Altarelli Prize, to be shared by two exceptional young scientists, one in theory and one in experiment, carrying out research in one of the subjects studied by Guido.

This book is meant as a tribute to Guido's vast accomplishments. It is a collection of original contributions, whose authors have been close to Guido, or have been influenced by him. Our main criterion in deciding on the authors, content, and style guidelines was to arrive at a book that, we hope, Guido would have enjoyed reading.

Stefano Forte, Aharon Levy and Giovanni Ridolfi

Contents

Chapter 1

A Giant of particle physics

Fabiola Gianotti
CERN Director General
1211 Geneva 23, Switzerland

Guido Altarelli is a giant of particle physics, and I use the present tense advisedly, as his outstanding contributions to the discipline make his marks everlasting and as relevant today as ever. The breadth and depth of his scientific output are evidenced by the range of topics covered by the many distinguished physicists writing in this volume. *From my vast repertoire* is indeed a fitting title, since Guido's work ranged from QCD to deep inelastic scattering and neutrino physics, and from HERA to LEP, the LHC and beyond. There were very few areas of particle physics that did not interest him, or to which he did not contribute.

Guido's scientific life undoubtedly coincided with an extremely vibrant era for the field. He lived and worked through the construction and consolidation of the Standard Model and the development of new theories addressing the Standard Model's inadequacies. *The excessive success of the Standard Model*, as he used to say, was a puzzle that accompanied him through the last years of his life.

Throughout his long career, Guido played a prominent role in bridging theory and experiment. A luminous example is his pioneering work in QCD in the 1970s, when he applied the newly minted theory to address a range of longstanding issues, notably through the equations he developed with Giorgio Parisi while they were both working in Paris, which describe the evolution of quark and gluon distributions inside hadrons. The Altarelli–Parisi equations have become a key tool in interpreting particle physics data. It's safe to say that there isn't a particle physics student anywhere who hasn't come across the DGLAP (Dokshitzer–Gribov–Lipatov–Altarelli–Parisi) equations and included a reference to them in their theses.

Guido's work was crowned with success in the form of a string of prestigious prizes, starting with the Julius Wess Prize in 2001 for "theoretical concepts in Quantum Field Theory and their connection to experiment, in particular his research on the consequences of supersymmetric models on physics at high energies". This was followed in 2012 by the award of the American Physical Society's J. J. Sakurai prize for "key ideas leading to the detailed confirmation of the Standard Model of particle physics, enabling high-energy experiments to extract precise information about quantum chromodynamics, electroweak interactions and possible new physics," and in 2015 by the European Physical Society's High Energy and Particle Physics Prize for "developing a probabilistic field theory framework for the dynamics of quarks and gluons, enabling a quantitative understanding of high-energy collisions involving hadrons". In all of these citations, the common thread is clear: Guido's theoretical work has been invaluable to the interpretation of experimental data.

Guido was a huge asset to CERN, where he spent much of his career, heading the Theory Division from 2000 to 2004. He was a prominent figure in the design of the LHC programme, participating from the very start in the first workshop on the feasibility of a proton-proton collider in the LEP tunnel, held in Lausanne in 1984. Following that meeting, he went on to provide a steady stream of guidance to the experimental programme as it evolved. Today, we still rely very much on Guido's contributions to the three main pillars of the CERN programme: full exploitation of the LHC, maintaining a diverse scientific programme, and preparing for a post-LHC high-energy accelerator.

Guido was everything a top physicist and a great human being should be. He was an inspiration to generations of scientists and a very generous person. He loved talking physics and he was a great teacher and communicator. His scientific stature, lucidity and clarity of expression, his intellectual integrity and innate charisma made him a universally acknowledged authority in the worldwide particle physics community, someone whose voice and opinion carried real weight. He was also something of an iconoclast, as the greatest scientists of our time have always been. While fully engaged in the hunt for new physics, he was always ready to dismantle overly-optimistic claims inevitably with the sharp sense of humour well known to his friends. On a more personal note, I miss running into him at CERN in the mornings. With a newspaper under his arm, he would slowly bend his head towards me, in that unique, avuncular style, and with his ceremonial tone would say: "Buongiorno, cara Fabiola!". There was no better way to start my day.

Guido's scientific and human legacy will be with us forever, and I invite you to dive into the pages of this volume and relish his exceptional contributions to physics as told through the words of his friends and colleagues.

Chapter 2

Guido, my dearest friend

Giorgio Capon

Guido was my dearest friend. Our friendship lasted more than 60 years.

Our families knew each other before we did; our fathers had met and become close friends during their military service. His father was best man at my wedding as I was for Guido and Monica 20 years after.

I first met Guido when still a child but our friendship actually began when we were teenagers (we were both born in Rome in 1941). In Rome we lived rather close; a 15 minute walk separated our houses.

We both attended the classical lyceum where Latin and Greek had equal or more importance than mathematics and physics, but in different institutes: I went to "Torquato Tasso" and Guido to "Giulio Cesare", and this was often an argument between us, each one inviting the other one to join their own school.

In those times we met nearly every day. In his home we spent the afternoon playing chess, monopoly and sometimes poker; Guido and I were mostly alone: his younger brother Massimo was often pushed away rather roughly.

In my home we had fierce ping-pong competitions in the garden with a group of other friends. Guido and the others didn't use to ring the bell; instead they entered directly in the garden by climbing over the surrounding wall and being welcomed more or less in a friendly manner by my wolfhound.

Often at the end of the day one of us accompanied the other back home, chatting about matters of the day and more serious issues. I remember one of his favourite quotations: "fortunate is the man who can live satisfying his manias" from Bertrand Russell.

When he was about twenty he read Freud's "The Interpretation of Dreams": then for a month he asked me about my night dreams trying to decipher them. I think that he also kept for some time a diary of his own dreams.

When I was 17 my father died suddenly and my friendship with Guido became even stronger: I was an only child, therefore he became like a brother for me.

We both did perform well at school in physics and mathematics but — unlike other more precocious colleagues — we did not go beyond the simple school texts of the classical lyceum.

At that time a science magazine "SAPERE" used to propose to its readers geometry or mathematical problems that were not simple but doable for us; we loved to search for solutions, in some cases we succeeded and we won physics books. At that time he greatly appreciated the book "The great mathematicians" by E. T. Bell and gave me a copy as a gift.

During the high school years Guido was not a particularly zealous student, he used to get good marks without much effort; he led the normal life of a boy his age: football, parties, flirts, not without transgressions: to use his father's car without permission, once he smuggled a copy of the car keys. He loved to play and watch football; I still remember the 1970 summer night when we watched on TV the famous match where Italy beat Germany 4 to 3.

Guido's mother, who came from a noble Sicilian family, had a strong personality and gave her sons a sound and rigorous education. Memorable was the day when she recalled Guido back home because he had gone out without combing his hair; Guido unwillingly obeyed to avoid further discussions. However Guido, notwithstanding his young age rebellions, cared to transfer the principles of such an education to his own children.

We were interested in physics but engineering did seem to us a safer profession. So at La Sapienza University of Rome, we started with engineering with the idea of a possible move to physics; at that time the first biennium courses were almost all in common.

We attended the university lectures together and sat generally side by side, but we did study each on their own. Among our lecturers of the first biennium I remember Amaldi, Salvini, Fichera (math analysis), Cattaneo (mechanics).

We discovered various errors in a textbook of physics exercises. We sent an errata-corrige to the authors, which was included in the subsequent edition with the authors' acknowledgments.

Afterwards in our scientific activity, given our different choices, theory and experiments respectively, we never published a paper together; only much later when Guido was invited to take care of an edition of Fermi's papers on statistical mechanics, he kindly invited me to join him in the task.

We took together the first university examinations. At the chemistry exam we were caught conferring between each other and with a third colleague; that was noted on our papers, but there was no consequence on the final mark.

At an analytical geometry test, we tried to communicate to each other the solution of the problem which ended with the equation of a hyperboloid: we obtained the same 2nd order terms but not the first order ones. So we were very worried that one of us was wrong; instead the cunning examiners had set slightly different problems according to the student locations in the classroom and we both were right.

We received very good marks in the first biennium; a colleague said: "they form a strong pair those two: Altarelli and the mute guy"; referring to me as being always rather taciturn. So at the end of first biennium we decided to switch to physics which appeared to us much more interesting than engineering.

In the second biennium we had Persico, Cini, Careri and Carlo Bernardini as main lecturers. After the degree we followed a beautiful course on statistical mechanics by Bruno Touschek.

The third year laboratory practicals were undertaken by teams of three students: so Guido, Maiani and I formed a team. Maiani was the most active of the three. At a practice on vacuum instruments we were repeatedly told to take care not to let the mercury enter in the rotary vane pump. But during the work we lost control of the various pipes and notwithstanding our frantic efforts we could not avoid the mercury eventually entering in the pump. So we had to switch off the whole apparatus and confess to the supervisor our ineptitude.

At the second biennium Guido opted without hesitation for a thesis in theoretical physics while I chose an experimental thesis at the Frascati Laboratories with Giordano Diambrini (Maiani also started his career with an experimental thesis in electronics).

At university Guido was hard working and competitive; in his family he had several examples of personal success. His father was a self-made high ranking civil servant while his mother's brothers were two brilliant professionals. Sometimes his father did appear on TV (in those years this was a family event); then Guido used to say with some compliance: "this evening I have to watch my father on TV".

He worked hard to respect the exams' tight schedule and to get the best marks. Once when he was given an exam mark below his expectations he was very upset and left the institute kicking his lined notebook.

Guido was ambitious but at the beginning he was not fully sure of his capabilities; once he said "I have just the threshold intelligence for this job". In fact at that time there were in our course many brilliant colleagues: Doplicher, Gallavotti, Maiani, Buccella, Greco, DeGasperis, Pacini, Cerdonio, Fontanesi etc.

Moreover we were pressed by the exuberant Preparata, who was one year younger. I remember Giuliano wandering in the institute corridors saying "have you read Herman–Weyl, how can you study Quantum–Mechanics without it...?".

During the university years we used to play tennis with other colleagues such as Mario Greco, Toni DeGasperis, Mario Spinetti, Luciano Maiani.

We also took holidays together. Once he was at my house at the seaside in Riccione; we were rather busy revising for an exam together, but exposed to frequent exhortations from other friends to join and be more prone to the amusements of the site.

The next summer we were in the mountains near Cortina; there we both courted a young French girl, and Guido's more brilliant and extroverted personality prevailed.

After our degree we went for a two week holiday to London with his brother Massimo and another friend. We lived in a small apartment and carefully visited monuments, museums and surroundings; there we also discovered and appreciated

Riccione (Italy), summer 1960

Riccione (Italy), summer 1960 with Giorgio Capon

the rich English breakfast. However Guido was not willing to behave as an ordinary tourist; he preached that we should lead in London the same life that we were used to in Rome.

Once we went to visit a famous country-house close to London, Hatfield House, but when we reached the site the appearance of the building was below his expectations and he refused to go in. So he stubbornly remained outside for almost two hours while I and the other two visited the very rich interiors and the beautiful gardens on the rear.

Guido's family owned a nice house with a large garden at the seaside near Rome (Fregene); there I spent some very pleasant days, initially just with Guido and later on with our wives and children.

Guido worked for his thesis with Gatto together with Buccella. Gatto, who often did not remember Buccella's name used to call him "l'altro laureando" i.e. "the other student". For the calculations needed for the thesis — on the single bremsstrahlung in e^+e^- collisions — he had to use the IBM machine of the Frascati Laboratories' computing center.

At that time there were no video-terminals and the programs were charged by using punched cards. Not only that, but the Fortran compiler was also on punched cards and had to be input before the user program; the output was sorted out on teletype. Often he told me how he had to struggle with all this machinery and with strongly varying functions to integrate. He had to choose carefully the integration steps to get reasonable results in reasonable time: he liked to mention this as "changing the gear" of the computer.

Cortina (Italy), summer 1961

After the degree, which he obtained in November 1963, Guido went to Florence, where he remained until 1968, to work with Gatto who had gathered around himself a group of brilliant young people as Maiani, Buccella, Preparata, Veneziano, Chiuderi, Ademollo forming the best school of theoretical physics in Italy.

Gatto was very appreciated by his pupils since he was able to quickly lead them to compete and publish at the physics frontiers, however he had a peculiar personality, shy and uneasy in human relationships, and a characteristic way of speaking. I heard several times Guido's imitations of Gatto's voice and tales of funny episodes in Florence.

When he left Rome for Florence our roads began to diverge and we did no longer meet as frequently as before. However we never lost contact and, whenever he was back in Rome, we resumed as usual our meetings, now also including our families.

He married in 1966 and I in 1968; so also our wives, Rita and Teresa, became involved in our friendship. From his first marriage he had two children: Claudia (1968) and Fabrizio (1973). He was a careful and demanding father regarding the

Cortina (Italy), summer 1961 with friends

education and the behaviour of his children; sometimes in our conversations he expressed concern when they did not follow the life path he had imagined for them.

We met at the Erice school in 1967 and 1968; there Guido and Buccella once got the "best student of the course" prize. I also remember a scene at the beach below Erice: Sidney Coleman cautiously plunging in the sea waters while Glashow was warning him "be careful of the algebra of the currents". We also attended in 1967 the Heidelberg Conference on Elementary Particles; on this occasion, I was offered through Maiani a contract to work with Ting in Hamburg, but I was reluctant to leave my routine in Rome. I think that Guido never forgave me for not accepting it; he reproached me for "throwing in the waste basket a unique opportunity".

Soon after leaving Florence he spent two years in New York with all his family, 1968–1970, first at the New York University then at the Rockefeller University.

In 1971 with him, my wife, Mario Greco and other colleagues we made a long trip in the USA on the occasion of the Cornell conference. We visited Chicago and the Fermi Lab. On one evening the whole group was welcomed at the house

London, summer 1964 with Giorgio Capon and brother Massimo

of my aunt Laura Fermi who lived in Chicago. Then we went to Washington and San Francisco.

For one year, 1976/77, he was in Paris, at the Ecole Normale Superieure, alone this time. There he wrote the famous Altarelli–Parisi paper. My wife and I visited him at his comfortable apartment in Bvd Saint Marcel. I remember a dinner in a restaurant near Les Halles: he was suffering from gastritis and was appalled at the prospect of leaving the day after for Germany (Bielefeld?) where he was invited for a seminar. Notwithstanding this inconvenience, given his high sense of responsibility, he went anyway.

In 1980 his achievements were recognised and he won the chair of theoretical physics at Rome La Sapienza. This was a long awaited goal of his and a well deserved reward for years of intense work.

However, soon after that his marriage went in crisis; we were very close witnesses of the events, as he told us about the evolution of the situation; he was very worried about the impact of the separation on his children, especially on Fabrizio who was still a child.

In 1983 he had his first contract with CERN (afterwards in 1987 he was awarded a permanent position). At that time I also was there with my family with a fellowship; there I met Monica for the first time; they had met a few weeks before at the Erice school. They first lived at Thoiry and afterwards bought the house at Sergy–Haut. They married in October 1988; I signed the paperwork as wedding witness in the small municipality of Sergy; then a lavish reception followed at Divonne castle. It was a radiant autumn day and both Guido and Monica were really very happy.

While at CERN he continued to keep the chair of theoretical physics at La Sapienza University of Rome but in 1992 he transferred his chair to Roma Tre. On the first year, given the recent foundation of the university, he had only one student for his course. This student had the privilege of individual lessons in theoretical physics from Guido who came expressly for him from Geneva!

When at CERN for my work in the ALEPH experiment, I often met Guido at the CERN cafeteria for lunch or at seminars and I often consulted him about the physics implications of the ALEPH results. I was also regularly invited by Guido and Monica at their Sergy house, where from year to year I saw Giulia (1990) and Marco (1994) grow up.

We also met at various international physics conferences; I remember the one in Bari in 1985, where amid various episodes of pickpocketing at the expenses of the participants, the sessions were also characterised by heated debates between Guido and Preparata about the validity of QCD.

In these years Guido was frequently in Rome for his university lectures, as well as conferences and visits to his family. For the Christmas holidays he also came

Guido with first wife Rita in Baden-Baden 1967

with Monica and their children; in these occasions we met either at home or for dinner at a restaurant. I remember Guido's tall figure arriving at our home for a drink and greeting us "carissimo Giorgio, carissima Teresa" with a bit of emphasis and then asking "where are we going to eat this evening?".

In 2004 we met in Milano for an Ukiyo-e exhibition; in 2008 we were at the party for Parisi's 60th birthday in the Roman countryside. In September 2010 his son Fabrizio married at a venue in Bracciano, near Rome. It was a beautiful day and a very pleasant gathering of his families, friends and colleagues. Guido was very happy; both for his son's choice and for seeing him, at 37, finally settling down to start a family.

In July 2011 I sent him by e-mail my best wishes for his 70th birthday and asked "what is your impression of reaching 70 years?" and he replied "a bad impression". Unfortunately he was prophetic: a few weeks later his illness was discovered.

I was among the very first to know. He told me in September 2011 at a meeting in Rome (Maiani's 70 years) with his typical indirect style: "the holidays went well notwithstanding the bad news about my health" (these were his exact words). He had prostate cancer; often these cancers are curable or have a very slow progression, but he told me that in his case it was "a fetid one".

For the first few years he continued to lead an apparently normal life: a journey to China with Marco as a prize for having passed his bac, trips to Rome to visit his daughter Claudia and the grandchildren, to give talks or seminars, or family matters (he also took care of his old and disabled uncle).

Guido with Teresa Capon at Giorgio Parisi 60° anniversary

When together, we did not ask about his health, but from time to time it was Guido himself who told us about his status. He carried the burden of the illness with coolness (in fact many people did not realise he was ill) and courage without losing heart or cultivating false illusions.

In the last years a frequent topic of our talks was the search for a new house in Switzerland. He was very fond of the Sergy house but Switzerland seemed to offer better opportunities to the children. Eventually they found a piece of land at Grand Saconnex where they would build the new house. His daughter Giulia — she is an architect — and a colleague of hers took care of the project.

He guided us to visit the building site; we followed the construction progress and were also involved in many related discussions; in fact he was not fully satisfied with some aesthetical solutions backed by Monica and Giulia.

2015 started with a complicated and tiring move from Sergy to Grand Saconnex; then he had to increase the therapy but he was still in a reasonably good state. "However, medical tools are running out," he warned me. Notwithstanding that he was able to travel to Greece, Normandie, Cogne and finally in July to Vienna to receive the EPS prize.

We met again in Geneva in August, he invited us to visit the completed new house, another time he came with Monica to our home in Ornex for dinner and before we left we had a dinner with friends at the Auberge de Satigny: that was the last time I saw him.

In late August his state worsened. In September we had long telephone conversations: he was melancholic, we chatted about physics and colleagues, about his children. He was always very concerned about them, proud of their achievements and worried about their problems or difficulties. In those days friends and relatives were guests at his house; "there is a continuous pilgrimage" he said. But he was weak: "I walk 100 metres and I am dead". When alone he spent time reading and listening to music "but I lack the drive" he told me.

On the evening of September 30th I received a phone call from Mario Greco announcing his sudden death. So I lost the friend and the person to whom, except my family, I was most attached and that had had such a relevant role in my life.

At the sober, moving ceremony for his funeral, the small church of Grand Saconnex was full of friends and colleagues from CERN, Rome and elsewhere. After his death Monica showed me hundreds of messages and letters, many not simply formal but proving his popularity and a sincere widespread regret for his death.

He now rests in Rome's cemetery; my parents' grave is close by; so when I go there I stop by for a brief pause and to deposit a flower.

Guido had a rich and complete personality: he had many interests beyond physics. He loved classical music; he had a large collection of CDs and often when coming to his house he showed me the more recent acquisitions; we also exchanged music records as Christmas gifts. He was also interested in fine arts, liked to visit museums and exhibitions, when abroad he took the chance to buy art objects and had a collection of Japanese prints.

He liked to travel, he accepted willingly some of the many invitations for talks or seminars. He was a much appreciated speaker for the clarity and the verve of his talks. He visited a lot of countries and when travelling, he cared to learn about the site (I remember in depth studies before a trip in Rajasthan, a vivid exhaustive description of Isfahan when he went in Iran).

His reading interests spanned a wide range of essays, novels and biographies: among the latest ones he loved the Dirac biography by Farmelo and warmly recommended it to me, so I bought the book and appreciated a thorough and accurate piece of work.

He loved skiing; Monica is an enthusiastic and expert skier; he gladly managed to follow her as far as possible. He also loved the Greek seaside, the islands close to Sicily; he enjoyed and valued the better moments of life.

He was generous, convivial and an exquisite host; when we were invited at his home he prepared personally "la pasta" or some other courses and he himself served us. In fact he did learn to cook and often we met him shopping at the Ferney market.

He had a brilliant conversation, full of irony and humour. He liked to distort words, for example "I took a Caravelle plane" became "I took a Caravaggio plane". As a matter of fact our conversation was mostly unidirectional; Guido spoke and I liked to listen to him.

At times his irony could become sarcasm and derision; I remember that once he mentioned two distinguished colleagues of mine as "X and Y seem two funeral directors". When irritated he could become rough; I saw sometimes this happening. However, later in his life, after his scientific success and the marriage with Monica he softened considerably these features of his personality.

I admired his bright intelligence, his rational and equilibrated character. He was not interested in abstract speculation, in physics as well he usually had a rather conservative attitude. For that and for his methodical approach to scientific issues once at a dinner he was dubbed by a renowned colleague as "il ragioniere (i.e. the accountant) Altarelli". Once he told me that what he mostly liked in his work was to do analytical computations; equally he liked also to play with the Mathematica package.

He was not one of those who are uncritically enthusiastic about particle physics; once when new experimental data and new ideas were lacking he told me "this physics is boring".

He was ironic and skeptical about Italian politicians and parties; he wished for an efficient and meritocratic society rather than an egalitarian one; he never embraced leftist views; he was always very critical about the influence of the trade unions in Italian schools and universities.

He was honest and rigorous and these qualities he expected in others; he despised sloppiness and mediocrity not to mention trickery.

A few days after his death, I noticed by chance the following verses on the Byron monument in Rome:

But I have lived, and have not lived in vain:
My mind may lose its force, my blood its fire,
And my frame perish even in conquering pain;
But there is that within me which shall tire
Torture and Time, and breathe when I expire;

Byron (Childe Harold; Canto iv. Stanzas 134–137.)

Chapter 3

Guido Altarelli: the person, the scientist

Guido Martinelli
CERN, Theoretical Physics Department
PH-TH, CERN, CH-1211, Geneva 23, Switzerland
and Physics Department and INFN Sezione di Roma La Sapienza
Piazzale Aldo Moro 5, 00185 Roma, Italy

3.1 Introduction

Guido Altarelli, a great physicist, an exceptional teacher, a man of the highest human profile, for many of us a dear friend. These notes are a tribute to the scientist and to the person.

On September 30th, 2015, Guido Altarelli, one of the protagonists of the development of the Standard Model and one of the most representative figures of the Italian and international physics community of the last fifty years, passed away in Geneva. The scientific contributions of this great scientist range over almost all the most important topics of the physics of the fundamental forces and constituents of nature, from strong and weak interactions to the properties of neutrinos, from theories beyond the Standard Model to precision measurements in particle physics.

Guido Altarelli has been a major figure among the scientists who led to the full development and to the understanding of the phenomenology and of the limitations of the Standard Model. In all the subjects on which he worked during his career, Guido left an important legacy. The relevance of his contributions to fundamental physics is highlighted by the fact that many of the results of his studies are already part of the textbooks on which young talents are formed. In recognition of his important contributions Guido Altarelli was awarded with three very prestigious prizes: the Julius Wess Award (2011), the J. J. Sakurai Prize of the APS (2012) and the High Energy and Particle Physics Prize of the EPS (2015). These very important awards only partially represent the variety of his activity, the relevance of the contributions that he gave to the development of the Standard Model, his strong imprint and lasting influence in the field of high energy physics. Guido Altarelli will be remembered by future generations for the extraordinary scientific heritage linked to the results of his research. After Nicola Cabibbo, with Guido Altarelli the historic research group of Rome lost another reference component, one of its most lucid and penetrating minds, a scientist who also created a school of younger scientists, a scientist with his feet firmly on the ground who did not refuse some daring flight.

The long and outstanding scientific career of Guido Altarelli can be grossly divided in four periods: the first period starting from Rome as a physics student to Florence and New York as a young researcher, the Rome period as a young professor, the CERN period and the last period of his career spent between CERN and the University of Roma Tre.

3.2 The early period

Born in Rome in July 1941, Guido Altarelli studied at the Physics Institute Guglielmo Marconi of the University La Sapienza. At that time he encountered several other young students who later became his colleagues and friends: among the others Giorgio Capon, Mario Greco and Luciano Maiani, but also Franco Buccella, Sergio Doplicher, Giovanni Gallavotti, Adalberto Giazotto, Francesco Melchiorri, Piergiorgio Picozza, and Piero Spillantini. Several of them became renowned scientists in particle physics, astrophysics or mathematical physics.

In those years, around 1960, Bruno Touschek proposed the construction of a "magnetic ring" — the forerunner of the modern e^+e^- accelerators — in which to run and subsequently collide electrons and positrons. So he set up a group of physicists, mostly Italians, intent on building the accelerator that would later be called "Storage Ring", in Italian *Anello di Accumulazione*, AdA, realised in 1961, with an energy of 250 MeV. Subsequently a larger accelerator, named ADONE, able to reach a centre of mass energy of about 3 GeV was built in Frascati. Because of the developments of the new e^+e^- accelerators an important theoretical effort was devoted to investigate the physics of electron-positron annihilation. Nicola Cabibbo and Raul Gatto aimed to compute the cross section for every possible process that could be observed [Cabibbo and Gatto (1961)]. Actually Guido made an attempt for his diploma thesis with Bruno Touschek, but this did not succeed. He graduated instead in Physics at the University of Rome in 1963 with Raul Gatto, with a thesis on the radiative corrections to the electron-positron diffusion ($e^+e^- \to e^+e^-$ and $e^+e^- \to e^+e^-\gamma$). For this thesis, which as usually happens in Italy was his first research paper, he strictly collaborated with Franco Buccella, another student of the Physics Department. They had to evaluate the cross section for this process, taking into account the infrared divergences appearing in the process $e^+e^- \to e^+e^-$. These divergences cancel out only when combined with the real emission cross section. Although the results that they derived were correct, unfortunately, for very small momentum transfer, due to rounding effects in the floating point arithmetics, the result looked rather weird giving an extremely large cross section that left everybody puzzled. They were however quick enough to understand the reason of the strange result and they finally wrote a successful thesis [Altarelli and Buccella (1964)].

After the thesis Guido followed Raul Gatto in Florence together with a group of brilliant young graduates who would become the protagonists of Italian theoretical physics from the 70's, among others Franco Buccella, Giovanni Gallavotti, Luciano Maiani and Giuliano Preparata. These young physicists were called "gattini" after the name of Gatto which in Italian means "cat". During these first years, together with Gatto, they had been working on advanced subjects which were relevant at that time, like applications, mainly to the phenomenology to hadron spectroscopy, of various unitary symmetries, $SU(6)$, $U(6)\times U(6)$ and $U(12)$ [Altarelli *et al.* (1965); Ademollo *et al.* (1965); Altarelli *et al.* (1966)] and strong interaction physics, like Regge poles [Altarelli *et al.* (1968)]. Guido also continued to work on the theoretical evaluations of the cross section of different reactions in electron-positron colliding beams.

The Florence school broke up in 1967–1968 when Gatto went to Geneva and then to Padova. In 1968–70 Guido spent two academic years at NYU and Rockfeller University working mainly with Hector Rubinstein on deep inelastic reactions and Regge pole predictions coming from the resummation of ladder diagrams, on S-Matrix duality and bootstrap [Altarelli and Rubinstein (1969)]. At that time the dominant school was asserting that strong interactions could be understood using only general analyticity and self-consistency properties of the S matrix (S matrix bootstrap). This phenomenological programme culminated in the duality and in the superconvergence sum rules. This line of research reached a definitive success with the Veneziano model [Veneziano (1968)] (and a few months later with the Virasoro model [Virasoro (1969)]). It is amusing to note that these two models were the starting point of the most sophisticated existing field theory, i.e. string theory; they correspond respectively to the open and to the closed string models.

Guido worked also on the light-cone expansion with Richard Brandt and Giuliano Preparata and on muon pair production in hadron-hadron collisions [Altarelli *et al.* (1971)]. The general physics context at that time was the approximate scaling properties shown by the SLAC deep inelastic scattering data, which had motivated the Feynman's parton model as well as the study of the properties of the commutator of the electromagnetic currents on the light cone. On the other hand Leon Lederman's group had presented their results on the production of muon pairs in hadronic collisions and immediately later S. Drell and T. Yan had proposed the quark-antiquark annihilation mechanism. At Frascati in the meantime ADONE was producing the first tests of QED in the GeV region and the early results on the large multi-hadron production. The phenomenological context for the rising of the theory of strong interactions was already there. The idea of the colour interaction of quarks was put forward by Fritzsch, Gell-Mann and Leutwyler [Fritzsch *et al.* (1973)] and soon afterwards came the discovery of asymptotic freedom of the Yang–Mills theory by David Gross, Franck Wilczeck and by David Politzer [Gross and Wilczek (1973b,a); Politzer (1973, 1974)].

3.3 The Rome period and the Altarelli–Parisi equations

After the period of two years spent at the University of New York and at Rockefeller University, Guido returned to Rome and became Assistant Professor at the Physics Department of the University La Sapienza where he started a very close collaboration with Luciano Maiani and Nicola Cabibbo. In 1980 he was promoted to full Professor of Theoretical Physics and from 1985 to 1987 he served as Director of the local INFN Section. During these years he also carried out research and teaching activities in various prestigious universities, including the Ecole Normale Superieure and Boston University.

With his return to Rome, the scientific interests of Guido moved toward the parton model and later were enlarged to cover essentially all the phenomenological aspects of what we now call the Standard Model. This move was in agreement with the strong tradition of the Rome school, led by Nicola Cabibbo, of using field theory to understand particle physics, an alternative viewpoint with respect to the S matrix approach to strong interactions. This period was also characterised by a close collaboration of Guido with many members of the theoretical group of Rome: besides Nicola Cabibbo and Luciano Maiani, already mentioned before, new younger people added to the group in the early seventies, enlarging the Altarelli–Cabibbo–Maiani collaboration. Let us remember Massimo

Testa and Giorgio Parisi, Cabibbo's graduate students, Keith Ellis, Roberto Petronzio and, later, Guido Martinelli, Cabibbo's graduate student also recruited by Guido.

In the '70s and '80s the scientific interests of Guido Altarelli focussed on the applications of the emerging theory of strong interactions, QCD, to weak interactions and to parton model processes namely to deep inelastic scattering, e^+e^- annihilation into hadrons at high energy, Drell–Yan processes and to fragmentation functions.

One of the first applications of QCD to weak interactions, and one of the most important, is the paper on the octet enhancement in collaboration with Luciano Maiani. In 1969 Ken Wilson had argued that strong interactions, although respecting isospin symmetry, could renormalise differently the $\Delta I = 1/2$ and $\Delta I = 3/2$ amplitudes, $A_{1/2}$ and $A_{3/2}$. Shortly after the discovery of asymptotic freedom and the proposal of QCD as the theory of strong interactions, Guido Altarelli and Luciano Maiani showed that indeed the dynamics of strong interactions could explain the large enhancement of the $\Delta I = 1/2$ amplitude, $A_{1/2}/A_{3/2} \sim 20$ [Altarelli and Maiani (1974)]. The calculations were later extended to charm decays [Altarelli *et al.* (1975)], and the dynamics of the enhancement at the level of the anomalous dimensions confirmed by next-to-leading order calculations of the effective weak Hamiltonian both in charm [Altarelli *et al.* (1981)] and in kaon (b-hadron) decays [Buras *et al.* (1993b,a); Ciuchini *et al.* (1994)]. It took more than thirty years, however, to prove the enhancement by combining the renormalisation of the relevant operators with the calculation of the relevant matrix elements in lattice QCD [Bai *et al.* (2015)].

In the same period, in collaboration with different combinations of other members of the Rome group, Guido had a quite intense and successful research activity related to several aspects of the decay processes of heavy flavours with seminal papers ranging from the enhancement of non-leptonic decays of charmed particles previously mentioned [Altarelli *et al.* (1981)], the first NLO computation of the strong interactions corrections in non-leptonic weak decays, to bound state corrections to inclusive charm and beauty decays [Altarelli *et al.* (1982)] and non-parsonic corrections to charmed particle decays [Altarelli and Maiani (1982)].

Among all the remarkable contributions to these issues, the evolution equations of the partonic densities, obtained together with Giorgio Parisi, really stand out [Altarelli and Parisi (1977)]. These equations, that are known throughout the world as the Altarelli–Parisi equations, in the following years have had an enormous influence on the whole theoretical and phenomenological development of perturbative QCD in high-energy processes and constitute a milestone in the history of the physics of strong interactions. The Altarelli–Parisi equations were in fact the starting point for the demonstration of the factorisation theorems and the extension of the QCD improved parton model to processes that are not light-cone dominated, essentially all the most important processes relevant for modern collider phenomenology like Drell–Yan processes, jet physics, Higgs production etc.

Before the AP equations, the applications of QCD to physical processes were quite complicated at least for two reasons. Using Guido's own words:

QCD is a theory of quarks and gluons while only hadrons are observable. Moreover perturbation theory can only be applied in those particular domains of the strong interaction where approximate freedom, which is only asymptotic, can be reached. In addition, in spite of the relative simplicity of the final results, their derivation, although theoretically rigorous, is somewhat abstract and formal, being formulated in the language of renormalisation group equations for the coefficient functions of the local operators which appear in the light-cone expansion for the product of two currents.

In the paper "The early days of QCD (as seen from Rome)" [Altarelli (2012)] Guido wrote:

In 1977 the well known work on the QCD evolution equations by myself and Parisi was published. In the academic year 76-77 both of us were on sabbatical in Paris. I was at the Ecole Normale Superieure (ENS) and Giorgio at the Institut des Hautes Etudes Scientifiques at Bures-sur-Yvette. Often Giorgio preferred to stay downtown, spending some time at the ENS to discuss with the people there and, in particular, with me on QCD phenomenology, a subject of great interest for both of us at that time. Out of these regular contacts our work on the evolution equations was developed. The main virtue of our approach was to formulate the evolution of parton densities as a branching process with probabilities determined (at leading order) by the splitting functions (proportional to the running coupling). In our paper a particular emphasis was devoted to prove that the splitting functions are a property of the theory and do not depend on the process (in particular the evolution does not apply only to deep inelastic scattering). This last point is of fundamental importance because it is on this basis that it is possible to generalise the concept of universal (QCD improved) parton densities to all hadronic processes at high energy.

On the genesis of the Altarelli–Parisi equations, which later became known as the DGLAP equations (Dokshitzer Gribov Lipatov Altarelli Parisi), we can add some further reference to Guido's own point of view:

The evolution equations are now often called DGLAP equations (Dokshitzer Gribov Lipatov Altarelli Parisi). The first article by Gribov and Lipatov was published in '72 [Gribov and Lipatov (1972a)] (even before the works by Gross and Wilczek and by Politzer!) and was followed in '74 by a paper by Lipatov (these dates correspond to the publication in Russian). All these articles refer to an abelian vector theory (treated in parallel with a pseudo-scalar theory). Seen from the point of view of the evolution equations, these papers, in the context of the abelian theory, ask the right question and extract the relevant logarithmic terms from the dominant class of diagrams. But from their formal presentation the relation to real physics is somewhat hidden (in this respect the '74 paper by Lipatov makes some progress and explicitly refers to the parton model). The article by Dokshitser was exactly contemporary to ours [Dokshitzer (1977)]. It now refers to the non abelian theory (with running coupling) and the discussion is more complete and explicit than in the Gribov–Lipatov articles. But, for example, the notion of the evolution as a branching process and the independence of the kernels from the process are not emphasized. An important point is also that the Gribov–Lipatov papers were known to Dokshitzer (while not to us). Their works were in fact his starting point and are quoted among the references given in his article.

Within the formulation of the general framework given by the DGLAP equations, the Rome group gave an important contribution to the understanding of Drell–Yan processes. In '78–'79 Guido, in collaboration with Keith Ellis and Guido Martinelli, made an important progress with the first calculation of the next to the leading (NLO) corrections to Drell–Yan processes [Altarelli *et al.* (1979a)]. After the calculation of the longitudinal structure function [Altarelli and Martinelli (1978)], this was one of the first calculations of NLO corrections in QCD. An important point was the definition of the parton densities beyond the leading order. Guido and collaborators fixed the definition of the quark parton densities (PDF) $q_f(x)$ at NLO in a precise, physically motivated, way by adopting at the NLO the structure function F_2 as the defining quantity. F_2 is defined in terms of the scale-dependent quark distributions $F_2(x, Q^2) = x \sum_f e_f^2 q_f(x, Q^2)$, where $q_f(x, Q^2)$ obey the DGLAP evolution equations. For the gluon PDF the definition can be postponed at higher orders since gluons already enter at NLO in Drell–Yan processes. Within their definition, the calculation of the $O(\alpha_s)$ diagrams for both deep inelatic scattering and the Drell–Yan process allowed to derive the correction to the NLO expression of the Drell–Yan cross section, as function of the invariant mass of the lepton-pair Q^2. The resulting correction turned out to be surprisingly large, casting doubts on the convergence of perturbation theory. It is also rather constant in Q^2 and in rapidity so that they called the ratio of the leading order to NLO cross section the K-factor. Indeed the main part of the large $O(\alpha_s)$ correction could be traced back to the QCD induced behaviour of the time-like and space-like quark form factors in Drell–Yan and deep inelastic scattering processes as $|Q^2| \to \infty$ and in the infrared behaviour of the soft gluon emission near the phase space boundary ($\tau = Q^2/s \to 1$ and $x \to 1$ respectively). Both these effects can be re-summed to all orders using methods developed within the context of QED. Thus several formulae to re-sum to all orders the large corrections, thus saving the perturbative behaviour of QCD, were proposed, for example by Mario Greco and collaborators and by G. Parisi [Curci and Greco (1978, 1981); Curci *et al.* (1979a,b); Parisi (1980)].

In the '80s, with the discovery of the W's and Z^0, a quite relevant aspect of the production, detection and mass measurements of the heavy vector bosons was the theoretical prediction of their transverse momentum (q_T) distribution. Guido Altarelli was a protagonist on this subject with a series of papers which determined with increasing precision the q_T distribution for Drell–Yan processes. In collaboration with Giorgio Parisi and Roberto Petronzio, he was the author of a first calculation of the transverse momentum distribution of lepton pairs produced in hadronic collisions [Altarelli *et al.* (1978b)]. The distribution and the increasing of the average q_T with the centre of mass energy s are an important check of the hard scattering of the partons in perturbative QCD, and also provide an independent method to measure α_S. As soon as the data on the W and Z production from UA1 and UA2 at CERN proton-antiproton collider were available, a theoretical prediction which combined the result of the Sudakov re-summation to the complete one-loop calculation [Ellis *et al.* (1981a)] was finally obtained in a paper by Guido, M. Greco, K. Ellis and G. Martinelli [Altarelli *et al.* (1984)]. This is how Guido commented on this paper:

This is an important paper, because it essentially contained all the crucial ingredients that describe the physics of this phenomenon. In the subsequent years the accuracy was much improved with the computation of sub-leading effects and with several different refinements, but the essential points were all present in our paper and the accuracy of our treatment was adequate for the quality of the first data. The same techniques are at present applied to the calculation of the q_T distribution of the Higgs boson produced by gluon fusion.

3.4 The CERN period

In 1987 Guido became Senior Staff Physicist at the Theory Division of CERN (where he remained up to the age of sixty-five in 2006). For a period of four years, 2000–2004, he was the Theory Division Leader. In his capacity he did everything to the best, always amiable and available to all members of the Division, always ready to discuss physics and help anyone who needed help. A real person, elegant in manners, a great gentleman, an example for all.

During the CERN years his scientific activity turned to the theoretical predictions concerning the experiments at the Hadronic Colliders and LEP at CERN in close collaboration with the major experimental groups. His interaction with CERN's environment strongly reflected his choice of research subjects. He coordinated several working groups and gave crucial contributions to the understanding of the physics of the colliders. The planning of experiments in CERN was strongly influenced by his theoretical analysis. His work has also been crucial in putting some order into the confrontation of the huge mass of experimental data with a plethora of theories and in identifying crucial experiments or theoretical analyses that could be done to remove the apparent inconsistencies, including fake signal of new physics mainly due to an incomplete understanding of the underlying phenomena in the Standard Model.

During the CERN period Guido worked on a large variety of subjects, ranging from neutral B meson mixing [Altarelli and Franzini (1988)] to inclusive decays of heavy flavours and heavy flavour lifetimes [Altarelli *et al.* (1996)], from the production of particles beyond the Standard Model [Altarelli *et al.* (1985b,c)] and searches of the Higgs meson [Altarelli and Franco (1986); Altarelli *et al.* (1987, 1989)], to extended gauge models, with Raoul Gatto and the Firenze group [Altarelli *et al.* (1990a,b, 1991a,b, 1993b)]. From this period it is also worth mentioning the very nice reappraisal of the bounds on the Higgs mass, work done in collaboration with Gino Isidori [Altarelli and Isidori (1994)]. All this activity shows the amplitude of his scientific interests and of his competence and his attention to the most advanced phenomenological issues in particle physics.

Besides subjects on which Guido worked quite occasionally, we may identify several main lines of research which he pursued very successfully over these years with several collaborators: QCD in all its relevant aspects, QCD at hadron colliders, the experimental determination of the strong coupling constants α_S [Altarelli *et al.* (1995b)], jet physics and vector boson production, the production of heavy flavours at hadron colliders [Altarelli *et al.* (1988)], QCD corrections to heavy quark asymmetries [Altarelli and Lampe (1993)], QCD renormalons, (polarised) structure functions [Altarelli *et al.* (1994b,a, 1997a, 1998a)] and partonic PDFs [Altarelli *et al.* (1998c, 2000a)]. The main collaborators on QCD starting in 1993 were R. Ball, S. Forte, M. Mangano, P. Nason and G. Ridolfi.

A related issue, very hot at that time, around 1988, was the interpretation of the experimental results by the EMC experiment concerning the polarised structure functions. Among the most significant results obtained by Guido in collaboration with Graham Ross were those related to the contribution of the QCD anomaly to the structure functions of the polarised protons and neutrons [Altarelli and Ross (1988)]. It turned out that, not only were the experimental results not in contradiction with the theoretical predictions, but they were a firm confirmation of QCD as the actual theory of strong interactions, including its more subtle quantum effects, namely the triangle anomaly. On the subject of polarised structure functions Guido worked over the years with accurate predictions and a continuous comparison with the experimental data, in collaboration with R. Ball, S. Forte, P. Nason, G. Ridolfi and J. Stirling.

One of the major contributions by Guido, related to this central period of his career, is the development of the model independent analysis of the electroweak data as a precision test of the Standard Model and of signals of new physics models [Altarelli and Barbieri (1991); Altarelli *et al.* (1992, 1993a, 1995a)]. Guido started being interested in the precision tests of the Standard Model when he contributed to the LEP yellow books which appeared in 1986. This interest generated another very important line of research, with a very strong impact on the LEP experimental program, constituted by the extraordinary series of analyses of precision measurements of the electroweak quantities and checks of the Standard Model, started in 1991, mainly in collaboration with Riccardo Barbieri. This approach was quite new and the methodology of this study became a standard for future analyses. Other collaborators on this subject were S. Jadach and F. Caravaglios. The precision analyses were subsequently extended to constraint extended gauge or supersymmetric models. This important tool is still used nowadays to simultaneous analyses of the precision LEP measurements with the LHC data, see for example [de Blas *et al.* (2016)] and references therein.

3.5 The last period, conclusion

In 1992 he (partially) moved to the newly founded University of Roma Tre, where he remained up to formal retirement, dividing his time between teaching (in Rome) and research (mostly at CERN). He continued his research activity up to his last months.

At the end of the 90's, beginning of 2000, without neglecting his previous interests, the focus of the research activity of Guido became neutrino physics in an attempt to decipher the information of the large harvest of newly incoming data on neutrino mixing. In particular he was attracted by models of mixing the leptonic flavour based on discrete symmetries, in collaboration with Ferruccio Feruglio [Altarelli and Feruglio (2005, 2006, 2010)].[1] He looked for fingerprints of symmetries; his old expertise in symmetry groups (from Florence times) was certainly very useful to him. He was fascinated by the elegance of the tribimaximal neutrino mixing, and many of his papers are dedicated to the search for the possible origins of this baffling symmetry. He also studied the detailed predictions of realistic SU(5) unification (also with supersymmetry) both for proton decay and for neutrino mixing. Although nowadays we know that neutrino mass matrix is not tribimaximal, it is still possible that deviations from a tribimaximalily come from a perturbative expansion in some small parameter. It is worth mentioning that Guido did not abandon his passion for QCD and all over these years he continued to work on the small x behaviour of the structure function mainly in very fruitful collaboration with R. Ball and S. Forte [Altarelli *et al.* (2002, 2003, 2006, 2008)].

Guido Altarelli worked on all the most relevant aspects of the Standard Model and studied in many different processes possible signals of new physics out of precise experimental measurements and accurate theoretical predictions. His lasting influence is pervasive and can be traced everywhere in the development of particle physics in the last fifty years. His most famous contribution, in collaboration with Giorgio Parisi, is the discovery of the equations generating the evolution of quark and gluon distribution functions inside the hadrons which have influenced all the development of the physics of strong interactions thereafter.

It is virtually impossible to make a list of all the collaborators that Guido had over the years. It is worth stressing, however, the very important role of educator he has played in initiating to the scientific career dozens of young people, some of whom have become famous scientists at the international level.

The wide range of interests and results achieved by Guido Altarelli, the very strong impact that his research has had on the development of modern elementary

[1]The number of papers by Guido Altarelli and collaborators on neutrino physics is very large. Here we only reference the three papers mostly cited in the literature.

particle physics, the prestigious international assignments and recognitions are a testament to the exceptional stature of the physicist. With the disappearance of Guido Altarelli we have not lost, however, only a great scientist, but also an extraordinary person, able to face and solve the problems of the scientific community with soundness and common sense, a point of reference for his Italian and CERN colleagues, an example of moral and scientific integrity. In the space of more than thirty years Guido has given a great deal to the scientific community, and in particular to CERN, engaging at first hand in the organization, operation and success of the study groups that have expanded the research strategies of the LEP collider and LHC. Many of his scientific works, both the most famous and the least known, were in fact born of a spirit of service to the community to which he belonged.

Guido was not only a great scientist and his figure cannot be separated from his human qualities. His most characteristic traits were the intellectual honesty, the moral integrity, the disinterest for his personal advantage compared to his dedication to the common interest (of the community, of the institution to which he belonged). He coupled a great and natural kindness with a great, ironic sense of humor and a rare quality, the ability to make self-irony. Among my most vivid memories, an episode happened while we worked together. We were damned to correct a mistake in the calculation of the differential Drell–Yan cross section, which we knew existed but which was difficult to find. Finally satisfied with the result Guido said: *Look, we never make computational errors, but only conceptual ones!* This sounds particularly strange since if there was a quality that everyone recognised to him was his logical rigour, the precision, the clarity in the derivation of the results, the overall vision of the theoretical context. Another episode worth mentioning is the renowned "Altarelli Cocktail" by which he demolished a too hasty "discovery" of supersymmetry at the proton-antiproton collider by using a cocktail, precisely, of Standard Model processes to explain the missing energy of the misunderstood events.

Guido was animated by an inexhaustible scientific curiosity in all the aspects of physics, from formal to phenomenological and followed and studied carefully the literature of all the subjects he was interested in. For this reason he had an extraordinary capability of synthesising all the different aspects of the fundamental questions in particle physics, combining all the pieces of a puzzle together and expressing the main issues in precise formulations which could be understood by all. Indeed one cannot forget his magisterial summaries on different topical subjects, from all aspects of QCD, to precision measurements in the electroweak theory, from Higgs searches to neutrino physics. His review talks were crucial not only because they were complete and clear cut, state-of-the-art in a given subject, but because of their clarity they indicated the new directions to take. It is difficult to think what the status of the different fields he worked on would be without his contributions.

Guido has been a reference figure in the development of particle physics in the last fifty years, and an example of a scientist and colleague to imitate for his moral and human qualities. We have all admired the dignity and courage with which

Guido left this world. Those who have had the privilege of collaborating or simply knowing Guido will continue to remember him with admiration and respect. We, who have been friends of Guido and have loved him, will not forget him and we will mourn for him for a very long time.

Acknowledgements

I am very grateful to Giorgio Parisi [Parisi (2017)], Mario Greco [Greco (2017)] and Luis Alvarez–Gaumé [Alvarez–Gaumé (2017)] from which I borrowed part of the material, including some personal recollection, presented in this article.

Chapter 4

Precision QCD, higher order corrections and resummation

R. Keith Ellis[1] and Giulia Zanderighi[2]
[1] *Institute for Particle Physics Phenomenology, Department of Physics Durham University, Durham DH1 3LE, UK*
[2] *CERN, Theoretical Physics Department, CH 1211 Geneva 23, Switzerland*
On leave from Rudolf Peierls Centre for Theoretical Physics, University of Oxford Oxford OX1 3NP, UK

Giulia Zanderighi and R. Keith Ellis.

4.1 Introduction

It is a privilege to contribute to this volume dedicated to the memory of Guido Altarelli, to pay tribute to his role in the development of perturbative QCD, and to acknowledge his insights so generously shared with us. Although the bulk of this article talks about the current status of QCD, and about ideas for the future, we shall devote a paragraph to describing Guido's role in the early years. This description will be brief, because the issues of 40 years ago are no longer the issues of today. It is the nature of progress that some of the problems that were struggled with in the early days of QCD are no longer as difficult, precisely because of the advances in the intervening years. The early days of QCD are described at length in refs. [Altarelli (2012); Ellis (2017)].

Prior to the discovery of asymptotic freedom, Guido had written a number of papers on deep inelastic phenomena and the parton model, not only the classic electron-proton (ep) scattering, but also muon pair production. The discovery of asymptotic freedom gave a calculable framework in which to pursue higher order corrections. The original applications were limited to deep inelastic processes where the light-cone expansion provided a tool to incorporate higher orders. There was initial diffidence in using perturbation theory for other processes. Subsequently, the factorisation ideas [Politzer (1977)] gave support to the idea that perturbation theory could also be used for high-energy processes involving two initial state hadrons.

The first calculation of strong radiative corrections to a hadronic process was the calculation of the radiative corrections to the Drell–Yan process [Altarelli *et al.* (1979a)]. The result of this calculation is shown in Fig. 4.1. This calculation showed that the QCD correction to the Drell–Yan process was large at the energies then being explored, and stimulated a detailed examination of radiative corrections to other hard processes in perturbative QCD.

4.2 NLO corrections

Since those early days the problem of calculating next-to-leading order (NLO) corrections has been systemised and techniques have evolved to make predictions for less inclusive quantities. This requires the combination of states with differing numbers of final state partons, which we shall now describe.

4.2.1 *Cancelling singularities*

The problem of assembling the final states of differing multiplicities which contribute beyond the leading order goes back to the very early days of QCD. At NLO, the cross section is given by

$$\sigma = \sigma^{LO} + \sigma^{NLO}, \quad \sigma^{LO} = \int_m d\sigma^B \,, \tag{4.1}$$

where σ^{LO} is obtained by integrating the Born-level cross section over the m final state partons. σ^{NLO} is given by two different contributions,

$$\sigma^{NLO} \equiv \int d\sigma^{NLO} = \int_{m+1} d\sigma^R + \int_m d\sigma^V \,. \tag{4.2}$$

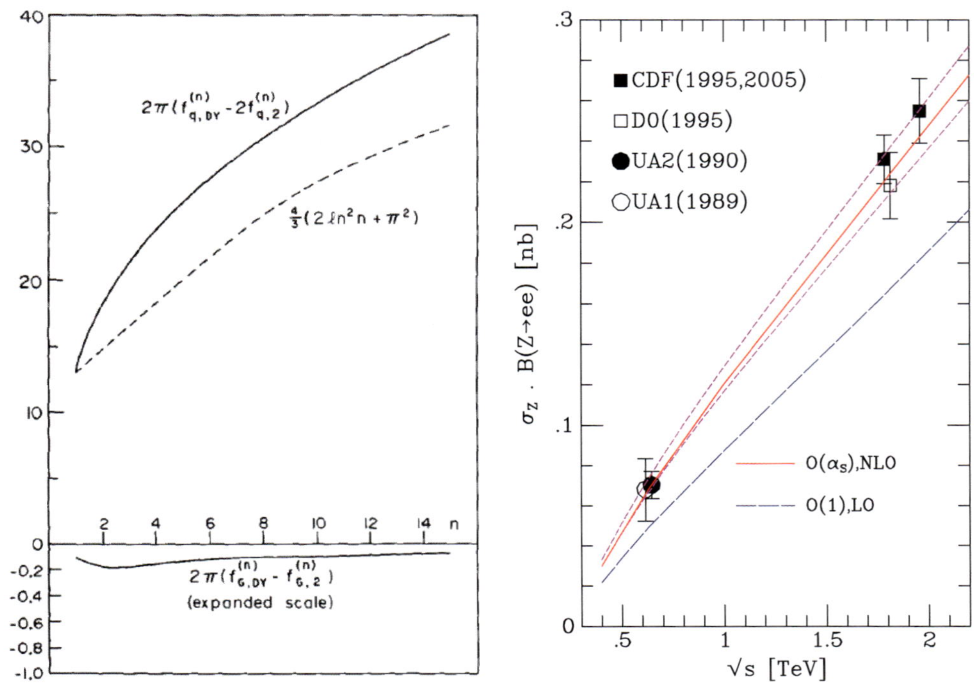

Fig. 4.1 The left-hand plot, taken from ref. [Altarelli *et al.* (1979a)] shows moments of the correction to the Drell–Yan process in units of α_S. For values of $\alpha_s \geq 10\%$ the correction is large. The contribution of the $q\bar{q}$ processes is positive, large and increasing with moment number. The dashed curve attempts to isolate the dominant terms responsible for this growth. The gluon induced processes (qg and $\bar{q}g$) are negative and small. The right-hand plot shows early data on Z production, showing that no agreement with the data can be achieved, using parton distributions measured in deep inelastic scattering, without inclusion of the $O(\alpha_S)$ correction. The outlying short-dashed pink curves give an estimate of the error on the theoretical NLO prediction.

These are the real emission contributions, involving $m+1$ partons in the final state and the virtual contributions involving m partons in the final state. Each one of these two contributions can contain infra-red divergences due to the emission of soft gluons, and mass singularities due to the emission of collinear partons, although, after integrating over phase space and factorisation of the initial state collinear singularities, the sum of the two pieces is finite. The divergences in the two separate pieces are best regulated by performing the calculation in $d = 4 - 2\epsilon$ dimensions. There is a tension between the need to cancel the infra-red singularities, which requires a suitably inclusive quantity and the desire to have predictions for a more exclusive quantity, so that, for example, experimental cuts can be applied to the theoretical predictions. The two methods of resolving this dilemma go under the names of subtraction [Ellis *et al.* (1981b); Kunszt *et al.* (1989)] and slicing [Fabricius *et al.* (1980); Giele *et al.* (1993)]. A brief description of methods at NLO will serve to introduce the more complex techniques needed at NNLO. General procedures have been presented in refs. [Catani and Seymour (1997)] and [Frixione *et al.* (1996)].

The difference between slicing and subtraction can be illustrated by considering a simple integral I [Kunszt and Soper (1992)],

$$I = \lim_{\epsilon \to 0} \left(\int_0^1 \frac{dx}{x} x^\epsilon F(x) - \frac{1}{\epsilon} F(0) \right). \tag{4.3}$$

In the slicing method I can be rewritten by dividing the region of integration and setting $F(x) = F(0)$ for $0 < x < \delta$,

$$\begin{aligned} I &= \lim_{\epsilon \to 0} \left(\int_0^\delta \frac{dx}{x} x^\epsilon F(0) + \int_\delta^1 \frac{dx}{x} x^\epsilon F(x) - \frac{1}{\epsilon} F(0) \right) \\ &\sim F(0) \ln \delta + \int_\delta^1 \frac{dx}{x} F(x) \,, \end{aligned} \tag{4.4}$$

because we have chosen $\delta \ll 1$. At small values of δ the full result should be independent of δ, until numerical instabilities set in.

In the subtraction method we write I as

$$\begin{aligned} I &= \lim_{\epsilon \to 0} \left(\int_0^1 \frac{dx}{x} x^\epsilon \Big[F(x) - F(0) \Big] + F(0) \int_0^1 x^\epsilon \frac{dx}{x} - \frac{1}{\epsilon} F(0) \right) \\ &= \int_0^1 \frac{dx}{x} \Big[F(x) - F(0) \Big]. \end{aligned} \tag{4.5}$$

In practical implementations it may be necessary to increase the lower limit of integration to a small postive value for reasons of numerical stability.

For the particular case of NLO corrections the subtraction method proceeds by constructing a series of counterterms, $d\sigma^A$, which have the same behaviour in the singular limits as the real emission graphs, but are simple enough that they can be integrated over the unresolved parton,

$$d\sigma^{NLO} = \left[d\sigma^R - d\sigma^A \right] + d\sigma^A + d\sigma^V. \tag{4.6}$$

Integrating over the final state momenta, the final result for the NLO cross section is,

$$\sigma^{NLO} = \int_{m+1} \left[(d\sigma^R) - (d\sigma^A) \right] + \int_m \left[d\sigma^V + \int_1 d\sigma^A \right]. \tag{4.7}$$

The first square bracket on the right-hand side of Eq. (4.7) is finite by construction, and can be evaluated in four dimensions; the second term is also finite for infra-red safe jet measures, and after integration over the unresolved parton, the limit $\epsilon \to 0$ can be taken.

The dipole method [Catani and Seymour (1997)] is a practical realisation of a subtraction method and relies on the identity,

$$\frac{1}{p_i \cdot k \, p_j \cdot k} = \frac{1}{(p_i \cdot k + p_j \cdot k)} \frac{1}{p_i \cdot k} + \frac{1}{(p_i \cdot k + p_j \cdot k)} \frac{1}{p_j \cdot k} \,, \tag{4.8}$$

which allows the eikonal soft singularity to be written as a sum of two terms, each with collinear singularities only along a single direction. In addition the method exploits simple factorised forms of the $(m \otimes 1)$-dimensional phase space. In light of Eq. (4.8) we see that the term dipole is really a misnomer.

The subtraction method of Frixione, Kunszt and Signer [Frixione *et al.* (1996); Frederix *et al.* (2009)] realises the subtraction in a different way, defining the subtraction terms in the centre-of-mass frame in terms of energies, and angles; these variables have a direct relationship with the form of the soft and collinear singularities, rendering the calculation of the integrals of the subtractions much easier. In addition, the final phase space is partitioned into sectors, in such a way that only a single soft singularity and/or a single mass singularity due to the collinearity of two partons can contribute. Thereafter it is easy to isolate the singularities in the subtractions, using formal identities such as

$$\theta\big(\xi(1-\xi)\big)\,\xi^{-1+\epsilon} = \frac{1}{\epsilon}\delta(\xi) + \left[\frac{1}{\xi}\right]_+ + \epsilon\left[\frac{\ln\xi}{\xi}\right]_+ + O(\epsilon^2), \tag{4.9}$$

where the "plus" distributions on the right-hand side are defined as

$$\begin{aligned} \int_0^1 d\xi\, f(\xi)\left[\frac{1}{\xi}\right]_+ &= \int_0^1 d\xi\left[f(\xi) - f(0)\right]\frac{1}{\xi}, \\ \int_0^1 d\xi\, f(\xi)\left[\frac{\ln(\xi)}{\xi}\right]_+ &= \int_0^1 d\xi\,\left[f(\xi) - f(0)\right]\frac{\ln\xi}{\xi}. \end{aligned} \tag{4.10}$$

4.2.2 *Current status*

One of the most exciting developments of the last few years has been the development of automatic procedures for the calculation of NLO corrections, such as aMC@NLO [Alwall *et al.* (2014)], Gosam [Cullen *et al.* (2014)], Recola [Actis *et al.* (2017)], and Openloops [Actis *et al.* (2017)]. In general these programs treat the virtual diagrams using semi-numerical procedures, performing the reduction of virtual amplitudes to standard integrals, which are then evaluated using analytic expressions. These reductions are either of a traditional sort [Passarino and Veltman (1979)], or make use of newer methods performing the reduction at integrand level [Ossola *et al.* (2007)]. These procedures all rely on tabulations of one-loop scalar integrals ['t Hooft and Veltman (1979); Ellis and Zanderighi (2008)] and the implementation of these integral results into reliable codes [van Hameren (2011); Carrazza *et al.* (2016); Denner *et al.* (2017)]. The methods are thus semi-numerical. Purely numerical methods formulating subtractions also for virtual integrals have been formulated in ref. [Becker and Weinzierl (2012)].

Although in principle the automatic procedures can handle any number of final state partons, in practice there is a limit to the number of final state partons which have been calculated using these automatic procedures. A number of projects have succeeded to go beyond the current reach of the automatic programs. Blackhat+Sherpa [Andersen *et al.* (2014); Maitre *et al.* (2016)] have produced results for

$W + n$ jets, $(n = 1, 2, 3, 4)$, $Z + n$ jets, $(n = 1, 2, 3, 4)$ and n-jets, for $(n = 1, 2, 3, 4)$. The program Njet handles 5-jet production in QCD [Badger *et al.* (2014a)], and the related process di-photon + 3 jets in ref. [Badger *et al.* (2014b)].

Despite the fact that α_S is small at high energy of order 10%, the calculation of NLO corrections shows that the uncertainty on the corrections is often much larger. This is a feature of renormalisation group improved perturbation theory, because all predictions depend on an unphysical scale, μ. As higher order corrections are included the dependence on this unphysical scale is formally less important, although this does not imply that this dependence is numerically small. The practical consequence of this is that if we wish to have theoretical predictions which are commensurate with other measurement errors, (for example, the luminosity which is measured at LHC with an error of 2.5% [CMS Collaboration (2017b)]), we have to proceed beyond NLO corrections.

4.3 NNLO

A necesssary prerequisite for the calculation of NNLO corrections is the calculation of two-loop virtual amplitudes. We shall now describe the status of these calculations.

4.3.1 *Loop amplitudes*

4.3.1.1 $2 \to 2$ *processes*

Advances in loop calculation methods [Laporta (2000); Gehrmann and Remiddi (2000)] have permitted the calculation of many two-loop diagrams, necessary for the calculation of next-to-next-to-leading order (NNLO) processes. The virtual two-loop $2 \to 2$ processes calculated first involved massless partons circulating in the loop and a limited number of massive external legs. As we shall see later, for many of the simplest $2 \to 2$ processes these ingredients have been assembled to produce physical cross sections. Processes involving a greater number of internal masses, massive external lines, or non-planar topologies are more challenging [Henn (2013)].

Although technically they are NLO processes, we mention here radiative corrections to processes which commence at one loop. The complexity of the calculation of these processes at NLO, is similar to NNLO calculations, since they require the inclusion of two-loop diagrams. Important examples of these processes are $gg \to H$ [Pak *et al.* (2010)] and approximate results for $gg \to H + j$ [Harlander *et al.* (2012)] where these processes are calculated in the full theory, rather than the effective theory valid as $m_t \to \infty$. Also noteworthy is the numerical calculation of double Higgs production [Borowka *et al.* (2016)] with full internal loop mass dependence, performed using a sector decomposition technique. Results have been obtained for the planar integrals relevant for the process $H \to 3$ partons [Bonciani *et al.* (2016)] including the full mass dependence of the fermions in the loop. This is an important step in the full NLO calculation of $H + 1$ jet at large p_T. The same amplitudes have been calculated in full in the limit of small mass [Melnikov *et al.* (2016, 2017)]. In addition other processes such as $gg \to ZZ$ [Caola *et al.* (2015a, 2016a); Campbell *et al.* (2016a)] have been calculated.

4.3.1.2 $2 \to 3$ *processes*

Only very few intermediate results exist for $2 \to 3$ processes. In ref. [Gehrmann *et al.* (2016)] analytic results for the planar 2-loop five-gluon amplitude for a particular helicity amplitude have been presented. These results build on earlier results of Badger *et al.* [Badger *et al.* (2015)]. This constitutes the current frontier for these NNLO calculations. However $2 \to 3$ processes are of great phenomenological importance, for example, Higgs + dijet and $WW+$ jets (as a background to Higgs production by vector-boson fusion), and $pp \to 3$ jets (for α_S measurements). It is therefore to be expected that great efforts will be devoted to these topics in the near future.

4.3.2 *Combination techniques*

The problem of assembling these amplitudes into physical cross sections follows the pattern described above for NLO, but with greater complexity because of the extra parton in the real emission diagrams and the presence of overlapping singularities. At NNLO for a process that begins at tree level, we need to consider two-loop amplitudes for processes with m partons in the final state. In addition the NNLO corrections involve one-loop processes with $m + 1$ final state partons, and tree-level processes with $m + 2$ partons. All of these processes contain soft and potentially also collinear singularities which only cancel when the final phase space is integrated over. As before this necessity to integrate over the final phase space, is in tension with the desire to produce a differential cross section.

The different combination strategies can be thought of as generalisations of the NLO Catani–Seymour and FKS procedures. A list of the various techniques used for NNLO calculations is given in Table 4.1, indicating whether it is a subtraction method (Su) or slicing method (Sl).

4.3.3 *Physical results*

As one can see from Tables 4.2 and 4.3 a large number of $2 \to 2$ processes have already been calculated at NNLO. The reason to pursue these calculations is to obtain a theoretical error commensurate with the errors in the experimental measurements, such as luminosity or statistical errors. In addition the error due to the limitation in the number of perturbative terms which have been calculated should be in line with error on other theoretical inputs, such as α_s and parton distribution functions.

Table 4.1 Subtraction (Su) and slicing (Sl) methods available at NNLO.

Name	Su/Sl	Features/Limitations
antenna subtraction [Gehrmann-De Ridder *et al.* (2005)]	Su	analytic counterterms
sector-improved residue subtraction [Czakon (2010)]	Su	local counterterms
nested subtraction [Caola *et al.* (2017)]	Su	reduced number of singular sectors
projection to Born [Cacciari *et al.* (2015)]	Su	local, limited types of processes
colourful subtraction [Somogyi *et al.* (2007)]	Su	local; currently final-state only
q_T subtraction [Catani and Grazzini (2007)]	Sl	Colour singlet final states only
N-jettiness subtraction [Gaunt *et al.* (2015); Boughezal *et al.* (2015c)]	Sl	

The fact that a large number of NNLO corrections have now been calculated, gives us the possibility to assess how large the theoretical errors are at NLO and NNLO, and in addition how well calculations (and their associated error estimates) performed at NLO, match with the observed situation at NNLO. This is demonstrated in Fig. 4.2 due to Gavin Salam [Salam (2016)]. We see that many processes have errors at the 2% level, although some of the most interesting quantities for precision Higgs physics do not. More troubling is that the NNLO central value lies within the NLO error band in only 4 out of the 17 cases shown.

We note parenthetically that if NNLO QCD corrections are included it is also necessary to include electroweak corrections since $\alpha \simeq \alpha_s^2$.

4.4 N^{3}LO

Only two hard hadronic processes have been calculated for the LHC at NNNLO. One of the currently most pressing problems is the characterisation of the Higgs boson, so it is not surprising that both calculations refer to Higgs processes. The gluon-gluon fusion process at NNNLO has been calculated in ref. [Anastasiou *et al.* (2015)] using the approximation that the top quark is infinitely heavy. Of course it is not just the NNNLO corrections to the cross section which are interesting, but rather the totality of all corrections of similar size (together with an estimate of the remaining errors, which contribute to the error budget). The final result for the inclusive correction at $\sqrt{s} = 13$ TeV is

$$\sigma = 48.58\,\mathrm{pb}^{+2.22\,\mathrm{pb}\,(+4.56\%)}_{-3.27\,\mathrm{pb}\,(-6.72\%)}\ (\text{theory}) \pm 1.56\,\mathrm{pb}\,(3.20\%)\ (\text{PDF}+\alpha_s)\,. \tag{4.11}$$

In the following we report the analysis, taken from ref. [Anastasiou *et al.* (2016)] which illustrates how this final value of 48.58 pb is made up:

$$\begin{aligned} 48.58\,\mathrm{pb} = &\quad 16.00\,\mathrm{pb} \quad (+32.9\%) \qquad (\text{LO, rEFT}) \\ &+20.84\,\mathrm{pb} \quad (+42.9\%) \qquad (\text{NLO, rEFT}) \end{aligned}$$

Table 4.2 Table summarising the main calculations completed recently at NNLO.

Process and Features	Application
W, Z, γ^* [Melnikov and Petriello (2006); Catani *et al.* (2009)] fully exclusive, NNLO	PDFs, Luminosity
H [Anastasiou *et al.* (2015); Anastasiou and Melnikov (2002)] [Harlander and Kilgore (2002)] HEFT, fully inclusive, N^{3}LO	Higgs couplings
H+1jet [Boughezal *et al.* (2015a,b); Caola *et al.* (2015b)] [Chen *et al.* (2016)] HEFT, fully exclusive	Higgs couplings, ggH vertex
$t\bar{t}$ pair [Czakon *et al.* (2016a, 2017)] fully exclusive, stable tops	top mass, p_T, asymmetry, PDFs
single top, t-channel [Brucherseifer *et al.* (2014); Berger *et al.* (2016b)] [Berger *et al.* (2017)] fully exclusive, narrow width approximation for the top	$V_{\rm tb}$, width, PDfs, BSM
VBF Higgs [Bolzoni *et al.* (2010); Cacciari *et al.* (2015)] [Dreyer and Karlberg (2016)] factorized approximation, exclusive VBF cuts; N^{3}LO p_t spectrum	Higgs couplings
$W+j$ [Boughezal *et al.* (2015c)] fully exclusive, decays	PDFs
$Z+j$ [Gehrmann-De Ridder *et al.* (2016b,c,a); Boughezal *et al.* (2016)] fully exclusive, off-shell effects	PDFs, Dark Matter
ZH [Ferrera *et al.* (2015); Campbell *et al.* (2016c)] Fully exclusive, with decay to $b\bar{b}$ at NLO	Higgs couplings
WH [Ferrera *et al.* (2014); Campbell *et al.* (2016c)] [Astill *et al.* (2016); Ferrera *et al.* (2017)] fully exclusive, Higgs decay to b *barb*	Higgs couplings
HH [de Florian *et al.* (2016a)] HEFT, fully exclusive	Higgs self-coupling
ZZ [Cascioli *et al.* (2014); Grazzini *et al.* (2015b)] [Heinrich *et al.* (2017)] fully exclusive, off-shell	trilinear gauge couplings, BSM

$$
\begin{array}{llll}
- & 2.05\,\text{pb} & (-4.2\%) & ((t,b,c),\ \text{exact NLO}) \\
+ & 9.56\,\text{pb} & (+19.7\%) & (\text{NNLO, rEFT}) \\
+ & 0.34\,\text{pb} & (+0.2\%) & (\text{NNLO}, 1/m_t) \\
+ & 2.40\,\text{pb} & (+4.9\%) & (\text{EW, QCD-EW}) \\
+ & 1.49\,\text{pb} & (+3.1\%) & (\text{N}^3\text{LO, rEFT})
\end{array}
\qquad (4.12)
$$

Table 4.3 Table summarising the main calculations completed recently at NNLO (continued).

Process and Features	Application
WW[Gehrmann *et al.* (2014); Grazzini *et al.* (2016a)], WZ[Grazzini *et al.* (2016b, 2017)] fully inclusive	trilinear gauge couplings, BSM
$W\gamma$, $Z\gamma$[Grazzini *et al.* (2015a); Campbell *et al.* (2017b)] fully exclusive	trilinear gauge couplings, BSM
$\gamma\gamma$[Catani *et al.* (2012); Campbell *et al.* (2016b)] fully differential	Background studies
γ+jet[Campbell *et al.* (2017a,b)] fully differential	PDFs
top decay[Brucherseifer *et al.* (2013); Gao *et al.* (2013)] exclusive	Top couplings
$H \to b\bar{b}$[Anastasiou *et al.* (2012); Del Duca *et al.* (2015)] exclusive, massless b-quark	couplings boosted
Single inclusive jet[Currie *et al.* (2014, 2017d)] Inclusive	strong coupling, PDFs
Di-jet[Currie *et al.* (2017b)] exclusive	strong coupling, PDFs
Direct photon[Campbell *et al.* (2017a)] Inclusive	PDFs, Monojets
Jet in DIS[Currie *et al.* (2017a)] Exclusive	PDFs
Charm in DIS[Berger *et al.* (2016a)] Exclusive	strange PDF
$e^+e^- \to 3j$ [Gehrmann-De Ridder *et al.* (2007); Weinzierl (2009); Del Duca *et al.* (2016b,a); Gehrmann *et al.* (2017)] Global event shapes, jet rates	α_S

Although the total Higgs production process may be special in this regard, because of the very large corrections at NLO, we note that the final error budget is roughly at the 5% level only after the calculation on N^3LO. This emphasises the point made earlier that the errors in QCD perturbation theory are far larger than the size of α_S would lead one to predict.

The inclusive cross section for vector boson fusion production of the Higgs at N^3LO has been considered in ref. [Dreyer and Karlberg (2016)] building on the work of ref. [Cacciari *et al.* (2015)]. Similar to the case cited above, this shows a nice apparent convergence of the perturbation series (i.e. NNNLO band inside NNLO band). In this case however the perturbative error is much smaller than for inclusive Higgs production with gluon-fusion, and follows the naive scaling of a power expansion in the coupling constant.

4.5 Resummation

Up to this point, we have discussed fixed-order perturbative expansions. While these calculations provide in general a good description of inclusive cross sections, it is well known that they fail in several cases, when the smallness of the coupling

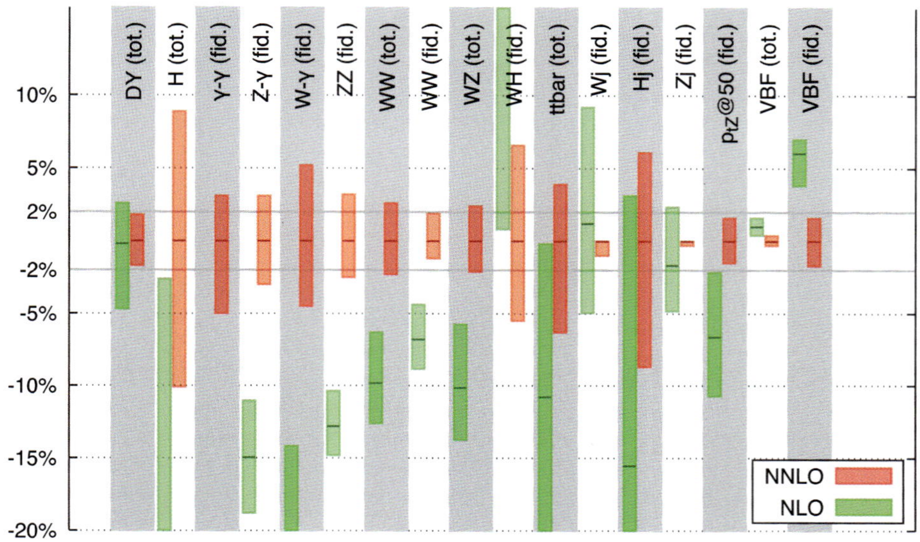

Fig. 4.2 Percentage deviations from the calculated NNLO central values, according to the NNLO error estimates (shown in pink) and the NLO error estimates (shown in green). The NLO (NNLO) central value is indicated by the dark green (red) horizontal mark.

constant is compensated by some large logarithm. This happens for instance when exclusive measurements are performed, since the phase space restrictions on additional radiation introduce a second (soft) scale in the problem, besides the typical hard scale of the scattering process. To obtain reliable predictions one then needs to resum, to all orders in perturbation theory, towers of terms that are logarithmically enhanced. These enhancements have different physical origins, most notably

- small-x resummation: This is the regime where the momentum fraction x carried by partons in protons becomes very small. This soft region is described by the BFKL (Balitsky–Kuraev–Fadin–Lipatov) equation which takes into account the small-x evolution of parton densities [Fadin *et al.* (1975); Balitsky and Lipatov (1978)] resumming to all-orders large $\ln(x)$ terms.
- threshold resummation: When a heavy object of mass M (a Drell–Yan pair, a top-quark pair, a Higgs boson, etc.) is produced in high-energy collisions, there is an enhancement when the partonic centre-of-mass energy is close to the mass threshold. Large logarithms of the form $\ln(1 - M/\sqrt{\hat{s}})$ appear as a result of a restriction of the phase space for additional soft and collinear radiation. In e^+e^- colliders, when the energy is close to the threshold no accurate description can be achieved without a calculation of all-order threshold logarithms. The situation is somehow different at hadron colliders, as one integrates over the partonic incoming energies, which dilutes

the effect of threshold logarithms. Still, calculations that include threshold effects are in general more accurate.

- soft and collinear effects: These are logarithms that appear in exclusive measurements whenever one restricts the phase space for additional soft- and collinear radiation explicitly through a measurement function. Observables where these logarithms have been studied in great detail include transverse momentum distributions, event shapes and jet rates. The requirement that the value of a transverse momentum or of an event-shape is below a fixed value imposes a restriction on the hardness and angle of real radiation, while no corresponding restriction is imposed on virtual radiation. The incomplete cancellation between real and virtual emissions gives rise to large logarithms. Observables have at most double logarithms, a soft and a collinear one for each power of the coupling constant. There are also observables that are not sensitive to collinear radiation, e.g. radiation away from jets, in which case only a single logarithm is present for each power of the coupling constant.

Guido performed a seminal work on small-x resummation. This topic is covered in some detail in Chapter 10 of this book. Guido achieved also ground-breaking work on the resummation of collinear logarithms through the DGLAP (Dokshitzer–Gribov–Lipatov–Altarelli–Parisi) evolution equation, which will be reviewed briefly in Section 4.5.1. A number of early papers of Guido are also devoted to the description of the transverse momentum of Drell–Yan pairs, and to the description of the hadronic radiation that recoils against the Drell–Yan pair [Altarelli *et al.* (1978b,a,c, 1979a, 1985a, 1984)]. In a modern approach this radiation is clustered into jets, and one talks about the resummation of the jet-veto for vector-boson production. In the following, we will discuss some key contributions to resummed calculations in the late 70's and will briefly review recent progress in the field.

4.5.1 *Collinear resummation*

Already very shortly after the formulation of QCD as an asymptotically free theory it was understood that scaling relations were broken by logarithmic terms. Experimental measurements seemed to confirm these findings and it was also known how to compute the deviations from scaling for deep inelastic scattering (DIS). The calculation of these logarithmic violations involved a renormalisation group equation for coefficient functions of local operators which appear in the light-cone expansion for the product of two currents. While the result had a simple physical interpretation, the derivation was rather technical. Instead, in ref. [Altarelli and Parisi (1977)] Altarelli and Parisi showed that one could obtain the Q^2 behaviour of DIS structure functions using only the parton language throughout. This led to the formulation of a system of integro-differential coupled equations that describes the evolution in

Q^2 of parton densities. These equations are now known as DGLAP evolution equations, as they were derived about at the same time also by Dokshitzer [Dokshitzer (1977)], based on earlier techniques of Gribov and Lipatov [Gribov and Lipatov (1971, 1972a)]. While the final results obtained were essentially the same, the language used in these papers didn't have the striking simplicity of the Altarelli–Parisi paper, which showed how to compute the Q^2 dependence of parton densities in a very intuitive and transparent way. The Altarelli–Parisi master equation, e.g. for the singlet $q^{\mathrm{NS}}(x,t)$, which is defined as the difference between quark and anti-quark distributions $q^{\mathrm{NS}}(x,t) = q(x,t) - \bar{q}(x,t)$, with a momentum fraction of the parent hadron, x, and at scale $t = Q^2$ is given simply by

$$\frac{t\, dq^{\mathrm{NS}}(x,t)}{dt} = \frac{\alpha_s(t)}{2\pi} \int_x^1 \frac{dy}{y} q^{\mathrm{NS}}(y,t) P\left(\frac{x}{y}\right), \tag{4.13}$$

where $P(z)$ is the so-called splitting kernel, whose interpretation is illustrated below. The above equation can be re-written as

$$\begin{aligned} &q^{\mathrm{NS}}(x,t) + dq^{\mathrm{NS}}(x,t) \\ &\quad = \int_0^1 dy \int_0^1 dz \delta(x - zy) q^{\mathrm{NS}}(y,t) \left(\delta(1-z) + \frac{\alpha_s(t)}{2\pi} P_{qq}(z) \frac{dt}{t}\right). \end{aligned} \tag{4.14}$$

The physical interpretation of the above equation was immediately clear: $\frac{\alpha_s(t)}{2\pi} P_{\mathrm{qq}}(z)$ is the variation per unit $\ln t$, at order α_s, of the probability to find a non-singlet quark inside another quark with fraction z of the parent momentum. Given this interpretation, the generalisation to include more flavours and gluons is straightforward:

$$\begin{aligned} \frac{t\, dq^i(x,t)}{dt} &= \frac{\alpha_s(t)}{2\pi} \int_x^1 \frac{dy}{y} \left[\sum_{j=1}^{2n_f} q^j(y,t) P_{q^i q^j}\left(\frac{y}{x}\right) + g(y,t) P_{q^i g}\left(\frac{y}{x}\right) \right], \\ \frac{t\, dg(x,t)}{dt} &= \frac{\alpha_s(t)}{2\pi} \int_x^1 \frac{dy}{y} \left[\sum_{j=1}^{2n_f} q^j(y,t) P_{g q^j}\left(\frac{y}{x}\right) + g(y,t) P_{gg}\left(\frac{y}{x}\right) \right], \end{aligned} \tag{4.15}$$

where the summations run over quarks and anti-quarks of all flavours. Again the physical interpretation is evident: a quark distribution can change because the quark loses energy when radiating a collinear gluon, or because a gluon splits to produce a new $q\bar{q}$ pair. Similarly, the gluon distribution changes because gluons can radiate collinear gluons or split into $q\bar{q}$ pairs, or because gluons are created through the collinear radiation from quarks. Ref. [Altarelli and Parisi (1977)] also showed how to compute the $\mathcal{O}(\alpha_s)$ splitting functions starting from the basic vertices of QCD making the process independence of the splitting functions manifest. After the Altarelli–Parisi paper, the higher order splitting functions were computed to $\mathcal{O}(\alpha_s^2)$ in ref. [Curci *et al.* (1980); Furmanski and Petronzio (1980)] to $\mathcal{O}(\alpha_s^3)$ in ref. [Vogt *et al.* (2004); Moch *et al.* (2004)], and to $\mathcal{O}(\alpha_s^4)$ in ref. [Moch *et al.* (2017)].

The Altarelli–Parisi evolution is driven by collinear emissions. The connection of the Altarelli–Parisi splitting kernels and the resummation of collinear logarithms is as follows. Going back for simplicity to the case of the non-singlet distribution, one defines its Mellin moment as

$$M_n^{\rm NS}(t) = \int_0^1 dx x^{n-1} q^{\rm NS}(x,t)\,. \tag{4.16}$$

Mellin moments are very useful since convolutions in x-space are just ordinary products in Mellin space. The LO evolution equation for the Mellin moment becomes

$$\frac{dM_n^{\rm NS}(t)}{d\ln t} = \frac{\alpha_s(t)}{2\pi} A_n^{\rm NS} M_n^{\rm NS}(t)\,, \tag{4.17}$$

with

$$A_n^{\rm NS} = \int_0^1 dz z^{n-1} P_{qq}^{(0)}(z)\,. \tag{4.18}$$

The latter quantity is now referred to as LO anomalous dimension. The solution to the differential equation (4.17) reads

$$M_n^{\rm NS}(t) = M_n^{\rm NS}(0) \left[\frac{\alpha_s}{\alpha_s(t)}\right]^{\frac{A_n^{\rm NS}}{2\pi\beta_0}}\,, \tag{4.19}$$

where β_0 is the first coefficient of the beta-function that governs the running of the strong coupling constant. This equation shows explicitly that the knowledge of $P_{qq}^{(0)}(z)$ allows one to compute the exponent necessary to describe the evolution of the non-singlet distribution. Today the evolution of parton distribution functions is known to NNLO, which is necessary to match the NNLO accuracy reached in the description of the hard partonic cross sections discussed in Sec. 4.3.

It is also interesting to point out that the collinear resummation of the DGLAP equations controls the parton showering in Monte Carlo programs. These tools are today absolutely essential for LHC studies (for a review on Monte Carlos see e.g. [Sjöstrand (2009)]). In fact, the Altarelli–Parisi equation in integrated form defines a Markov chain, hence is suitable for Monte Carlo implementation. Similar methods of collinear resummation have been used also for the resummation of $\log(R)$ effects, i.e. to resum collinear logarithms that appear in cross sections for jet production when the jet-radius R is small [Dasgupta *et al.* (2015)].

4.5.2 *Transverse momentum resummation*

In the late 70's there were a number of reasons for studying Drell–Yan collisions at small transverse momenta. First, the study of the small transverse momentum region could reveal whether an intrinsic transverse momentum component in the parton distribution function was required to describe data. Second, QCD in contrast to the naive parton model, predicted an approximately linear rise of the average transverse momentum $\langle q_T^2\rangle$ with the centre-of-mass energy s and the invariant mass Q^2 at fixed $\tau = Q^2/s$. Experimental data on the other hand, pointed to a rather flat behaviour for $\langle q_T^2\rangle$ as a function of Q^2 at fixed s. The question was then whether such a behaviour could be explained in perturbative QCD. In fact, at that time asymptotic freedom had just been established and since quarks and gluons were not observed as free objects, it was unclear to which extent calculations performed in perturbative QCD could describe measured quantities. In this sense, Drell–Yan was one of the cleanest available laboratories where QCD could be tested.

It was soon clear that the region of low transverse momentum could not be described by fixed-order calculations, and a description that included all-order terms in the coupling constant was necessary. This was first studied in ref. [Dokshitzer *et al.* (1978b)], which pointed out the fact that the transverse momentum distributions of lepton pairs could be described without the need to introduce a large transverse momentum for the constituent quarks. The result relied on the well known DDT formula [Dokshitzer *et al.* (1978a)] which resums enhanced terms in the so-called leading logarithmic approximation (LLA). This approximation accounts for terms $(\alpha_s \ln^2(q_T^2/\mu^2))^n$ for all $n \geq 1$. The DDT formula for $a + b \to l^+l^- + X$ is given by

$$\frac{d\sigma}{dq^2 dq_T^2 dy} = \frac{4\pi\alpha^2}{9sq^2q_T^2}\frac{\partial}{\partial \ln q_T^2} \times \left\{ \sum_{f=q,\bar{q}} D_{H,f}\left(x_1, \ln\frac{q_T^2}{\mu^2}\right) D_{H,\bar{f}}\left(x_2, \ln\frac{q_T^2}{\mu^2}\right) T^2(q_T^2, q^2) \right\}, \quad (4.20)$$

where $D_{H,i}\left(x, \ln\frac{q_T^2}{\mu^2}\right)$ denotes the probability to find a parton i in a hadron H with momentum fraction x at scale μ. This probability could be obtained from experimental measurements of the cross section for electro-production. The double-logarithmic Sudakov form factor $T(q_T^2, q^2)$, which we will specify below, instead guarantees that, in the region of small q_T, regardless of the value of q^2, no gluon is emitted with transverse momentum larger than q_T. Shortly after this result was published, a resummation procedure for q_T was suggested in a seminal paper of two leading physicists of the Rome group, Parisi and Petronzio [Parisi and Petronzio (1979)], based on the Bloch–Nordsieck method, which was known to resum soft photon effects in QED. First, Parisi and Petronzio confirmed the observation made in the DDT paper that even the region of $q_T \to 0$ could be computed by means of perturbative QCD, something that was not clear at the time. In fact, one could have naively thought that the region of very small transverse momenta is dominated by effects of the unknown, intrinsic k_T distributions in the hadrons. The crucial element was that the probability to have a lepton pair accompanied by the emission of no gluons with transverse momentum larger than q_T vanishes exponentially, hence faster than any power of q_T. As a consequence, the intercept $d\sigma/dq_T^2$ can be computed by means of perturbation theory at $q_T = 0$. The resummation was performed using factorisation properties of the matrix elements, namely the fact that precisely as in QED, in the soft limit, i.e. when the energy of the quarks is much larger than that of the secondary gluons, the QCD matrix element for the emission of n gluons could be written in a factorised form.

For the emission of n photons in QED with momenta k_i and polarisation ϵ_i the matrix element can be written as

$$\begin{aligned}&\alpha^n \frac{p\cdot\epsilon_1}{p\cdot k_1}\frac{p\cdot\epsilon_2}{p\cdot(k_1+k_2)}\cdots\frac{p\cdot\epsilon_n}{p\cdot(k_1+\ldots+k_n)}+(n-1)\,\text{permutations}\\ &=\alpha^n\prod_i w(k_i)\,, \quad \text{with} \quad w(k_i)=\frac{p\cdot\epsilon_i}{p\cdot k_i}\,.\end{aligned} \tag{4.21}$$

The factors $w(k_i)$ are denoted as eikonal factors. The constraint that the transverse momentum of the lepton pair is given by the sum of the transverse momenta of individual soft emissions can also be factorised by taking the Fourier transform,

$$\delta^2\left(\vec{q}_T+\sum_i \vec{k}_{t,i}\right)=\frac{d^2b}{(2\pi)^2}e^{-i\vec{b}\cdot\vec{q}_T}\prod_i e^{-i\vec{b}\cdot\vec{k}_{t,i}}\,, \tag{4.22}$$

which allowed one to write the differential distribution as

$$\frac{1}{\sigma_0}\frac{d\sigma}{d^2q_T}=\frac{1}{4\pi^2}\int d^2b\, e^{-i\vec{b}\cdot q_T}\,\tilde{\sigma}(b)\,, \tag{4.23}$$

with

$$\tilde{\sigma}(b)=\exp\left[\int d^2k_t\frac{\alpha}{\pi}\exp[i\vec{k}_t\cdot\vec{b}\,(w(k_i)-1)]\right]\,, \tag{4.24}$$

where the -1 in the exponent includes the virtual corrections that can be included by unitarity.

Performing the Fourier transform, in the double logarithmic approximation one obtains for the integrated cross section

$$T_{QED}(q_T^2,s)=\exp\left[-\frac{\alpha}{2\pi}\ln^2\frac{s}{q_T^2}+\ldots\right]. \tag{4.25}$$

The discussion up to this point applies to photons in QED. To extend the argument to gluons in QCD Parisi and Petronzio proposed to start from a one-gluon independent emission formula, at $k_T^2\neq 0$, given by

$$\frac{1}{\sigma_0}\frac{d\sigma^{(1)}}{dk_T^2}=\frac{C_F\alpha_s(k_T^2)}{\pi}\frac{1}{k_T^2}\ln\frac{Q^2}{k_T^2}+\ldots\;, \qquad C_F=\frac{4}{3}, \tag{4.26}$$

which extends the QED result and suggests that the argument of the strong coupling constant is the relative transverse momentum of the emitted gluon. It is then conjectured that one can obtain the leading logarithmic result by just exponentiating the first order result, precisely as in the case of QED:

$$T(q_T^2,s)\sim\exp\left[-C_F\frac{\alpha(q_T^2)}{2\pi}\ln^2\frac{s}{q_T^2}\right]. \tag{4.27}$$

This very simple idea set the basis for all resummations of final states performed in the following decades. The resulting distribution agrees to order $\mathcal{O}(\alpha_s)$ with the DDT result, but not at higher orders. In fact according to the DDT formula the leading logarithms do not exponentiate, while the Parisi–Petronzio paper demonstrated exponentiation of soft gluon radiation for the first time.

One of the first papers that aimed at an accurate description of the whole Drell–Yan spectrum both at low and high transverse momenta is ref. [Altarelli *et al.* (1984)]. The requirement imposed on the resulting "matched" distributions were

(1) at large q_T the fixed-order result is recovered (NLO in the case of ref. [Altarelli *et al.* (1984)]);
(2) at small q_T all double logarithmic enhanced terms are included;
(3) only terms for which the exponentiation is justified are exponentiated;
(4) the integral reproduces the known fixed-order result;
(5) the average q_T is identical to the fixed-order result;
(6) the treatment is such that it can be extended to higher order terms when these become available.

It is quite remarkable, that after more than thirty years since ref. [Altarelli *et al.* (1984)] the criteria (1-2) and (4-6), that a matched distribution should satisfy, are essentially still the same. It is also interesting to note that the debate about whether or not it is a better approximation to resum terms for which the exponentiation is not fully justified is still an open one. For instance in POWHEG the whole real radiation is exponentiated, while in approaches like MC@NLO only the soft contributions are exponentiated.

Ref. [Altarelli *et al.* (1984)] also contained a detailed discussion of the treatment of the theoretical uncertainties. In fact, while in fixed order calculation the uncertainty is usually assessed by varying renormalisation and factorisation scales, in matched calculations many more handles are available, e.g. the choice of resummation scale, of the precise argument of the logarithm and of the treatment of subleading logarithms that are beyond the formal accuracy. The question of assigning a realistic, but not too conservative uncertainty to theoretical predictions is still an open one. Beyond varying parameters in the calculations, recent suggestions to estimate the uncertainty due to higher orders include exploiting the convergence of the perturbative expansion (see e.g. [Baglio and Djouadi (2010)]) or adopting a Bayesian approach to uncertainties of perturbative calculations [Cacciari and Houdeau (2011)].

Sample Drell–Yan distributions are shown in Fig. 4.3. The left-hand side shows data for W production from the UA1 and UA2 experiments taken in the 70's, the time of the first pioneering QCD calculations of the Drell–Yan spectrum. The right-hand side shows current high-precision data from the CMS experiment for Z production. Both plots show a comparison to the best available theory predictions, at the time of data taking. While fixed order predictions for the transverse momentum rise towards small transverse momenta, it is clear from data that they instead tend to zero for $q_T \to 0$, precisely as suggested by the first resummed calculations. Since the pioneering paper of Parisi and Petronzio, resummed predictions for the transverse momentum spectrum reached a high level of sophistication and

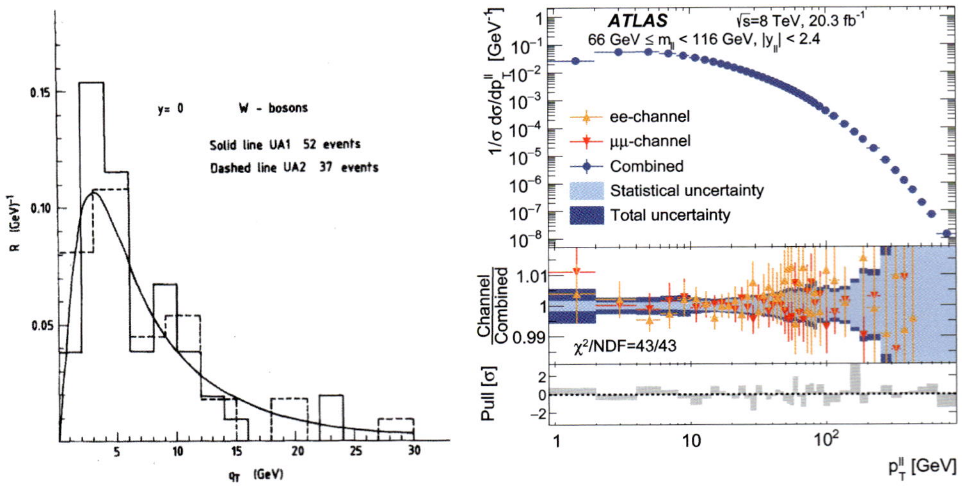

Fig. 4.3 Predictions for the normalised transverse momentum distribution of the W boson from [Altarelli *et al.* (1984)] compared to date from the UA1 and UA2 experiments (left). Modern LHC data [Aad *et al.* (2016b)] on the normalised Z transverse momentum distribution (right).

accuracy. Next-to-leading logarithmic (NLL) predictions were first computed in ref. [Catani *et al.* (1988); Kauffman (1992)], NNLL predictions were first computed in refs. [Bozzi *et al.* (2003)]. Very recently even N^3LL predictions are available and matched to NNLO Z+jet predictions [Bizon *et al.* (2017)].[1] At variance with previous calculations, the latter reference performs the resummation directly in q_T space, rather than in b-space, as in Eq. (4.23).

4.5.3 *Recent progress in resummations*

Since the pioneering work on resummed calculations performed by the Rome group, the field of resummation has made an enormous progress. In the 80's seminal articles by Collins, Soper and Sterman [Collins *et al.* (1984, 1985, 1988)] established the formalism to extend the logarithmic accuracy to higher orders. In particular the mechanism by which soft gluons factorise, decouple from the incoming hadrons and cancel in hard inclusive hadron-hadron cross sections was studied. In ref. [Catani and Trentadue (1989)] a general method to exponentiate leading and next-to-leading logarithms to all orders in perturbation theory was developed. A complete formula for the large-n moments was given and shown to agree with previous two-loop calculations.

[1] The only missing ingredient being the four-loop anomalous dimension, which is set to zero in ref. [Bizon *et al.* (2017)].

Since then, a large number of different observables have been resummed at NLL level (for a review see e.g. [Dasgupta and Salam (2004)]). Resummation to NLL for suitable observables has been automated in the CAESAR approach in the case of suitable observables [Banfi *et al.* (2005)]. In fixed order calculations the choice of observables plays a modest role and, provided an observable is infra-red safe, predictions for it can be made, as long as the hard process has been computed to the given perturbative order. The situation is different for resummations. Here observables play a primary role, since the appearance and the structure of the logarithms is intimately connected to the properties of the observable. It was shown in ref. [Banfi *et al.* (2005)] that a NLL resummation can be automated if the observable has a suitable scaling in the presence of soft-collinear emissions, is recursively infra-red-collinear safe, and is continously global (for a precise definition of these properties see ref. [Banfi *et al.* (2005)]).

Recently the accuracy was pushed to NNLL for e^+e^- event shapes and jet rates [Banfi *et al.* (2015)]. The idea underlying these calculations is to isolate different sources of NNLL contributions, and to compute them as finite corrections numerically using a Monte Carlo numerical simulation. Such an approach opens up the question of whether it is possible to improve the parton showers that underlie exclusive Monte Carlo simulations (Herwig, Pythia, Sherpa, ...), so as to guarantee NLL or even NNLL accuracy after the parton shower, at least for a given class of observables. Indeed, while Monte Carlo simulations are used ubiquitously at the LHC, a major open question is what is the accuracy of the parton shower simulation. This question is particularly timely since currently the precision of fixed order calculations is improving quickly and one can expect $(2 \to 3)$-scattering process to be also known up to NNLO in the coming five years. Techniques to match NLO calculations to parton showers have been developed more than ten years ago now [Frixione and Webber (2002); Frixione *et al.* (2007)] and matching to NNLO is also possible in the cases of simple final states [Hamilton *et al.* (2013); Karlberg *et al.* (2014); Alioli *et al.* (2015); Höche *et al.* (2015, 2014); Astill *et al.* (2016)]. It is natural to expect that NNLO+parton shower accurate predictions will soon be the state-of-the-art. In such a situation, it is clear that controlling the all-order accuracy of the parton shower will be necessary.

An approach that is somehow similar to the one of ref. [Banfi *et al.* (2015)], but is based on effective field theory (EFT) methods, relies on setting up calculations using an EFT designed to describe QCD in the regime dominated by soft-collinear emissions. This EFT was formulated in at the beginning of this millenium [Bauer *et al.* (2001)] and is referred to as soft-collinear-effective theory (SCET). Here again higher-order corrections are classified into different contributions, but the classification is inherited from the EFT operators, rather than from an expansion of the phase space and the observable as in ref. [Banfi *et al.* (2015)]. A lot of progress has been made recently in SCET-based resummation (for a pedagogical reference on the subject see ref. [Becher *et al.* (2015)]). One notable calculation is that of the resummation of thrust in e^+e^- at N^{3}LL [Becher and Schwartz (2008)]. Another example is the N-subjettiness observable τ_N which was suggested in ref. [Stewart *et al.* (2010)] and a resummed prediction for the 1-jettiness using SCET methods was presented in ref. [Boughezal *et al.* (2015a)]. The importance of τ_N is twofold. We have already presented the use of this variable for slicing methods in NNLO calculations. On the other hand, it has been used for phenomenological studies, as a discriminating variable between QCD backgrounds and new physics searches since the distribution of e.g. τ_2/τ_1 is very different for BSM/EW signals and QCD backgrounds in the boosted regime [Altheimer *et al.* (2014)].

More generally, resummation techniques play an important role in other emergent sub-fields of QCD, most notably boosted methods and jet substructure techniques. The main idea behind these approaches is that it is possible to disentangle small signals from large QCD background thanks to the fact that the jets produced in the two cases have different properties. These differences originate from the fact that QCD matrix elements have soft-collinear singularities, hence QCD branchings are typically asymmetric, while when a colourless particle branches, the splitting is typically symmetric. This results in different properties of the structure of the particles in a jet. A large number of tools have been suggested recently to exploit this difference, such as the modified-Mass Drop tagger, Y-splitters, trimming, and pruning techniques (for a review see e.g. [Salam (2010); Altheimer *et al.* (2014)]). Considerable effort has been devoted to validate, compare and improve different methods using Monte Carlo simulations. A more recent activity has started recently to perform analytic calculations for these observables, both at fixed order, but also to all orders [Dasgupta *et al.* (2013)]. Understanding all-order properties would open up the possibility to design even better tools in a targeted manner. Furthermore, contrary to Monte Carlo simulations, it is possible to understand what are the properties of the signal (mass, quantum numbers, 2-prong or 3-prong decay) that are crucial for a given tagger to perform well. As a consequence, one can obtain robust conclusions about the performance of a tagger without running Monte Carlo simulations for every single interesting case.

Another subject that attracted considerable attention in recent years is the one of non-global (NG) logarithms. Their existence was first pointed out in ref. [Dasgupta and Salam (2001)] which showed that a number of NLL resummed calculations for NG observables (those sensitive to emissions only in a restricted phase space region, like the light-jet mass or the narrow jet broadening) were in fact not correct at NLL. The simple observation here is that all NLL calculations are based on using the eikonal approximation of Eq. (4.21), which implies that only emissions from the primary hard partons must be accounted for, while non-abelian emissions from secondary partons cancel at NLL when one includes real and virtual contributions. For NG observables instead emissions from a secondary soft, large-angle gluon can affect the distributions at NLL. This happens typically when the radiating secondary gluons is not observed, but the radiation from it is resolved. Progress has been made recently in the description of these logarithms. In ref. [Becher *et al.* (2016)] an equation is written whose solution allows to resum these logarithms to all orders, however no closed-form solution for the equations has been found yet. It is interesting also to point out that there is a connection between non-global and BFKL logarithms studied by Guido, which was formulated recently in ref. [Caron-Huot (2015)].

4.6 Epilogue and Conclusions

Precision QCD, which really started for hadron-hadron processes in 1979 [Altarelli *et al.* (1979a)], has turned out to be a fundamental tool in interpreting the data from the LHC. Its importance stems from small statistical uncertainty on Standard Model processes, because of their large cross sections at $\sqrt{s} = 13$ TeV, and the necessity of precise estimates of backgrounds in searches for new processes.

A personal note from G. Zanderighi

Writing this contribution has been a spur to look back at a number of papers written about forty years ago. Many older papers are hard to decipher, despite the fact that the topic discussed is by now well-understood, or even textbook material. This is because many papers rely on a language and concepts that are no longer in use, they lack vision of the big picture and focus on details that are not relevant. In the case of papers co-authored by Guido this is never the case. Guido had the ability to go directly to the heart of a problem, and to discuss its essence with a remarkable simple language. Still, Guido was not a person of hand-waving arguments. He moved on firm grounds, relied on logical proofs and proceeded with solid arguments. As a true and pure phenomenologist, Guido focused his attention on what experimental data are telling us, rather than on following beautiful theoretical constructs that lack experimental confirmation. This is what makes his contributions to particle physics timeless. Despite Guido being a true giant in physics, I like to remember my most human encounters with him. As a CERN fellow, I used to go to the Ferney–Voltaire market on Saturday mornings and I would buy a "sacristain", a huge puff-pastry covered in powdered sugar and almonds. Walking through the market I would eat it bite after bite out of the paper-bag, and when my face and clothes would be covered in sugar, I would hear Guido's warm and friendly voice telling me "Buon appetito!". Apparently, we liked to visit the market at about the same time. I could never tell whether he enjoyed or even noticed my slight embarrassment, but he certainly never looked the other way.

Acknowledgements

It is a pleasure to thank Andrea Banfi, Sophia Borowka, Fabrizio Caola and Claude Duhr for discussions on fixed order and resummed calculations.

Chapter 5

Precision physics at LEP

Siegfried Bethke
Max-Planck-Institut für Physik
Föhringer Ring 6
80805 Munich, Germany

Scientifically, the notion of *precision* relates to the reproducibility and repeatability of a measurement or a prediction. In practical terms, it relates to the degree of consistency of repeated measurements, i.e. to the spread of such results around their average or mean value. The term *precision measurements* usually implies that one expects significantly *higher* precision, i.e. smaller spread, than previous measurements. In absolute terms, the attribute *precise* is often used for results with precisions below the per-cent level, but this boundary also depends on the current and commonly accepted level of consistency. For instance, measurements or predictions in the field of Quantum Chromodynamics, the gauge field theory of the strong interaction, are regarded to be *precise* if they reach precisions of 1% or less, while in Quantum Electrodynamics, "precision" would require levels of at least sub-per-mille to even parts-per-million.

The electron-positron collider LEP, operated at CERN from 1989 to 2000, at e^+e^- centre-of-mass energies between 89 and 209 GeV, fulfilled the definition of "precision" in all technological and scientific respects, from the layout of the machine, its performance and stability, to the detectors, the data analysis to — last not least — the corresponding theoretical calculations, predictions and interpretations which are indispensable to extract physical quantities and conclusions from measurements.

Guido Altarelli was a great admirer of the precision reached at LEP, as is evident from the fact that he himself is author of many summaries and reviews of "Electroweak Precision Tests" at LEP, see e.g. [Altarelli *et al.* (1998b); Altarelli and Grunewald (2004); Altarelli (2011a)]. Guido also very actively prepared the grounds for performing electroweak precision tests at LEP, see e.g. [Altarelli (1982a, 1989b)], and he proposed model independent methods to detect effects of new physics beyond the Standard Model, based on vacuum polarisation effects and precise measurements at LEP [Altarelli and Barbieri (1991)].

5.1 The LEP collider

The **L**arge **E**lectron **P**ositron Collider was the largest and highest energy accelerator colliding electrons and positrons ever built. Located in a ring tunnel of 26.67 km circumference about 100 m below surface, LEP operated from 1989 to 2000, providing collision energies from 89 to 209 GeV, at 4 interaction points equipped with large multi-purpose particle detectors. The layout of LEP is depicted in Fig. 5.1.

Planning, design, approval and construction of LEP commenced from 1975 to 1989, followed by 11 years of successful operation until LEP's close-down in the fall of 2000. Initially, 3 phases of operation were planned: LEP-1 to run at c.m. energies around the rest-mass of the Z^0 boson, $M_Z \approx 91$ GeV, LEP-2 to run at energies at and above the W^+W^- production threshold, $2M_W \approx 160$ GeV, and finally LEP-3 as an electron-proton or proton-(anti-)proton collider after installing, additionally, proton beam acceleration and storage in the LEP tunnel.

Phase 2 was realised after the development and industrial production of superconducting RF cavities, providing nominal acceleration gradients of 6 MeV/m, and their staged installation from 1995 to 1999. An overview of the maximum energy, maximum luminosities and integrated luminosities provided by LEP between 1989 and 1999 is given in Fig. 5.2. Phase 3 was altered and finally realised as the Large Hadron Collider (LHC), after the shutdown of LEP in late 2000, removing the LEP

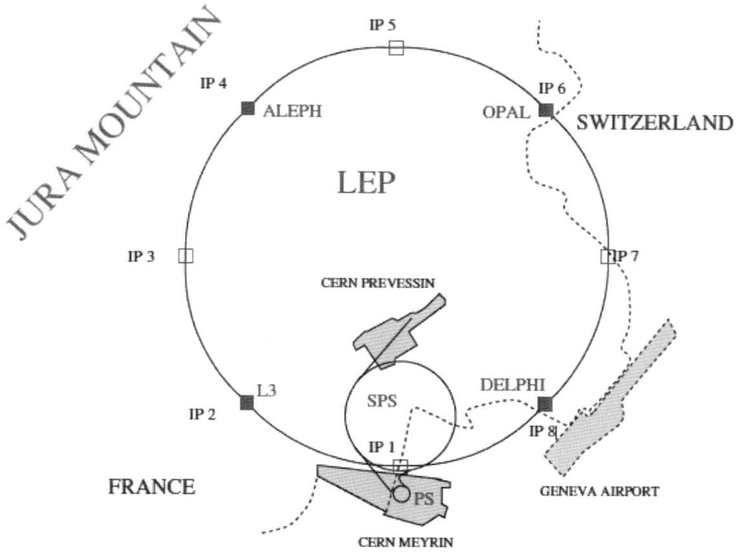

Fig. 5.1 Layout of the LEP ring and the 8 access points IP1-8, on the border of France and Switzerland. The four LEP experiments L3, ALEPH, OPAL and DELPHI are installed at the even-numbered access points. Positrons travel clockwise, electrons anticlockwise. The location of two LEP injectors, the SPS (Super Proton Synchrotron) and the PS (Proton Synchrotron), is also indicated. From [Brandt *et al.* (2000)].

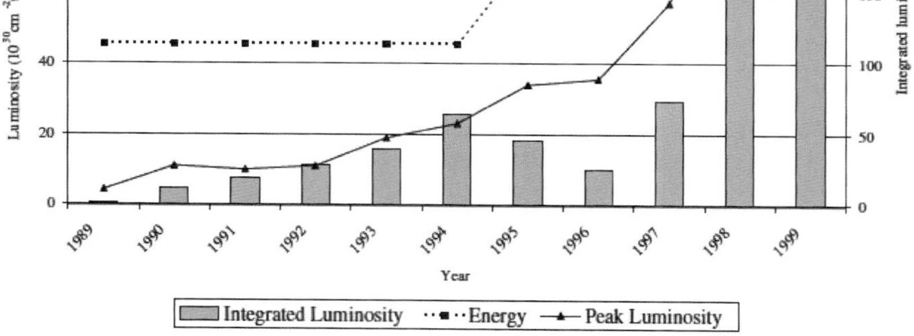

Fig. 5.2 Maximum energy, maximum luminosity and integrated luminosity per experiment of LEP for each year between 1989 and 1999. From [Brandt *et al.* (2000)].

electron-positron infrastructure altogether. LHC started first operation in 2008, with first proton-proton collisions in 2009.

The history, design, construction and operation of LEP is documented in a large number of publications, see e.g. [Brandt *et al.* (2000); Assmann *et al.* (2002); Hubner (2004); Schopper (2009)] and references quoted therein. Some of the many *highlights* of this machine shall be reviewed here.

The most general demands on LEP were to reach and provide highest energy, luminosity and precision. The quest for the highest possible beam *energies* and collision *luminosities* is obvious in terms of determining the parameters of the Standard Model (SM) of electroweak and strong interactions, i.e. the masses and couplings of gauge bosons (W and Z) and of the fundamental fermions (quarks and leptons), but also for direct searches of new particles and phenomena (e.g. Supersymmetry) and missing ingredients of the SM (the Higgs boson). *Precision* is a key issue for most of these aspects, because precise measurements of parameters constitute important consistency checks of the underlying theories, and they provide possibilities to test energy scales much higher than the c.m. energies of the collisions, through the effects of virtual processes like vacuum polarisation and vertex corrections.

Among other specifications of the machine, precision knowledge of the actual beam energy at the collision points was one of the key issues for LEP operation, most important for the precise determination of the masses of the Z boson and the W-bosons. The initial goal of determining M_Z to 20 MeV (corresponding to about 0.2 per-mille accuracy) was finally surpassed by a factor of 10 to 20 parts-per-million, due to the very elaborate and efficient energy calibration and operation of LEP phase-1.

This precision of beam energy calibration was achieved by resonant depolarisation [Assmann *et al.* (1999)] of the transverse self-polarisation of the beams that reached, at beam energies around 45 GeV, up to 60%. This procedure provides accuracies of single measurements to the level of 1 MeV, however it then unveiled a number of rather subtle and largely unexpected effects which had to be accounted and corrected for. Of such, energy deviations due to variations of the circumference of LEP, caused by geological movements due to earth tides and hydrological strains, and effects caused by leakage currents from the French Train Grande Vitesse (TGV), DC-operated electrical trains circulating on nearby railway lines, are perhaps the most remarkable and "entertaining" systematic effects, demonstrating the enormous size of precision achieved in operating such a large and delicate machine [Brandt *et al.* (2000)]. These efforts and achievements finally reduced the contribution of the average beam energy to the Z-mass uncertainty to 1.7 MeV (i.e. 18 parts-per-million) and to its overall decay width, Γ_Z, to 1.3 MeV [Assmann *et al.* (1999)].

At the higher c.m. energies of LEP phase 2, self-polarisation of the beams could not be established due to many depolarising effects, so the calibration of beam energy was performed at lower energies and then transferred "upwards" using

Table 5.1 Basic parameters of LEP.

	LEP-1	LEP-2
beam energy up to	55 GeV	104.5 GeV
magnetic field (bending dipoles)	0.065 T	0.111 T
accel. voltage per turn	260 MV	2700 MV
clystron power	16 MW	16 MW
RF cavities	Cu (normal conducting)	Cu-Nb (super conducting)
	128 at P2 and P6	272 at P2,4,6,8
beam currents	3 mA	5 mA
number of e^+e^- bunches	4×4	4×4 ($\times$ 2 bunchlets)
max. luminosity	$1.6 \times 10^{31}\ cm^{-2}\ s^{-1}$	$5 \times 10^{31}\ cm^{-2}\ s^{-1}$
c.m. energy spread	60 MeV	140 - 260 MeV
c.m. energy uncertainty	1.4 MeV	20-40 MeV
beam lifetime	$\sim$ 6–8 h	$\sim$ 5 h

nuclear magnetic resonance (NMR) and flux-loop measurements in the LEP dipole magnets [Assmann *et al.* (2005)]. This provided contributions to the uncertainty of the W mass as low as 10 MeV.

A summary of the most important parameters of LEP operation (phases 1 and 2) is given in Table 5.1.

5.2 The LEP detectors

Four large general-purpose particle detector systems had been developed, approved, constructed and operated at LEP. In general, they all shared common features like hermeticity, precision charged particle tracking and momentum measurement using silicon vertex detectors, gaseous tracking detectors and solenoidal magnetic fields, particle identification, electromagnetic and hadron calorimeters, muon tracking and identification, however the four detectors used different technologies for their sub-detectors, magnet systems and detection strategies:

- the ALEPH detector [Decamp *et al.* (1990)] with its (at that time) novel and sophisticated double-sided silicon vertex detector,
- the DELPHI detector [Aarnio *et al.* (1991)] with emphasis on particle identification by dE/dx sampling in a large TPC with RICH detectors,
- the L3 detector [L3 Collaboration (1990)] with a large BGO electromagnetic calorimeter and all detector components placed inside the world's largest conventional 0.5T magnet, and
- the OPAL detector [Ahmet *et al.* (1991)] which was based on largely conventional technologies and anticipated as a safe bet to fully function right from the start of LEP.

As an example for the size and general structure of the LEP experiments, a sketch of the DELPHI detector is displayed in Fig. 5.3. A description of all four LEP detector systems would exceed the scope of this contribution. Overviews of the LEP

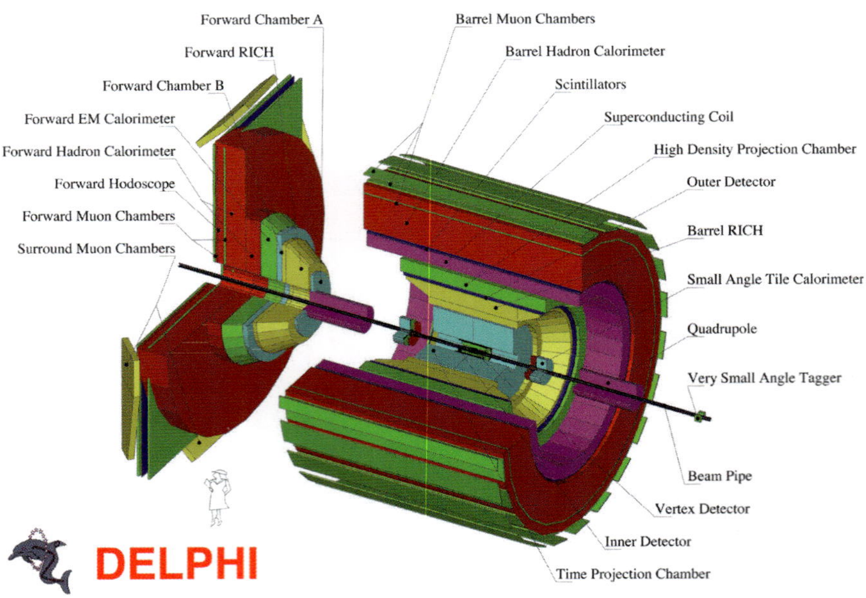

Fig. 5.3 The DELPHI detector at LEP.

detectors can be found e.g. in [Schopper (2009)] and [Sauli (2004)]. Here, and owing to the subject of *precision* at LEP, only two of the many innovative technologies and subdetectors shall be mentioned in more details: the OPAL Silicon-Tungsten (SiW) luminometer and the ALEPH double-sided silicon vertex detector.

For precision measurements of production cross sections and (partial) decay widths of physical objects like the Z boson, measurements must be compared to well known "standard" cross sections that are theoretically and experimentally well defined, that can be measured and monitored with high accuracy, and that are largely independent from any of the anticipated new physics measurements. At LEP, the measurement of small angle Bhabha (electron-positron) scattering was chosen to precisely determine the luminosity of the LEP beams and of recorded data samples, which — at very small scattering angles — is basically determined by precision calculations of Quantum Electrodynamics, and is very little affected by effects of the weak interaction.

In the initial phase of LEP-1 running, all 4 experiments had luminometers which determined beam luminosities to accuracies of about 1%. They were soon replaced by more sophisticated and precise devices which achieved accuracies to about 10^{-4}. One of those second-generation luminometers was the silicon-tungsten luminometer of OPAL [Abbiendi *et al.* (2000)]. It was composed of two calorimeters encircling the LEP beam pipe, on opposite sides of the interaction point, at approximately 2.5 m from the interaction point, detecting electrons from small angle Bhabha scattering at angles between 25 and 58 mrad w.r.t. the beam pipe. Each calorimeter was a

Fig. 5.4 One of the calorimeters of the OPAL SiW luminometer during construction. The diameter of the central hole for the beampipe was 10 cm. The anticipated precision of the luminometer required an accuracy of the radial positioning of 4.4 μm.

stack of 19 layers of silicon sampling wafers interleaved with 18 tungsten plates. Electromagnetic showers generated in a stack of 1 radiation length of tungsten absorber plates were sampled by 608 silicon detectors with 38,912 radial pads of 2.5 mm width. A photograph of one of the calorimeters is represented in Fig. 5.4.

The fine segmentation of the detector, combined with the precise knowledge of its physical dimensions, allowed the trajectories of incoming 45 GeV electrons or photons to be determined with a total systematic error of less than 7 microns. The total systematic measurement uncertainty was 3.4×10^{-4}, and therefore smaller than the theoretical error of 5.4×10^{-4} that was assigned to the QED calculation of the Bhabha acceptance at that time. It thus contributed negligibly to the total uncertainty in the OPAL measurement of the relative invisible decay width of the Z boson, $\Gamma_{inv}/\Gamma_{\ell^+\ell^-}$.

The ALEPH Silicon Vertex Detector [Mours *et al.* (1996)] was the first detector operating in a colliding beam environment that used silicon strip detectors which provide readout on both sides, allowing a three-dimensional point measurement for the trajectory of charged particles. Since then double-sided detectors have been applied in other experiments as well, and have become a standard technology. Two layers of silicon strip detectors were arranged in two concentric barrels around the beam pipe with an average radius of 6.3 cm for the inner layer and 10.8 cm for the outer layer. The strip detectors had readout strips on both sides. The strips on one side were parallel to the beam direction while the strips on the other side were perpendicular to the beam.

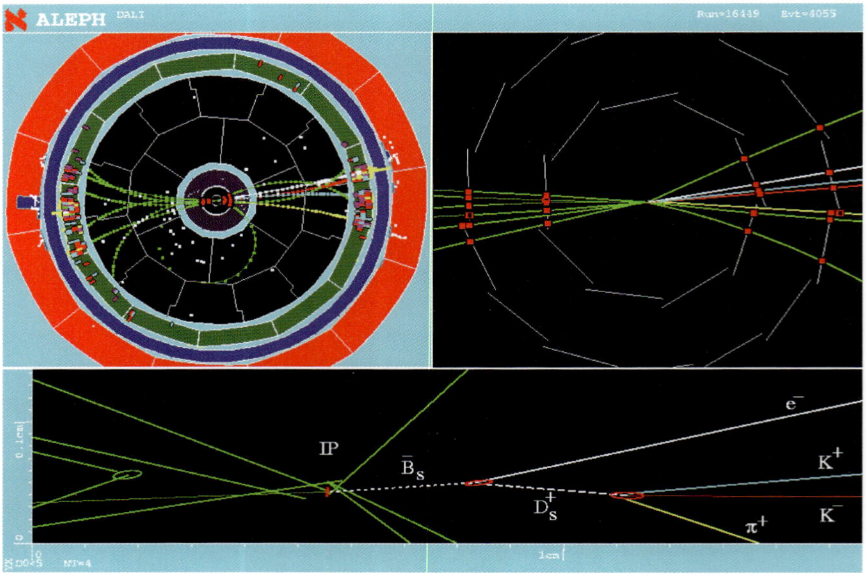

Fig. 5.5 Candidate for an event $e^+e^- \to Z^0 \to b\bar{b}$ as recorded by the ALEPH detector: full detector view perpendicular to the beam axis (top left), silicon vertex detector with fitted tracks (top right), and blow-up of reconstructed tracks and decay vertices close to the collision vertex (bottom).

The detector system was fully commissioned during 1991 while taking data at the Z^0 resonance. The achieved spatial resolution of the complete 73 728-channel device was 12 μm in the plane perpendicular to the beam, and 12 μm in the direction of the beam. This precision, close to the primary collision vertex of LEP, allowed the measurement of the lifetimes and decay topologies of short-lived particles with typical decay lengths down to a few tenths of a millimetre. An e^+e^- collision event with secondary decay vertices of hadrons containing b-quarks, recorded by ALEPH, is shown in Fig. 5.5.

Until 1995, the initial ALEPH silicon vertex detector played a vital role in the measurement of B-hadron and τ-lepton lifetimes, of direct B_d^0 mixing and limits on B_s^0 mixing, of $\Gamma_{b\bar{b}}$, and of limits on the existence of new particles. For the further running of ALEPH at LEP phase 2, the device was replaced by an advanced version of silicon strip detector, with larger spacial acceptance, increased radiation hardness and smaller strip widths.

5.3 Recap of Standard Model terminology

The minimal Standard Model (SM) of electroweak interactions in lowest order ("Born Approximation") describes processes like $e^+e^- \to f\overline{f}$ using only 3 free parameters:

- α (the electromagnetic coupling *or* fine structure constant),
- G_F (the Fermi constant), and
- $\sin^2\Theta_W$ (the weak mixing angle),

which are fundamental parameters whose numerical values are not given by theory but must be determined by experiment. For instance, G_F and $\sin^2\Theta_W$ are known from the muon lifetime and from deep inelastic neutrino-nucleon scattering experiments, respectively. Alternatively, α, G_F and M_Z, the mass of the Z^0 boson, can be chosen as the 3 free and fundamental parameters of the SM, since

$$\sin^2\Theta_W\ \cos^2\Theta_W = \frac{\pi\alpha}{G_F\sqrt{2}}\frac{1}{M_Z^2}.$$

The process $e^+e^- \to f\overline{f}$, where f can be any of the fundamental fermions of matter, i.e. $(f\overline{f}) \equiv (e^+e^-)$, $(\mu^+\mu^-)$, $(\tau^+\tau^-)$, $(\nu_e\overline{\nu_e})$, $(\nu_\mu\overline{\nu_\mu})$, $(\nu_\tau\overline{\nu_\tau})$ or (quark-antiquark) like $(u\overline{u})$, $(d\overline{d})$, $(s\overline{s})$, $(c\overline{c})$, $(b\overline{b})$ and $(t\overline{t})$, is described by s-channel annihilation diagrams like Fig. 5.6(a), which proceed via an intermediate photon (γ) or a Z^0 boson which then branches into the final state fermion-antifermion pair. For the electron-positron final state, also the t-channel scattering diagram given in Fig. 5.6(b) contributes. Neutrino final states proceed via s-channel Z^0 (all neutrino flavours) and t-channel W exchange (electron-neutrinos only).

Further diagrams and final states are possible for e^+e^- annihilation above the W-boson and Z-boson pair production thresholds, see e.g. Fig. 5.7, where W bosons instantly decay into either a lepton and its associated (anti-)neutrino or into a quark-antiquark pair as depicted in Fig. 5.8, and the Z boson into a fermion-antifermion pair (charged leptons, neutrinos or quarks).

Further processes and final states which were relevant and produced at LEP were e.g. two-photon scattering processes and initial state photon radiation, see Fig. 5.9, and higher order radiation processes like gluon emission off quarks, final state photon radiation off charged final state fermions, and virtual processes like

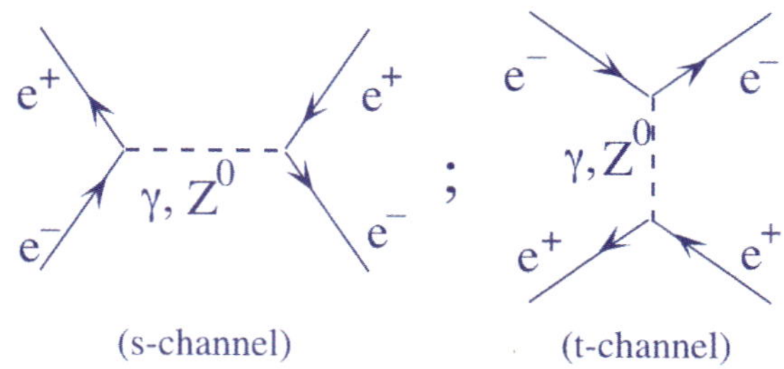

(a) (b)

Fig. 5.6 s-channel (a) and t-channel (b) Feynman diagrams for the process $e^+e^- \to e^+e^-$, also called "Bhabha scattering". For other charged leptons in the final state, like $(\mu^+\mu^-)$, $(\tau^+\tau^-)$ or (quark-antiquark), only the s-channel annihilation process exists.

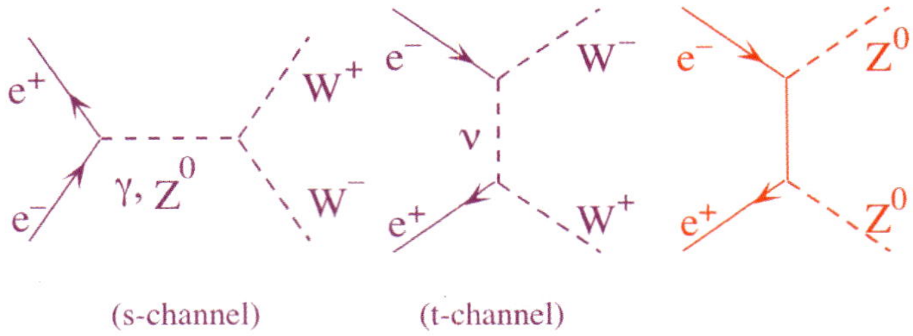

Fig. 5.7 Additional processes and final states in e^+e^- annihilations for phase 2 of LEP running, i.e. at c.m. energies above the W^+W^- production threshold.

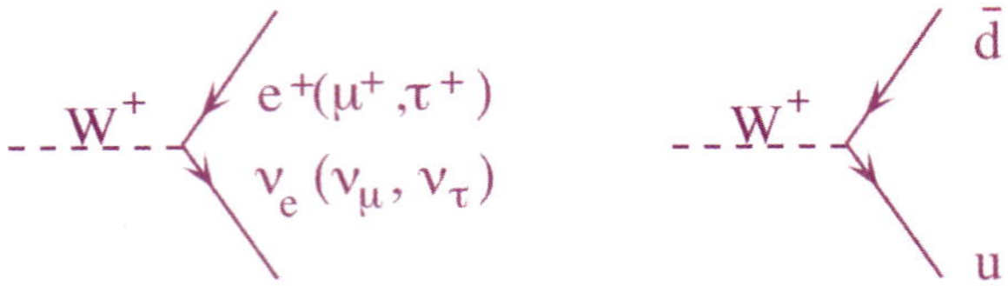

Fig. 5.8 Decay modes of W bosons.

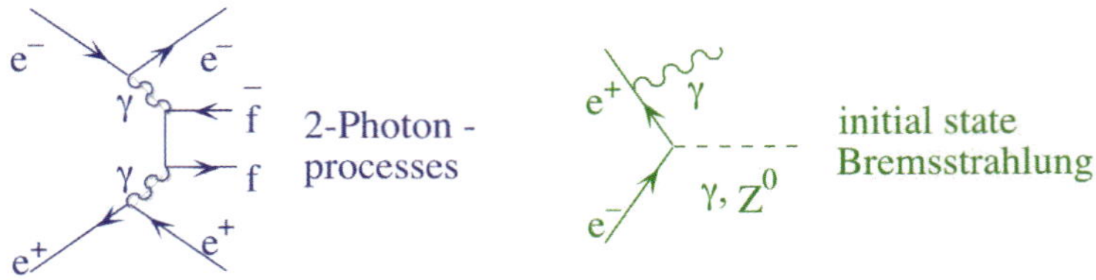

Fig. 5.9 Further processes relevant and produced at LEP.

vacuum polarisation and vertex corrections, some of which are depicted in Figs. 5.10 and 5.11.

The production cross sections for fermion-antifermion pairs f in e^+e^- annihilations around c.m. energies of the mass of the Z boson, $\sqrt{s} = E_{cm} \approx M_Z$, can be written as

$$\sigma_f(s) = \sigma_f^0 \cdot \frac{s\Gamma_Z}{(s - M_Z^2)^2 + M_Z^2\Gamma_Z^2} + \text{“}\gamma\text{”} + \text{“}\gamma Z\text{”},$$

where the first term parametrises the resonant cross section around the Z mass, the second term, “γ” is the electromagnetic cross section for point-like annihilation processes — a function falling like $1/s$ — and “γZ” denotes electroweak interference terms. σ_f^0 are the pole cross sections for fermions f,

$$\sigma_f^0 = \frac{12\pi}{M_Z^2}\frac{\Gamma_e\Gamma_f}{\Gamma_Z^2},$$

and Γ_Z, Γ_f and Γ_e are the total width and the partial widths of the Z into fermions f and electrons, respectively.

Measurements of s-dependent cross sections around the resonance of the Z boson provide *model independent* results for M_Z, Γ_Z, Γ_f and σ_f^0. In the SM, Γ_f are no free parameters, but are parametrised as functions of the so-called vector and axial-vector constants $g_{a,f}$ and $g_{v,f}$:

$$\Gamma_f = \frac{G_f M_Z^3}{6\pi\sqrt{2}} \left[g_{a,f}^2 + g_{v,f}^2\right] N_{c,f},$$

where $N_{c,f} = 3$ for quarks and 1 for all other fermions; $g_{a,f} = I_{3,f}$ and $g_{v,f} = I_{3,f} - 2Q\sin^2\theta_w$ with $I_{3,f}$ the 3rd component of the weak isospin ($= \pm 1/2$) and Q the electric charge of the fermion.

Radiation corrections in the SM are treated in two different classes:

- Photonic corrections, cf. Fig. 5.10, may be large and cause corrections up to $\mathcal{O}(100\%)$; they may depend on selection criteria and their effects can thus be reduced; and they are factorisable, i.e. their contributions can be factored out by expressions like $(1 + \delta_{rad})$.
- Non-photonic corrections, c.f. Fig. 5.11, involving loops, vertex corrections etc. from gluons, fermions, W, Z and Higgs bosons, even if they are too heavy to be produced as "free" particles at the given collider energies. These corrections are in general smaller than the photonic ones, of $\mathcal{O}(10\%)$, but they are independent of selection criteria. However, they can be absorbed in running couplings.

Higher order radiation corrections render couplings and parameters of the SM to become energy dependent, and the common notion is to deal with an effective weak mixing angle, $\sin^2\theta_{eff}(s)$, with the electromagnetic coupling $\alpha(s) = \alpha/(1 - \Delta\alpha)$ where $\Delta\alpha = 0.064$ at $\sqrt{s} = M_Z$, with $N_{c,f}\left(1 + \frac{\alpha_s}{\pi} + 1.4\left(\frac{\alpha_s}{\pi}\right)^2 + ...\right)$ for quarks, modified by the running strong coupling constant, $\alpha_s(s)$, and with $M_W^2/M_Z^2 = \rho\cos^2\theta_w$ with $\rho = 1/(1 - \Delta\rho)$, whereby

$$\Delta\rho = 0.0026\frac{M_t^2}{M_Z^2} - 0.0015\ln\left(\frac{M_H}{M_W}\right).$$

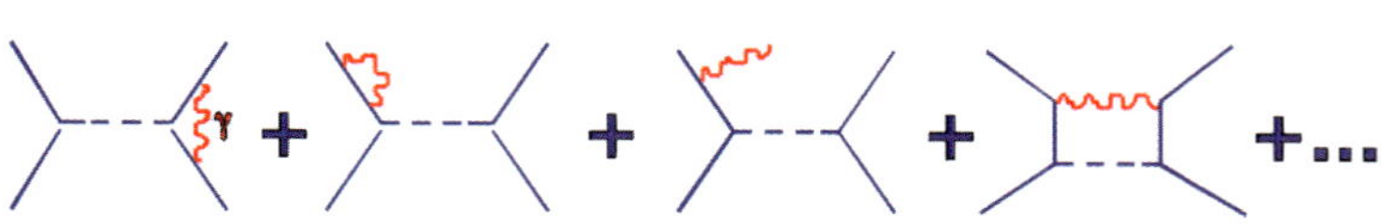

Fig. 5.10 Photonic processes and corrections.

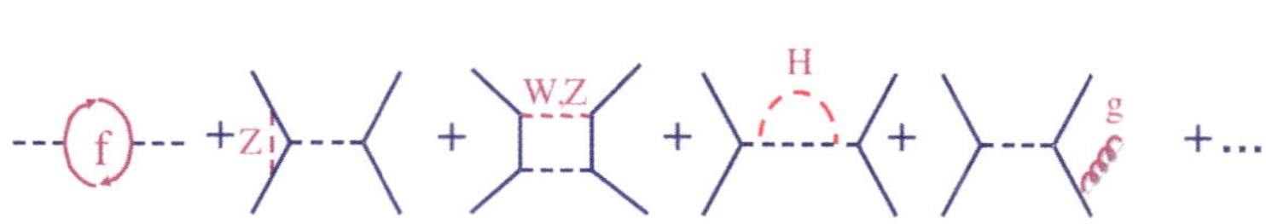

Fig. 5.11 Non-photonic processes and corrections.

Inserting the running couplings into the "Born" approximation, the partial Z decay widths Γ_f, the cross sections and other observables acquire dependences on

- the top quark mass M_t,
- the Higgs boson mass M_H, and
- the strong coupling parameter α_s.

This then provides the possibility for indirect determinations (i.e. fits) of M_t, M_H and α_s from a combination of measurements of electroweak observables. From the numerical structure of the $\Delta\rho$ parameter given above it is quite obvious that such determinations, for them to provide significant results, must be based on *very precise* measurements: taking the top-quark and Higgs-boson masses as we know them today, from direct production and measurements of these objects, $M_t = 173.1 \pm 0.6$ GeV and $M_H = 125.09 \pm 0.24$ GeV [Patrignani *et al.* (2016)], leads to contributions to $\Delta\rho$ of 0.009 and -0.0012, respectively. Aiming, for example, at relative uncertainties of 5% on M_t and 10% on M_H, in indirect determinations from fits to precision electroweak data alone, requires control of all statistical and systematic errors to levels corresponding to 10^{-3} and 5×10^{-4}, respectively.[1]

A more detailed account of the basic theory, calculations of higher order QED, electroweak and QCD (strong interaction) corrections, implementations into Monte-Carlo programs, fitting algorithms and analysis procedures is given in the final reports of the LEP Electroweak Working Group [Schael *et al.* (2006, 2013)] and respective web updates [LEP electroweak working group (2012)], as well as in many summaries and reviews such as those given by Guido himself, e.g. [Altarelli *et al.* (1998b); Altarelli and Grunewald (2004)], and references quoted therein.

5.4 Precision electroweak results of LEP

One of the main tasks of the LEP phase-1 experimental programme was the measurement and determination of the essential parameters of the Z resonance, its mass, its width, its branching fractions, and the angular distribution of its decay products. About 17 million Z decays were accumulated by the four experiments at LEP. In most of the combined and global analyses of these data, LEP data were supplemented by the results from the SLD detector at the Stanford Linear Accelerator (SLC). SLD collected about 600,000 Z decays overall, however with electron beam polarisation of up to about 80%, adding additional means for electroweak precision tests that LEP could not provide.

The final results of electroweak precision measurements around the Z pole were presented in [Schael *et al.* (2006)]. Owing to the precision of the LEP beam energies, the mass and total width of the Z boson were determined with unprecedented

[1] These numbers are given for illustration only.

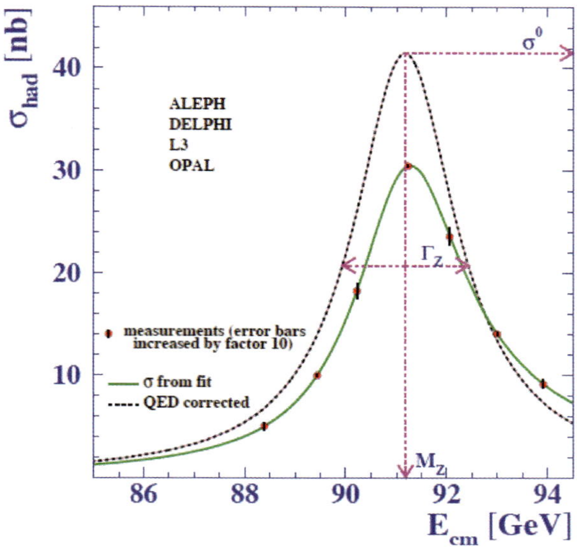

Fig. 5.12 Average of measurements of the hadronic cross section as a function of e^+e^- centre of mass energy. The full line is a model independent fit to the measurements; the dashed line is obtained by correcting the former for QED photonic effects. From [Schael *et al.* (2006)].

precision:

$$M_Z = 91.1875 \pm 0.0021 \text{ GeV}$$
$$\Gamma_Z = 2.4952 \pm 0.0023 \text{ GeV},$$

corresponding to a precision of 23 parts-per-million for M_Z and 0.9 per-mille for Γ_Z.

One of the key measurements, the cross section for $Z^0 \to$ hadrons,[2] is given in Fig. 5.12, from which the basic parameters of the Z boson are being derived. A particular important derivation of these data, with global and even cosmological relevance, is the determination of the number of light neutrino flavours coupling to the Z boson, N_ν. It is derived from the Z invisible partial decay width, Γ_Z^{inv}, which is determined from the measurements of visible Z decays into charged leptons and quarks, and results in

$$N_\nu = 2.9841 \pm 0.0083,$$

assuming lepton universality and excluding other contributions to Γ_Z^{inv} other than SM neutrinos. It is therefore concluded that only the three known sequential generations of fermions exist in nature.

The combined precision measurements of cross sections, masses, ratios of decay widths, $R_\ell = \Gamma_\ell/\Gamma_{had}$, asymmetries of angular distributions of the final state

[2]About 70% of all Z bosons decay hadronically, into a quark-antiquark pair.

Fig. 5.13 Comparison of measurements with the expectation of the SM, calculated for the five SM input parameter values in the minimum of the global χ^2 of the fit. The electroweak precision fit includes results from LEP (upper 13 observables), from SLD (A_l) and from Tevatron (M_W, Γ_W, M_t). From [LEP electroweak working group (2012)].

fermions from Z-boson decays (including forward-backward asymmetries like $A^{0,\ell}_{f,b}$), and from LEP-external measurements like the left-right asymmetry from SLD, the top-quark mass from the Tevatron proton-antiproton collider and the mass of the W boson from LEP phase-2 and from Tevatron, are listed in Fig. 5.13. They are compared to the results of a global fit of the SM predictions to all these measurements [LEP electroweak working group (2012)]. To the far right, the "pull"-factors, i.e. the deviation between the measurement and the fit result, in terms of the number of standard deviations, are also shown. The measurements agree with the SM expectations within their assigned uncertainties and with a "normal" distribution of deviations; only three out of 18 measurements deviate by more than 1 standard deviation, and one by more than 2 (and less than 3) standard deviations.

Even though the overall picture does not indicate the need for or presence of new physics beyond the SM, the largest deviation, of the forward-backward asymmetry of b-quarks, $A^{0,b}_{f,b}$, gave and still gives rise to speculations about possible reasons

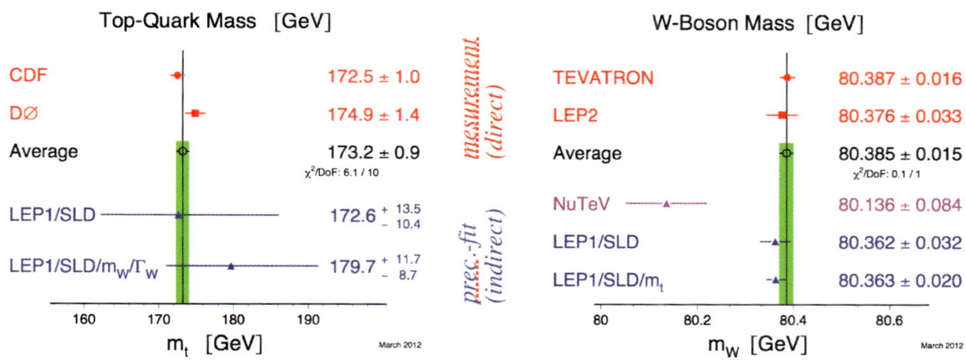

Fig. 5.14 Results for the top-quark and the W boson mass, from direct measurements (M_t from Tevatron experiments, M_W from Tevatron and from LEP-2) and indirect determinations from electroweak fits to the precision data from LEP-1 and the SLD detector at the Stanford Linear Collider (SLC). From [LEP electroweak working group (2012)].

for it to occur. No compelling explanation was found so far, and the origin of this deviation is likely to be due to a statistical fluctuation or to unknown (measurement or theoretical) systematic effects.

The precision of these measurements manifests in the possibility to determine quantities like the mass of top-quark (discovered at the Tevatron in 1995, at the end of LEP phase-1 running), and the mass of the Higgs boson (discovered at the LHC in 2012, long after LEP's active phases), from their effects through radiation corrections and quantum loop effects, see the previous section of this article. This can be admired by the results shown in Fig. 5.14, where the direct measurements of the masses of these objects, shown on top, are compared with the results of the indirect results obtained from the precision fits. While the uncertainties of the indirect fits are larger than those of the direct measurements, the agreement between both are breathtaking — considering the smallness of the radiative corrections and therefore the precision needed to obtain these results.

Similarly as for indirect determinations of the W boson and the top-quark mass, significant limits for the mass of the SM Higgs boson emerged from the electroweak precision fits, and from direct (but unsuccessful) searches for Higgs boson production at LEP. The direct searches resulted in a lower limit of the Higgs boson mass of 114.4 GeV [Barate *et al.* (2003)], while

$$M_H = 129^{+74}_{-49}|\mathrm{GeV}$$

was obtained from the precision fits [Schael *et al.* (2006)] — a value that was frequently updated and changed within the uncertainties, but was amazingly close to the actual value of the Higgs boson, as known today, of $M_H = 125.09 \pm 0.24$ GeV [Patrignani *et al.* (2016)].

While it was important to obtain early indications about the possible mass range of the missing, at that time, ingredients and objects of the SM, these results today,

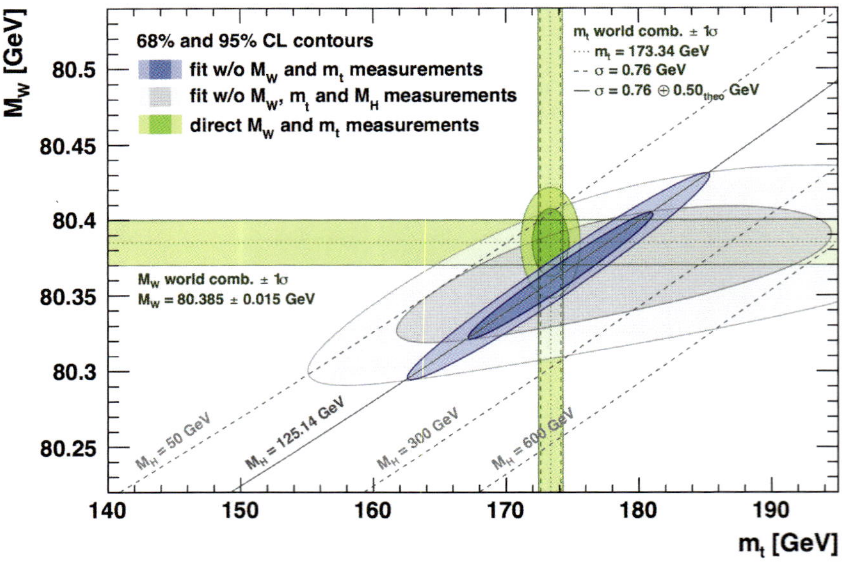

Fig. 5.15 Comparison of direct measurements of the mass of the top-quark and of the Higgs boson (vertical and horizontal green bands) with indirect determinations (fits to precision data) excluding direct measurements (grey shaded contours) and including the measured Higgs-boson mass (blue shaded contours). From [Baak *et al.* (2014a)].

with the top-quark and the Higgs boson discovered and directly measured, still serve as important checkmarks for the consistency and reliability of the SM. This consistency is also obvious from Fig 5.15, where the more recent status of indirect precision fits and direct mass measurements, now also including the value of the Higgs boson mass from the experiments at the LHC, are displayed [Baak *et al.* (2014a)]. It is intriguing to see that the SM indeed still is in excellent shape!

The success of the SM continued at LEP phase-2, at c.m. energies above the Z-boson resonance [Schael *et al.* (2013)] already became obvious from the results, including the LEP measurement of the W-boson mass discussed above. The full beauty of agreement of s-dependent, measured cross sections of virtually all processes predicted by the SM, in the accessible range of c.m. energies, is further demonstrated in Fig. 5.16. Note that the processes $e^+e^- \to e^+e^- q\overline{q}$ (c.f. Fig. 5.9) and $e^+e^- \to \gamma\gamma$ do not see the Z-boson resonance. Also note that, unfortunately for those days' LEP-enthusiasts, the Higgs-boson — now known to have a mass of $M_H = 125.09 \pm 0.24$ GeV — was just outside of the reach of LEP.[3] The measured shape of the W-boson pair cross section directly proved the existence of the

[3]The main production mode of Higgs bosons would have been Higgs-Strahlung off the intermediate Z boson, $e^+e^- \to ZH$, which requires c.m. energies of at least $M_Z + M_H \sim 216$ GeV to obtain viable cross sections, while the ultimate LEP energy was 209 GeV.

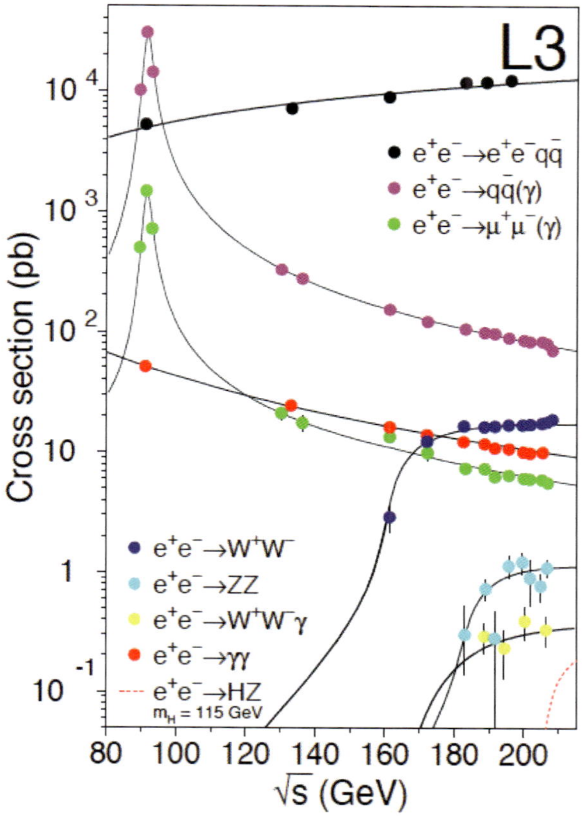

Fig. 5.16 Cross sections of electroweak SM processes observed at LEP phase-1 and phase-2. The curves show the theoretical predictions based on the SM. From [Schael *et al.* (2013)].

electroweak triple-gauge couplings (c.f. the left-most diagram of Fig. 5.7), and the mass and width of the W-boson was precisely measured, at LEP phase 2, to

$$M_W = 80.376 \pm 0.33 \text{ GeV, and}$$
$$\Gamma_W = 2.195 \pm 0.83 \text{ GeV.}$$

Further measurements and combinations of results obtained at LEP phase-2 are summarised in [Schael *et al.* (2013)].

5.5 Precision QCD results from LEP

LEP data also were a source for precision studies of the strong interactions and its underlying theory, Quantum Chromodynamics (QCD), which is regarded to be part of the Standard Model of particle physics. QCD also and very specifically was at the heart of Guido Altarelli. In fact he belongs to the "fathers" of this theory with

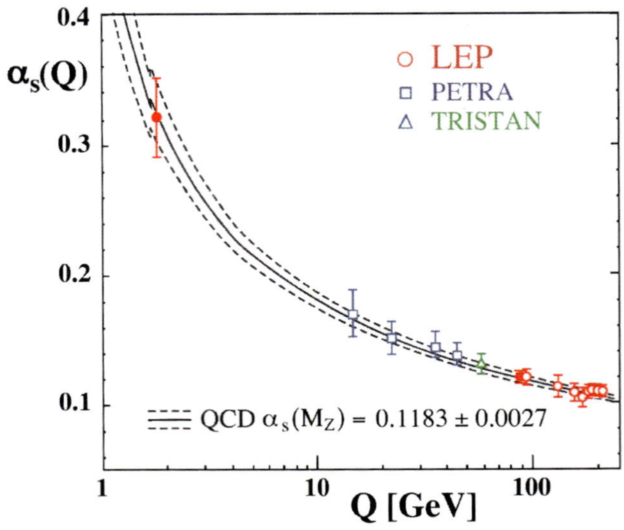

Fig. 5.17 Summary of measurements of $\alpha_s(Q \equiv E_{cm})$ from LEP. Results from previous e^+e^--colliders are also included. Open symbols are from event shapes in resummed NLO, filled symbols from τ and Z-boson hadronic decay widths, in full NNLO QCD. The curves represent the QCD predictions of the running coupling for the world average value of $\alpha_s(M_Z)$ at that time. From [Bethke (2004)].

epoch-making publications like [Altarelli and Parisi (1977)], and he summarised and reviewed the theoretical and experimental status of QCD at many instances, like [Altarelli (1989a)] (before the advent) and [Altarelli (2002)] (after the shutdown of LEP).

The spectrum of QCD studies at LEP was summarised e.g. in [Bethke (2004)]. Here, only precision determinations of the strong coupling α_s shall be briefly reviewed. At LEP, the strong coupling was determined from a variety of observables and their respective QCD predictions. The most significant determinations were obtained from measurements of hadronic event shapes and jet-rates, from the hadronic width and the spectral functions of τ-lepton decays, and from the hadronic decay width of the Z boson or the fits of the electroweak precision data. A summary of the results available in 2004 is reproduced in Fig. 5.17.

Almost all QCD calculations and predictions that were available at the time of LEP running have been significantly enhanced, from next-to-leading order in perturbation theory (NLO, $\mathcal{O}(\alpha_s^2)$) at the beginning of LEP, to next-next-to-leading order (NNLO, $\mathcal{O}(\alpha_s^3)$) plus resummation in next-to-leading-logarithmic approximation (NLLA), and even to N^{3}LO in some cases, at the time thereafter until today. Since then, the results summarised in [Bethke (2004)] have been reanalysed and significantly improved, see the QCD review in [Patrignani *et al.* (2016)] for the most recent update of α_s determinations at LEP and elsewhere. From this, the following main results of α_s, based on LEP data only, can be derived:

from τ decays: $\alpha_s(M_Z) = 0.1192 \pm 0.0018$ (in N^{3}LO),

from electroweak precision fits: $\alpha_s(M_Z) = 0.1196 \pm 0.0030$ (in N^{3}LO),

from hadronic event shapes and jets: $\alpha_s(M_Z) = 0.1196 \pm 0.0036$ (in NNLO).

These results are remarkably consistent with each other, with practically no correlations between the three different classes. They are also consistent with the current world average, $\alpha_s(M_Z) = 0.1181 \pm 0.0011$ [Patrignani *et al.* (2016)], although systematically lying on the higher side. This indicates that other results that enter the determination of the world average, like e.g. from deep inelastic scattering, but also from other studies involving a broader range of (LEP and non-LEP) e^+e^- data and alternative analysis methods, partly arrive at smaller α_s values than those from LEP given above; see the discussion in [Patrignani *et al.* (2016)].

Note that the most precise results obtained at LEP, from hadronic τ- and Z-boson decays, at the smallest energy scale of M_τ and the overall reference scale of M_Z, provide a very significant proof of the running of α_s as predicted by QCD, and thus, of asymptotic freedom of quarks and gluons, excluding a hypothetical assumption of an energy *in*dependent coupling by more than 13 standard deviations (6 standard deviations at the time of 2004). These results were a decisive input for granting the Physics Nobel Prize of 2004, "for the discovery of asymptotic freedom in the theory of the strong interaction", to David J. Gross, H. David Politzer and Frank Wilczek [Nobel Committee (2004)].

5.6 Summary

The LEP collider provided high statistics of electron-positron collisions at centre-of-mass energies around the mass of the Z boson, from 1989 to 1995, and continued — after upgrades to using superconducting accelerator cavities — at energies above the Z boson, crossing the W-boson pair production threshold at 160 GeV, up to its highest c.m. energy of 209 GeV. The LEP stopped operation in the year 2000, and was replaced by the Large Hadron Collider being installed in the same tunnel, starting to deliver proton-proton collisions in the (multi-)TeV c.m. energy range in 2009.

Based on the precise energy calibration and efficient operation of LEP, the four large general purpose detector systems ALEPH, DELPHI, L3 and OPAL, established precision measurements of Standard Model observables and parameters, most of which exceeded initial expectations and are unsurpassed in testing the underlying theory and in determining its free parameters, to the level of its radiative corrections.

The LEP era manifested an overwhelming degree of glory, success and consistency of the Standard Model of electroweak and the strong interactions. This review could only summarise and point out the few most important measurements and determinations. The story of success of the Standard Model continues with the results from LHC, demonstrating its validity up to the multi-TeV range and possibly even beyond, perhaps reaching out and up to the Planck scale. This success goes together with the unexpected absence of any significant signal for new physics beyond the Standard Model, which already started to emerge at LEP, too. Such a situation was considered "*possible but not natural*" by Guido Altarelli [Altarelli (2011c)]. In the absence of direct new physics signals at current and future collider projects, precision measurements and calculations that can reach to energy scales far beyond those realised in the laboratory will remain to be indispensable tools to further approach our ultimate understanding of the dynamics and origin of matter and forces.

Chapter 6

Looking inside the proton at HERA: investigating the fundamental forces and structure of matter

Matthew Wing

Department of Physics and Astronomy
University College London
Gower Street
London WC1E 6BT, UK

Soon after Guido Altarelli's passing away, the H1 and ZEUS collaborations published their final paper combining all measurements of inclusive deep inelastic scattering cross sections into a coherent data set. These data have a richness of physics, providing results for student textbooks and will provide the backbone of parametrisations of the structure of the proton for years to come. There is no doubt that Guido's influence on these final HERA results was far reaching; as such the paper was dedicated to his memory. A summary of the highlights of this paper form the bulk of the work presented in this chapter on HERA physics and is a fitting tribute to Guido's contribution to science.

6.1 Introduction

HERA was the world's first and so far only lepton-hadron collider, extending the kinematic range for studying the fundamental structure of matter by many orders of magnitude compared to fixed-target experiments. As with all general-purpose colliders and detectors, the H1 and ZEUS experiments have studied a wide range of physics, ranging from precise measurements and investigation of new phenomena in QCD, to investigation of the electroweak sector as well as searches beyond the Standard Model. The HERA collider operated from 1992 to 2007 and both experiments continue to release new papers on a broad array of physics, with plans to continue for several years to come. Perhaps the most significant publication from the HERA experiments is the recent combination of all data on inclusive deep inelastic scattering [Abramowicz *et al.* (2015)]. The paper presents many textbook results such as the demonstration of the unification of the electromagnetic and weak forces as well as providing the most detailed view of the structure of the proton. With the large kinematic range measured and high precision achieved, the inclusive cross sections in deep inelastic scattering can be used not only to extract the partonic content of the proton, but also measure electroweak couplings and constrain models of physics beyond the Standard Model. Given its importance, this flagship paper forms the basis of this chapter; the paper was also dedicated to Guido Altarelli and as such seems an appropriate choice of topic.

In such a short chapter, it is impossible to cover all of HERA physics and the reader is referred to the hundreds of papers produced by the H1 and ZEUS collaborations.[1] Alternatively, various reviews have been written which summarise certain areas. Other reviews on the structure of the proton as seen at HERA can be found at [Abt (2016); Perez and Rizvi (2013)]. The search for physics beyond the Standard Model is discussed in [South and Turcato (2016)]. Reviews have been written on the hadronic final state and diffraction [Newman and Wing (2014)], as well as dedicated articles on diffraction [Wolf (2010)] and on heavy quarks [Behnke *et al.* (2015)]. A general review of HERA physics was also written [Klein and Yoshida (2008)], although an update of this would now be timely.

6.2 Deep inelastic scattering

6.2.1 *Kinematics*

Deep inelastic scattering [Devenish and Cooper–Sarkar (2004)] can be classified as either neutral current (NC) or charged current (CC), depending on whether the

[1]See `http://www-h1.desy.de/publications/H1publication.short_list.html` and `http://www-zeus.desy.de/zeus_papers/zeus_papers.html`.

exchanged boson is a photon or Z^0 boson (NC) or a $W^\pm$ boson (CC). The NC reaction in electron-proton collisions can be written as $e^-(k)+p(p) \to e^-(k')+X(p')$ with the four vectors given in the brackets and X referring to a hadronic final state. In the case of CC, the final state electron is replaced by a neutrino. The four-momentum of the exchanged boson, q, is given by $q = k - k'$. The events can be described by the following quantities: the squared centre-of-mass energy, s, is given by

$$s = (k+p)^2\,; \tag{6.1}$$

the virtuality, Q^2, of the exchanged boson is given by

$$Q^2 = -q^2\,; \tag{6.2}$$

the Bjorken-x variable, interpreted as the fraction of proton's momentum carried by the struck quark in the proton's infinite-momentum frame, is given by

$$x = \frac{Q^2}{2p \cdot q}\,; \tag{6.3}$$

and the inelasticity, y, interpreted in the proton rest frame as the fraction of energy transferred from the lepton to the proton, is given by

$$y = \frac{p \cdot q}{p \cdot k}\,. \tag{6.4}$$

The above variables are related, $Q^2 = s\,x\,y$, where particle masses can be ignored at the very high energies considered here. Another quantity of importance is the exchanged boson-proton centre-of-mass energy, W, which is given by

$$W^2 = (q+p)^2\,. \tag{6.5}$$

6.2.2 *Cross sections and structure functions*

6.2.2.1 *Formalism for unpolarised lepton beams*

The neutral current reduced cross section, $\sigma^\pm_{\mathrm{r,NC}}$, and double-differential cross section for unpolarised $e^\pm p$ scattering, $\frac{d^2\sigma^{e^\pm p}_{\mathrm{NC}}}{dx_{\mathrm{Bj}}\,dQ^2}$, can be written in terms of the kinematic variables and generalised structure functions of the proton as

$$\sigma^\pm_{\mathrm{r,NC}} = \frac{d^2\sigma^{e^\pm p}_{\mathrm{NC}}}{dx_{\mathrm{Bj}}\,dQ^2} \cdot \frac{x_{\mathrm{Bj}}\,Q^4}{2\,\pi\,\alpha^2\,Y_+} = \tilde{F}_2 \mp \frac{Y_-}{Y_+} x\tilde{F}_3 - \frac{y^2}{Y_+}\tilde{F}_{\mathrm{L}}\,, \tag{6.6}$$

where α is the fine structure constant and $Y_\pm = 1 \pm (1-y)^2$. The generalised structure functions, $\tilde{F}_2$, $\tilde{F}_3$ and $\tilde{F}_{\mathrm{L}}$, are sums of the structure functions, $F_{2,3,\mathrm{L}}$, $F^{\gamma Z}_{2,3,\mathrm{L}}$ and $F^Z_{2,3,\mathrm{L}}$, relating to photon exchange, photon-Z interference and Z exchange and depend on the electroweak parameters as [Patrignani *et al.* (2016)]:

$$\begin{aligned}
\tilde{F}_2 &= F_2 - \kappa_Z v_e \cdot F_2^{\gamma Z} + \kappa_Z^2 (v_e^2 + a_e^2) \cdot F_2^Z\,, \\
\tilde{F}_{\mathrm{L}} &= F_{\mathrm{L}} - \kappa_Z v_e \cdot F_{\mathrm{L}}^{\gamma Z} + \kappa_Z^2 (v_e^2 + a_e^2) \cdot F_{\mathrm{L}}^Z\,, \\
x\tilde{F}_3 &= -\kappa_Z a_e \cdot xF_3^{\gamma Z} + \kappa_Z^2 \cdot 2 v_e a_e \cdot xF_3^Z\,,
\end{aligned} \tag{6.7}$$

where v_e and a_e are the vector and axial-vector weak couplings of the electron to the Z boson, $\kappa_Z(Q^2) = Q^2/[(Q^2+M_Z^2)(4\sin^2\theta_W\cos^2\theta_W)]$ and M_Z and θ_W are the mass of the Z boson and the electroweak mixing angle, respectively. The structure function F_2 is sensitive to the quark and anti-quark distributions in the proton and dominates at low Q^2. At low Q^2, $Q^2 \ll M_Z^2$, the contribution from Z exchange is negligible and

$$\sigma^{\pm}_{\mathrm{r,NC}} = F_2 - \frac{y^2}{Y_+}F_{\mathrm{L}}\,. \tag{6.8}$$

The longitudinal structure function F_{L} is sensitive to the gluon distribution in the proton and becomes important at high values of y.

The structure function xF_3 is sensitive to the difference in the quark and anti-quark distributions in the proton and arises due to the interference of the photon- and Z-exchange contributions at Q^2 values around the mass of the W and Z bosons. Using Eq. (6.6), the structure function xF_3 can be determined from the difference between the e^-p and e^+p reduced cross sections as

$$xF_3 = \frac{Y_+}{2Y_-}(\sigma^-_{\mathrm{r,NC}} - \sigma^+_{\mathrm{r,NC}})\,. \tag{6.9}$$

The charged current reduced cross section, $\sigma^{\pm}_{\mathrm{r,CC}}$, and double-differential cross section for unpolarised $e^{\pm}p$ scattering, $\frac{d^2\sigma^{e^{\pm}p}_{\mathrm{CC}}}{dx_{\mathrm{Bj}}\,dQ^2}$, can also be written in terms of the kinematic variables and the structure functions of the proton as

$$\sigma^{\pm}_{\mathrm{r,CC}} = \frac{2\,\pi\,x_{\mathrm{Bj}}}{G_F^2}\left[\frac{M_W^2+Q^2}{M_W^2}\right]^2 \frac{d^2\sigma^{e^{\pm}p}_{\mathrm{CC}}}{dx_{\mathrm{Bj}}\,dQ^2} = \frac{Y_+}{2}W_2^{\pm} \mp \frac{Y_-}{2}xW_3^{\pm} - \frac{y^2}{2}W_{\mathrm{L}}^{\pm}\,, \tag{6.10}$$

where G_F is the Fermi constant and M_W is the mass of the W boson.

6.2.2.2 *Formalism for polarised lepton beams*

In deep inelastic scattering, the forms of the cross sections and structure functions become modified when considering polarised lepton beams. In the NC reaction, $e^{\pm}(P_e)+p \to e^{\pm}+X$, where the incoming lepton has polarisation, P_e, and the proton is unpolarised, the generalised structure functions (see Eq. (6.7)) are modified as

$$\begin{aligned}\tilde{F_2}^{\pm} &= F_2 - \kappa_Z(v_e \pm a_e P_e)\cdot F_2^{\gamma Z} + \kappa_Z^2(v_e^2 + a_e^2 \pm 2v_e a_e P_e)\cdot F_2^Z\,,\\ x\tilde{F_3}^{\pm} &= -\kappa_Z(a_e \pm v_e P_e)\cdot xF_3^{\gamma Z} + \kappa_Z^2(2v_e a_e \pm (v_e^2 + a_e^2)P_e)\cdot xF_3^Z\end{aligned} \tag{6.11}$$

and

$$P_e = \frac{N_R - N_L}{N_R + N_L}\,, \tag{6.12}$$

where N_R and N_L are the number of right- and left-handed leptons in the beam, respectively. For the high-Q^2 measurements with polarised lepton beams, the contribution from $\tilde{F_L}$ is negligible and so is not considered in this section.

The coupling in the CC process is intrinsically chiral and so the cross section for $e^{\pm}p$ scattering of polarised leptons with unpolarised protons is given by

$$(1 \pm P_e)\, \sigma^{\pm}_{\mathrm{r,CC}} = \frac{2\,\pi\, x_{\mathrm{Bj}}}{G_F^2} \left[\frac{M_W^2 + Q^2}{M_W^2}\right]^2 \frac{d^2\sigma^{e^{\pm}p}_{\mathrm{CC}}}{dx_{\mathrm{Bj}}\, dQ^2}\,, \tag{6.13}$$

in which the cross section has a linear dependence on the polarisation and is identically zero for a fully right-handed e^- beam ($P_e = 1$) or a fully left-handed e^+ beam ($P_e = -1$).

6.2.3 *Distribution of partons in the proton*

In order to illustrate the sensitivity of the data to the different partons in the proton, it is useful to consider the cross sections and structure functions in terms of the quark-parton model (QPM) where gluons are not present and $\tilde{F}_{\mathrm{L}} = 0$ [Callan and Gross (1969)]. In the QPM, x_{Bj} is equal to x.

In the QPM, the NC structure functions in Eq. (6.7) become

$$\begin{aligned}\left(F_2, F_2^{\gamma Z}, F_2^Z\right) &\approx \left[\left(e_u^2, 2e_u v_u, v_u^2 + a_u^2\right)\left(xU + x\bar{U}\right)\right.\\ &\qquad \left. + \left(e_d^2, 2e_d v_d, v_d^2 + a_d^2\right)\left(xD + x\bar{D}\right)\right],\\ \left(xF_3^{\gamma Z}, xF_3^Z\right) &\approx 2\left[\left(e_u a_u, v_u a_u\right)\left(xU - x\bar{U}\right) + \left(e_d a_d, v_d a_d\right)\left(xD - x\bar{D}\right)\right],\end{aligned} \tag{6.14}$$

where e_u and e_d denote the electric charge of up- and down-type quarks and v_u, v_d and a_u, a_d are the vector and axial-vector weak couplings of the up- and down-type quarks to the Z boson. The variables xU, xD, $x\bar{U}$ and $x\bar{D}$ denote the sums of parton distributions for up-type and down-type quarks and anti-quarks and are related to the u, d, s, c quark distributions as:

$$xU = xu + xc, \quad x\bar{U} = x\bar{u} + x\bar{c}, \quad xD = xd + xs, \quad x\bar{D} = x\bar{d} + x\bar{s}\,. \tag{6.15}$$

The valence-quark distributions can be written as

$$xu_v = xU - x\bar{U}, \quad xd_v = xD - x\bar{D} \tag{6.16}$$

assuming symmetry between quarks and anti-quarks in the sea.

The $x\tilde{F}_3$ structure function is related to the valence-quark distributions and in the HERA kinematic range this is dominated by photon-Z interference, such that

$$xF_3^{\gamma Z} \approx \frac{x}{3}(2u_v + d_v)\,. \tag{6.17}$$

In the QPM, $W_{\mathrm{L}}^{\pm} = 0$, the CC structure functions (see Eq. (6.10)) can be written in terms of quark and anti-quark distributions as

$$W_2^+ \approx x\bar{U} + xD, \quad xW_3^+ \approx xD - x\bar{U}, \quad W_2^- \approx xU + x\bar{D}, \quad xW_3^- \approx xU - x\bar{D}\,, \tag{6.18}$$

for e^+p or e^-p interactions. It then follows that the reduced cross sections can be written as

$$\sigma^+_{\mathrm{r,CC}} \approx \left(x\bar{U} + (1-y)^2 xD\right), \quad \sigma^-_{\mathrm{r,CC}} \approx \left(xU + (1-y)^2 x\bar{D}\right). \tag{6.19}$$

Within the framework of perturbative QCD (pQCD), the structure of the proton is described in terms of parton density functions, $f(x)$, which are the probability to find a parton (gluon or quark) with a fraction x of the proton's momentum. The probability is predicted to depend on the so-called factorisation scale, μ_f, at which the proton is probed; in inclusive deep inelastic scattering, this is usually taken to be Q^2. The functions are usually presented as parton distribution functions (PDFs), $xf(x)$. Perturbative QCD does not predict the absolute values of the PDFs, but does provide the framework to evolve the PDFs with the scale of the process. Therefore the PDF must be determined in fits to data. The PDFs can then be convoluted with the short-distance scattering cross section to calculate a given cross section with particular incoming particles such as *ep* and *pp*.

Several groups, JR [Jimenez-Delgado and Reya (2014)], MSTW/MMHT [Martin *et al.* (2009); Harland–Lang *et al.* (2015a)], CT/CJ [Dulat *et al.* (2016); Accardi *et al.* (2016a)], ABM/ABMP [Alekhin *et al.* (2014, 2017)] and NNPDF [Ball *et al.* (2015)] determine PDFs using a variety of data (e.g. fixed-target and hadron-collider data), with the HERA data discussed in this chapter forming the backbone of the QCD analyses. The H1 and ZEUS collaborations have also determined PDF sets, called HERAPDF [Aaron *et al.* (2010a); Abramowicz *et al.* (2015)], based solely on the HERA data. Although the global PDF sets have a wider kinematic coverage and more data, the advantage of the HERAPDF approach is that a single coherent high-precision data set containing NC and CC cross sections is used as input. This allowed rigorous statistical procedures to be applied to the determination of the experimental uncertainties in the HERAPDF fits; they are also independent of heavy nuclear (or deuterium) corrections through the use of *ep* data only. The QCD analyses are performed within the DGLAP [Gribov and Lipatov (1972a,b); Lipatov (1975); Dokshitzer (1977); Altarelli and Parisi (1977)] formalism pioneered by Guido Altarelli and others. The PDFs are determined using the DGLAP equations at leading order (LO), next-to-leading order (NLO) and next-to-next-to-leading order (NNLO) in pQCD.

6.3 The HERA collider and H1 and ZEUS experiments

The Hadron-Elektron Ring Anlage (HERA) [Voss and Wiik (1994)] was the first and, to date, only lepton-hadron collider. It was located in a tunnel of 6.3 km at the DESY laboratory in the west of Hamburg, Germany. The collider ran from 1992 – 2000 (HERA I) and 2002 – 2007 (HERA II) with most data taken with proton energies of $E_p = 920$ GeV and electron energies of $E_e = 27.5$ GeV giving a centre-of-mass energy of $\sqrt{s} = 318$ GeV. Such a centre-of-mass energy is equivalent to a lepton beam of energy ~ 50 TeV for a fixed-target experiment which had dominated the world of deep inelastic scattering before HERA. HERA was initially run at somewhat lower energies with the next largest data sample taken with $E_p = 820$ GeV giving $\sqrt{s} = 300$ GeV. At the end of 2007, data was also taken with $E_p = 575$ and 460 GeV, yielding $\sqrt{s} = 251$ and 225 GeV, respectively, in order to measure $F_{\rm L}$.

During the HERA II period, the electron or positron beams were longitudinally polarised with an average polarisation of about ±30%.

Experiments were located at up to four points around the ring of which this chapter is concerned with the two, H1 [Abt *et al.* (1997a,b); Appuhn *et al.* (1997)] and ZEUS [Holm (1993); Derrick *et al.* (1991); Foster *et al.* (1994); Polini *et al.* (2007)], where the electrons and protons were brought to collision. Due to the much larger proton energy, both detectors were asymmetric with significantly more instrumentation in the direction of the outgoing proton beam.[2] Other than the forward-backward asymmetry in instrumentation, the detectors were similar to other general-purpose detectors at high energy colliders such as those at LEP, Tevatron and the LHC. Both H1 and ZEUS detectors had microvertex detectors for measuring weak decays, surrounded by a central drift chamber for the precise tracking of charged particles. Forward and rear tracking detectors gave additional charged-particle information with the rear detectors particularly useful for measuring low-angle scattered electrons at small Q^2 where the deep inelastic scattering cross section is largest. Beyond the tracking detectors were calorimeters in which a single technology was used for the ZEUS detector with the H1 detector also including a dedicated higher-precision electron calorimeter in the rear direction. Finally, large muon detectors surrounded the calorimeters. Further, smaller detectors with dedicated purposes, such as for electron or proton tagging, were placed along the beampipe in both directions. The different technologies used by the two experiments often led to different reconstruction techniques, analysis methods and hence systematic uncertainties. This effective cross calibration was exploited when combining measurements from the two collaborations, as will be presented in later sections.

6.4 Investigation of the structure of the proton

The structure of the proton has principally been investigated at HERA through inclusive deep inelastic scattering and these results make up the bulk of this section. Much can be learnt on QCD (and electroweak physics, see Section 6.5) by just considering the cross-section measurements and the basics of the Standard Model. Jet and heavy-quark production also provide valuable information on the structure of the proton, in particular the gluon content of the proton as well as on the strong coupling, α_s, and the heavy-quark masses. The extraction of the parton densities and parameters such as α_s and the heavy-quark masses comes from detailed QCD analyses and comparison with the data.

---

[2] A right-handed Cartesian coordinate system is used throughout this chapter, with the Z axis pointing in the proton beam direction, referred to as the "forward" direction, and the X axis pointing towards the centre of HERA. The coordinate origin is at the nominal interaction point. The pseudorapidity is defined as $\eta = -\ln(\tan\frac{\theta}{2})$, where the polar angle, θ, is measured with respect to the proton beam direction.

6.4.1 *Inclusive deep inelastic scattering and F_2*

The first measurements of the proton structure were performed at HERA using the 1992 data sample with each experiment having only just over $20\,\text{nb}^{-1}$. Fixed-target experiments were restricted to low Q^2, below about $100\,\text{GeV}^2$, and high x_{Bj}, above about 10^{-2}. Even with the small data samples from the 1992 running period, the HERA experiments extended the reach in Q^2 beyond $1000\,\text{GeV}^2$ and in x_{Bj} down to 10^{-4}. First measurements are shown in Fig. 6.1, with the ZEUS results (left) [Derrick *et al.* (1993)] displaying F_2 versus x_{Bj} for fixed Q^2, and the H1 data (right) [Abt *et al.* (1994)] displaying F_2 versus Q^2 for fixed x_{Bj}. The data demonstrate a clear and strong rise to low x_{Bj} and this discovery was one of the major achievements of the HERA experiments. The predictions from different PDF groups [Gluck *et al.* (1993); Botts *et al.* (1993); Martin *et al.* (1993)] have a large spread, with some below and some above the data. This indicates how little the fixed-target data constrain the PDFs at this low x_{Bj} and, already with these small data samples, how well the HERA data can constrain them. The value of F_2 rises with increasing Q^2 ever more steeply for decreasing x_{Bj}, showing scaling violations in contrast to early measurements at higher x_{Bj} where the structure function was independent of Q^2. Simple linear fits describe the data well.

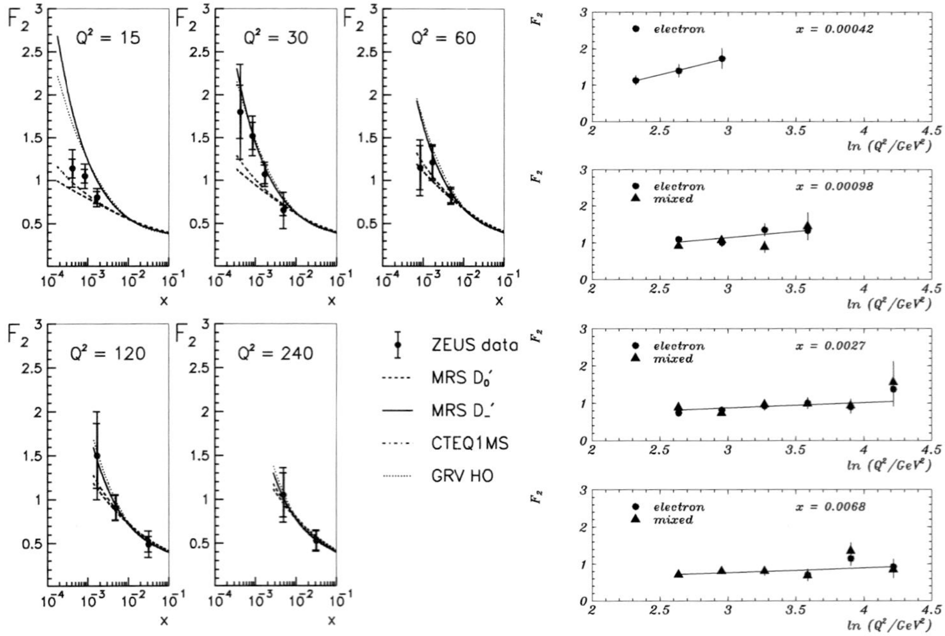

Fig. 6.1 First measurements at HERA of the proton structure function F_2 shown (left) versus x_{Bj} at fixed Q^2, measured by ZEUS [Derrick *et al.* (1993)] and (right) versus Q^2 at fixed x_{Bj}, measured by H1 [Abt *et al.* (1994)]. Predictions from (left) different PDF groups and (right) linear fits to the data are shown.

These and other phenomena were investigated in the recent combination of all H1 and ZEUS data on inclusive deep inelastic scattering [Abramowicz *et al.* (2015)]. The combination used data collected over the 15-year lifetime of HERA, corresponding to a total luminosity of $1\,\text{fb}^{-1}$. The data were combined separately for the e^+p and e^-p runs and for the different centre-of-mass energies. Overall, 41 separate data sets were used in the combination, spanning $0.045 < Q^2 < 50\,000\,\text{GeV}^2$ and $6 \times 10^{-7} < x_{\text{Bj}} < 0.65$, i.e. six orders of magnitude in each variable. The initial measurements consisted of 2937 published cross sections in total, which were combined to produce 1307 final combined cross-section measurements. The procedure for combining the data involved a careful treatment of the various uncertainties between all of the data sets. In particular, the correlations of the various sources were assessed, and those uncertainties deemed to be point-to-point correlated were accounted for as such in the averaging of the data based on a χ^2 minimisation method. The resulting χ^2 is 1687 for 1620 degrees of freedom, demonstrating excellent compatibility of the multitude of data sets. Figure 6.2 illustrates the power of the data combination. It displays a selection of the data in bins of the photon virtuality, Q^2, and for fixed values of x_{Bj}, showing separately individual data sets

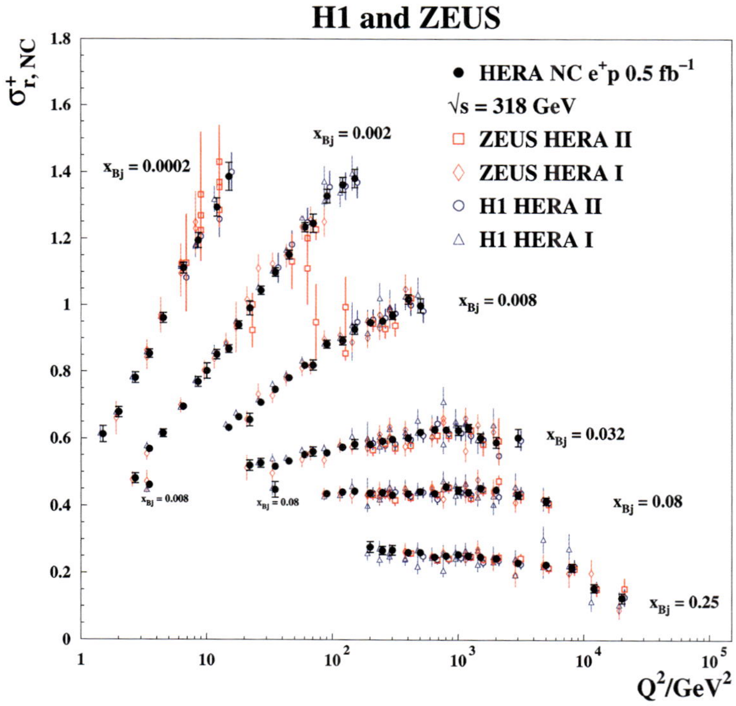

Fig. 6.2 Neutral current reduced cross section, $\sigma^+_{\text{r, NC}}$, versus Q^2 for H1 and ZEUS data sets (open points) and after the HERA combination (solid points). A selection of data for fixed x_{Bj} is shown. Figure taken from [Abramowicz *et al.* (2015)].

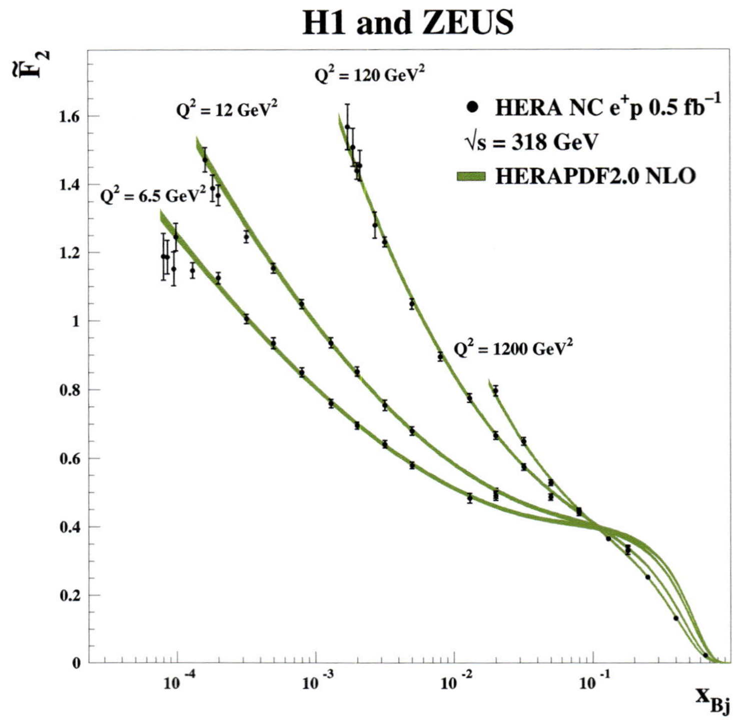

Fig. 6.3 The structure function $\tilde{F}_2$ extracted from the measured reduced cross sections for four values of Q^2. The predictions from HERAPDF2.0 NLO with their total uncertainties (bands) are also shown. Figure taken from [Abramowicz *et al.* (2015)].

from several different analyses. A combined data point can be the combination of up to eight individual measurements. The improvement in precision is striking; an indication of the precision of the combined data is that the total uncertainties are close to 1% for the bulk region of $3 < Q^2 < 500\,\mathrm{GeV}^2$.

Examples of the HERA combined data are shown as $\tilde{F}_2$ as a function of x_{Bj} at fixed Q^2 in Fig. 6.3, displayed in a similar fashion to the first HERA results in Fig. 6.1 (left). Comparison of the two figures shows the tremendous improvement in precision through the larger data sample and the detailed understanding of the detectors and systematic uncertainties. The combined HERA data show the clear increase in $\tilde{F}_2$ to decreasing low x_{Bj}, with the gradient increasing with increasing Q^2. At higher Q^2, the exchanged boson effectively probes smaller distances, and so can see more of the inner structure of the proton and hence resolves more and more gluons.

The final HERA results are plotted for the reduced cross section as a function of Q^2 for fixed x_{Bj}, shown separately for e^-p and e^+p data at $\sqrt{s} = 318\,\mathrm{GeV}$ in Fig. 6.4. Again, the comparison to the first HERA results (Fig. 6.1 (right)) is striking. For $x_{\mathrm{Bj}} \sim 0.08$, the cross-section dependence is reasonably flat as a

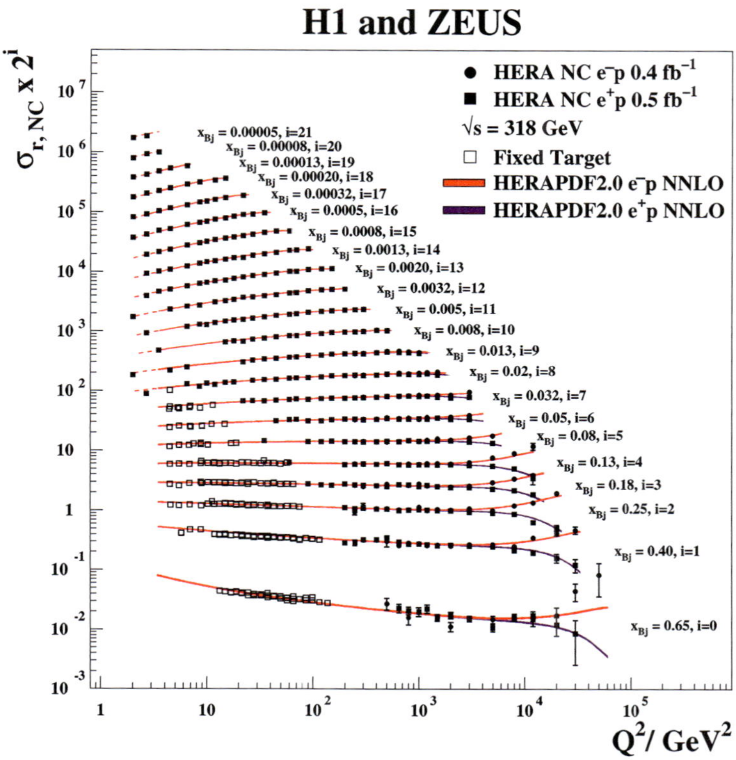

Fig. 6.4 Neutral current reduced cross sections as a function of Q^2 and at fixed x_{Bj}, shown for combined HERA data for e^-p (circular points) and e^+p (solid square points) data, together with fixed-target data (open square points). The predictions from HERAPDF2.0 NNLO with their total uncertainties (bands) are also shown separately for e^-p and e^+p collisions. The dashed lines represent extrapolations of HERAPDF2.0 NNLO to a region not included in the fits. Figure taken from [Abramowicz *et al.* (2015)].

function of Q^2; this is known as Bjorken scaling and is expected from the simple parton model in which inelastic electron-proton scattering is viewed as a sum of elastic electron-parton scattering where the partons are free point-like objects. At larger x_{Bj}, $\sigma_{\mathrm{r,NC}}$ falls with increasing Q^2, reflecting the valence structure of the proton and the reduced phase space for gluon emission. At lower x_{Bj}, $\sigma_{\mathrm{r,NC}}$ rises with increasing Q^2 (also seen in Fig. 6.2), reflecting the increased phase space for gluon emission and hence the increased gluon density, as also discussed for Fig. 6.3. In summary, Fig. 6.4 gives a beautiful demonstration of scaling violations, made possible through the high precision of the data and the large kinematic range in Q^2 and x_{Bj} of the measurements.

The difference in e^-p and e^+p cross sections at high Q^2, also seen in Fig. 6.4, reveals information on the valence structure of the proton and is discussed later in Section 6.5.2. In both Figs. 6.3 and 6.4, the PDF fit to the HERA data alone,

HERAPDF2.0, gives a good description of the data, including those from fixed-target experiments, and is discussed in more detail in Section 6.4.4.

6.4.2 *Longitudinal structure function*

As outlined in Section 6.3, in the final months of data taking HERA ran with different proton beam energies, and so data at $\sqrt{s}$ of 318, 300, 251 and 225 GeV were used to extract the longitudinal structure function, F_L. By measuring the cross section at different centre-of-mass energies at fixed values of x_{Bj} and Q^2, but at different values of y, an experimental separation of F_2 and F_L is possible. A measurement of F_L provides a powerful vindication of QCD by allowing a comparison of the gluon density extracted mainly from scaling violations with an observable that is directly sensitive to the gluon density.

The extraction of F_L is experimentally challenging and is also the result of extracting a small number from two large numbers, the cross section and F_2 structure function. To be able to measure events at such high y, scattered electrons need to be reconstructed at lower energy, down to 3 GeV, rather than the more regular minimum requirement of ~ 10 GeV for deep inelastic scattering. The final measurements of F_L from both H1 [Andreev *et al.* (2014)] and ZEUS [Abramowicz *et al.* (2014a)] collaborations are shown in Fig. 6.5 as a function of Q^2 for fixed x_{Bj}. It can be seen that the uncertainties on both measurements are relatively large and it should be noted that within a given measurement the uncertainties are correlated. Although the H1 and ZEUS measurements appear systematically different,

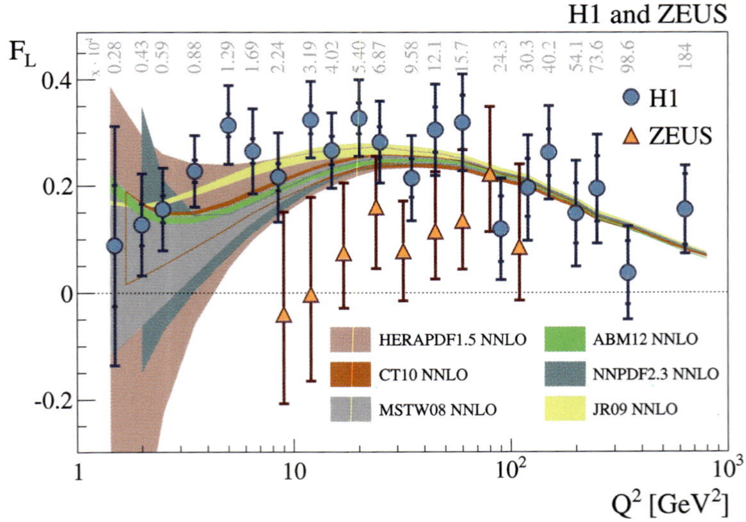

Fig. 6.5 The longitudinal structure function, F_L, as a function of Q^2 and at fixed x_{Bj} measured by the H1 (circles) and ZEUS (triangles) experiments. A selection of QCD predictions at NNLO is shown.

taking into account the correlations, the significance is about 1.5 standard deviations. The data are reasonably well described by various PDF predictions [Alekhin *et al.* (2014); Ball *et al.* (2013c); Jimenez–Delgado and Reya (2009); Lai *et al.* (2010); Martin *et al.* (2009)], although the uncertainties on the data mean they lack real constraining power. Although the combined HERA data could be used to extract a common HERA value for $F_{\rm L}$, the procedure needs optimised binning and so a dedicated analysis would be needed to have the ultimate measurement from HERA. As well as increased statistics through a larger data sample, any measurement of $F_{\rm L}$ at a future facility should pursue a larger lever arm in centre-of-mass energy in order to achieve more precise measurements of the longitudinal structure function.

6.4.3 *Heavy-quark contribution to the structure of the proton*

Heavy-quark production at HERA is dominated by boson-gluon fusion and so is directly sensitive to the gluon density in the proton. As with the measurement of $F_{\rm L}$, a measurement of heavy-quark production provides a valuable verification of the gluon density extracted from scaling violations in inclusive deep inelastic scattering data. The measurements of the heavy-quark cross sections are also sensitive to the heavy-quark masses. The charm cross section is significantly higher than the beauty cross section due to the lower mass and higher charge.

Many measurements of heavy-quark production have been made using different techniques such as the reconstruction of specific charm hadrons, the decays to electrons or muons and the reconstruction of a displaced vertex due to the weak decay of the heavy quark. The different charm measurements from the H1 and ZEUS collaborations were combined into a coherent HERA result [Abramowicz *et al.* (2013a)], much as the inclusive measurements were, as described in Section 6.4.1. The data spans a range of $2.5 \leq Q^2 \leq 2000\,{\rm GeV}^2$. An example range for the data combination is shown in Fig. 6.6; the combined data provide a precise and coherent data set, significantly better than any individual data set.

All of the HERA charm cross sections are shown in Fig. 6.7 in comparison with the HERAPDF1.5 set which was a precursor to HERAPDF2.0 discussed elsewhere in this chapter. A central charm quark mass parameter of $M_c = 1.4\,{\rm GeV}$ was used and the uncertainty bands reflect a variation to 1.35 GeV and 1.65 GeV. The predictions describe the data well.

An NLO QCD analysis was performed to determine the $\overline{\rm MS}$ running charm quark mass, $m_c(m_c)$, based on the inclusive neutral and charged current HERA I deep inelastic scattering data and the reduced charm cross sections. The strong coupling constant is evolved setting the number of active flavours to $n_f = 3$, using $\alpha_s^{n_f=3}(M_Z) = 0.105$. The fit gives a value of

$$m_c(m_c) = 1.26 \pm 0.05_{\rm exp} \pm 0.03_{\rm model} \pm 0.02_{\rm param} \pm 0.02_{\alpha_s}\,{\rm GeV}\,, \tag{6.20}$$

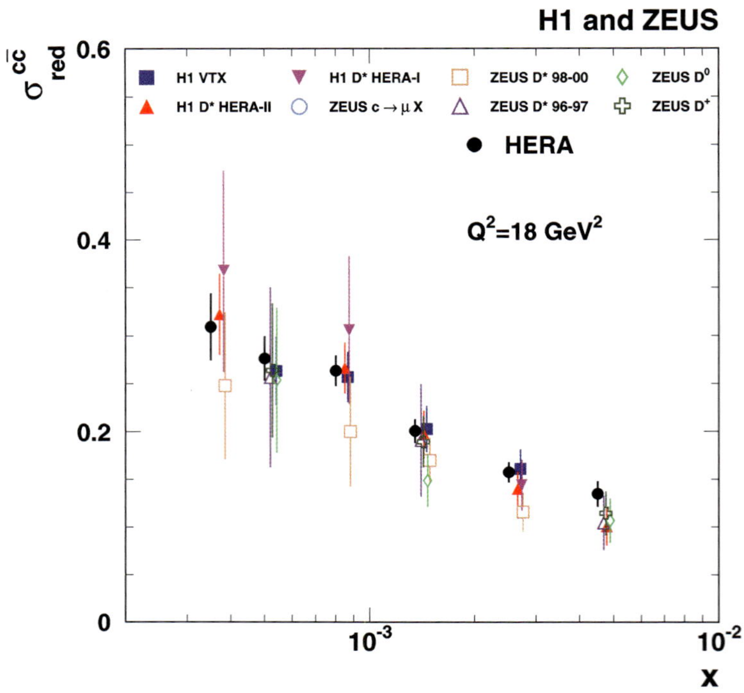

Fig. 6.6 Reduced charm cross sections, $\sigma^{c\bar{c}}_{\rm red}$, as a function of $x_{\rm Bj}$ for an example Q^2 of 18 GeV^2. The original measurements by H1 and ZEUS are shown as the coloured points and the combined HERA result as the solid black circles. Figure taken from [Abramowicz *et al.* (2013a)].

where the uncertainties correspond to the experimental, model, parametrisation and α_s-dependent uncertainties. The measured value of the running charm quark mass is consistent with the world average [Olive *et al.* (2014)], based on lattice calculations and measurements of time-like processes.

Since the publication of the combined charm data, several new measurements of charm production in deep inelastic scattering have been made [Abramowicz *et al.* (2013c,b, 2014c)]. At the time of writing, a new and final HERA combination of charm cross sections and a first and final combination of beauty cross sections [Abramowicz *et al.* (2010, 2011, 2014c); Chekanov *et al.* (2010); Aaron *et al.* (2010b)] is being finalised. These data can be used in future PDF fits and allow extractions of the masses of both quarks to be made.

6.4.4 *Comparisons with QCD and PDF fits*

The data in the previous sections have been compared with various predictions of QCD, in particular PDF fits up to NNLO accuracy in QCD. The combined HERA data on inclusive deep inelastic scattering were used to determine new sets of PDFs,

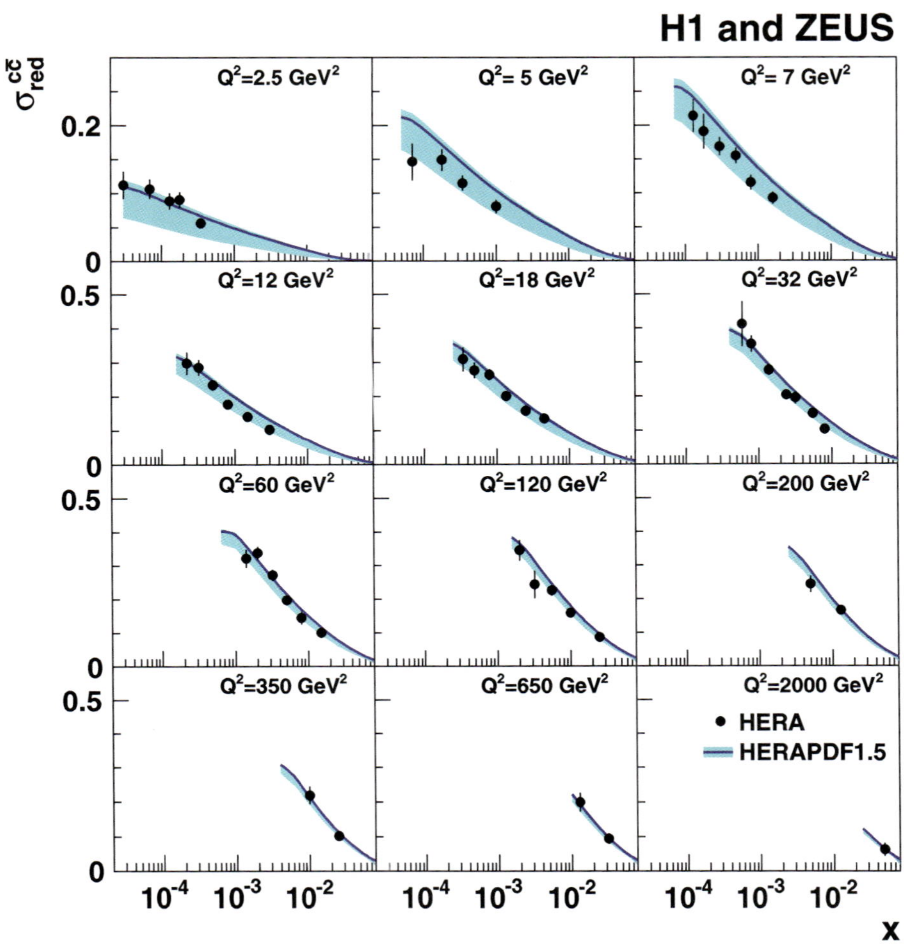

Fig. 6.7 Combined reduced beauty cross sections, $\sigma_{\rm red}^{c\bar{c}}$, as a function of $x_{\rm Bj}$ and at fixed Q^2. The predictions from HERAPDF1.5 NLO with their total uncertainties (bands) are also shown. Figure taken from [Abramowicz *et al.* (2013a)].

HERAPDF2.0, which used only HERA data so as to have well-controlled and well-understood data sets. Details of the HERAPDF2.0 set of PDFs and the QCD analysis are given in the publication [Abramowicz *et al.* (2015)]. The analysis was done and PDFs extracted at LO, NLO and NNLO and examples of the comparisons were shown already in Figs. 6.3 and 6.4. A further detailed comparison of the NNLO set of HERAPDF2.0 is shown in Fig. 6.8 in comparison to NC reduced cross sections. On a global level, HERAPDF2.0 describes the data well, similarly well at NNLO, as shown here, and NLO.

On closer inspection of the description of the data, however, and a study of the quality of the fit, interesting features are revealed. A minimum Q^2 value was chosen

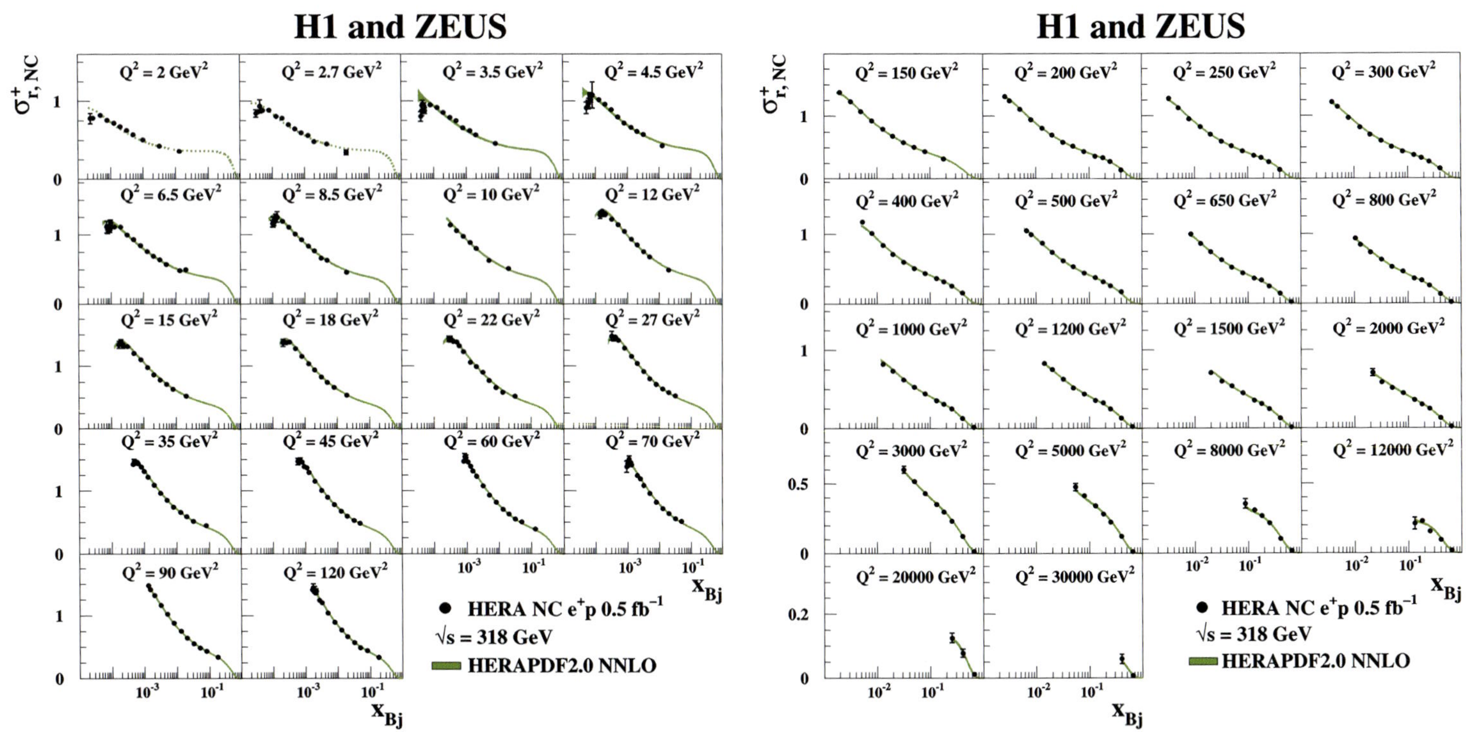

Fig. 6.8 Neutral current reduced cross sections as a function of x_{Bj} and at fixed Q^2, shown for combined HERA e^+p data. The predictions from HERAPDF2.0 NNLO with their total uncertainties (bands) are also shown. The dotted lines represent extrapolations of HERAPDF2.0 NNLO to a region not included in the fits. Figure taken from [Abramowicz *et al.* (2015)].

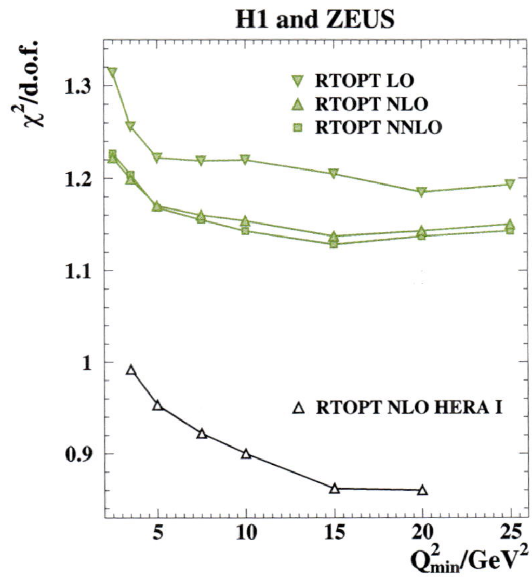

Fig. 6.9 Dependence of χ^2/d.o.f. on the minimum Q^2 of the LO, NLO and NNLO fits to the HERA combined inclusive data. The values for the NLO fit to HERA I data are also shown. The fits were performed using the RTOPT heavy-flavour scheme [Thorne and Roberts (1998); Thorne (2006, 2012)]. Figure taken from [Abramowicz *et al.* (2015)].

such that pQCD was expected to be applicable; the range included in the fits was $3.5 \leq Q^2 \leq 50\,000\,\text{GeV}^2$, corresponding to an x_{Bj} range of $0.651 \times 10^{-4} \leq x_{\text{Bj}} \leq 0.65$. The typical χ^2/d.o.f. for NLO and NNLO fits was ~ 1.2 for about 1130 degrees of freedom (d.o.f.). The dependence of χ^2/d.o.f. as a function of the minimum Q^2 of the data used in the fit is shown in Fig. 6.9 for the three default LO, NLO and NNLO PDFs as well as the fit to HERA I data. The χ^2/d.o.f. of the NLO and NNLO fits are observed to fall with increasing Q^2 until about $10\,\text{GeV}^2$ where they are χ^2/d.o.f. ~ 1.13. The NLO fit to HERA I data has χ^2/d.o.f. values closer to 1, but shows the same trend with Q^2.

The χ^2/d.o.f. values rise again for a minimum Q^2 of $15\,\text{GeV}^2$ due to high-Q^2 data. If the data between 15 and $150\,\text{GeV}^2$ only is fit, the χ^2/d.o.f. value is close to unity. The poor χ^2/d.o.f. values could not be attributed to a particular region of x_{Bj} or to a particular NC or CC process. These results indicate that pQCD cannot fully describe the structure of the proton and its internal dynamics. Although a poor description of data at low Q^2 and low x_{Bj} may be expected and alternative evolution schemes beyond DGLAP could be applicable, this cannot also explain the poor χ^2/d.o.f. values at high Q^2.

The detailed QCD analysis led to the release of the HERAPDF2.0 set of PDFs. These were produced at LO, NLO and NNLO in QCD, along with several other variants such as using an alternative gluon parametrisation, imposing a higher minimum Q^2 cut on the data used in the fit, including data on jet and charm

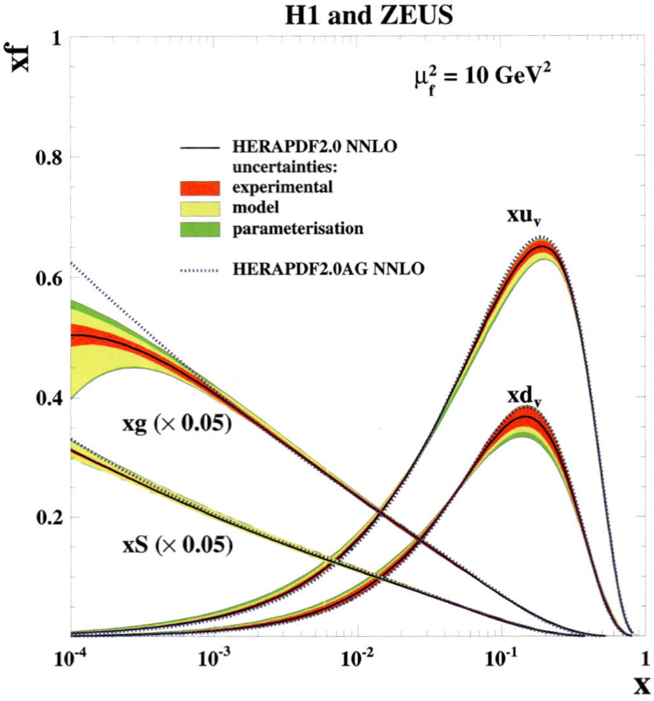

Fig. 6.10 Parton distribution functions, xu_v, xd_v, $xS = 2x(\bar{U} + \bar{D})$ and xg of HERAPDF2.0 at NNLO for scale $\mu_f^2 = 10\,\text{GeV}^2$. The experimental, model and parametrisation uncertainties are shown. The sea and gluon distributions are scaled down by a factor of 20. The dotted line represents the HERAPDF2.0 NNLO set with an alternative gluon parametrisation. Figure taken from [Abramowicz *et al.* (2015)].

production, etc. A rigorous statistical analysis was performed and an experimental uncertainty determined on the PDFs; model and parametrisation uncertainties were also assessed. The PDF set then provides a numerical value for the density of partons in the proton for given values of Q^2 and x and these can, in principle, be used for predictions whenever a proton is one of the incoming particles. Example distributions of HERAPDF2.0 at NNLO QCD are shown in Fig. 6.10 showing the expected distributions for the valence quarks peaking at relatively high x and gluon and sea distributions which rise rapidly and dominate at low x.

The QCD analysis was also extended to include data from the production of charm quarks (see Section 6.4.3) and jets at HERA. Jet production was measured in the Breit frame, where jets with non-zero transverse momentum are expected from hard QCD processes only. By including the charm and jet data, the analysis becomes particularly sensitive to the strong coupling constant, $\alpha_s(M_Z)$, whereas without jet data the coupling constant is strongly correlated with the normalisation of the gluon density. The combined analysis of inclusive data, charm data and jet data at NLO results [Abramowicz *et al.* (2015)] in an experimentally very precise measurement of the strong coupling constant, $\alpha_s(M_Z) = 0.1183 \pm 0.0009\,(\text{exp.})$, with significantly larger uncertainties of $(^{+0.0039}_{-0.0033})$ related to the model and theory.

Within the last year, a major new development is the release of NNLO calculations of jet production in deep inelastic scattering and in hadron-hadron collisions [Currie *et al.* (2016, 2017a,d,b)]. The extra order in the perturbation series should lead to more reliable calculations with reduced uncertainties, in particular from those due to the estimation of effects from missing higher orders. This allows a high precision comparison with the various jet measurements at HERA, which have previously been compared to calculations at NLO in QCD for the last 25 years. In particular this allows the first extraction of α_s at NNLO accuracy which can then be compared with extractions at the same order from other processes such as in e^+e^- reactions. In a new measurement from the H1 collaboration [Andreev *et al.* (2017)], these comparisons and extractions of α_s are performed. When using the jet measurements alone, which span $5 < Q^2 < 15\,000\,\text{GeV}^2$, the value of α_s extracted is $\alpha_s(M_Z) = 0.1157 \pm 0.0020(\text{exp}) \pm 0.0029(\text{theo})$. When the jet data is used in combination with inclusive deep inelastic scattering data from H1 and a simultaneous PDF and α_s fit performed, the extracted value is $\alpha_s(M_Z) = 0.1142 \pm 0.0028(\text{tot})$. The value extracted from jet data alone is shown in Fig. 6.11 as a function of the

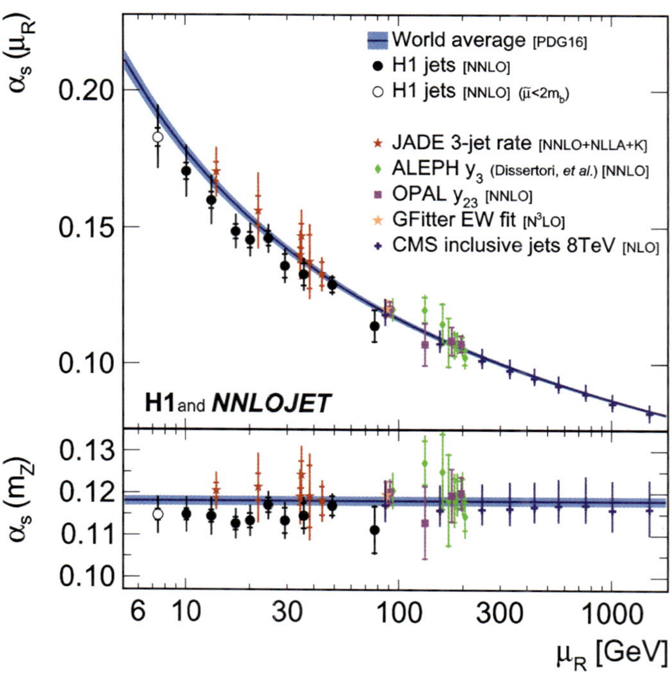

Fig. 6.11 Extractions of $\alpha_s(\mu_R)$ and $\alpha_s(M_Z)$ where μ_R represents the energy scale for H1 jet data compared to extractions from other experiments. Figure taken from [Andreev *et al.* (2017)].

energy scale and compared with other data. The H1 extractions are lower than the world average but are also consistent with other extractions and the world average. The HERA values of α_s have a precision competitive to all other extractions.

6.5 Electroweak and high-Q^2 physics

Although low-Q^2 NC events dominate the data sample at HERA, measurements extend up to about 50 000 GeV^2. Charged current cross sections were generally measured with a minimum Q^2 of about 200 GeV^2 and also extend up to statistics allowed, $\sim$30 000 GeV^2. A significant number of events was acquired for $Q^2 >$ 10 000 GeV^2 in both NC and CC interactions such that clear effects of the weak force were observed as summarised in this section.

6.5.1 *Unification of the electromagnetic and weak forces*

The differential cross sections at high Q^2 are shown in Fig. 6.12 for NC and CC e^-p and e^+p scattering. At the lowest Q^2 shown, $\sim$200 GeV^2, the NC cross section is significantly higher than the CC cross section due to the dominance of photon exchange. At $Q^2 >$ 10 000 GeV^2, the cross sections become similar, due to

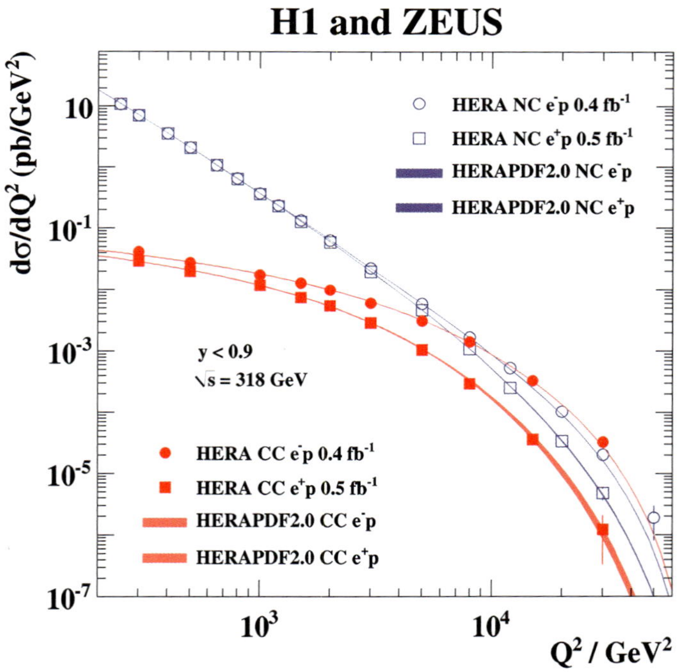

Fig. 6.12 Differential cross section $d\sigma/dQ^2$ for NC and CC processes for both e^+p and e^-p collisions. The data are compared with Standard Model predictions using the HERAPDF2.0 set of PDFs. Figure taken from [Abramowicz *et al.* (2015)].

the exchange of massive vector bosons, providing a beautiful demonstration of the unification of the electromagnetic and weak forces. Predictions based on the Standard Model and using the HERAPDF2.0 set of proton PDFs provide an excellent description of the data.

6.5.2 *Valence structure of the proton*

The data also clearly demonstrate that the NC e^+p and e^-p cross sections are the same at low Q^2 when photon exchange dominates, but differ at high Q^2, again around $10\,000\,\text{GeV}^2$, due to $\gamma{-}Z$ interference effects. This effect was also clearly seen in the reduced cross sections shown in Fig. 6.4. By subtracting the NC e^+p and e^-p cross sections, the structure function $xF_3^{\gamma Z}$ can be extracted, which directly probes the valence structure of the proton (see Section 6.2.2.1). The extracted values of $xF_3^{\gamma Z}$ are shown in Fig. 6.13 and are well described by the HERAPDF2.0 NLO prediction. An integration of $F_3^{\gamma Z}$ was performed over the full range in x_{Bj} and the values for data and theory obtained were:

$$\text{Data}: 1.790 \pm 0.078\,(\text{stat.}) \pm 0.078\,(\text{syst.}), \qquad \text{HERAPDF2.0 NLO}: 1.588^{+0.078}_{-0.100}. \tag{6.21}$$

The value from HERAPDF2.0, as well as the simple prediction from the QPM of 5/3, agree with the data.

6.5.3 *Chiral structure of weak interactions*

Using polarised lepton beams, the chiral structure of the NC electroweak exchange was investigated. In contrast to the CC cross section which has a linear dependence on the polarisation, the NC cross sections have a weaker dependence which

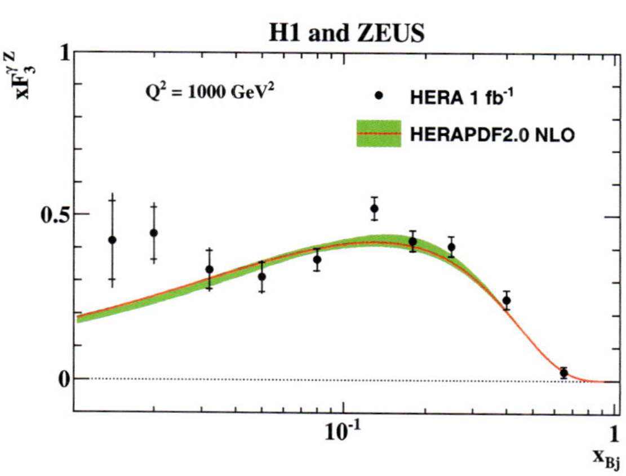

Fig. 6.13 The structure function $xF_3^{\gamma Z}$ at the scale $Q^2 = 1000\,\text{GeV}^2$ together with the prediction from HERAPDF2.0 NLO. Figure taken from [Abramowicz *et al.* (2015)].

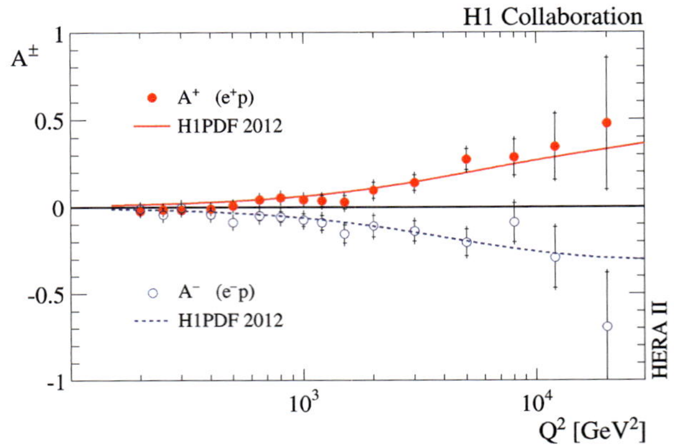

Fig. 6.14 Dependence on Q^2 of the polarisation asymmetry, $A^{\pm}$, for e^+p and e^-p NC interactions. The data are compared with Standard Model expectations using the H1PDF 2012 set of PDFs. Figure taken from [Aaron *et al.* (2012)].

is expected to manifest itself at higher Q^2 (see Section 6.2.2.2). The polarisation asymmetry, $A^{\pm}$, is defined as

$$A^{\pm} = \frac{2}{P_L^{\pm} - P_R^{\pm}} \cdot \frac{\sigma^{\pm}(P_L^{\pm}) - \sigma^{\pm}(P_R^{\pm})}{\sigma^{\pm}(P_L^{\pm}) + \sigma^{\pm}(P_R^{\pm})}, \tag{6.22}$$

where $P_L^{\pm}$ and $P_R^{\pm}$ are the left-handed and right-handed longitudinal lepton beam polarisations for $e^{\pm}p$ data. At leading order, $A^{\pm}$ can be written in terms of the structure functions as $A^{\pm} \approx \mp \kappa_Z a_e F_2^{\gamma Z}/\tilde{F}_2$ which is proportional to $a_e v_q$ and so is a direct measure of parity violation with minimal assumptions of the proton structure. The asymmetry for e^+p and e^-p collisions is shown in Fig. 6.14 measured by the H1 collaboration [Aaron *et al.* (2012)]. The magnitude of the asymmetry increases with increasing Q^2 and is positive for e^+p and negative for e^-p scattering; the magnitude is below 10% for $Q^2 < 1000\,\text{GeV}^2$ and is up to $30-50\%$ for highest $Q^2 \sim 20\,000\,\text{GeV}^2$. The data are well described by the Standard Model prediction; similar results and conclusions were also observed by the ZEUS collaboration [Chekanov *et al.* (2009); Abramowicz *et al.* (2013d)]. These results from HERA confirm parity violation effects in electroweak interactions at large Q^2.

The helicity structure of the $W^{\pm}$ exchange and the different quarks being probed also leads to the observed difference in Fig. 6.12 for the e^+p and e^-p data in CC interactions. However differences are more striking when the initial leptons are polarised. The dependence of the total CC cross section on the polarisation of the lepton beam, at high Q^2, is shown in Fig. 6.15. The data cross sections show a clear and strong linear dependence on the polarisation, as expected from Eq. (6.13). Linear fits to the data demonstrate that the cross section is consistent with zero in e^-p collisions at $+100\%$ electron polarisation, i.e. fully right-handed and in e^+p collisions at -100% positron polarisation, i.e. fully left-handed. This is consistent with

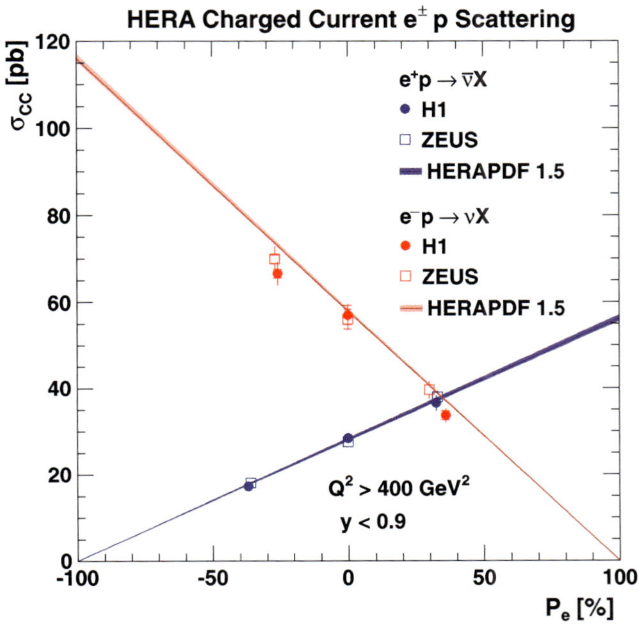

Fig. 6.15 The total CC cross section for $Q^2 > 400\,\text{GeV}^2$ and $y < 0.9$ measured as a function of the lepton polarisation by H1 (solid points) and ZEUS (open points) for e^-p (red points) and e^+p (blue points). The predictions from HERAPDF1.5 NLO with their total uncertainties (bands) are also shown.

the Standard Model in which the W boson couples only to left-handed fermions and right-handed anti-fermions. The results exclude the existence of charged currents involving right-handed fermions mediated by a right-handed W boson below about 200 GeV.

6.5.4 *Fits of electroweak parameters*

Given the sensitivity the data exhibit to electroweak effects and their wide kinematic range, they can be used to constrain the parameters that govern electroweak interactions.

A publication by the H1 collaboration [Aktas *et al.* (2006)] used data on inclusive deep inelastic scattering from HERA I in which the incoming leptons were not polarised. Nevertheless, through the measurement of NC and CC cross sections over the widest possible kinematic range, a combined fit of QCD and electroweak parameters, which accounts for the correlations between the electroweak parameters and the proton PDFs, was possible. The axial (a_u and a_d) and axial-vector (v_u and v_d) weak couplings of u and d quarks to the Z^0 boson were extracted. A recent ZEUS analysis also extracted these couplings, this time using the unpolarised data

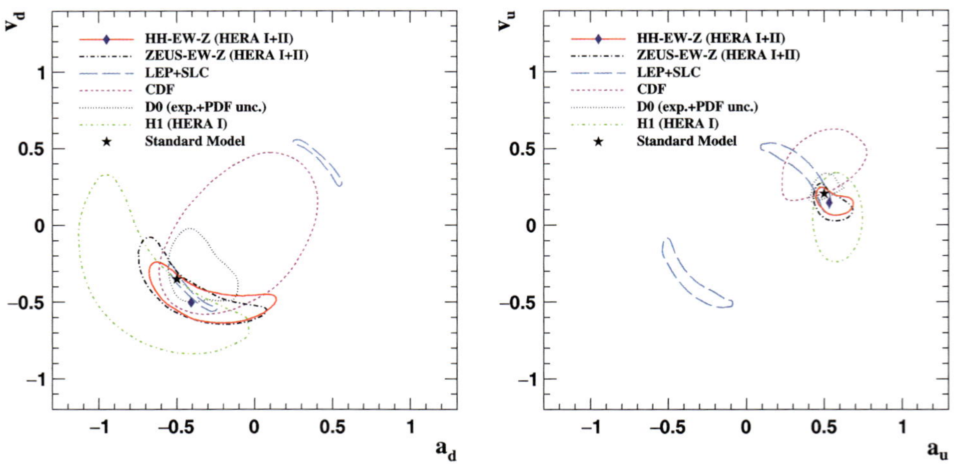

Fig. 6.16 The 68% confidence level contours for (a_d, v_d) and (a_u, v_u) obtained from the HH-EW-Z fit compared with results from ZEUS-EW-Z, H1 (HERA I), LEP and SLC combined, and CDF and D0.

which fed into the HERA combination of all unpolarised deep inelastic scattering [Abramowicz *et al.* (2015)] and the data from the ZEUS experiment in which the incoming leptons were longitudinally polarised. This analysis was extended to also include H1 data in which the incoming leptons were longitudinally polarised [Abt *et al.* (2016)].

The results of the extractions of (a_d, v_d) and (a_u, v_u) from these three analyses are shown in Fig. 6.16 compared to results from a LEP and SLC combined extraction [Schael *et al.* (2006)] and from CDF [Acosta *et al.* (2005)] and D0 [Abazov *et al.* (2011)]. The results from HERA show that the Z^0 couplings to the u quark are better constrained than those to the d quark, due to the high u valence density and higher electric charge. All results for each of the couplings are shown in Fig. 6.17, compared with the expectations from Standard Model and the PDG 2014 average [Olive *et al.* (2014)], which does not include the HH-EW-Z and ZEUS-EW-Z values. All extractions are compatible with each other and agree with the Standard Model prediction. The results of HH-EW-Z are the most precise values for v_u and a_u extracted from a single process. Comparing the three HERA results shows the importance of the inclusion of data using polarised lepton beams on the extractions of the axial-vector couplings, where the latest result, HH-EW-Z, uses the most such data.

The above analyses of HERA data also extracted values for the mass of the W boson, M_W, and $\sin^2 \theta_W$. Although the resulting values of M_W are less precise than the world average, they are determined at HERA for t-channel processes, rather than s-channel processes which dominate the world average. The agreement in the different determinations is a valuable consistency check of the Standard Model.

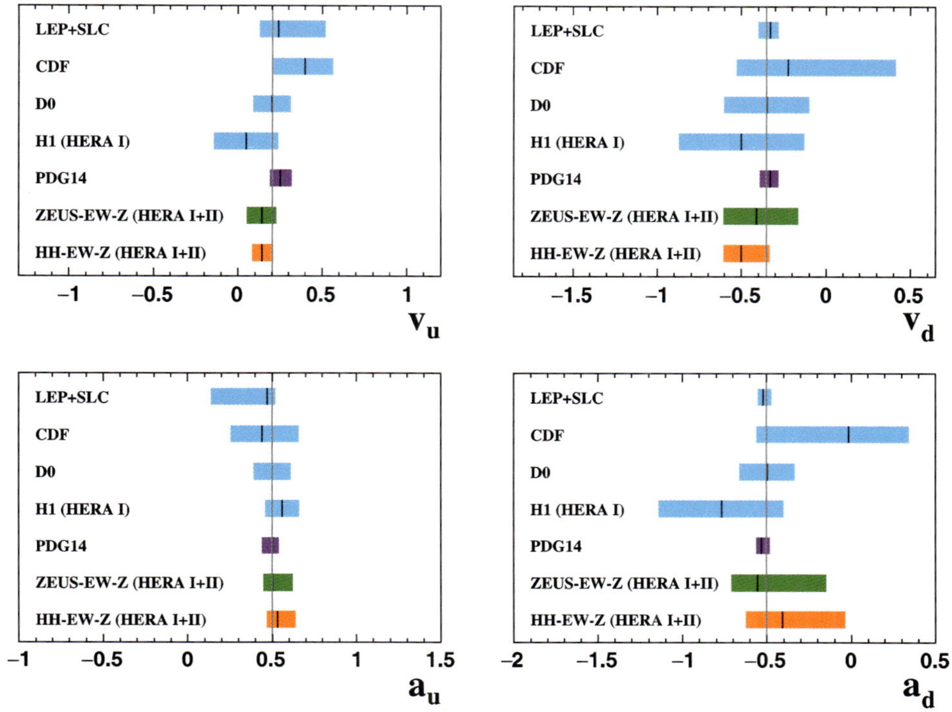

Fig. 6.17 The values for v_u, v_d, a_u, a_d obtained from the HH-EW-Z fit compared with results from ZEUS-EW-Z, H1 (HERA I), LEP and SLC combined, and CDF and D0. The PDG 2014 world average, which does not include the HH-EW-Z and ZEUS-EW-Z values, is also shown. The black line in each box represents the central value and error bars on the value represent the total uncertainties. The grey vertical line represents the SM prediction.

The values extracted for $\sin^2\theta_W$ are also consistent with other measurements and the world average and of a precision competitive with those in the neutrino sector. It should be noted that the analyses done here use leading order expressions for electroweak effects and next-to-leading order expressions for QCD effects in the combined fit of both theories. Future fits encompassing QCD at next-to-next-to-leading order and electroweak effects at next-to-leading order are most desirable given the precision being achieved at the Large Hadron Collider and the need for precise Standard Model predictions.

6.5.5 *Constraints on physics beyond the Standard Model*

Inclusive data on deep inelastic scattering from HERA has also been used to search for physics beyond the Standard Model. In particular, quark substructure has been investigated looking for deviations of the Standard Model theory from the measured cross sections. This follows a similar procedure to the investigation of the structure of the proton and indeed the work done by Rutherford in elucidating the structure of matter. No quark substructure was observed and a 95% confidence level upper limit on the quark radius of $R_q < 0.43 \cdot 10^{-16}$ cm was extracted. The limit is numerically similar to that extracted at LEP by the L3 collaboration ($R_q < 0.42 \cdot 10^{-16}$ cm) [Acciarri *et al.* (2000)], and as the beyond-the-Standard-Model physics can be very different, the results are largely complementary. Determination of the quark radius is just one example of contact interactions which are general models [Eichten *et al.* (1983); Rückl (1983, 1984); Haberl *et al.* (1991)] for interactions beyond the Standard Model and also encompass large extra dimensions and leptoquarks. A general search for contact interactions was performed by the H1 collaboration [Aaron *et al.* (2011b)] and a wide range of limits placed: lower limits on the compositeness scale were set in the multi-TeV range and lower limits for the mass scale of leptoquarks and extra dimensions set at the ~1 TeV scale.

6.6 Summary and perspective

In summary, the HERA collider experiments H1 and ZEUS have produced a wealth of results which have deepened our understanding of the fundamental particles and forces. In particular, the collaborations combined their data on deep inelastic scattering, reaching a precision of almost 1% in the double-differential cross-section measurements. It is the largest coherent data set on proton structure, spanning six orders of magnitude in the kinematic variables x_{Bj} and Q^2. A QCD analysis of the HERA data alone results in a set of parton distribution functions, HERAPDF2.0, without the need for data from other experiments. Also using HERA jet and charm data, the strong-coupling constant is measured together with the structure of the proton. QCD and electroweak effects are probed at high precision in the same data set, providing beautiful demonstrations of the validity of the Standard Model. An underpinning thread of all this work and our view of QCD is the pioneering work and insight of Guido Altarelli.

HERA was the world's first and so far only lepton-hadron collider; it is hoped that this situation will be remedied soon. Future projects such as the electron-ion collider (EIC) [Abelleira Fernandez *et al.* (2012a)], large hadron-electron collider (LHeC) [Accardi *et al.* (2016b)] and the very high energy electron-proton (VHEeP) [Caldwell and Wing (2016a)] collider would push the study of *ep*/*eA* physics to new boundaries. The different proposed facilities, i.e. EIC, LHeC and VHEeP, have much in common, but also significant differences. All have exciting and rich physics programmes and these should be continually developed, in particular highlighting where a given facility has a unique possibility. There is a lot of complementarity between the low energy, high luminosity polarised physics, such as the 3D mapping of the proton, and the physics at high energy, such as saturation. These facilities are discussed elsewhere in this volume and it is hoped that one or more of them is realised soon to continue the legacy of the likes of Guido Altarelli.

Acknowledgements

The work summarised in this chapter is due to many people such as current and former members of the H1 and ZEUS collaborations, the HERA machine group and staff at the DESY laboratory and other institutes who contributed but were not authors of the papers. The work that went in to producing the final combined data on deep inelastic scattering was crucial, forming the backbone of the chapter, and deserves special acknowledgement. The author also acknowledges support from DESY and the Alexander von Humboldt Stiftung.

Chapter 7

The strong coupling: a theoretical perspective

Gavin Salam
CERN, Theoretical Physics Department,
1211 Geneva 23, Switzerland

The strong coupling, α_s, is one of the fundamental parameters of the Standard Model. It enters into all cross-section calculations for processes at the Large Hadron Collider (LHC), whether directly at leading order, or through higher order QCD calculations. It also enters indirectly through the evolution of parton distribution functions (PDFs) and their correlation with the strong coupling. Consequently, as the LHC experiments' work evolves towards precision physics, accurate knowledge of the strong coupling is becoming increasingly important. The value of the coupling matters also for the question of gauge coupling unification at high scales and for the stability of the universe in any given particle-physics scenario.

The question of the value of α_s is hotly debated, with a range of discussions in the literature [Bethke *et al.* (2011); Pich (2013); Moch *et al.* (2014); d'Enterria (2015); Patrignani *et al.* (2016)]. It's a subject that Guido had an active interest in [Altarelli (2013b)]. Since, with Siggi Bethke and Günther Dissertori, I'm one of the authors of the PDG review chapter on QCD, which also includes a discussion and average of α_s, we often had exchanges on the subject. As I'll explain below, Guido was rather critical of our PDG approach to the question.

To illustrate the problem of establishing the value of α_s, consider the two following determinations: one, from a lattice-QCD calculation of Wilson loops, quotes $\alpha_s(m_Z) = 0.11840 \pm 0.00060$ [McNeile *et al.* (2010)]; another, from a fit to the thrust distribution in e^+e^- collisions [Abbate *et al.* (2011)], yields 0.1135 ± 0.00105. The two determinations are four standard deviations apart from each other, and there is no single value of α_s that isn't at least three standard deviations from one or the other of them. Both determinations pay extensive attention to the question of potential systematic uncertainties, yet in all likelihood, at least one of the two has underestimated them.

A first question is what accuracy do we need for the strong coupling. Consider the case of LHC phenomenology. The change induced in key LHC cross sections, e.g. the top-quark cross section or the Higgs-boson cross section, in going from $\alpha_s = 0.118$ to $\alpha_s = 0.113$ is a reduction of about 8–9%.[1] This is larger than the total theoretical uncertainties on these cross sections from missing higher order corrections (about 4–6% [de Florian *et al.* (2016b); Anastasiou *et al.* (2016)]) and larger also than current or foreseen experimental uncertainties: the top cross section is measured to about 3–4% uncertainty [Aad *et al.* (2014b); Khachatryan *et al.* (2016d)], and in the long term the Higgs cross section should also reach a similar or better precision. Even a 1% uncertainty on α_s leads to effects that are comparable to any other single theoretical uncertainty on the Higgs cross section, i.e. at the 2% level. Therefore, only for a determination of the coupling with a precision comfortably below the percent level can α_s uncertainties largely be ignored for extracting fundamental information from the LHC.[2]

To understand the limitations that arise in determining the strong coupling, it is useful to keep in mind the essence of any determination of the coupling. Some quantity V is measured experimentally, giving a result $V_{\text{exp}} \pm \delta V_{\text{exp}}$. This needs to be related to a theoretical prediction for the same quantity in terms of powers of the coupling,

$$V_{\text{th}}(\alpha_s(\mu), \mu) = \sum_{n}^{N} c_n(\mu)\alpha_s^n(\mu) + \mathcal{O}(\alpha_s^{N+1}) + \mathcal{O}\left(\frac{\Lambda^p}{Q^p}\right), \tag{7.1}$$

[1]This is based on the CT14nnlo PDF set [Dulat *et al.* (2016)], for a centre-of-mass energy of $\sqrt{s} = 13$ TeV, taking into account the correlation of the PDFs with α_s. Cross sections were evaluated with the ggHiggs code [Bonvini *et al.* (2016)] (at N3LO [Anastasiou *et al.* (2016)]) and the top++ code [Czakon and Mitov (2014)] (at NNLO). In the case of $t\bar{t}$ production with ABMP16 PDFs [Alekhin *et al.* (2017)], the effect is larger, a 14% reduction.

[2]One may also consider how α_s impacts vacuum stability estimates, assuming validity of the Standard Model up to high scales. A 1% increase in $\alpha_s(m_Z)$ has roughly the same impact on the stability criterion as a 0.4 GeV decrease in the top-quark mass, while a 5% increase in the value of $\alpha_s(m_Z)$ would ensure stability, rather than just metastability, of the Standard Model vacuum [Espinosa *et al.* (2015)].

where the c_n factors are the coefficients of the perturbative series, which can in practice be calculated up to some finite order $n = N$. They depend on the choice of renormalisation scale μ, as does the coupling itself. The quantity Λ is the non-perturbative scale of QCD and Q is the order of magnitude of the momentum transfer in the process used for measuring V. The term $\frac{\Lambda^p}{Q^p}$ reflects the inevitable existence of non perturbative contributions. Our understanding of its structure and relevance is, in some contexts, the subject of debate, though in most cases at the least the value of the power p is known.

Requiring $V_{\rm th}(\alpha_s(\mu), \mu) = V_{\rm exp}$ in Eq. (7.1) fixes α_s. The uncertainty on the $\alpha_s(m_Z)$ determination then has three sources: (i) the extent to which V can be measured precisely, i.e. the size of the $\delta V_{\rm exp}$ uncertainty; (ii) the estimated impact of terms c_n beyond those that can be calculated with today's technology, of order α_s^{N+1}, e.g. as found by varying μ; and (iii) the size of the "power correction" terms, Λ^p/Q^p and the degree to which they can be reliably understood. There may also be missing higher order electroweak terms, or uncertainties associated with other fundamental parameters. The discussion of different determinations will essentially be a discussion of the relative sizes of each of these sources of uncertainty, and the degree of consensus on our understanding of them.

The Particle Data Group (PDG) [Patrignani *et al.* (2016)] world average (in the QCD review chapter) limits its inputs to cases where the perturbative series is known at least to next-to-next-to-leading order (NNLO), and in the interests of brevity the discussion below will similarly concentrate on those cases. In the PDG we have taken the approach that we should be as neutral as possible with regard to disputes in the community about different determinations, with uniform prescriptions applied to all reasonable determinations. That is motivated in part by a desire to minimise any risk of bias in the outcome of the average. I think it's fair to say that Guido wasn't impressed by this approach. In our discussions on the subject he would insist that one should attempt to bring a theorist's critical view to each determination. He argued

> [...] one should select few theoretically simplest processes for measuring α_s and consider all other ways as tests of the theory.

In the chapter I'll give my take on what is needed to make a "clean" determination of the strong coupling: clean (and "transparent") were words regularly used by Guido in this context.

7.1 Jet rates and event shapes and in e^+e^- collisions

One natural way of thinking about the meaning of the QCD coupling is that it governs the probability of emitting a gluon. Gluons, of course, cannot be directly observed, but jets of hadrons can be used as a stand-in for hard gluons, i.e. for gluons that are energetic and radiated at large angles with respect to their emitter.

The simplest environment in which to study jets is $e^+e^- \to$ hadrons reactions. Perturbatively, the lowest order process is $e^+e^- \to q\bar{q}$, i.e. a two-jet event. There is a probability α_s to radiate a hard gluon, giving $e^+e^- \to q\bar{q}g$, i.e. a three-jet event. By measuring the fraction of three-jet events one can determine α_s.

There is considerable freedom in how one carries out the extraction: there are different algorithms for defining the jets, including the Durham [Catani *et al.* (1991c)] and Cambridge [Dokshitzer *et al.* (1997)] algorithms. Each algorithm comes with a parameter to define how energetic the emission should be in order to be considered a jet. For the Durham and Cambridge algorithms, the parameter is called y_{cut}, and corresponds to the squared transverse momentum of the emission, normalised to the squared centre-of-mass energy.

To illustrate some of the characteristics and challenges of extractions of the coupling from jet rates, Fig. 7.1 (top-left) shows the Durham 3-jet rate as a function of y_{cut}, as measured by the ALEPH collaboration at particle (i.e. hadron) level. It is compared to the pure NNLO prediction in 5-flavour massless QCD at parton level and to the NNLO result multiplied by an additional hadronisation correction. The upper axis shows the effective transverse momentum (p_t) cut that a given value of $\ln y_{\text{cut}}$ corresponds to. For a typical choice $\ln y_{\text{cut}} = -4.0$, one hovers close to a p_t cut of 10 GeV. If one considers that a jet's energy may change by an amount of the order of a GeV due to the parton-to-hadron transition, then the fact that the jet p_t is just 10 GeV becomes a concern. That parton-to-hadron transition is believed to be the main reason the NNLO (parton) theory doesn't immediately agree with the hadron-level data. To extract a value of the coupling it is mandatory to apply a hadronisation correction to the NNLO prediction (and also corrections for b-quark mass effects). This was done for the analysis in the top right-hand plot of Fig. 7.1, using Monte Carlo event generators to estimate the hadronisation correction (about a 5% effect). The plot shows the ALEPH collaboration's resulting extraction of the strong coupling as a function of y_{cut} [Dissertori *et al.* (2010)]. For $\ln y_{\text{cut}} > -4$ the extracted α_s value is fairly independent of y_{cut}, a sign of robustness of the analysis. Below that, however, the result depends substantially on the choice of y_{cut}. One can attribute that feature to a breakdown of fixed order perturbation theory, associated with logarithmically enhanced terms of the perturbative series, which go as $(\alpha_s \ln^2 y_{\text{cut}})^n$.

The final extraction from the ALEPH collaboration involves a choice of $\ln y_{\text{cut}} \simeq -3.9$ that remains within the plateau and minimises the final error. It gives:

$$\begin{aligned}\alpha_s(m_Z) &= 0.1175 \pm 0.0004(\text{stat.}) \pm 0.0016(\text{det.}) \pm 0.0011(\text{exp.})\\ &\quad \pm 0.0006(\text{had.}) \pm 0.0002(\text{mass}) \pm 0.0014(\text{pert.}),\\ &= 0.1175 \pm 0.0025(\text{total}),\end{aligned}$$

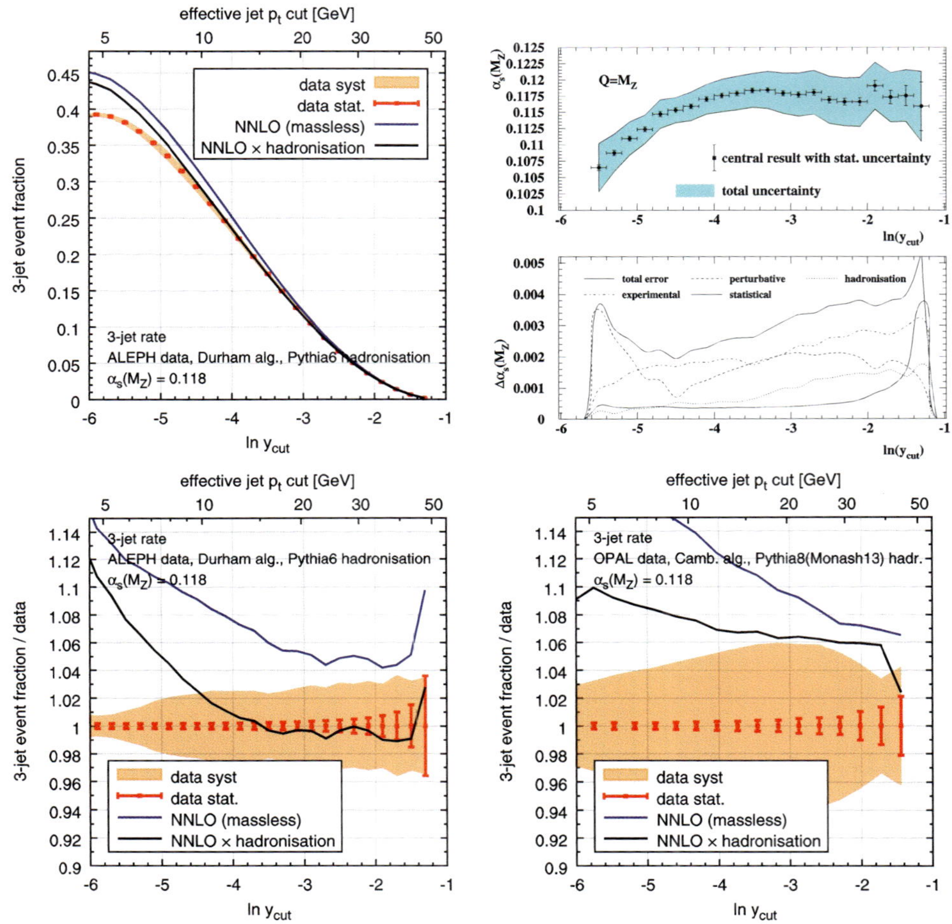

Fig. 7.1 Top left: comparison of ALEPH jet rate data [Heister *et al.* (2004)] for the Durham algorithm with the pure NNLO result for $\alpha_s(m_Z) = 0.118$ [Weinzierl (2011)] (see also [Gehrmann-De Ridder *et al.* (2008); Del Duca *et al.* (2016b)]), and the NNLO result multiplied by hadronisation corrections estimated with Pythia 6.428 [Sjöstrand *et al.* (2006)] with the DW tune. Top right: strong coupling (top) and the uncertainty breakdown from the ALEPH jet rate extraction (Fig. 1 of ref. [Dissertori *et al.* (2010)]). Bottom left: ratio of NNLO theory to data for $\alpha_s(m_Z) = 0.118$, with the same data and simulation choices as for the top-left plot. Bottom right: ratio of NNLO theory to data for $\alpha_s(m_Z) = 0.118$, but now with the Cambridge jet algorithm, OPAL data and Pythia 8.223 [Sjöstrand *et al.* (2015)] (Monash 2013 tune [Skands *et al.* (2014)]) for the hadronisation.

for which the dominant quoted error sources are detector and other experimental systematics as well as missing high order contributions (assessed through variation of the renormalisation scale).

The aspect of this kind of determination that is perhaps most called into question is the hadronisation correction, because the way non-perturbative effects arise in

Monte Carlo simulations (through a cutoff) does not match the way they need to be applied to perturbative calculations. Rather than applying a cutoff at some scale of order 1 GeV, perturbative calculations integrate down to zero momentum, but using an integrable, perturbative expansion of the coupling. The overall size of the hadronisation correction is 4–5%, as is visible from the ratio plot in the bottom-left of Fig. 7.1. The definition of how hadronisation interfaces with a perturbative calculation could conceivably modify this by a couple of percent.

A further question with jet-rate data is the stability of the determination. The bottom-right plot of Fig. 7.1 is analogous to the bottom-left one except that the OPAL data replaces the ALEPH data, the jet algorithm has been switched to the Cambridge algorithm and hadronisation corrections have been evaluated with the Pythia 8 generator instead of Pythia 6. The ratio of theory to data goes up by over 6%, with each of the three changes accounting for about one third of that.[3] Since the OPAL data have larger systematic uncertainties than the ALEPH data, the results appear to be still (just barely) compatible with $\alpha_s(m_Z) = 0.118$.[4] Nevertheless the difference between the bottom-left and right-hand plots of Fig. 7.1 could be taken as a more conservative estimate of the possible size of theoretical uncertainties in α_s determinations from e^+e^- jet rates.

The ALEPH result is reproduced in Table 7.1 together with a range of other strong-coupling determinations (all at least at NNLO) from hadronic final-state measurements at LEP and earlier e^+e^- colliders. As well as jet rates, they also make use of "event shapes" such as the thrust [Brandt *et al.* (1964); Farhi (1977)], C-parameter [Ellis *et al.* (1981b)] or jet broadenings [Catani *et al.* (1992)]. Each

Table 7.1 Determinations of the strong coupling from jet rates and event shapes in e^+e^- collisions.

Determination	Data and procedure	Reference
0.1175 ± 0.0025	ALEPH 3-jet rate (NNLO+MChad)	[Dissertori *et al.* (2010)]
0.1199 ± 0.0059	JADE 3-jet rate (NNLO+NLL+MChad)	[Schieck *et al.* (2013)]
0.1224 ± 0.0039	ALEPH event shapes (NNLO+NLL+MChad)	[Dissertori *et al.* (2009)]
0.1172 ± 0.0051	JADE event shapes (NNLO+NLL+MChad)	[Bethke *et al.* (2009)]
0.1189 ± 0.0041	OPAL event shapes (NNLO+NLL+MChad)	[Abbiendi *et al.* (2011)]
$0.1164\,^{+0.0028}_{-0.0026}$	Thrust (NNLO+NLL+anlhad)	[Davison and Webber (2009)]
$0.1134\,^{+0.0031}_{-0.0025}$	Thrust (NNLO+NNLL+anlhad)	[Gehrmann *et al.* (2013)]
0.1135 ± 0.0011	Thrust (SCET NNLO+N^{3}LL+anlhad)	[Abbate *et al.* (2011)]
0.1123 ± 0.0015	C-parameter (SCET NNLO+N^{3}LL+anlhad)	[Hoang *et al.* (2015)]

[3] The OPAL collaboration had already noted the smaller value of α_s for the Cambridge algorithm than for the Durham algorithm in fits at NLO+NLL [Abbiendi *et al.* (2006)].

[4] Note that in the full OPAL fits of ref. [Abbiendi *et al.* (2006)], which use slightly different jet variables than the 3-jet rate discussed here, the experimental systematic error on α_s is substantially smaller than one might deduce based on the size of the band in the bottom-right plot of Fig. 7.1.

event shape provides a continuous measure of the extent to which an event's energy flow departs from a pure back-to-back (i.e. leading order $e^+e^- \to q\bar{q}$) structure. One sees that the central values and uncertainties for α_s vary quite substantially between determinations, even though they all relate to the same underlying phenomenon, the probability of gluon emission from a quark-antiquark system. This is perhaps unsurprising given what we've seen in Fig. 7.1.

The result with the highest quoted precision is that for the thrust (SCET),

$$\alpha_s(m_Z) = 0.1135 \pm 0.0002_{\text{exp.}} \pm 0.0005_{\text{hadr.}} \pm 0.0009_{\text{pert.}} = 0.1135 \pm 0.0011, \tag{7.2}$$

closely followed by the C-parameter determination. A first point to be aware of is that in the 2-jet limit the thrust and C-parameter are highly correlated, both in the data and in the structure of the theoretical calculations. The two extractions are very similar in approach: they supplement fixed order perturbation theory with the resummation of enhanced logarithmic contributions, specifically accounting for terms ranging from $\alpha_s^n \ln^{n+1}$ down to $\alpha_s^n \ln^{n-2}$, i.e. N^{3}LL accuracy. Furthermore they use an analytic estimate of hadronisation corrections with a fitted free parameter, which to a first approximation corresponds to a shift of the distribution by a shift of the thrust (specifically $1-T$) or C-parameter by an amount proportional to a "power correction" Λ/Q (see the reviews [Beneke (1999); Dasgupta and Salam (2004)]). An advantage of analytic hadronisation estimates is that they can address the concern of matching Monte Carlo hadronisation corrections to perturbative calculations.

Guido's comment about the thrust result was

> I think that this is a good example of an underestimated error which is obtained within a given machinery without considering the limits of the method itself.

What might these methodological limitations be? One is that the formalism of resummation holds only for $C, 1-T \ll 1$, where every emission is so soft and collinear that one can effectively neglect the kinematic cross-talk (e.g. energy-momentum conservation) that arises when there are multiple emissions. There is almost no quantification in the literature of the corrections induced by such cross-talk, so they potentially represent a neglected systematic error.

Another limitation is that the power correction that is used holds in the 2-jet limit, i.e. $1-T \ll 1$. However the fits extend into the full 3-jet region, e.g. $(6\text{ GeV})/Q < 1-T < 0.33$ for the thrust, keeping in mind that $1-T = \frac{1}{3}$ is the largest value that can be obtained with 3 partons. One way of viewing the problem is that the power correction in this region should be governed by a different operator and simply taking the 2-jet result is risky. A different way of phrasing this is that if the average effect of hadronisation is essentially a shift of $1-T$ by an amount $\delta \sim \Lambda/Q$, then δ itself is likely to be a non-trivial function of the parton-level value of $1-T$. Only for small values of $1-T$ may one neglect that functional dependence.

A sign that this might be causing problems comes from the dependence of the thrust fit results on the choice of limits. Figure 17 of ref. [Abbate *et al.* (2011)] shows that modestly restricting the fit range to $(8\ \mathrm{GeV})/Q < 1 - T < 0.25$ increases the central fit value to about 0.1151, an increase of 0.0016 relative to Eq. (7.2), which is larger than the overall quoted error of 0.0011.[5]

Yet another concern relates to the treatment of experimental systematic errors. The ALEPH fit had total detector and experimental systematic errors of 0.0019, to be compared to those quoted in the SCET thrust fit of 0.0002. Even if the latter puts together several experimental results, such a large reduction is puzzling.

What do I conclude about event shapes? Ultimately I have doubts about how realistic it is to extract $\alpha_s(m_Z)$ with 1% precision from data whose underlying physical scale is 10–20 GeV, using observables that have $\sim 1\ \mathrm{GeV}/Q$ power corrections. Perhaps, today, the value of such fits should instead be seen in terms of what they might teach us about the limits of our understanding of hadronisation corrections and also of resummation.

7.2 LEP electroweak fits

Rather than looking for the signature of actual gluon emission, electroweak (EW) fits for α_s rely on the slight non-cancellation between higher order real and loop graphs in a range of EW observables, many of them connected with Z production at LEP and SLC. For example the quantity $R = \frac{\sigma(e^+e^- \to \mathrm{hadrons}, Q)}{\sigma(e^+e^- \to \mu^+\mu^-, Q)} \equiv R(Q) = R_{\mathrm{EW}}(Q)(1 + \delta_{\mathrm{QCD}}(Q))$ is sensitive to α_s through the δ_{QCD} term

$$\delta_{\mathrm{QCD}}(Q) = \frac{\alpha_s(Q)}{\pi} + \dots, \tag{7.3}$$

where the series is known up to $\mathcal{O}(\alpha_s^4)$ [Baikov *et al.* (2008, 2012)]. Substituting $\alpha_s(m_Z) = 0.118$ shows that $\delta_{\mathrm{QCD}}(m_Z)$ is about a 4% effect, so the roughly per mil measurement accuracy for R [Schael *et al.* (2006)] leads to a 2.5% uncertainty on α_s. The actual fits for α_s, carried out in the context of a global electroweak fit, are from the GFitter group [Baak *et al.* (2014b)],

$$\alpha_s(m_Z) = 0.1196 \pm 0.0028_{\mathrm{exp}} \pm 0.0009_{\mathrm{th}} = 0.1196 \pm 0.0030, \tag{7.4a}$$

and from the PDG electroweak chapter

$$\alpha_s(m_Z) = 0.1203 \pm 0.0028\,. \tag{7.4b}$$

[5] A full interpretation would require an analysis of the correlation of the errors of the results with different choices of limits. Interestingly, the corresponding danger sign is not there for the C-parameter results. However the absence of an obvious danger sign in the numerical fit should not be taken as an indication of the absence of danger. In particular, with Gionata Luisoni and Pier Monni, we have started investigating the impact on $\alpha_s(m_Z)$ of different forms of C-dependence for the power correction. Our preliminary finding is that the effect is substantial, potentially on the order of 5%.

The two fits differ in the details of which EW input variables are used and in their error treatment, with the GFitter group conservatively taking the theoretical uncertainty to be the size of the last term in the perturbative series.

From the point of view of QCD, the EW fits are arguably the most robust. One reason is that the perturbative series is under good control: even the conservative theory uncertainty from GFitter is below a percent. The other reason is that non-perturbative corrections are also small, with $(\Lambda/Q)^4$ (suppressed) or $(\Lambda/Q)^6$ corrections. Any reasonable estimate of their numerical impact gives contributions that are much below the experimental uncertainty. Finally, the extraction is conceptually straightforward, (mostly) satisfying Guido's criterion of transparency.

A concern that is sometimes raised about the EW extraction of α_s (in particular by Guido) is the potential impact of new physics contributions, such as non-universal vertex corrections, for example in the $Zb\bar{b}$ vertex. For this reason, even if a future collider were to bring improved experimental accuracy, one might not wish to rely on electroweak fits alone to extract a precise value for the strong coupling.

7.3 Tau decays

The hadronic branching ratio of τ leptons is an observable that is sensitive to the strong coupling in a way that is very similar to the Z width to hadrons, i.e. through the slight non-cancellation of QCD real and virtual graphs. One practical difference, aside from the much lower momentum scale, is that in the decay $\tau \to \nu_\tau W^*(\to \text{hadrons})$, one should integrate over all allowed virtualities for the off-shell W. Schematically, following the simplified notation used by Guido, this gives the following relation for the hadronic branching ratio of the τ:

$$R_\tau = N \int_0^{m_\tau^2} \frac{ds}{m_\tau^2} \left(1 - \frac{s}{m_\tau^2}\right)^2 \operatorname{Im} \Pi_\tau(s) \,, \tag{7.5}$$

where N is an electroweak normalisation factor and $\operatorname{Im}\Pi_\tau(s)$ is the QCD spectral function.

Experimentally, both R_τ and the detailed spectral function can be measured, and typically ALEPH [Davier *et al.* (2014); Schael *et al.* (2005)] and OPAL [Ackerstaff *et al.* (1999)] data are used, with the former having somewhat smaller uncertainties.

Theoretically, Eq. (7.5) involves integrating over squared hadronic momenta s in a region that has resonance structure and where perturbative QCD is clearly not applicable. However, analyticity means that it is possible to rewrite the integral as

$$R_{\tau,h} = \frac{N}{2i} \oint_{|s|=m_\tau^2} \frac{ds}{m_\tau^2} \left(1 - \frac{s}{m_\tau^2}\right)^2 \Pi_\tau(s) \,. \tag{7.6}$$

The fact that $|s| = m_\tau^2$, together with the $(1 - s/m_\tau^2)^2$ factor, ensures that the integral stays away from the most dangerous, resonance regions. The theoretical prediction gets evaluated in two main ways: (1) it can be written as a series in powers of $\alpha_s(m_\tau)$, referred to as fixed order perturbation theory (FOPT); or (2) one can use the perturbative expression for $\Pi_\tau(s)$ in the integrand and use the full renormalisation-group equation for the evolution of $\alpha_s(-s)$ around the contour, called contour-improved perturbation theory (CIPT).[6] For some time the choice of FOPT v. CIPT was a hotly-debated one. Brief and quite approachable explanations of the different points of view are given in ref. [Bethke *et al.* (2011)]. Nowadays, most groups tend to quote both and give an average of them for the final results.

Currently the most contentious issue in the literature concerns non-perturbative corrections. The two main lines of thought have been exposed recently in refs. [Pich and Rodríguez-Sánchez (2016)] (PRS) and [Boito *et al.* (2017)]. Table 7.2 shows results[7] from the Boito *et al.* paper, with their favoured non-perturbative choice, which includes "duality violations" (DV), i.e. an attempt to allow for differences that may arise between an operator-product expansion (with quarks, no hadronic resonances) and real data (with hadrons and corresponding resonances). It also includes a truncated OPE result, which corresponds to the approach favoured by PRS, with a more minimal set of non-perturbative corrections.

Table 7.2 Extractions of $\alpha_s(m_\tau)$ using different choices for both the perturbative (PT) evaluation and non-perturbative (NP) contributions, based on ALEPH and OPAL data. The results are taken from ref. [Boito *et al.* (2017)], but the central values are indicative also of results from other groups when they make similar perturbative and non-perturbative data choices.

PT choice	NP choice	$\alpha_s^{(n_f=3)}(m_\tau)$
FOPT	DV	0.303 ± 0.009
CIPT	DV	0.319 ± 0.012
FOPT	trunc-OPE	0.321 ± 0.009
CIPT	trunc-OPE	0.339 ± 0.011

[6]Yet another scheme is discussed in ref. [Abbas *et al.* (2013)].

[7] Extractions of α_s from the τ data are usually quoted at the scale m_τ with three light flavours. To convert $\alpha_s^{(n_f=3)}(m_\tau)$ to $\alpha_s^{(n_f=5)}(m_Z)$, an approximation that is good to within about two per mil over the relevant range is

$$\alpha_s^{(n_f=5)}(m_Z) \simeq 0.1180 + 0.125\,[\alpha_s^{(n_f=3)}(m_\tau) - 0.314]\,, \tag{7.7}$$

based on 5-loop running [Baikov *et al.* (2017); Herzog *et al.* (2017)] and four-loop flavour thresholds [Chetyrkin *et al.* (2006)]. A given absolute error on $\alpha_s(m_\tau)$ goes down by a factor of 8 when translating it to an error on $\alpha_s(m_Z)$, in reasonable accord with the leading-order expectation that it should scale as α_s^2. This is part of the rationale of extracting α_s at such a low scale, because the impact of non-negligible non-perturbative effects may be still be compensated by the reduction of the error when evolving up to m_Z.

Table 7.3 Recent determinations of $\alpha_s(m_Z)$ from τ decays that take into account the updated ALEPH spectral-function data [Davier *et al.* (2014)] as well as the highest order theoretical predictions [Baikov *et al.* (2008)].

Determination	Reference
0.1199 ± 0.0015	[Davier *et al.* (2014)]
0.1197 ± 0.0015	[Pich and Rodríguez-Sánchez (2016)]
0.1175 ± 0.0018	[Boito *et al.* (2015)]
$0.1174 {}^{+\,0.0019}_{-\,0.0017}$	PDG EW 2016 [Patrignani *et al.* (2016)]

PRS argue their approach is justified in part based on the observation of stability of the extracted α_s when replacing the m_τ^2 upper limit of Eq. (7.5) with an arbitrary limit s_0 and then varying s_0. The PRS final result, averaging the FOPT and CIPT determinations, is $\alpha_s^{(n_f=3)}(m_\tau) = 0.328 \pm 0.013$. However they have considered a range of analyses of non-perturbative contributions and in some cases found the overall uncertainty can go up to 0.020 depending on the precise procedure used (cf. their summary table 11).

Different results from τ determinations are summarised in Table 7.3, now showing the values of $\alpha_s(m_Z)$. In the case of the Boito *et al.* result, the numbers are taken from an earlier publication of theirs, however the results are essentially the same as in their most recent work.

Guido's discussion of α_s came before the latest iteration in the debate about non-perturbative contributions. Nevertheless he expressed clear concerns, for example about possible $\mathcal{O}(\Lambda^2/m_\tau^2)$ contributions. These are absent in the limit of massless quarks, while quark-mass effects themselves contribute as m_q^2/m_τ^2. Guido highlighted that if one considers a constituent quark mass, $m_q \sim 0.3$ GeV, then the resulting $\mathcal{O}(0.3\ \text{GeV}/m_\tau)^2$ correction to R_τ would have a significant impact on the extracted α_s value. Another potential concern is that it is commonplace in τ-based determinations to vary the renormalisation scale μ_R by a factor of $\sqrt{2}$ rather than the more canonical factor of 2 (though perhaps this is covered by the CIPT/FOPT difference).

My inclination is to share Guido's caution about determinations of α_s from τ decays, especially in view of the ongoing debates on the subject. One question is whether to view τ decays as a source of precise information about α_s or rather a unique window into physics close to the edge of the perturbative regime.

7.4 PDF determinations

PDF fits are sensitive to the strong coupling in various ways. One way is through the Q^2 dependence of Deep Inelastic Scattering (DIS) structure functions, itself driven by the DGLAP [Gribov and Lipatov (1972a); Altarelli and Parisi (1977); Dokshitzer (1977)] evolution of the underlying PDFs, which is proportional to the

Table 7.4 Determinations of the strong coupling in the context of PDF fits. The first uncertainty is the statistical uncertainty of the fit as defined by the fit authors (scaled from 90% to 68% confidence level in the case of CT14), the second is a theory uncertainty estimated as half the difference between the NLO and NNLO fit results. In the case of ABMP16, which carried out only a NNLO fit, the NLO−NNLO difference is taken from the earlier, conceptually similar, ABM12 fit [Alekhin *et al.* (2012)]. The JR result corresponds to their "standard" fit variant, which has minimal assumptions about the structure of parton distributions at low Q^2.

Determination	PDF fit	Reference
$0.1141\,^{+\,0.0022}_{-\,0.0020} \pm 0.0003$	BBG06 non-singlet	[Blumlein *et al.* (2007)]
$0.1173 \pm 0.0007 \pm 0.0009$	NNPDF21	[Ball *et al.* (2012)]
$0.1162 \pm 0.0006 \pm 0.0014$	JR14	[Jimenez-Delgado and Reya (2014)]
$0.1147 \pm 0.0008 \pm 0.0023$	ABMP16	[Alekhin *et al.* (2017)]
$0.1150\,^{+\,0.0036}_{-\,0.0024} \pm 0.0010$	CT14	[Dulat *et al.* (2016)]
$0.1172 \pm 0.0013 \pm 0.0014$	MMHT2014	[Harland–Lang *et al.* (2015b)]

coupling,

$$\frac{dq(x,Q^2)}{d\ln Q^2} = \alpha_s(Q^2)\int_x^1 \frac{dz}{z}(P_{qq}(z)q(x/z,Q^2) + P_{qg}(z)g(x/z,Q^2)) + \mathcal{O}(\alpha_s^2)\,, \tag{7.8}$$

where P_{ij} are splitting functions and q and g are quark and gluon distributions. Another source of sensitivity is that some cross sections are proportional to α_s or α_s^2, e.g. jet cross sections, which overlaps with the question of collider determinations below.

A summary of extractions of α_s carried out in the context of PDF fits is given in Table 7.4, taking only the most recent published result from any given fitting group and/or approach. One element to note is that in contrast to almost all other classes of strong coupling determination, PDF fits don't usually quote a theory uncertainty. Partly this is because it is not straightforward to extend the standard method for theory uncertainty estimation, scale variation, to PDF fits: many different processes come into play and in each one scale variations effectively play a different role. One then ends up with hard-to-answer questions of whether scale variations should be correlated across processes and even across different regions of x and Q^2. Neglecting theory uncertainties in the PDF fit means that if two processes or kinematic regions have similar statistical constraining power but one has much larger theory uncertainties, the region with larger theory uncertainties will get more weight than is appropriate.[8]

A poor-man's approach to estimating theory uncertainties is to take half the difference between fits at NNLO and NLO, as adopted by the NNPDF collaboration.

[8]As far as I'm aware, the extent to which this situation occurs in practice hasn't been studied in detail.

This is the basis of the theory uncertainties shown in Table 7.4.[9] Formally it's a conservative approach (the theory uncertainty estimated in this way is of the same order as the NNLO corrections), though in practice it might underestimate the error if NNLO and NLO results just happen, numerically, to be close.

Taking into account both the experimental and the estimated theory uncertainties, the different PDF determinations are largely consistent with each other. This has not, however, prevented heated debate between groups. For example if one leaves aside the theory uncertainty estimate and assumes the experimental errors to be uncorrelated, the relatively recent MMHT2014 and ABMP16 results are 1.7σ apart. In reality the experimental errors should be correlated, since much of the underlying data is the same. So the disagreement must be ascribed to systematic differences between the fit procedures. These include: the treatment of heavy flavour, whether a fixed-flavour number scheme as in ABMP16 or a general-mass variable flavour number scheme (GM-VFNS) as in MMHT2014 (as well as CT14 and NNPDF21); the treatment of higher twist effects, with ABMP16 explicitly including higher-twist terms within their cross-section calculations, while many other groups don't; the inclusion of collider jet data only at NLO, which is inconsistent with the rest of the fit being NNLO, an issue whose importance was debated given the non-negligible experimental uncertainties of the data sets. Each group argues that its results are robust (see e.g. refs. [Alekhin *et al.* (2014); Thorne (2014)]).

Guido expressed a preference for the results based on global fits, i.e. using the largest available data sets, which today corresponds to the CT, MMHT and NNPDF results. Those also represent my first choice, mainly because of their use of a GM-VFNS which is relevant for the moderate and high Q^2 DIS data. They also represent the widely adopted choice of the LHC community for the PDFs themselves, notably through the PDF4LHC15 combined PDF set [Butterworth *et al.* (2016)] (which does not involve an α_s fit).

Looking to the future, there are prospects for significant theoretical improvements in such fits, for example from the full NNLO jet cross sections [Currie *et al.* (2017d)] already recently included into a first PDF fit [Harland–Lang *et al.* (2017)] or the inclusion of Z p_t distribution data [Boughezal *et al.* (2017)].[10] Other advances

[9]The BBG06 result is based on N^{3}LO coefficient functions and in that case the table shows half the difference between N^{3}LO and NNLO. At the time of its publication only NNLO splitting functions were available, however the authors argue that the uncertainty from the N^{3}LO splitting function was small relative to other uncertainties and that this statement is further supported by the recent calculation of the exact N^{3}LO non-singlet splitting functions [Moch *et al.* (2017)]. I am grateful to Johannes Blümlein for correspondence on this point.

[10]Beware, however, of the effect of Λ/Q power corrections for these data sets. In particular the pattern of soft gluon emission from a Z+jet event is azimuthally asymmetric: there is more radiation away from the Z than in the same direction as the Z. As a result one may expect non-perturbative modifications of the pattern of emission to also be azimuthally asymmetric, resulting in a net average shift of the Z p_t by an amount of order Λ. A 0.5 GeV shift would translate to a 1.5% change in the Z p_t distribution around $p_t = 100$ GeV, which is significant compared to the sub-percent accuracies of some measurements [Aad *et al.* (2016b)]. This type of effect cannot be

that could be of benefit to all fits include small-x resummation (as used in ref. [Ball *et al.* (2017a)]) and recent progress towards $\mathrm{N^3LO}$ splitting functions [Moch *et al.* (2017)].

However, I do have concerns about potential fundamental limits in PDF fits as they are carried out currently. In the case of τ decay we saw there is considerable debate about non-perturbative corrections, for a kinematic region $s = m_\tau^2 = 3.16\ \mathrm{GeV}^2$. DIS fits extend to a comparably low Q^2, but power corrections rather than being Λ^p/Q^p with $p = 4$ or 6, have $p = 2$, i.e. they are potentially larger. Additionally, charm production is a relevant contribution to the F_2 structure function and it is not clear how reliably it can be predicted near threshold (even within a minimal "fitted" charm framework), given the hundreds of MeV difference between a charm quark mass and charm-meson masses and the significant mass-dependence of the cross sections.

How could these problems be addressed? Some PDF fits already explore the use of higher Q^2 cutoffs on the data being used and I think this is an avenue that deserves to be pursued further. For example, one might argue that a relative precision ϵ on PDFs, and the associated α_s, is to be trusted only insofar as the DIS data being used satisfies $Q^2 > Q^2_{\min} \gtrsim \Lambda^2/\epsilon$. The choice of Λ^2 would need to be debated, but could be of the order of $0.5\ \mathrm{GeV}^2$.

7.5 Collider determinations

By "collider determinations" I mean determinations of α_s based on cross sections measured at hadron-hadron and hadron-lepton colliders that are used to constrain the strong coupling independently of a PDF fit.

With the recent advances in NNLO calculations (see a recent review [Dittmaier (2017)]), a significant number of new processes is becoming available for collider-based NNLO strong-coupling determinations.

In practice two processes have been used so far, $t\bar{t}$ production [Chatrchyan *et al.* (2014); Klijnsma *et al.* (2017)] together with the calculation of ref. [Czakon *et al.* (2013)], giving

$$\alpha_s(m_Z) = 0.1177 \pm 0.0010\,(\text{exp.})\,{}^{+0.0020}_{-0.0024}\,(\text{PDF})\,{}^{+0.0021}_{-0.0021}\,(\text{scale}) = 0.1177^{+0.0034}_{-0.0036}\,, \tag{7.9}$$

which combines a number of top-production cross section measurements from ATLAS, CMS and the Tevatron; and jet production in DIS [Andreev *et al.* (2017)] by the H1 collaboration, together with the calculation ref. [Currie *et al.* (2016)],

$$\begin{aligned}\alpha_s(m_Z) &= 0.1157 \pm 0.0020(\text{exp}) \pm 0.0006(\text{had}) \pm 0.0005(\text{PDFs}) \pm 0.0027(\text{scale}),\\ &= 0.1157 \pm 0.0034,\end{aligned} \tag{7.10}$$

straightforwardly estimated by turning hadronisation and multiple-parton-interaction effects on and off in Monte Carlo simulations.

where the PDF uncertainty includes several sources: the actual PDF uncertainty, the dependence of the PDF on α_s and differences between PDF sets. These uncertainties are all small, perhaps because the jet production kinematic region that was used is dominated by quark-induced processes.

One potential question about individual collider determinations is why one would go down this route at all: isn't it better simply to include the collider data in a global PDF fit and extract α_s that way? One answer to this is that it is much simpler to properly account for scale and other theoretical uncertainties when considering a single observable than when considering many different observables and kinematic ranges. What's more, one sees that the theory uncertainties tend to be as large as any other uncertainty: neglecting them, as is common in global PDF fits, is clearly not justified.[11]

Overall, even if I've been involved in them myself, I am inclined to approach collider α_s fits with some caution. Ultimately, insofar as they rely on knowledge of PDFs, they inherit the same drawbacks as PDF fits, notably the potential sensitivity to low Q^2 non-perturbative effects. Only if one can devise sets of collider observables where the sensitivity to PDFs is mostly eliminated can one evade this problem. To some extent this sensitivity to PDFs is eliminated in the H1 study. However the use of a cut of $\tilde{\mu}^2 \simeq Q^2 + p_{t,jet}^2 > (28 \text{ GeV})^2$ means that some fraction of the jets will have a transverse momentum that is sufficiently low that (Λ/p_t) hadronisation effects could be a concern beyond the quoted hadronisation uncertainty, as was the case for event-shape fits.

7.6 Lattice QCD

Whereas most α_s determinations involve comparing a perturbative calculation directly with an experimental observable, lattice QCD approaches use a somewhat different methodology: firstly, lattice parameters are tuned so as to reproduce suitably chosen low-energy hadronic data (e.g. pion and kaon decay constants); then the same lattice simulation is used to calculate some observable at a perturbative scale that is also amenable to calculation in perturbation theory; finally α_s is determined by requiring agreement between the lattice and perturbative predictions for that observable.

The main recent lattice results for α_s (using at least $2+1$ flavours) are summarised in Table 7.5. They differ from each other both in the type of lattice simulations used (e.g. the treatment of light quarks) and in the choice of observable used to match with perturbation theory.

---

[11] The H1 paper also shows the result of extracting α_s within a PDF fit that incorporates the H1 structure function and jet data and find $\alpha_s(m_Z) = 0.1142 \pm 0.0028$. Interestingly this fit includes scale variations both for the structure functions and the jet cross section and it is the scale variation uncertainty (± 0.0026) that dominates the final α_s uncertainty. The fit also restricts its attention to $Q^2 > 10 \text{ GeV}^2$, which is further from the dangerous non-perturbative region than the $Q^2_{\min}$ value used in many global fits.

Table 7.5 Selected determinations of the strong coupling within lattice QCD.

Determination	Approach	Reference
0.1184 ± 0.0006	HPQCD Wilson loops	[McNeile *et al.* (2010)]
0.1192 ± 0.0011	Maltman-HPQCD Wilson loop	[Maltman *et al.* (2008)]
0.1182 ± 0.00074	HPQCD heavy quark current correlator	[Chakraborty *et al.* (2015)]
0.1177 ± 0.0026	JLQCD charmonium correlators	[Nakayama *et al.* (2016)]
0.1162 ± 0.00084	MP charmonium correlators	[Maezawa and Petreczky (2016)]
0.1196 ± 0.00108	ETM ghost-gluon coupling	[Blossier *et al.* (2014)]
$0.1166 {}^{+\,0.0012}_{-\,0.0008}$	QCD static energy	[Bazavov *et al.* (2014)]
$0.1205 {}^{+\,0.00094}_{-\,0.00197}$	PACS-CS step scaling	[Aoki *et al.* (2009)]
0.1185 ± 0.00084	ALPHA step scaling	[Bruno *et al.* (2017)]

The first high-precision α_s results with reasonable dynamical quark masses were those from the HPQCD collaboration [Davies *et al.* (2008); McNeile *et al.* (2010); Chakraborty *et al.* (2015)], with precisions approaching 0.5%. The solidity of this precision claim has been widely discussed in the literature. Guido's comment was

> With all due respect to lattice people I think this small error is totally [i]mplausible.

The lattice part of the calculation uses the staggered fermion approach for the light quarks, a method that is the subject of misgivings by part of the lattice community. The perturbative correspondence is made using Wilson loops and heavy-quark correlators. Concerns have been raised as to whether perturbation theory is sufficiently precise for these observables at the scale where α_s is extracted. The HPQCD authors argue that they have been conservative in their estimate of perturbative matching systematics. However the method involves fitting higher order coefficients in the perturbative series, an approach that is not widely used in other α_s determinations. With a subset of the same lattice data, but different assumptions in using it, ref. [Maltman *et al.* (2008)] found a slightly different result with almost double the uncertainty. The FLAG working group [Aoki *et al.* (2014)], representing part of the community, argued that a larger error uncertainty should be assigned to the HPQCD results, notably associated with the matching with perturbation theory and its final estimate for $\alpha_s(m_Z)$ was 0.1184 ± 0.0012.

The question of the reliability of perturbation theory applies to essentially all of the determinations shown in Table 7.5, and in many cases there is also a concern about the potential impact of discretisation errors. This is because of the fundamental computational limitation, within any single lattice calculation, on the ratio of the smallest length scale to the longest length scale. The former needs to be as small as possible for good perturbative matching and small discretisation errors at the perturbative scale, while the latter needs to be large to minimise finite-volume effects in low-energy observables. The issue of the perturbative scale being insufficiently high leads to a situation where different groups, using broadly similar methods, obtain substantially different error estimates. This is the case for the JLQCD v. MP charmonium correlator results: the JLQCD result uses scale variation to estimate the perturbative uncertainty and this dominates the 2.2% uncertainty. In contrast the MP result varies an unknown higher order term in the perturbative series, within some estimated reasonable range, and finds a negligible contribution to the overall 0.7% uncertainty, which is instead dominated by continuum extrapolation and statistical components.

A notable advance in this respect has come recently from the ALPHA collaboration [Bruno *et al.* (2017)]. Here the matching is performed at a scale of 70 GeV, where α_s is sufficiently small that systematic errors from neglected higher order terms are convincingly subdominant [Dalla Brida *et al.* (2016)]. The authors obtain lattice results at this high scale with the help of the step scaling method [Luscher *et al.* (1991)], i.e. by using a series of lattice simulations with progressively smaller sizes, i.e. higher momentum scales, non-perturbatively matching each lattice to the previous one.[12] Their final error of 0.7% is dominantly statistical.

A demonstration of the perturbative robustness of the ALPHA determination is given in Fig. 7.2, which shows the extraction of the QCD scale Λ (defined in terms

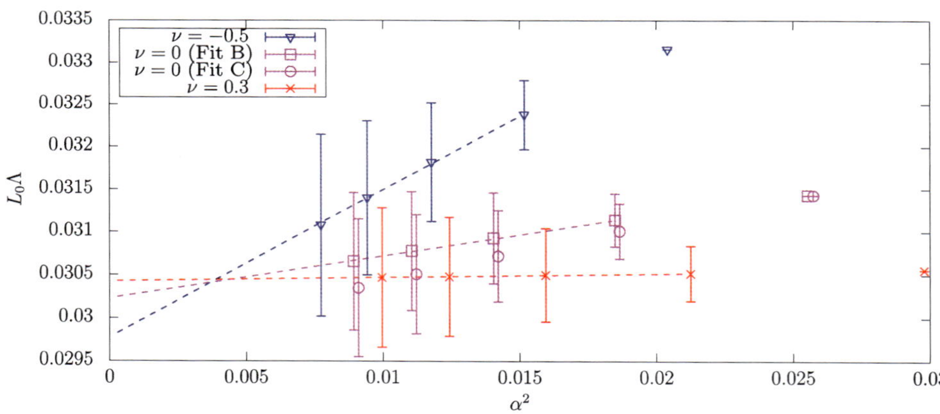

Fig. 7.2 Adaptation of Fig. 2 of ref. [Dalla Brida *et al.* (2016)], kindly provided by Alberto Ramos. The figure shows the extraction of the QCD Λ parameter, in units of $L_0 \simeq 4.22$ GeV, in the Schrödinger Functional scheme. Different lattice sizes (or correspondingly scales $\mu_{\rm PT}$ in the step scaling approach) each lead to one $\alpha_s(\mu_{\rm PT})$, the value of which is shown squared on the x axis (labelled α^2). For each $\alpha_s(\mu_{\rm PT})$ result, the authors deduce a value for the QCD Λ parameter (y axis). The parameter ν in the underlying Schrödinger Functional approach can be thought of as a renormalisation scheme choice (Λ is always converted back to a unique scheme). The two $\nu = 0$ sets of results (labelled Fit B, C) correspond to different correction schemes for lattice discretisation artefacts.

[12] An earlier step-scaling analysis was carried out by the PACS-CS collaboration [Aoki *et al.* (2009)] and while it was relatively far from the chiral limit, its results are consistent with those from the ALPHA collaboration.

of $\alpha_s(\mu)$ in Eqs. (1,2) of ref. [Dalla Brida *et al.* (2016)][13]), based on a number of differently sized lattices, covering a factor of 16 in scales. For each lattice size, they determine a value of α_s at a scale $\mu_{\rm PT}$ associated with the lattice size. Then using purely perturbative running they can convert it to a value for Λ. This is done for a variety of schemes, corresponding to the choices of ν in the figure (the Λ value shown is defined in a scheme invariant way). With a β-function known to all orders and in the absence of power corrections, the extracted Λ would be independent of $\mu_{\rm PT}$ or correspondingly of the $\alpha_s^2(\mu_{\rm PT})$ value shown on the x axis. In practice, missing higher order terms for the β-function in the Schrödinger Functional scheme should correspond to a residual offset for the extracted Λ value that scales as $\alpha_s^2(\mu_{\rm PT}^2)$, with a ν-dependent coefficient. This is observed in the lattice data over the last three iterations of step scaling, insofar as the four left-most points (covering a factor of 8 in $\mu_{\rm PT}$) are consistent with a linear dependence of the extracted Λ parameter on α_s^2. An extrapolation of Λ to zero coupling is always consistent with the result at the lowest available value of α_s, within its statistical errors. Furthermore, at that lowest value of α_s, the different schemes are all in agreement. This is a powerful cross-check of the stability of the perturbative side of the extraction over a broad range of scales. The final uncertainty of 2.6% on $L_0\Lambda$ that is quoted by the authors corresponds to about 0.54% uncertainty on $\alpha_s(m_Z)$. The conversion of these results to an $\overline{\rm MS}$ Λ (or coupling) in physical units, in particular the determination of L_0 in GeV, corresponds to the work of refs. [Dalla Brida *et al.* (2017); Bruno *et al.* (2017)].

One issue raised by the ALPHA collaboration is that their results (as most others) are for 2+1 flavours, and so require perturbative matching at the charm mass in order to be interpreted as a 5-flavour coupling at m_Z. However, the 2015 HPQCD result with $2+1+1$ flavours [Chakraborty *et al.* (2015)] is very close to the earlier result with 2+1 flavours and a perturbative charm threshold [McNeile *et al.* (2010)]. This suggests that non-perturbative charm-threshold effects are small. A further potential concern is that QED and isospin breaking effects are not accounted for in the lattice simulations of the low-energy hadronic quantities, however these are believed to be subdominant compared to the current uncertainties. Nevertheless, such points may need to be addressed in future higher-accuracy determinations of the strong coupling.

---

[13] In most circumstances, the PDG recommends against referring to values for Λ. This is because in the collider community Λ is often used as parameter in closed-form formulae for α_s that do not exactly satisfy the renormalisation group equation to some truncated order in the beta-function. A same value for Λ can also lead to different α_s values depending on the specific closed-form formula used. In contrast the definition of Λ that is exploited in ref. [Dalla Brida *et al.* (2016)] is defined through an implicit equation for α_s that does exactly satisfy the renormalisation group equation.

7.7 Concluding remarks

The many different (NNLO and better) determinations are summarised in Fig. 7.3. A wide range of methods can deliver results for α_s with an accuracy of a few percent. However such an accuracy is not sufficient if one is to fully exploit the results that are coming and still to come from the LHC. The question then for the field is which, if any, of the percent-level determinations to trust given that some of them are mutually incompatible. A few years ago, Guido's answer to that question was, essentially, none!

The key issues that recur are the estimate of uncertainties from missing higher orders and the problem of non-perturbative corrections. Many of the determinations based on τ decays, event shapes, PDF fits and lattice methods are either directly determining α_s at a scale in the range of 1–3 GeV or are in some other way sensitive to the physics occurring on those scales. Most of them give detailed reasoning as to why they believe they are able to control the problems that might arise from

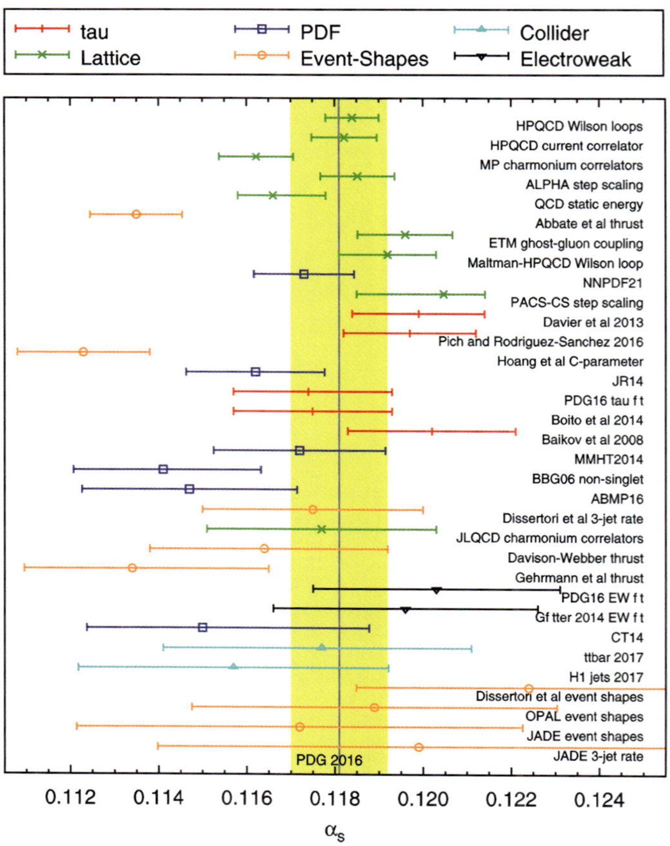

Fig. 7.3 The α_s determinations discussed in this review, in order of increasing quoted uncertainty.

proximity to the non-perturbative region and the poor convergence of the series when α_s is not so small. However the accessible range of scales over which one can rigorously test these statements tends to be limited, leaving the door open to sometimes heated debate about the degree of control over systematic uncertainties.

In this respect the recent ALPHA lattice result is potentially a breakthrough. Together with EW precision fits, it is the only approach where the connection with perturbation theory is made at a genuinely high scale, with almost no assumptions needed about low-scale physics. Its precision of 0.7% is almost three times better than the result from EW fits, thanks in part to the fact that it is not limited by LEP statistics. Furthermore the method can provide an extraction over a wide range of momentum scales and the pattern of results obtained in that way gives strong support to the authors' arguments that the perturbative extraction is robust.

What would be needed to consolidate the ALPHA advance? Any determination of α_s that reaches percent-level accuracy tends to be a complicated endeavour, with aspects that can be adequately judged only by experts in that sub-field. That goes against the criterion of transparency called for by Guido. In such situations, sometimes the best way of judging a result is to attempt to reproduce it, ideally making complementary choices where relevant. A second, independent, up-to-date lattice step-scaling determination would thus be important to help establish this approach as the reference method for α_s determinations.

In the meantime, what global average should one use? For the 2017 PDG update we decided to maintain the 2016 average, $\alpha_s(m_Z) = 0.1181 \pm 0.0011$ in order to allow more time to collect community input on the latest α_s determinations. This world average remains consistent with, if somewhat more conservative than, a simple weighted average of the two determinations that appear today to be the cleanest, namely the ALPHA and EW determinations, which gives $\alpha_s(m_Z) = 0.1186 \pm 0.0008$.

Acknowledgements

Aside from those with Guido, I have benefited from discussions on α_s with many colleagues and friends over the years. In particular I would like to thank Siggi Bethke and Günther Dissertori for the past decade of collaboration on the PDG QCD review chapter, including discussions about all aspects of α_s determinations (and together with Thomas Klijnsma also our work on α_s from top-quark cross sections), as well as Jens Erler for regular discussions about aspects that overlap with the Electroweak review; Martin Lüscher, Luigi del Debbio and Agostino Patella, for numerous discussions about lattice approaches and Alberto Ramos for extensive exchanges on the ALPHA determination and for supplying Fig. 7.2; Martin Beneke and Toni Pich for their insights on τ determinations (if my conclusions differ from theirs it may simply be because I still have more to learn), as well as Andreas Höcker and Zhiqing Zhang for clarifications; Johannes Blümlein, Stefano Forte, Sven Moch, Juan Rojo and Robert Thorne for discussions about PDF fits; Andrea Banfi, Stefan Kluth, Gionata Luisoni, Pier Monni and Giulia Zanderighi for their insights about

e^+e^- jet rates and event shapes and their input for Fig. 7.1 as well as Günther Dissertori for permission to reproduce the top-right panel of Fig. 7.1; and, finally, Zoltán Trócsányi for discussions concerning the interplay of the strong coupling and vacuum stability. I would also like to thank Siggi Bethke, Michelangelo Mangano, Pier Monni, Alberto Ramos and Giulia Zanderighi for helpful comments on the manuscript.

Chapter 8

The top mass in hadronic collisions

Paolo Nason
INFN, Sezione di Milano Bicocca
Dipartimento di Fisica "G. Occhialini"
Piazza della Scienza, 3
20126 Milano, Italy

8.1 Introduction

There is something strange about the current status of top mass measurements at hadron colliders. Regarding the most precise measurements, performed by kinematic reconstruction of the full $t\bar{t}$ event, and commonly referred to as "Standard Measurements", often it is not stated clearly what is measured, in contrast with the very small error that is proudly quoted by the experimental collaborations. In some circumstances, it is claimed that what is measured is a "Monte Carlo" mass, i.e. just a parameter in the generator used to perform the analysis. But yet extensive variations of Monte Carlo parameters are performed in order to estimate errors, in a clear effort to reach something more fundamental. Other measurements, for example from the total cross section, or from some kinematic distributions, are presented collectively as "Pole Mass Measurement", that is (in the context of perturbation theory) a theoretically well defined parameter. No attempt is made to combine these pole mass measurements with the direct measurements.[1]

I remember having a discussion on this issue with Guido Altarelli, many years ago, at an Italian conference on LHC physics. Much time has gone by, and I do not remember the exact year and conference. A recurring argument was circulating among theoretical physicists, stating that top mass measurements at hadron colliders were just extracting the Monte Carlo mass, a parameter that did not have a well defined field theoretical definition. The argument went as follows: the top mass is extracted by fitting Monte Carlo templates to measured distributions; Monte Carlo generators have just leading order accuracy; different mass renormalisation schemes cannot be distinguished at leading order, since renormalisation enters at least at one loop; hence the Monte Carlo mass is not in a definite mass scheme. One-line arguments like this are quite catchy, and tend to spread very quickly, since they do not require any specialised knowledge to be understood. I strongly disagreed with it. The measurements were clearly aiming at observables that were strongly correlated to the mass of the system of decay products, i.e. were aiming at the pole mass. And the error associated with the LO accuracy of the Monte Carlo was estimated by several methods, typically by varying parameters in the Monte Carlo, and by comparing its output to more precise next-to-leading order calculations. I expressed my disappointment about the diffusion of this "Monte Carlo mass" concept to Guido Altarelli. I told him that the argument was just used to scare the experimentalists with theoretical concepts that they were not very familiar with, which was in fact an aspect of the issue that I was finding most irritating. Guido's answer was that, rather than scaring them, the argument should have made them think. I could not reply to that, since for an instant I thought that perhaps he was subtly implying that I myself should have done more thinking. This was quite typical of my exchanges with Guido. They were never free of some sort of friction. And that was certainly not the only occasion when I departed from a discussion with him with a slightly sore feeling. I must admit that I was never great at verbally arguing about physics issues, while Guido was clearly excellent at it. Yet, I was always eager to discuss with him. I had the perception (that I believe was shared by many others) that in order to settle an issue you had to convince him, since he was *the* uncontested authority in our field, and with the good reasons that we all know. And discussing with him always helped me to acquire a more detached and focused view of the problems. It is unfortunate that this discussion stopped at that point, since I didn't worry about this issue any more until recent times.

In this work I express my personal views on this subject, that at the moment is undeniably very controversial.

---

[1] I have heard a presentation where the speaker has jokingly introduced the subjects saying that the top is the only Standard Model particle with more than one mass.

In the first section I will illustrate how and why the current status of the theoretical interpretation of top mass measurements has been reached. I will not try to fully reconstruct the historical development of all related arguments, and thus I will certainly omit quoting more than one important paper on this topic. I will rather focus on the arguments and papers that have had more influence in determining the current status. The message that I would like to convey is that, rather than finding or denying some relation of the Monte Carlo mass parameter to some well defined field theoretical parameters, we should find ways to estimate the error on the extraction of a theoretically well defined mass parameter when using a Monte Carlo generator.

In the second part of this work I will illustrate what I believe has been a small but useful progress, i.e. an improved understanding of the relation of the pole mass to the $\overline{\text{MS}}$ mass. It is a self-contained subject where analytic calculations, not requiring modelling of non-perturbative and hadronisation effects, can lead to partial progress in our understanding of the top mass issues. I will expose this topic, avoiding technicality as much as possible, yet discussing what I believe are its most relevant aspects.

Needless to say, I am trying to convince Guido.

8.2 The top mass problem

The top quark mass is a key parameter of the Standard Model. Its large size, of the order of the electroweak scale, is associated with a Yukawa coupling of order 1, that gives important contributions, via radiative corrections, to Standard Model observables. After the Higgs boson discovery and the accurate measurement of its mass, the allowed values of the W-boson and top-quark masses have become strongly correlated, so that an accurate determination of both would lead to a SM test of considerable precision [Patrignani *et al.* (2016); Baak *et al.* (2014b)]. At present there is some tension, at the $1.6\,\sigma$ level, between the indirect top-mass determination from electroweak precision data (177 ± 2.1 GeV) and the combination of direct measurements at the Tevatron and at the LHC [ATLAS Collaboration (2014a)], that yields 173.34 ± 0.76 GeV. More recent determinations [Aaboud *et al.* (2016); Khachatryan *et al.* (2016e); CMS Collaboration (2017c); ATLAS Collaboration (2017c)] favour an even lower value, close to 172.5 GeV.

The top mass value is also critical in the issue of vacuum stability in the Standard Model [Degrassi *et al.* (2012); Buttazzo *et al.* (2013); Andreassen *et al.* (2017)]. At high scales, the Higgs quartic coupling λ evolves to increasingly small values as m_t grows, and above $m_t = 171$ GeV, i.e. very close to the present world average, λ becomes negative at the Planck scale, rendering the electroweak vacuum metastable, while for $m_t > 176$ GeV the electroweak vacuum becomes unstable. The only conclusion that can be drawn from this result is that the current value of the top and Higgs masses are such that no indication of new physics at any scale can be inferred by imposing the stability or metastability of the vacuum. On the other hand, the fact that the Higgs quartic coupling nearly vanishes at the Planck scale may have some deep meaning that we are now unable to unveil.

The abundant production of top pairs at the Large Hadron Collider (LHC) provides an opportunity for accurate top mass measurements, that are generically performed by fitting m_t-dependent kinematic distributions to Monte Carlo predictions. The most precise ones rely upon the full or partial reconstruction of the system of the t and $\bar{t}$ decay products. The CMS measurement of ref. [Khachatryan *et al.* (2016e)], yielding the value $m_t = 172.44 \pm 0.13$ (stat) ± 0.47 (syst) GeV, falls into this broad category.[2]

In contrast with the increasing experimental precision with which the top mass is measured at the LHC, the theoretical interpretation of the measurements seems to be in a questionable state. The so called "direct measurements", i.e. those that rely upon the reconstruction of the kinematics of the $t\bar{t}$ system, have been heavily criticised by some theorists as not possessing a clear relation between the extracted mass and a well defined quantum field theory parameter of the underlying theory.

As a consequence of that, in many experimental papers and talks, it is preferred not to qualify the measured mass parameter with a precise field-theoretical attribute, such as the "pole mass" or the "$\overline{\text{MS}}$ mass", and, at times, even to qualify it as a "Monte Carlo" mass [Patrignani *et al.* (2016); ATLAS Collaboration (2014a); CMS Collaboration (2015)]. This has led to the paradoxical situation that the most precise measurement available is not receiving the attention that it deserves from the theoretical community, and the interpretation of the measurement is left in a "limbo", with the hope that some theoretical work may clarify it in the future. At the same time, several theorists have suggested observables that allegedly overcome the problems of the direct measurements. Although all these suggestions are valuable, since they can provide alternative determinations and consistency checks, none of them seem to possibly lead to an accuracy comparable with the direct determination. Furthermore, these alternative measurements are grouped in a different category by the experiments, that do not attempt to combine them with the standard measurements.

The aspect of the top mass measurement that is mostly puzzling the theoretical community, is that the top mass cannot be defined in terms of the mass of the system of its decay products. Since top is a coloured object, no final-state hadronic system can be unambiguously associated with it. On the other hand, the top mass bears certainly some relation to the mass of the system of objects arising from its decay (i.e. leptons, neutrinos and hadronic jets). The mass distribution of this system can be computed, and the top mass enters as a parameter in this computation. This is in fact the case for many parameters that are measured in high energy physics

---

[2] A similar measurement performed at 13 TeV [CMS Collaboration (2017c)] yields a value of $172.25 \pm 0.08(stat + JSF) \pm 0.62(syst)$ GeV.

experiments. In the case of top, however, the computation in question is performed by a parton shower generator, thus the idea that what is measured is a "Monte Carlo" mass. This idea has been expressed by several authors, but not always in the same sense: one line of thinking has to do with the perturbative accuracy of Monte Carlo generators, while others worry about the non-perturbative effects that they model. As far as the "perturbative accuracy" is concerned, the argument is essentially as follows: since shower Monte Carlo are only accurate at leading order, they cannot possibly distinguish mass parameter definitions that differ only at next-to-leading order, like the pole mass and the $\overline{\text{MS}}$ mass, whose difference amounts to several GeV's.

In the following I will examine more carefully the arguments and the studies that have led to the current situation. I will show that the most disturbing aspect of the "Monte Carlo mass" concept is its ambiguous role. In other words, it has acquired a different significance according to different authors, up to the point where it is used to support conflicting points of view.

8.2.1 *The "Perturbative" argument*

Fixed order theoretical calculations of (infra-red safe) final state distributions can be catalogued according to their perturbative accuracy as Leading Order (LO), next-to-leading order (NLO) and so on. The top mass parameter in a theoretical calculation must be defined within a given renormalisation scheme, since (divergent) perturbative corrections arise order by order in perturbation theory. For the top mass parameter, one such scheme is the pole mass scheme, that prescribes to subtract the divergent mass corrections in such a way that the pole in the quark propagator remains fixed order by order in perturbation theory. Alternatively, the $\overline{\text{MS}}$ scheme prescribes to subtract the pure $1/\epsilon$ pole in the divergent mass correction. Thus, in the $\overline{\text{MS}}$ scheme the position of the pole in the top propagator receives corrections at all orders in perturbation theory. The two definitions lead to perturbatively equivalent theories, in the sense that there is a perturbative expression of the pole mass in terms of the $\overline{\text{MS}}$ mass that allows one to translate a physical result in the pole mass scheme into the corresponding physical result in the $\overline{\text{MS}}$ scheme. We notice that the $\overline{\text{MS}}$ mass begins to differ from the pole mass at first order in perturbation theory, that contributes about 7.5 GeV to the total difference.

We report below the relation between the top pole mass m_p and $\overline{\text{MS}}$ mass m up to the four loops level [Marquard *et al.* (2015)],

$$m_p = m + \underbrace{7.557}_{\text{NLO}} + \underbrace{1.617}_{\text{NNLO}} + \underbrace{0.501}_{\text{N}^3\text{LO}} + \underbrace{0.195 \pm 0.005}_{\text{N}^4\text{LO}}\,\text{GeV}, \tag{8.1}$$

for $m = 163.643\,\text{GeV}$ and $\alpha_s^{(6)}(m) = 0.1088$, in order to show the typical numbers that come into play. Thus, the "perturbative argument", if pushed to its extreme, would lead us to conclude that the relation of the extracted mass parameter to some physically well defined one has an ambiguity on the order of 10 GeV.

The earliest written record of this argument that I found is in ref. [Alioli *et al.* (2013)], where it is stated quite clearly that "the top-quark mass derived from the kinematical reconstruction does not correspond to a well defined renormalisation scheme leading to a theoretical uncertainty in its interpretation. Nevertheless it is usually interpreted as the "top-quark pole mass". In the same work it was proposed to extract the top mass by comparing kinematic observables in $t\bar{t}$ events involving one additional jet. Such a process can be computed at NLO level, and so it must use a well defined mass renormalisation scheme.[3]

This argument has also been used to support mass measurement based upon the $t\bar{t}$ cross section, that is now computed at NNLO order [Czakon *et al.* (2013)].

Sometimes, all the mass measurement techniques that refer to a fixed order calculation beyond the Leading Order are collected together and presented as "measurements of the top pole mass", in contrast with direct measurement where the mass scheme is never specified (see `https://twiki.cern.ch/twiki/pub/CMSPublic/PhysicsResultsTOP/pole_mtop.pdf`).

8.2.2 *The "Non-Perturbative" argument*

There are claims in the literature that the difference between the top pole mass and the mass extracted in direct methods arises due to non-perturbative effects, that are only modelled in a shower Monte Carlo. This claim is certainly true, although the difficult bit in it is to quantify this difference. Again, this difference is used to motivate a "Monte Carlo mass" concept, that cannot be related to a well defined theoretical parameter because of these non-perturbative effects. It does not have the same meaning as the "Perturbative" difference discussed previously. But also if we restrict ourselves to this "Non-Perturbative" meaning, different authors interpret it differently.

In ref. [Kawabata *et al.* (2015)], the direct measurements are criticised on the basis that they reconstruct the top mass making use of jets, and jet reconstruction is affected by hadronisation effects that are modelled by the shower Monte Carlo. In contrast, they propose using observables that only depend upon the lepton kinematics and that are insensitive to production dynamics, and thus should not (in their opinion) be affected by hadronisation effects.

Other kinds of lepton observables have also been advocated in ref. [Frixione and Mitov (2014)], There, however, an attempt is made to single out those observables that seem less affected by shower and hadronisation effects, on the basis of Monte Carlo simulation study. In other words, the authors rightfully recognise that also

---

[3] It should be recalled, however, that the difference between the pole and the $\overline{\text{MS}}$ mass receives contributions beyond the NLO order that amount to more than 2 GeV. Thus, one may also argue that in an NLO calculation the pole mass parameter used there is related to a fundamental parameter of the theory with an uncertainty of at least 2 GeV.

leptonic observables may be subject to hadronisation corrections, and try to quantify these effects by Monte Carlo studies.

A considerably more elaborated argument was first put forward in ref. [Hoang and Stewart (2008)], where it is stated quite explicitly that "it is not $m_t^{\rm pole}$ that is being measured by the Tevatron analyses". The argument goes through several steps, that I can only summarise here. First of all, it is argued that the pole mass scheme is a poor choice, because of the presence of an ambiguity of order $\Lambda_{\rm QCD}$ associated with the mass infra-red renormalon (that I will discuss later in this work), and that the so called MSr scheme avoids this problem. The MSr mass is a mass parameter that depends upon a scale R. It is defined using the same self-energy diagrams that contribute to the pole mass, but their contribution is limited to loop momenta of order larger than R. Thus, as R becomes small, the MSr mass approaches the pole mass. It is argued in [Hoang and Stewart (2008)] that an MSr mass, with R taken of the order of 1 GeV, should be used as mass parameter.

The argument is further developed by considering top production at high momentum in e^+e^- collisions at energies much larger than the top mass. It is argued that the hemisphere masses (defined by dividing the event with a plane orthogonal to the trust axis) are calculable using SCET techniques, up to the inclusion of power suppressed effects of order Λ, provided the mass parameter is the MSr mass. From this framework, implications for the standard measurements at the Tevatron are drawn. However, they suffer for the fact that in standard measurement the top is not ultrarelativistic, and the SCET factorisation cannot be applied.

It must be stressed that in ref. [Hoang and Stewart (2008)] it is recognised that the Monte Carlo mass parameter should be quite close to the pole mass. More specifically, it is claimed that it should be close to the MSr mass evaluated at a small scale, that in turn is close to the pole mass. This point of view is very different from the "perturbative" one, that leads to differences of order $\alpha_s m_t$ between the pole and the Monte Carlo mass.

There are several followups of the work in ref. [Hoang and Stewart (2008)]. In ref. [Hoang (2014)], it was argued that, in general, an additional uncertainty of 1 GeV should be accounted for in top mass measurements at hadron colliders. It is also argued that the top pole mass in general cannot be determined with a precision better than 1 GeV because of the mass renormalon problem. Although it is stated somewhere in this paper that 1 GeV "is the energy value I use in this talk for what theorists call hadronisation scale", this value, quoted in the abstract and in many places in the paper, has been echoed as is in several other publications and talks, to be taken as a serious hard limit on the precision that can be achieved in the measurement of the top mass by direct methods. In subsequent works [Butenschoen *et al.* (2016)], Hoang and collaborators attempt to quantify a relation between the so called Monte Carlo mass and an MSr mass evaluated at 1 GeV. Essentially, always in the framework of highly boosted tops, they compute the jet mass using both an NNLL SCET calculation, and the Pythia8 [Sjöstrand *et al.* (2015)] shower Monte

Carlo. They absorb the difference in the two results by a shift in the Pythia8 mass parameter with respect to the MSr mass used in SCET, and argue that this shift is the difference between the Monte Carlo and the MSr mass. Indeed, they find that the Monte Carlo mass exceeds the MSr mass by about 200 MeV with an uncertainty of about the same size. On the other hand, also the pole mass exceeds the MSr mass by a similar amount. Whatever method one adopts to resum the divergent series that gives the pole mass in terms of the MSr mass, it seems that the difference between the pole mass and the Monte Carlo mass should be significantly less than 1 GeV.

We also observe that in [Butenschoen *et al.* (2016)], effects that are accounted for in the SCET results (that is supposed to be NNLL accurate) and are not present in Pythia8 (that certainly cannot claim this accuracy) are absorbed into a shift in the mass. It thus seems that the mass shift they find, rather than having a universal character, should be dependent upon the chosen observable.

8.2.3 *Summary of the issue*

We have seen at least three arguments that are used to criticise the standard top mass measurements:

(1) The perturbative argument (i.e., MC are only LO accurate)
(2) A non-perturbative argument: jets have non-perturbative corrections, thus we should avoid top mass observables that involve jets, (i.e., use leptonic observables).
(3) A deeper non-perturbative argument: since MC's implement ad hoc models of non-perturbative effects, we should not rely upon them, since they imply an irreducible non-perturbative error. Rather, we should resort to methods that allow to treat non-perturbative corrections from first principles.

These arguments have different meanings, and in particular the first one is independent of the remaining two, since it deals with perturbative effects, while the other two address non-perturbative issues. Yet, for example, in ref. [Kieseler *et al.* (2016)] it is stated that the translation from the Monte Carlo mass to a short distance mass can only be estimated within a 1 GeV accuracy, quoting in particular [Hoang and Stewart (2008)],[4] and this statement is used to advocate that it is preferable to extract the top mass by considering observables that can be calculated in QCD at least at NLO. It is unclear why NLO precision should get rid of non-perturbative effects.

Regarding the first of the three arguments listed above, it should be reminded that one cannot simply say that a Monte Carlo has only LO accuracy. The accuracy

[4]A few months later, this 1 GeV accuracy would have gone down to 200 MeV in ref. [Butenschoen *et al.* (2016)].

of a Monte Carlo depends upon the observable, and there is no simple statement that can qualify it for all observables.

The production and decay dynamics of a coloured resonance in the narrow-width limit factorise in perturbation theory, and become independent, as a consequence of the fact that the resonance can propagate a long time before decaying. Thus, in the narrow width limit and in perturbation theory, there is an unambiguous definition of the system of the resonance decay products, whose mass coincides with the resonance pole mass to all orders in perturbation theory if the pole mass scheme is used. The factorisation feature of perturbation theory is also implemented in shower Monte Carlo generators, where radiation in production and decay are developed independently, and preserve the mass of the decaying resonance. Thus, up to non-perturbative effects and soft radiation of energy comparable to the resonance width (that can violate factorisation because of interference between production and decay) the resonance mass appearing in the Monte Carlo can be identified with the pole mass.

One could still argue that the accuracy of the shower Monte Carlo generator does play a role in the mass determination. Typically, in top production, the radiation generated at the production stage can enter the b-jet cone, or radiation from the b-quark can escape the jet cone, thus altering the mass of the reconstructed decay system. These effects can lead to an error on the extracted mass. This error, however, can be reduced by improving the perturbative accuracy of the Monte Carlo, or by tuning the Monte Carlo in such a way that it better describes radiation in production and the structure of the b-jet. In other words, these errors are modelling errors, and their impact can be estimated (and in fact is estimated by the experiments) by varying suitable parameters in the shower Monte Carlo. But none of these errors is associated to problems in the perturbative definition of the top mass.

Fixed-order calculations of kinematical distributions involving top quarks are also subject to soft radiation and non-perturbative effects, and thus cannot be considered as privileged observables for pole mass determinations, to be presented in isolation from those obtained with direct measurements. If the former can be considered pole mass determinations, the same holds for the latter. Notice that this also holds for total cross section. Even if one is willing to believe that the total cross section has no linear power corrections (i.e. a correction suppressed by a single power of a hadronic scale divided by the top mass), one should remember that total cross-section measurements requires extrapolations outside of a fiducial region, and fiducial regions have cuts that can introduce linear power corrections. Associated non-perturbative effects must thus be estimated also in this case.

We can also add that the accuracy of the Monte Carlo generators currently used in standard measurements make use of `POWHEG` or `MCNLO` generators, including NLO corrections in production. Most modern Monte Carlos, like `Pythia8` [Sjöstrand *et al.* (2015)] and `Herwig7` [Bellm *et al.* (2016)], include matrix element corrections to top decay, that makes them essentially (up to an irrelevant normalisation factor) NLO accurate in decay. Herwig7 also implements its own POWHEG implementation of NLO corrections to top decay. Furthermore, there are recent NLO+PS generators that implement top production and decay including finite width, non-resonant effects and interference of radiation in production and decay [Ježo *et al.* (2016)].

The second argument is based upon the general feeling that leptonic observables should not be affected by hadronisation effects. It is easy, with a simple example, to convince ourselves that this may not be the case. Consider the decay of the top into a W and a b-jet. The emission of an extra soft pion at large angle with respect to the b-jet can represent a typical non-perturbative effect. Such emission would always subtract an energy of the order of few hundred MeV's from the W, thus reducing also the lepton energy by few hundred MeV's. Thus, it is likely that linear power corrections are also present in leptonic observables, and a corresponding uncertainty should be estimated and associated with them.

The third item in the list should be regarded more carefully. It is certainly undeniable that shower Monte Carlos do not implement a solid theory of the leading power suppressed effects. It would certainly be desirable to have such a theory, or to perform top-mass measurement in frameworks where such a theory is available. This seems to be the case for top-mass measurements performed by a threshold scan of the e^+e^- cross section (see [Beneke *et al.* (2017a)] and references therein), or for the recently proposed mass measurement based upon the shape of the $\gamma\gamma$ mass spectrum at the LHC [Kawabata and Yokoya (2017)]. While the first possibility is conditioned by the future developments of high energy physics experiments, the second one seems to be limited by statistics even at the high luminosity LHC phase.

We have seen that there is a sequel of papers that advocate the use of highly boosted top-quark jets, claiming that they can be computed up to the leading power suppressed correction. These approaches require validation against data,[5] full scrutiny and criticism by the theoretical community, and, most important, should be demonstrated to be practically useful, i.e. they should lead to top-mass measurements with an an error smaller than the leading power suppressed effects that they claim to model correctly. At the moment, these conditions do not seem to be fulfilled.

So, the question remains about what to do with top-mass measurements by direct methods at hadron colliders. It is obvious that we have an error due to non-perturbative effects that should be estimated, but it is not enough to state that it is an error of the order of 1 GeV. We should understand whether it is 1 or more, 0.5, 0.2 or less GeV's. Some authors, typically those advocating the use of boosted tops, have argued in the past for uncertainties of non-perturbative origin, of magnitude near 1 GeV, affecting the relation of the Monte Carlo mass, and also of the pole mass, to some short distance mass. As we have seen, however, the 1 GeV figure is

[5]We should remember that this is always the case in perturbative QCD, that is a consistent framework, but is based upon some unproven assumptions.

scarcely motivated as a hard limit. One also gets the impression that these "near a GeV" uncertainties keep decreasing with time, as more thorough analyses are performed, leaving us wondering why such a large value was ever quoted in the first place.

8.2.4 *What to do*

We have seen that the theoretical problems related to the determination of the top mass at hadron colliders boil down to the problem of quantifying how power suppressed corrections affect the measurement. More specifically, it has to do with the uncertainties associated with how Monte Carlos implement power suppressed effects, and how they match them to perturbative effects. These Monte Carlo uncertainties have been translated by some authors into the concept of a Monte Carlo mass.

Translating a Monte Carlo inaccuracy into a Monte Carlo mass concept has also led to the equation: the Monte Carlo implements perturbation theory at leading order, thus its mass parameter cannot be associated with a well defined mass scheme. In this second case, it is quite clear that the right question to ask should have been: what is the error on the measurement due to the fact that the Monte Carlo is only accurate at leading order. This question has in fact been asked by the experimentalists, that have studied the errors associated with the MC perturbative inaccuracies with several techniques, and by theorists, that have developed improvements in the generators to promote the accuracy of the Monte Carlo to the NLO level.

The question now is whether uncertainties due to power suppressed effects can be reliably estimated in similar ways, i.e. by varying Monte Carlo parameters, or by including plausible alternatives in the model of hadron formation that they implement. This is currently done by the experimental collaborations and by Monte Carlo developers.[6] Yet, with these methods the doubt remains that we may be missing something, and investigations aiming at a better understanding of power suppressed effects, and their eventual interplay with renormalons, would be welcome. At the moment, however, there is no sound argument that suggests that the Monte Carlo mass is related to some well defined mass parameter via a systematic shift. It has been argued by some authors that, since shower Monte Carlo have a 1 GeV cutoff on soft radiation,[7] their mass parameter should be identified with an MSr mass at a scale $R = 1$ GeV, rather than a pole mass, which yields a difference of a few hundred MeV's. It should be recalled, however, that Monte Carlos implement soft

[6] As an example, the increase in the error in the 13 TeV top mass measurement of CMS reported in [CMS Collaboration (2017c)] relative to the previous 8 TeV measurement is due to the introduction of alternative models of colour reconnection in `Pythia8` (see [Argyropoulos and Sjöstrand (2014); Christiansen and Skands (2015)]).

[7] We note again the loose use of the 1 GeV figure. In Pythia8, such cutoff is by default 500 MeV.

radiation by ensuring that virtual effects cancel completely the real emission corrections in inclusive quantities, via a mechanism known as "shower unitarity". This mechanism implements in practice the cancellation of soft singularities. However, also finite, non-singular soft effects are cancelled out in this way, and we should remember that self-energy mass corrections are non-singular. Thus, the "shower cut off" argument does not track down Monte Carlo effects that can convert a pole mass to an MSr one.

Notice that it is also easy to set up simple and catchy arguments to prove that the Monte Carlo mass parameter *is* the top pole mass. One could argue, for instance, that, since Monte Carlos implement a top propagator with a complex pole at a fixed position, that pole must be at the (complex) top pole mass, since the use of a short distance mass would yield a pole position that is blurred by the mass renormalon effects. Needless to say, also this argument is not very convincing, and it is clear that these issues require more serious thinking. However, it doesn't seem appropriate to tell the experimental community that until such thinking is done they are not allowed to tell what they are actually measuring. Rather than telling the experimentalists to wait for new theoretical developments to interpret their results, it would be far more constructive to point out to them methods to estimate the uncertainties associated with the limitations of the generator they use. For example, if the lack of radiation below the Monte Carlo cutoff scale is a concern, one should recall that this cutoff scale can be changed. One could set up a different tune of the Monte Carlo, using a different cutoff scale, and examine whether it leads to relevant differences in the extracted top mass. I would also like to stress that this work should not be necessarily done by the experimentalists. It can be done by theorists as well, using simplified versions of top mass observables and of detector effects.

8.3 The renormalon problem

If the renormalon ambiguity on the top pole mass was as large as 1 GeV, it would make no sense to measure the top pole mass with the experimental precision that is quoted today. Fortunately, it turns out that this ambiguity is in fact much smaller. In refs. [Beneke *et al.* (2017b)] and [Hoang *et al.* (2017b)] the relationship of the pole mass to the $\overline{\text{MS}}$ mass for the top quark is studied. They find compatible results for the central values, while for the renormalon ambiguity they quote values of 110 MeV and 250 MeV respectively, both safely below currently quoted systematic errors.[8]

[8] In fact, values in this range were obtained much earlier in refs. [Pineda (2001); Bali *et al.* (2013)], mostly in a bottom physics context, but since the renormalon ambiguity does not depend upon the heavy quark mass, they also apply to top.

8.3.1 *A simplified description of the renormalon problem*

The quark mass parameter m that we introduce in our Lagrangian does not necessarily coincide with the quark pole mass, which is the position of the pole in the quark propagator. More precisely it coincides with it only at zeroth order in perturbation theory. At higher orders the mass parameter requires renormalisation, but this can be carried out without ever referring to the pole position, like for example when using the $\overline{\text{MS}}$ prescription.

In the following I will denote with m the mass parameter renormalised in the $\overline{\text{MS}}$ scheme.[9] The $\mathcal{O}(\alpha_s)$ correction to the position of the pole in the heavy quark propagator turns out to have a linear infra-red sensitivity to the scale of the momentum flowing in the loop, i.e. to yield a contribution of the form

$$\int_0^m \mathrm{d}l\, \alpha_s, \tag{8.2}$$

where l is the scale of the loop momentum. The upper cutoff arises since we have assumed that ultraviolet divergences related to the large l region have been subtracted by renormalisation. Here we assume that the renormalisation scale is taken equal to m. Thus, the pole position is given by an expression of the form

$$m_{\mathrm{P}} = m + N \int_0^m \mathrm{d}l\, \alpha_s + \mathcal{O}(\alpha_s^2), \tag{8.3}$$

where N is a suitable normalisation factor.

We now assume heuristically that we can obtain the higher order terms by simply writing

$$m_{\mathrm{P}} = m + N \int_0^m \mathrm{d}l\, \alpha_s(l), \tag{8.4}$$

with

$$\alpha_s(l) = \frac{1}{b_0 \ln\, l^2/\Lambda^2} = \frac{\alpha_s(m)}{1 - \alpha_s(m)\, b_0 \ln\, m^2/l^2} = \sum_{n=0}^{\infty} \alpha_s^{n+1}(m)\, b_0^n \ln^n \frac{m^2}{l^2}\,. \tag{8.5}$$

Inserting (8.5) in (8.4), and recalling that

$$\int_0^m \mathrm{d}l\, \ln^n \frac{m^2}{l^2} = m\, 2^n \int_0^1 \mathrm{d}x\, \ln^n \frac{1}{x} = m\, 2^n n! \tag{8.6}$$

we get

$$m_{\mathrm{P}} = m + N \sum_{n=0}^{\infty} m\, (2b_0)^n\, \alpha_s^{n+1}(m)\, n!\,. \tag{8.7}$$

[9]It is common practice to use the term "$\overline{\text{MS}}$ mass" (that we call m in this paper) to denote the running $\overline{\text{MS}}$ mass, $m(\mu)$, that depends upon a scale μ, evaluated self-consistently at the scale $\mu = m(\mu)$.

The factorial growth of the coefficients of the perturbative expansion is what is called "renormalon", where the name is suggested by the fact that it arises due to the renormalisation group evolution of the coupling constant. It leads to a power expansion that has zero radius of convergence. This is related to the fact that the integral in Eq. (8.4) runs over the Landau Pole, i.e. the divergence of $\alpha_s(l)$ when $l = \Lambda$. The terms in the sum (8.7) initially decrease for $\alpha_s(m)$ small enough, but at some value of n they start growing again. The value at which the minimum is attained is easily obtained by use of the Stirling approximation $n! \approx \sqrt{2\pi n}\exp(n\ln n - n)$. The term of order $n+1$ in Eq. (8.7) can be written as

$$\delta m^{(n+1)} = Nm\alpha_s(m)\sqrt{2\pi}\exp\left[\left(n+\frac{1}{2}\right)\ln n - n + n\ln(2b_0\alpha_s(m))\right], \tag{8.8}$$

that has a minimum when the derivative of the exponent with respect to n vanishes:

$$\ln n + \frac{1}{2n} + \ln(2b_0\alpha_s(m)) = 0\,. \tag{8.9}$$

Under this condition the minimal term can be written as

$$\delta m_{\min} = Nm\alpha_s(m)\sqrt{2\pi}\exp\left[\frac{1}{2}\ln n_{\min} - n_{\min} - \frac{1}{2}\right]. \tag{8.10}$$

Solving Eq. (8.9), we get the value of n at the minimum

$$n_{\min} = \frac{1}{2b_0\alpha_s(m)} - \frac{1}{2} + \mathcal{O}(\alpha_s(m)) \tag{8.11}$$

so that

$$\delta m_{\min} \approx Nm\sqrt{\frac{\pi\alpha_s(m)}{b_0}}\exp\frac{1}{2b_0\alpha_s} = N\sqrt{\frac{\pi\alpha_s(m)}{b_0}}\Lambda. \tag{8.12}$$

A rough procedure to sum divergent series is to sum up the terms as long as they decrease, stopping at the minimum, that therefore gives a first indication of the uncertainty in the result.

An alternative way to deal with factorially divergent expansion is by use of the Borel transform. Given the power series

$$f(\alpha_s) = \sum_{n=0}^{\infty} c_n\alpha_s^{n+1}, \tag{8.13}$$

the corresponding Borel transform is defined as

$$B[f](t) = \sum_{n=0}^{\infty} c_n\frac{t^n}{n!} \tag{8.14}$$

and we have formally

$$f(\alpha_s) = \int_0^{\infty} \mathrm{d}t\, e^{-t/\alpha_s}B[f](t). \tag{8.15}$$

In the case of Eq. (8.7), where $c_n = N(2b_0)^n n!$, we get

$$\begin{aligned} m_{\rm P} - m &= Nm \int_0^\infty {\rm d}t\, e^{-\frac{t}{\alpha_s(m)}} \sum_{n=0}^\infty (2b_0)^n t^n \\ &= Nm \int_0^\infty {\rm d}t\, e^{-\frac{t}{\alpha_s(m)}} \frac{1}{1-2b_0 t}. \end{aligned} \tag{8.16}$$

The presence of the pole along the real axis is again a manifestation of the renormalon problem.

The Borel procedure gives an alternative, more educated method for the summation of a factorially divergent power expansion: one takes as result the principal value of the integral, and as uncertainty something proportional to the (absolute value) of the imaginary part that arises when the contour of integration is distorted above or below the singularity (at $t = 1/(2b_0)$ in our case) in the complex plane. In ref. [Beneke (1999)] it is proposed to take the absolute value of the imaginary part divided by π as the one-sided ambiguity, that in our case leads to

$$Nm\frac{1}{2b_0} e^{\overline{\frac{1}{2b_0\alpha_s(m)}}} = \frac{N}{2b_0}\Lambda. \tag{8.17}$$

In the following we will call this method for estimating the renormalon ambiguity the "Im/Pi method". In ref. [Beneke *et al.* (2017b)] the Im/Pi method is adopted, with the motivation that it works well in contexts where the renormalon effect can be related to some physical observable [Beneke (1999)], and one can check with data that it gives reliable results.

Notice that the form of the minimal term in Eq. (8.12) differs parametrically by the one in Eq. (8.17), due to the presence of the extra factor $\sqrt{\alpha_s(m)}$. In fact, in order to use the minimal term to estimate the ambiguity of the result we should also account for the fact that several terms of similar size may lie near the minimum. Starting again from Eq. (8.7), we can expand the exponent around the minimum to get

$$\delta m^{n+1} \sim \delta m_{\rm min} \exp\left[\frac{(n-n_{\rm min})^2}{2n_{\rm min}}\right] \sim \delta m_{\rm min}\left[1+\frac{(n-n_{\rm min})^2}{2n_{\rm min}}\right]. \tag{8.18}$$

It is clear now that the size of the region where the terms of the series have similar size is proportional to $\sqrt{n_{\rm min}}$. For definiteness, let us define this region by requiring

$$\left[1+\frac{(n-n_{\rm min})^2}{2n_{\rm min}}\right] < f, \quad \text{with } f > 1. \tag{8.19}$$

This leads to

$$|n - n_{\rm min}| < \sqrt{2(f-1)n_{\rm min}} \tag{8.20}$$

and by multiplying the size of this region by the minimal term we get

$$\sqrt{2(f-1)n_{\rm min}}N\sqrt{\frac{\pi\alpha_s(m)}{b_0}}\Lambda = N\frac{\sqrt{(f-1)\pi}}{b_0}\Lambda. \tag{8.21}$$

This has the same parametric form of the ambiguity determined with the Im/Pi method, and is in fact identical to it if we choose $f = 1 + 1/(4\pi)$.

In the context of this oversimplified illustration of the mass renormalon problem, we also introduce the concept of the so called MSr mass [Hoang *et al.* (2017a)]. We define it as

$$m_{\rm MSr}(R) = m + N\int_R^m {\rm d}l\ \alpha_s(l) \tag{8.22}$$

or, equivalently

$$m_{\rm P} = m_{\rm MSr}(R) + N\int_0^R {\rm d}l\ \alpha_s(l)\,. \tag{8.23}$$

Comparing Eq. (8.23) to Eqs. (8.4) and (8.7), we see that the difference between the pole mass and the MSr mass is given by the same power expansion as the difference between the pole mass and the $\overline{\rm MS}$ mass, with the only difference that α_s is evaluated at the scale R rather than the scale m.

The MSr mass looks like a formal interpolation between the $\overline{\rm MS}$ mass m (when $R = m$) and the pole mass (when $R = 0$). However, at the low end, when $\mu = l$ the integral does not exist. For $\mu > l$, on the other hand, the whole series is convergent, and does not manifest any factorial growth for large n. In fact, the coefficients grow factorially as long as $n < \log(m/\mu)$. For larger n, the factorial growth is damped out.

Since $m_{\rm P}$ is μ independent, we can also derive from (8.23) an evolution equation

$$\mu\frac{{\rm d}m_{\rm MSr}(\mu)}{{\rm d}\mu} = -N\mu\alpha_s(\mu), \tag{8.24}$$

that, unlike the typical renormalisation group evolution equations, has also a linear dependence upon the scale μ. This allows for the solution to have a well defined perturbative expansion in terms of the coupling evaluated at some fixed scale, relative to its value at $\mu = 0$. On the other hand, it is also associated with the factorial growth of its coefficients. Notice also that if we use the notation $m_{\rm P}(m)$ to denote the pole mass, and $m_{\rm MSr}(m, R)$ the MSr mass of a heavy quark with $\overline{\rm MS}$ mass m, we have

$$m_{\rm P}(m) - m_{\rm MSr}(m, R) = m_{\rm P}(R) - R\,. \tag{8.25}$$

In fact, the left-hand side is equal to

$$N\int_0^R {\rm d}l\ \alpha_s(l)\,, \tag{8.26}$$

according to Eq. (8.23), and the right-hand side is equal to the same quantity according to Eq. (8.4) with m replaced by R, and m_P taken as the pole mass corresponding to a $\overline{\mathrm{MS}}$ mass equal to R.

As a last point, we can ask what happens if there are other heavy flavours (typically bottom and charm) below the top mass. The answer is quite obvious: we just replace $\alpha_s(l)$ in Eq. (8.4) with a variable flavour $\alpha_s^{\mathrm{vf}}(l)$ that solves the evolution equation

$$\mu^2 \frac{\mathrm{d}}{\mathrm{d}\mu^2} \frac{1}{\alpha_s^{\mathrm{vf}}(\mu)} = -b_0^{\mathrm{vf}}(\mu), \quad b_0^{\mathrm{vf}}(\mu) = \frac{11C_a - 4T_f[3 + \theta(\mu - m_c) + \theta(\mu - m_b)]}{12\pi}. \tag{8.27}$$

We can immediately anticipate that the change in the mass relation will not be dramatic, since it is induced by a relatively small change in α_s in a relatively small region of the integration domain. However, the renormalon uncertainty will be determined by Eq. (8.17) with Λ corresponding to the 3-flavour Λ. This is easily understood, since the factorial growth at large order is controlled by increasingly small momenta, and thus cannot be sensitive to flavours with large mass.

8.3.2 *The full story*

The pole mass m_P is given in terms of the $\overline{\mathrm{MS}}$ mass m by an expansion of the form

$$m_P = m(\mu_m) \left\{ 1 + \sum_{n=0}^{\infty} c_n(\mu, \mu_m, m(\mu_m)) \alpha_s^{n+1}(\mu) \right\}, \tag{8.28}$$

where $m(\mu_m)$ is the $\overline{\mathrm{MS}}$ mass evaluated at the scale μ_m, and μ is the renormalisation scale. The coefficients have been evaluated up to the fourth order in α_s [Marquard *et al.* (2015, 2016)]. For the moment we assume that we have only one massive flavour, and all others are massless. Formula (8.28) was originally presented in the 6-flavour scheme, where also divergences arising from the heavy flavour loops are subtracted according to the $\overline{\mathrm{MS}}$ prescription. The coupling constant is in this case the $(n_l + 1)$-flavours (where n_l is the number of light flavours) coupling constant. The same result can be expressed in the so called decoupling scheme, which is the scheme where divergences caused by the heavy flavour loops are subtracted at zero momentum [Collins *et al.* (1978)], in which case the coupling constant is the n_l flavours one.

In order to obtain the decoupling scheme formula, it is enough to express the $(n_l + 1)$-flavours strong coupling in terms of the n_l one, according to the standard matching formulae (see the QCD review in ref. [Patrignani *et al.* (2016)]) and expand to the relevant order. Since the matching conditions are known at four loops [Schroder and Steinhauser (2006); Chetyrkin *et al.* (2006)]), the full four loop accuracy of Eq. (8.28) can be retained. From now on, we assume that formula (8.28) is expressed in the decoupling scheme. We stress that also in this case heavy fermion loops do enter in the calculation, but they are renormalised by zero momentum subtraction.

The leading IR renormalon divergence implies the following large-n behaviour of the perturbative coefficients [Beneke (1995)] and [Beneke (1999)][10]

$$c_n(\mu, \mu_m, m(\mu_m)) \underset{n\to\infty}{\longrightarrow} N \frac{\mu}{m(\mu_m)} c_n^{(\mathrm{as})}, \tag{8.29}$$

where

$$c_{n+1}^{(\mathrm{as})} = (2b_0)^n \frac{\Gamma(n+1+b)}{\Gamma(1+b)} \left(1 + \frac{s_1}{n+b} + \frac{s_2}{(n+b)(n+b-1)} + \cdots\right). \tag{8.30}$$

The b and s_i coefficients can be found in ref. [Beneke *et al.* (2017b)]. The s_i coefficients of the sub-leading $\mathcal{O}(1/n^i)$ behaviour can all be given in terms of the coefficients of the beta-function [Beneke (1995)]. We notice that $m(\mu_m)$ cancels when inserting (8.29) in (8.28).

Although the proof of the form of the asymptotic expansion is non-trivial, it is not difficult to understand its properties. First of all, the leading asymptotic behaviour arises from the region of small momenta running in the loops. Since we are using a decoupling scheme, it is then natural that the heavy quark mass and the scale at which it is evaluated drop out. So, the only scale that can appear in the mass correction is μ. If we neglect for the moment the mass of the light flavours, the form in Eq. (8.29) is the only one allowed by dimensional analysis. The form of Eq. (8.30) then follows by imposing that in the derivative of the mass correction with respect to μ the factorially growing terms should cancel at high orders, consistently with the fact that factorial growth should arise from the low momentum region.

The coefficient N cannot be computed exactly from first principles at the moment. In the large n_l limit (both positive or negative) it assumes the form [Beneke and Braun (1994)]

$$\lim_{|n_l|\to\infty} N = \frac{C_f}{\pi} \times e^{\frac{5}{6}}, \tag{8.31}$$

which equals 0.97656 for $N_c = 3$ ($C_f = 4/3$). In ref. [Nason (2016)] it was noticed that the third and fourth known terms of the mass conversion formula (8.28) already show the renormalon asymptotic behaviour, and this can be used to infer higher order terms in the top pole mass relation. This was used in ref. [Beneke *et al.* (2017b)] to extract the value of N by fitting the third and fourth order coefficient of the exact calculation and give an improved determination of the mass relation including the resummation of the asymptotic terms.[11] The mass relation was determined by using the Borel prescription, illustrated earlier. The asymptotic expansion

[10] Here we follow the notation of ref. [Beneke *et al.* (2017b)].

[11] Yet again, the observation of ref. [Nason (2016)] was made much earlier in a bottom physics context, and fits to extract the value of N were performed even before the fourth order term was known, see refs. [Pineda (2001); Ayala *et al.* (2014)].

was evaluated by taking the principal value in the inverse Borel transform formula. Then, its first four terms were subtracted and replaced with the exact ones. The renormalon ambiguity, obtained according to the Im/Pi prescription, was determined to be 70 MeV, in the case in which the bottom and charm quark mass effects are neglected.

The procedure is modified when the charm and bottom mass effects are included. I will not describe it here in detail. I only wish to remark that in this case the renormalon ambiguity is obtained from the asymptotic expansion of the pole mass relation with three light flavours. In other words, the leading renormalon ambiguity only depends upon the number of light flavours, and nothing else. It is thus the same for the top, bottom and charm pole masses, and it was determined to be 110 MeV.

Adopting $\alpha_s(M_Z) = 0.1181 \pm 0.0013$ and $m_t = 163.508$ GeV (that yields $\alpha_s(m_t) = 0.108531$ for the 5-flavour coupling constant, the results can be summarised by the formulae

$$m_c,\, m_b \text{ massless:} \qquad \frac{m_{t,\mathrm{P}}}{m_t} = 1.06164^{+0.00086}_{-0.00089} \pm 0.00043 \tag{8.32}$$

$$m_c = 1.3,\, m_b = 4.2 \text{ GeV:} \qquad \frac{m_{t,\mathrm{P}}}{m_t} = 1.06213^{+0.00088}_{-0.00096} \pm 0.00066, \tag{8.33}$$

where the last error is the irreducible ambiguity in the Borel integral, obtained with the Im/Pi prescription, and the upper and lower errors on the central value are the sum in quadrature of all (reducible) errors in the procedure, that are largely dominated by the uncertainty in α_s.

In ref. [Hoang *et al.* (2017b)], the same task was dealt with a rather different method. Before discussing it, it is useful to compare the final result with that of ref. [Beneke *et al.* (2017b)]. For a top $\overline{\mathrm{MS}}$ mass of $m_t = 163$ GeV used in [Hoang *et al.* (2017b)], formulae (8.32) and (8.33) yield $m_{t,\mathrm{P}} = 173.047$ and 173.127 respectively, to be compared with the 173.086 and 173.165 values reported on the last column of Table 5 in ref. [Hoang *et al.* (2017b)], in the row corresponding to the R value equal to m_t. The final value they quote at the end of Section 4.4 is 173.186 GeV, about 60 MeV larger than the one obtained here using the result of ref. [Beneke *et al.* (2017b)], and thus well within the uncertainty. On the other hand, they quote a larger uncertainty of 180 MeV when the bottom and charm masses are neglected, and of 250 MeV when they are taken into consideration. It is instructive to examine the source of the differences. For simplicity, I will focus upon the case of massless bottom and charm quarks.

In ref. [Hoang *et al.* (2017b)], an extensive use is made of the MSr mass. In the case of massless bottom and charm, this is defined as

$$m_P = m_{\mathrm{MSr}}(R) + R \sum_{n=0}^{\infty} c'_n(\mu, R, R)\, \alpha_s^{n+1}(R), \tag{8.34}$$

where α_s is defined in the 5-flavours scheme, and the coefficients c'_n are almost the same as those in Eq. (8.28), since they are given by the same set of Feynman diagrams except for those involving top loops. Were it not for this difference, we would have $m_{\rm MSr}(m_{\rm MSr}) = m(m)$. This is the first of a number of subtleties involved in the definition of the MSr mass in ref. [Hoang *et al.* (2017b)]. More subtleties come into play when the effects of the bottom and charm masses are included. For the present discussion, however, these subtleties are irrelevant. The important point is that we can express $m_{\rm MSr}(R)$, for R near the top mass, in terms of the $\overline{\rm MS}$ mass m, and, as shown in our elementary example, the MSr mass obeys an evolution equation free of renormalons, so that it can be computed at any scale R below the top mass, at least for not too low values of R. Thus, we have the freedom to evaluate the pole mass in Eq. (8.34) at a scale R of our choice, with the first term evaluated using the evolution equation for $m_{\rm MSr}(R)$, while the second term is evaluated using some prescription to handle the factorial growth of the coefficients, that is again given by Eq. (8.29).

The method used in [Hoang *et al.* (2017b)] to estimate the asymptotic sum is as follows:

(1) Define: $m_{\rm P}(n)$ to be given by Eq. (8.34) truncated up to (and including) the $n^{\rm th}$ order, and define $\Delta(n) = m_{\rm P}(n) - m_{\rm P}(n-1)$.
(2) Find the minimal term $\Delta(n_{\rm min})$ and the set $\{n\}_f = \{n : \Delta(n) \leq f \times \Delta(n_{\rm min})\}$. The factor f is defined to be "larger but close to unity", and the value $f = 5/4$ is adopted, without further justification, in ref. [Hoang *et al.* (2017b)].
(3) The midpoint of the covered range of values in $m_{\rm P}(n)$ for $n \in \{n\}_f$ is used as central value, and the error is taken as half the size of the covered range. Scale variation is applied to the results within the range and included in the error.

First of all, we notice that this method resembles the one presented in the previous section, Eq. (8.19), except for point 3. In fact, rather than taking half the size of the range of values covered by $m_{\rm P}(n)$, it seems more natural to take the sum of all $\Delta(n)$ divided by two.[12] In the following we will adopt this amendment for item 3, and furthermore we will not consider scale variation effects, that we will discuss further on, and refer to this procedure as "method D", where D stands for "discrete". According to the discussion following Eq. (8.17), the Im/Pi method should roughly correspond to method D, with the choice $f = 1 + 1/(4\pi)$.

The situation is illustrated in Fig 8.1, where the size of the terms for the perturbative expansion of the pole mass relation[13] are shown as a function of the order in the left plot. In the right plot, we show the ambiguity in the top pole mass as a function of the top mass m, taking $f = 5/4$ or $f = 1 + 1/(4\pi)$. As suggested by Eq. (8.25), a plot analogous to the right plot in Fig. 8.1 could be shown representing the renormalon ambiguity, for a fixed $\overline{\rm MS}$ mass, of the second term of Eq. (8.34) as

[12] Following literally the procedure of item 3 one would get a null error if the set $\{n\}_f$ consisted of a single element.

[13] They are taken from ref. [Beneke *et al.* (2017b)], table 2, multiplied by 163 GeV and by the factor $N = 0.4616$, according to the same reference.

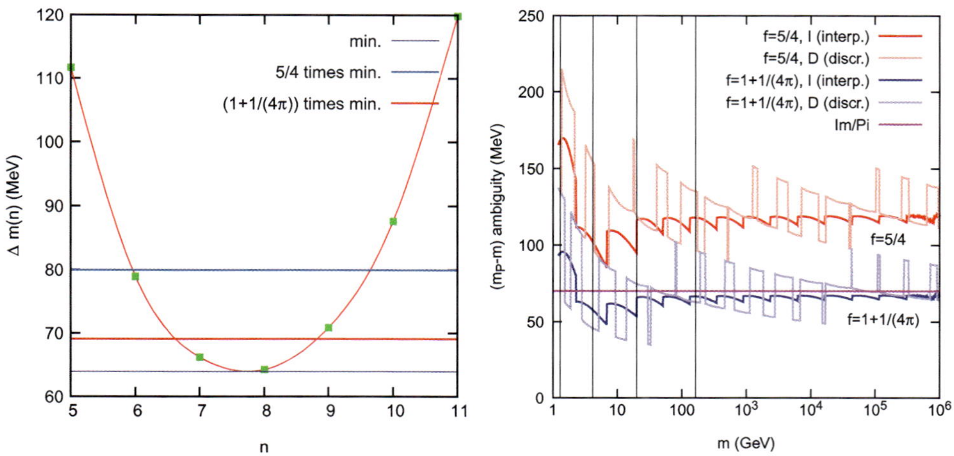

Fig. 8.1 In the left figure, the terms in the perturbative expansion for the pole mass relation are plotted as a function of the order. In the right figure, the pole mass ambiguity as a function of the pole mass, determined using the methods D and I described in the text are plotted as a function of the $\overline{\rm MS}$ mass m for two choices of the parameter f. The value corresponding to the Im/Pi method is also shown. The vertical lines correspond to $m = 1.3$, 4.2, 20 and 163 GeV.

a function of R. Such plot is easy to obtain, and it is indistinguishable from the one shown in Fig. 8.1. Thus I will not include it here. The reader should thus keep in mind that the plot shown in Fig. 8.1 is also representative of the renormalon ambiguity as determined in ref. [Hoang *et al.* (2017b)] as a function of R instead of m, and that the vertical black lines in the figure correspond to the specific values considered there.

The pole mass ambiguity should not depend upon m, since it is determined by the small momentum region. However, because of the "jumpy" nature of the D prescription, we see considerable variations in the error estimate, that get worse as the mass decreases. In order to remedy this problem, we can improve the procedure by extrapolating it to non-integer values of n. This approach was also used in ref. [Beneke *et al.* (2017b)] in order to give an alternative estimate of the sum of the asymptotic series, and turned out to give a result that exceeded the one obtained with the Borel prescription by only 22 MeV. Here I adopt a related procedure that has the advantage of requiring a shorter and more transparent explanation:

- Calling $m_{\rm P}(n)$ the sum of the asymptotic expansion up to (and including) the $n^{\rm th}$ term, one finds the value of n_0 such that $m_{\rm P}(n_0) - m_{\rm P}(n_0 - 1)$ is at a minimum. In other words, the $n^{\rm th}$ term of the series is the smallest one.
- One finds a cubic polynomial $P(n)$ such that $P(n) = m_{\rm P}(n)$ for $n = n_0 - 2$, $n_0 - 1$, n_0 and $n_0 + 1$. The polynomial $P(n)$ is taken as the extension of the

series to non-integer values of n. In particular, the value of the terms of the series at any n are now given by $\mathrm{d}P(n)/\mathrm{d}n$.

- One finds $n_{\rm min}$ such that $\mathrm{d}^2P(n)/\mathrm{d}n^2 = 0$ for $n = n_{\rm min}$, and interprets it as the location of the minimal term. $P(n_{\rm min})$ is taken as the central value for the resummed result.
- The error is taken equal to $\mathrm{d}P(n)/\mathrm{d}n$ evaluated at $n = n_{\rm min}$, times half the range in n such that $\mathrm{d}P(n)/\mathrm{d}n$ does not exceed its minimal value by more than a factor f.

We call the above procedure "method I", where I stands for "interpolation". The pole mass ambiguity obtained with this method is also shown in Fig. 8.1, where it is seen to stabilise considerably the result down to masses of a few GeV. It can also be noticed that, in an average sense, it seems consistent with the discrete method. At very low values of the mass, however, it also becomes unreliable. This is not unexpected: at low value of the mass the minimal term occurs very early, when the series has not reached its asymptotic form. Therefore, the use of a method based upon the size of the minimal term cannot be recommended in this case. On the other hand, the Im/Pi method always yields a constant ambiguity of 70 MeV, irrespective of the value of the mass. We notice that method I, with $f = 1 + 1/(4\pi)$, is fairly consistent with the Im/Pi method, as the simple argument following Eq. (8.17) suggests.

A point that needs discussion is the method adopted in ref. [Hoang *et al.* (2017b)] to perform scale variation. The minimal term and the set $\{n\}_f$ is found for the central value of the scale, and then the scale variation is performed on the result. However, one should keep in mind that also the set $\{n\}_f$ can be affected by scale variation, and its change should be included, since it leads to scale compensation. This is an important point that deserves a more detailed discussion.

If we knew the whole perturbative expansion for the pole mass relation, it would be formally independent of the renormalisation scale. The series, however, is only asymptotic, and we should devise a method to sum it up. Such a method had better be scale independent too, so that the result of the sum is also scale independent. It is easy to show with an example, borrowed from our oversimplified model, that this is the case for the Borel method. Let us assume that the top mass relation is exactly given by the formula

$$m_{\rm P} = m + \int_0^m \mathrm{d}l\, \alpha_s(l), \tag{8.35}$$

where $\alpha_s(l)$ is the one-loop strong coupling constant, and we assume that the only massive flavour is the top. We know how to express the above mass relation in terms of $\alpha(m)$, by using the Mellin transform technique, that we now interpret as a formal, order-by-order procedure:

$$m_{\rm P} - m = \int_0^m \mathrm{d}l\, \alpha_s(l) = m \int_0^\infty \mathrm{d}t\, e^{-t/\alpha_s(m)} \frac{1}{1-2b_0 t}. \tag{8.36}$$

We now want to express the same relation using $\alpha(\mu)$, with μ of order m, but different from it. This can be done as follows:

$$
\begin{aligned}
m_{\rm P} - m &= \int_{\mu}^{m} {\rm d}l\, \alpha_s(l) + \int_{0}^{\mu} {\rm d}l\, \alpha_s(l) \\
&= \int_{\mu}^{m} {\rm d}l\, \alpha_s(l) + \mu \int_{0}^{\infty} {\rm d}t\, e^{-t/\alpha_s(\mu)} \frac{1}{1-2b_0 t} \,.
\end{aligned} \tag{8.37}
$$

The first term can be expressed as a function of $\alpha_s(\mu)$ as long as $\mu > \Lambda_{\rm QCD}$, and in fact it has a convergent expansion in terms of $\alpha_s(\mu)$, while the second term has the divergent, asymptotic expansion, still in terms of $\alpha_s(\mu)$. We now show that the two expressions, if the Borel integral is evaluated according to the principal value prescription, are identical, i.e. the scale variation of the series resummed according to the principal value in the Borel integral is zero. In order to prove it, we take the difference of the two formulae both regulated according to the principal value prescription. We get

$$
\int_{\mu}^{m} {\rm d}l\alpha_s(l) + \int_{0}^{\infty} {\rm d}t\, \left[\mu e^{-t/\alpha_s(\mu)} - m e^{-t/\alpha_s(m)}\right] \frac{1}{1-2b_0 t} = 0\,. \tag{8.38}
$$

But now we see that the principal value prescription is no longer needed, since both terms in the square bracket become equal to $\Lambda_{\rm QCD}$ for $1 - 2b_0 t = 0$. At this point we have to show that the expression vanishes. It certainly vanishes for $\mu = m$, and its derivative with respect to μ, given by

$$
\begin{aligned}
& -\alpha_s(\mu) + \int_{0}^{\infty} {\rm d}t\, \left[e^{-t/\alpha_s(\mu)} \left(1 + \mu \frac{\partial}{\partial\mu} \frac{-t}{\alpha(\mu)}\right)\right] \frac{1}{1-2b_0 t} \\
&= -\alpha_s(\mu) + \int_{0}^{\infty} {\rm d}t\, \left[e^{-t/\alpha_s(\mu)} (1 - 2b_0 t)\right] \frac{1}{1-2b_0 t} \\
&= -\alpha_s(\mu) + \int_{0}^{\infty} {\rm d}t e^{-t/\alpha_s(\mu)} \; = 0,
\end{aligned} \tag{8.39}
$$

vanishes also, and thus we have full scale independence of the result.

In resummation methods that rely upon the minimal term, scale independence is no longer exact, since a truncation of the series is involved. But it is quite clear that the location of the minimal term must be kept variable with the scale, since it leads to scale compensation. This is illustrated in Fig. 8.2. The figure was obtained from Eq. (8.4), and its expansion given by the terms in Eq. (8.7), by expressing $\alpha_s(m)$ as a function of $\alpha_s(\mu)$ (using the leading order formula) and expanding again in powers of $\alpha_s(\mu)$, a procedure that can be easily performed using algebraic manipulation programs. The scale μ is taken equal to $2m$ and m. The series can then be resummed using our prescriptions D and I that rely upon the minimal term. We consider the two approaches: one where we recompute the location of the minimal term and of the range for each scale choice, and one where we do not, this last one corresponding to ref. [Hoang *et al.* (2017b)]. It is clear that in both the D and I methods, further scale compensation takes place if we recompute the location

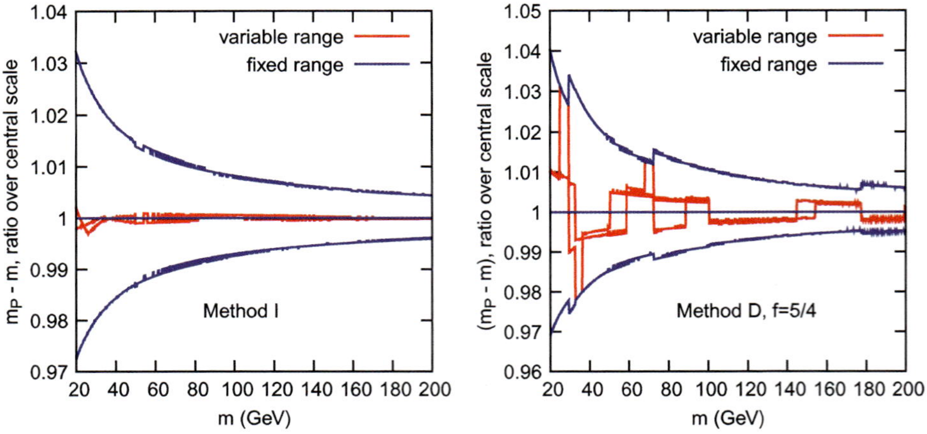

Fig. 8.2 Scale dependence of the pole — $\overline{\text{MS}}$ mass difference. The scale is varied by a factor of two above and below m. The curves labelled as "fixed range" are obtained by holding the position of the minimal term and of the range fixed while varying the scale; the curves labelled as "variable range" are obtained by recomputing the minimal term and the range for each scale choice. In the case of method I the result does not depend upon f, while in the case of method D a mild dependence upon f is found, as can be easily understood from the definition of the method.

of the minimal term and the range when we vary the scale. When using method I, the scale variation almost disappears, which is consistent with the observation that this method is fairly close to the Im/Pi prescription. Again, in the D method, due to its jumpy nature, scale compensation, although considerable, is less effective. When going at very low value of the mass, both methods become quite unstable, again suggesting that it is better to avoid the use of minimal term based prescriptions in this case.

In summary, although some freedom in the determination of the renormalon ambiguity cannot be avoided, one can identify three aspects in the procedure of ref. [Hoang *et al.* (2017b)] that should be amended: the choice of the parameter f; the insistence in evaluating the sum of the asymptotic expansion at low scales using a minimal term based, discrete method; and the procedure adopted to determine scale variation uncertainty. As far as the choice of f is concerned, some justification for the chosen value is clearly missing. As far as going to low scales for determining the renormalon ambiguity, it is clear that when using resummation methods based upon the minimal term there is essentially no limit to the size of the ambiguity that one can get, since at sufficiently low scales the first term of the expansion is the minimal term, and it can become as large as one pleases because of the Landau pole. Finally, the method adopted for the scale variation uncertainty does not include effects (i.e. the displacement of the minimal point and the range) that lead to scale compensation.

As a conclusion of this discussion, we may wonder whether the issue of the size of the mass renormalon ambiguity will ever have any importance at the LHC.

If we are arguing about whether the ambiguity is 110 or 250 MeV, this seems, at the moment, very unlikely. While the mass renormalon is now the only non-perturbative ambiguity that we can discuss to such detail, there are certainly other non-perturbative effects, very likely associated with renormalon uncertainty, that are much less understood, and that pose a much more serious threat to the accuracy of the top mass determination at the LHC.

8.4 Conclusions

The problem of the top mass measurement at hadron colliders is certainly a very subtle one, and has received the attention of many researchers. At the moment, there is no wide consensus in the theoretical community on how it should be dealt with.

In the present work I have reviewed the current theoretical status, highlighting several issues that need to be resolved. In the following, I summarise my view on the subject.

In Monte Carlo generators the top decay products are distributed with a Breit Wigner distribution. Thus, the corresponding mass parameter should be qualified as the top pole mass. It is clear that Monte Carlos represent physics processes only in an approximate sense, and thus, if we use them to extract the pole mass by fitting mass sensitive distributions, the result will be subject to errors due to Monte Carlo inaccuracies. However, this is true of *any* calculation of a physics process. Thus, I do not accept the view that, because of these inaccuracies, one should qualify the fitted mass parameter as a "Monte Carlo mass" rather than the top pole mass. What one should instead do is to ask how the approximations and inaccuracies of the Monte Carlo propagate into mass sensitive observables, thus leading to an error on the extracted top pole mass. Observe that this is the same approach that we adopt when we perform fixed order calculations, or resummed calculations at various levels of precision. There is no reason why Monte Carlo generators should be treated differently.

We can loosely separate the approximations and inaccuracies of the Monte Carlo into two categories: those that affect the perturbative part, and those that affect hadronisation phenomena and their matching to the perturbative part.

In relation to the Monte Carlo inaccuracies and approximations in the perturbative part, much has been done in order to understand their impact. We have now at our disposal generators of increasing perturbative accuracy, that include only the shower, the shower with the inclusion of matrix element corrections (MEC), or the shower matched to an NLO calculation (NLO+PS) of top production and decay. By comparing them, we can assess the errors associated with the less accurate generators, and at the same time get an estimate of the remaining inaccuracies.

It is clear that for the standard measurements of the top mass, the most relevant component of the perturbative model is the formation of the b-jet. Thus, a shower

Monte Carlo that handles production with only leading order accuracy, but performs matrix element corrections to top decay, may be quite adequate for this purpose. Thus, it is important to examine the Monte Carlo accuracy for the observable at hand. By relying upon the generic statement that Monte Carlos have only leading order accuracy, we simply miss the point.

Notice that in higher order calculations to be used for top mass extraction in standard measurements, such as those that enter the NLO+PS generators, we *must* use a mass scheme such that the top mass is close enough to the pole mass (i.e. that differs from it by less than the top width), otherwise an important shift in the mass of the reconstructed top[14] at the next perturbative order would systematically arise. Thus, we might as well use the pole mass scheme in these contexts.

More subtle effects, not correctly modelled with the Monte Carlo, have to do with interference of radiation in production and decay. It is unclear in this case whether the radiated parton should be considered part of the system of the top decay products (this phenomenon is a perturbative precursor of the non-perturbative colour reconnection problem). The worry is that these emissions, if not implemented correctly, may induce systematic shift in the mass of the reconstructed top that would be difficult to characterise.

In recent times, techniques to deal with interference effects in production and decay within the framework of NLO+PS generators have become available [Jezo and Nason (2015); Frederix *et al.* (2016a)], and have been used to build a $t\bar{t}$ generator [Ježo *et al.* (2016)]. This has been compared to a generator that does not include these features [Campbell *et al.* (2015)], and it was found to yield very similar results. This can be taken as evidence that interference in production and decay has little impact on top mass measurements.

In a recent study [Ravasio *et al.* (2018)] it has been shown that the new generator of ref. [Ježo *et al.* (2016)], interfaced with Pythia8, yields results on the mass of the reconstructed top that differ from the ones obtained with `POWHEG-hvq` interfaced with Pythia8 (that is now the default of the LHC experimental collaborations) by less than 50 MeV, if one assumes to have full access to the "particle truth" level, or less than 200 MeV, if the experimental resolution is taken into account. Results like this give us confidence that we have control over the perturbative side of our generators. They should be challenged and scrutinised by the theoretical community, also by building and trying different generators with the same or better accuracy.

Monte Carlo inaccuracies due to the modelling of non-perturbative effects, and to their interplay with renormalons, are certainly harder to examine. Even there,

[14] Here, by "reconstructed top" I mean a system defined at the particle level that is likely to have originated from the decay of a top quark. We may define it, for example, as the hardest b-jet, the hardest positive lepton and the hardest neutrino of matching flavour.

however, some recent progress has taken place. The calculation of the fourth order term in the relation of the pole mass to the $\overline{\text{MS}}$ mass, yielding a contribution of 200 MeV, has allowed to make reliable projections on the size of higher order terms, to the point where one can be confident that the pole mass can be safely used for the LHC top mass measurements. I recognise, however, that this is just one piece of the puzzle. Other power suppressed effects may be more tricky to estimate.

Notice that estimating a "perturbative precursor" of a non-perturbative effect does not necessarily yield an upper bound on the latter. For example, the fact that at leading order interference between radiation in production and decays is small does not imply that colour reconnection effects are small. This counter-intuitive fact may also be understood if we assume that non-perturbative effects are associated to renormalons, in which case a contribution at a given perturbative order may be followed by others that decrease less than geometrically and have all the same sign. Under these conditions, the given contribution cannot be considered a good estimate of the error due to the missing terms of higher order.

It is clear that more theoretical work is badly needed in this framework, with new ideas on how to estimate linear power suppressed effects and their interplay with renormalons. The works of ref. [Hoang and Stewart (2008)] and its followups go in the right direction, but should be extended to cover top quarks of moderate transverse momenta in order to be useful for standard top mass measurements, and should focus upon assessing the errors associated with direct measurement techniques.

There are other kinds of studies that are, in my opinion, even more urgent. Since Monte Carlos do fit the data, and since there is much arbitrarity in the way they model non-perturbative effects, studies that exploit this arbitrarity to estimate the associated uncertainties are badly needed. The inclusion of alternative models of colour reconnection in Pythia8 is one such example. These alternative models have been used by CMS, and have led to an increase of the systematic error of the top mass determination in the recent 13 TeV measurement. Needless to say, these models should undergo further scrutiny by the theoretical community. We should examine and criticise them, and reach a consensus on whether they are acceptable, too extreme or too narrow.

I firmly believe that more studies of this sort are needed. They should proceed as follows:

(1) Choose a parameter that is a matter of concern for the top mass measurement. Rather than varying a parameter, one may also consider more drastic variations, like changing a full component or the whole generator with a different one.
(2) Vary the parameter and re-tune the generator (or change the generator). Restrict the range of variation of the parameter so that an acceptable tune

can be achieved (or make sure that the new generator gives a good description of the data).

(3) Set up a simplified framework to assess the impact on top mass measurement, and determine how much the extracted mass changes with the new setting.

Notice that this implies a lot of work, and furthermore, by proceeding in this way we may end up uncovering problems that we have not yet foreseen. In other words, it takes courage to do it. However, I believe that this is the only way to set up a reliable method to assess the uncertainties, and gain an increased confidence on the reliability of our result.[15]

As an example, in the work of ref. [Ravasio *et al.* (2018)] it was found that by using Herwig7 rather than Pythia8 as shower generator, the reconstructed top peak according to the particle level truth (i.e. assuming that we can detect and identify all final state particles including neutrinos and b hadrons) is displaced by less than 100 MeV when using the new generators of ref. [Ježo *et al.* (2016)] and [Campbell *et al.* (2015)], and by less than 250 MeV when using the old `POWHEG-hvq` one. This result is quite remarkable, since the shower models (and, as a consequence, their interface to hadronisation) and the hadronisation models themselves are totally different in the two generators, and yet they cause a displacement in the top reconstructed peak that is significantly below the 1 GeV level. The same study also finds important differences, of the order of 1 GeV, if experimental resolution effects are taken into account. This should be considered, however, a less severe problem, since in principle this error could be reduced by increasing the resolution.[16]

To conclude, let me say again that I am very aware that this is a controversial subject. The overall view that I have expressed should be taken as my personal one. However, several points that I made are shared by many theorist colleagues, and my current opinions have been deeply influenced by discussing with them. None of these views have been expressed in writing so far, so I have just done it here, in the hope that a more transparent debate on this subject will be started, and some wider consensus may be reached in the future.

8.5 A personal note

My last exchange with Guido was by e-mail, on the 16th of April 2015, when I learned that he had been awarded the EPS High Energy and Particle Physics Prize. I did not know of his disease at that time. I sent him a very synthetic e-mail, that had a 'CONGRATULAZIONI PER L'EPS!' in the subject, and "Era ora ... ciao, Paolo" in the body. His reply was particularly warm: "Caro Paolo, ti ringrazio

---

[15] This can also guide us in improving the measurement method, along lines similar to those explored in ref. [Andreassen and Schwartz (2017)].

[16] Furthermore, in the same study some evidence was found that it may not be possible to fit the same data set with both generators. In other words, it may not be possible to satisfy item (2) in the above list with both generators.

moltissimo. Sei un caro amico. Saluti G". I now find this exchange particularly moving.

Acknowledgements

I wish to thank Giulia Zanderighi, Gavin Salam, Matthias Steinhauser and Michelangelo Mangano for reading the manuscript, and for providing useful suggestions.

Chapter 9

Standard Model physics at the LHC

Jan Kretzschmar

University of Liverpool

I discuss selected topics of Standard Model (SM) physics at the LHC and their connection to the work of Guido Altarelli. An overarching connection is given by Guido's fundamental contributions to our understanding of the strong force, parton distribution functions and their evolution, and techniques to calculate theoretical predictions through perturbative Quantum Chromodynamics. The increasingly precise LHC measurements of diverse processes can be confronted with SM predictions relying on these foundations to test the SM and contribute to the knowledge of the proton structure. The LHC experiments have also started to perform precise measurements of SM parameters, such as the mass of the W-boson, and have measured a variety of processes sensitive to interactions of multiple electroweak bosons. The discovery of the Higgs boson in 2012 and the measurements of its properties are

crucial to establish the nature of the mechanism of electroweak symmetry breaking and the status of these studies is summarised.

9.1 Introduction

Before the year 2009, the start of the Large Hadron Collider (LHC) was eagerly anticipated in the whole particle physics community. As Guido Altarelli put it: "physics [was] in a deadlock and [needed] to be restarted" [Altarelli (2011b)]. The Standard Model of particle physics (SM) was by many seen as just a low-energy effective version of the ultimate theory. The LHC was expected to take our understanding to the next level. In the following I attempt to discuss where the endeavour to explore the SM at the energy frontier has led us in the last seven years.

The LHC [Evans and Bryant (2008)] is the most powerful accelerator ever built. From 2010 to 2012 it provided proton-proton (pp) collisions at an unprecedented centre-of-mass energy of $\sqrt{s} = 7$–$8\,\text{TeV}$ in Run 1. In 2015 the LHC started the Run 2 operation at $\sqrt{s} = 13\,\text{TeV}$, close to the design value of 14 TeV. With this, it exceeds the typical energy scale of collisions at the previous generation of colliders, LEP, HERA and TeVatron, by about an order of magnitude. The LHC now regularly delivers an instantaneous luminosity higher by 50% than the original design goal of $10^{34}\,\text{cm}^{-1}\,\text{s}^{-1}$ to the ATLAS and CMS experiments [ATLAS Collaboration (2008); CMS Collaboration (2008)], exceeding previous colliders by about two orders of magnitude.

The SM describes the interactions of three generations of quarks and leptons through the gauge fields of the electroweak and strong interactions, with the electroweak symmetry being spontaneously broken by the Englert–Brout–Higgs mechanism. The unprecedented collision energy and large data sets provided by the LHC allow the probing of the fundamental building blocks of matter and their interactions at a new level. It was certainly one of the "major physics discoveries of our time" [Altarelli (2013a)] when the LHC experiments discovered "a" Higgs boson in 2012 [Aad *et al.* (2012c); Chatrchyan *et al.* (2012)]. The Higgs boson was the final SM particle and was seen as "directly related to most of the major open problems of particle physics" [Altarelli (2011b)].

It is important to remember, that the pp collisions delivered by the LHC are complex interactions of composite, strongly-interacting particles. Without the framework provided by the SM, we would hardly be in a situation to exploit the data at such a detailed level as we do it these days: decades of efforts in theory and phenomenology enable us to connect the basic Lagrangian to Feynman diagrams and those to quantitative predictions of cross sections and decay rates. A striking feature of the LHC data is the vast spectrum of measurements (and searches) that are possible, thereby testing a wide diversity of processes as well as very different energy scales.

9.2 The role of QCD in LHC physics

The protons collided by the LHC are strongly-interacting particles composed of quarks and gluons. It is thus somewhat obvious, that the theory of strong interactions, Quantum Chromodynamics (QCD), is of the highest importance for LHC physics. This is also true, if the aim is to study electroweak interactions. Or, as Guido Altarelli put it: "the understanding QCD processes is an essential prerequisite for all possible discoveries" at the LHC [Altarelli (2011c)]. Thanks to the work of Guido Altarelli and many others at the end of the 1970s, QCD was established as the theory of strong interactions and the framework of perturbative calculations was worked out. At the level of the Lagrangian "QCD is a 'simple' theory, but actually this simple theory has an extremely rich dynamical content" [Altarelli (2011c)].

The calculation of a cross section can be factorised into a part describing the structure of hadrons in terms of the Parton Distribution Functions (PDFs) and another part describing the interactions of the fundamental quarks and gluons. The PDFs are not static objects, but evolve with the energy scale of the interaction, a property studied by Dokshitzer, Gribov and Lipatov [Gribov and Lipatov (1972a,b); Dokshitzer (1977)], and then re-derived by Altarelli and Parisi "in a direct way from the basic vertices of QCD" [Altarelli and Parisi (1977)]. The resulting evolution equations are known as DGLAP equations and form one of the cornerstones of our understanding of QCD. The formulation given by Guido Altarelli has allowed the precision to be systematically improved. The splitting functions that describe the evolution are nowadays known to the three-loop level [Vogt *et al.* (2004); Moch *et al.* (2004)] and partially beyond this.

The elementary cross sections can be systematically calculated and similarly improved by an expansion in the strong coupling constant α_S. Guido Altarelli contributed to one of the first higher order calculations [Altarelli *et al.* (1979a)]: the production of di-leptons in hadron collisions through the Drell–Yan process [Drell and Yan (1970)]. The Drell–Yan process will be further discussed in Section 9.6. The mechanism to improve the calculations by considering diagrams higher-order in α_S has been critical. For LHC physics purposes it has turned out, that the next-to-leading order (NLO) is the minimum to have a reliable prediction, but often the next-to-next-to-leading order (NNLO) still provides an important correction.

Guido Altarelli's contributions have as such paved the way to quantitative predictions for QCD processes. The exploitation of the large LHC data sets keeps increasing the demands of experimentalists on the theory accuracy, which typically reaches the 5–10% percent level and may reach below 1% for simple processes. Still, the precision in the measurements is often better than the theory uncertainties, limiting the interpretation of the data.

9.3 A broad overview of SM processes studied at the LHC

The diversity of processes and energy scales accessible at the LHC is exceptionally wide. A very much simplified summary of cross-section measurements is given in Fig. 9.1, where each measured point often stands for a complex, multi-differential analysis and typically 10–1000 differential measurements. The range of cross sections studied ranges from 100 mb for the total *pp* cross section to 1 fb for the rarest processes, thus spanning 14 orders of magnitude.[1]

The theoretical predictions for the vast majority of the processes displayed in Fig. 9.1 use the techniques of perturbative QCD. The necessary PDF information is largely derived from deep inelastic lepton-nucleon scattering (DIS), although more recently the information from LHC measurements have started to become useful to

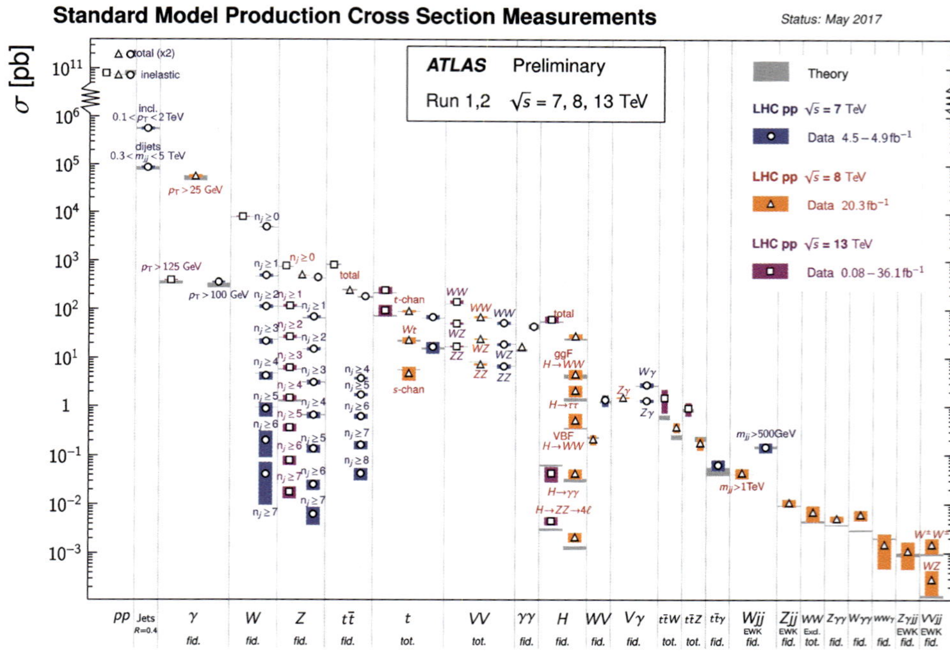

Fig. 9.1 A summary of cross-section measurements in *pp* collisions at $\sqrt{s} = 7, 8, 13\,\text{TeV}$ for a variety of SM processes by the ATLAS collaboration [ATLAS Collaboration (2017a)]. The measurements are compared to the respective SM theory predictions. A similar summary is available for results by the CMS collaboration [CMS Collaboration (2017a)].

[1] In addition, this overview does not even attempt to capture many other important SM measurements performed with LHC data, as for example the study of rare or CP-violating *B*-hadron decays with the LHCb [Alves *et al.* (2008)], CMS and ATLAS experiments or the study of Heavy Ion collisions and the quark-gluon plasma, where the major contributions of the ALICE experiments lie [Aamodt *et al.* (2008)].

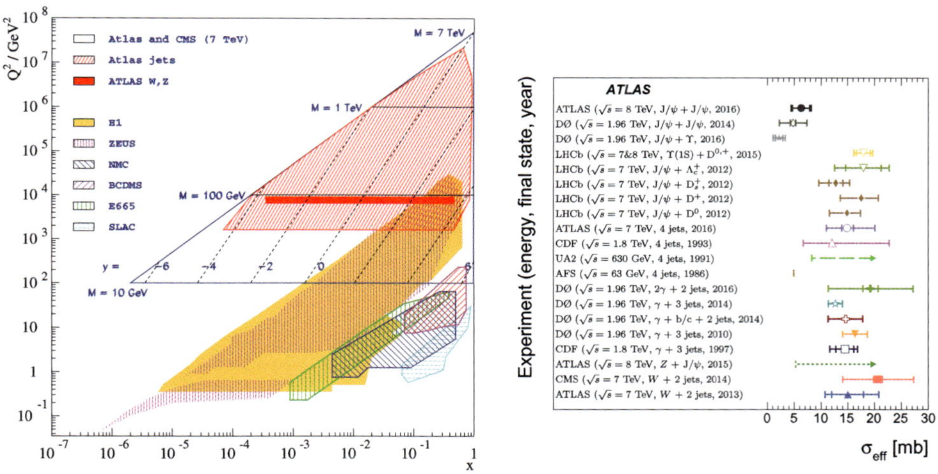

(a) (b)

Fig. 9.2 (a) Illustration of the plane in Bjorken-x and scale Q^2 relevant for the determination of PDFs from DIS data by the HERA experiments (H1, ZEUS) and a selection of fixed-target experiments (NMC, BCDMS, E665, SLAC). The kinematic region accessed by the LHC in terms of mass M and rapidity y of the produced system is shown for $\sqrt{s} = 7\,\text{TeV}$ together with example coverage by specific processes (jet and W,Z production). (b) A summary of effective DPI cross sections σ_{eff} derived from a variety of processes studied at different hadron-collider experiments [Aaboud *et al.* (2017f)].

constrain PDFs further. Figure 9.2(a) illustrates the relevant range of the Bjorken-x momentum fraction for LHC physics, which is roughly from 10^{-4} to 1. Indeed DIS fixed-target experiments and the HERA ep collider cover this range, however a significant evolution in energy scale is required to reach the LHC region. The broad agreement of the measurements with the predictions is thus a major triumph, both for the SM in general and more specifically for QCD and the DGLAP formalism.

In the following, I will discuss a selection of recent measurements, starting with the most abundant processes involving only the strong interaction, moving on to processes involving the production of single electroweak bosons and a small selection of multi-boson processes before concluding with the current status of the studies of the Higgs boson.

9.4 Soft QCD processes

The majority of pp interactions take place at small energy scales, where α_S is large and non-perturbative effects dominate. Effects of this "soft regime", such as inelastic minimum-bias interactions, often cannot be systematically computed through the usual perturbative QCD techniques. We thus need to resort to phenomenological models integrated in Monte Carlo (MC) tools like Pythia, Herwig or Epos [Sjöstrand *et al.* (2008); Bellm *et al.* (2016); Pierog *et al.* (2015)] that incorporate aspects of

QCD and require tuning to data. Despite the fact that these MC tools are not directly connected to the SM Lagrangian, they are indispensable to the work of experimentalists. MC simulations are used to model the soft effects and their influence on measurements of processes at large energy scales, such as hadronisation of quarks and gluons and additional interactions beyond the hard scattering ("underlying event"). It remains a challenge to develop and constrain these models further such that non-perturbative effects will not limit the interpretation of LHC measurements in the future. Detailed measurements often show significant discrepancies to the model predictions. For example, the ALICE collaboration recently reported an enhanced production of strange hadrons in high-multiplicity pp collisions [Adam *et al.* (2017)], suggesting a lack of understanding of the mechanism in the current MC models.

Another interesting aspect is the existence of hard double-parton interactions (DPI), which, despite taking place at perturbative energy scales, point beyond our typical pQCD framework. In a simple model, the DPI interaction may factorise into two effectively independent parton-parton scatters within the same pp interaction, however the probability for the second interaction is adjusted by a factor $\sigma_{\rm inel}/\sigma_{\rm eff}$. Here $\sigma_{\rm inel}$ refers to the total inelastic interaction cross section and $\sigma_{\rm eff}$ is the so-called effective DPI cross section, a parameter to be determined in the experiment. In Fig. 9.2(b) an overview of measurements of $\sigma_{\rm eff}$ from both TeVatron and LHC analyses is given. The current data set is still sparse and does not allow one to determine a clear dependence on the centre-of-mass energy or the final states considered. Most analyses find a value for $\sigma_{\rm eff}$ that is a significant factor ~ 5 below $\sigma_{\rm inel}$. This possibly suggests a "hot-spot" interpretation, where the existence of a first hard interaction increases the probability to find a second hard scatter. Interestingly the values of $\sigma_{\rm eff}$ extracted from double quarkonia final states feature further reduced $\sigma_{\rm eff}$ values, suggesting an enhanced correlation in these cases.

9.5 Production of jets and heavy-flavour jets

In the perturbative regime, the strong interaction of quarks and gluons leads to an abundant production of hadronic jets with high transverse momentum. The ATLAS and CMS collaborations have performed a series of measurements at energies of $\sqrt{s}$ = 2.76–13 TeV [Khachatryan *et al.* (2016c, 2017a); Sirunyan *et al.* (2017b); Aaboud *et al.* (2017b,e)]. Figure 9.3(a) shows an overview of a typical measurement of inclusive jet cross sections double-differentially in the transverse momentum p_T and rapidity y of the jet together with a QCD prediction at NLO accuracy, supplemented by corrections for electroweak and non-perturbative effects, where the latter include corrections of underlying event and hadronisation. Inclusive jet measurements have the potential to cover almost the full kinematic plane of accessible parton dynamics, as can be appreciated in Fig. 9.2(a). The data typically agrees with the QCD predictions to 5–10% as shown in Fig. 9.3(b), which is truly remarkable given the wide kinematic range and many orders of magnitude in the

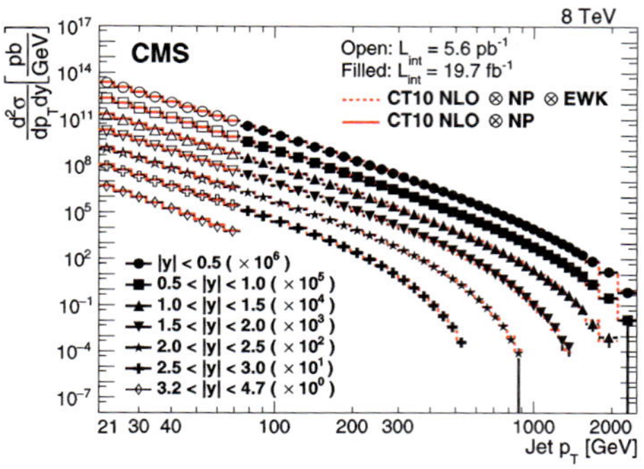

(a)

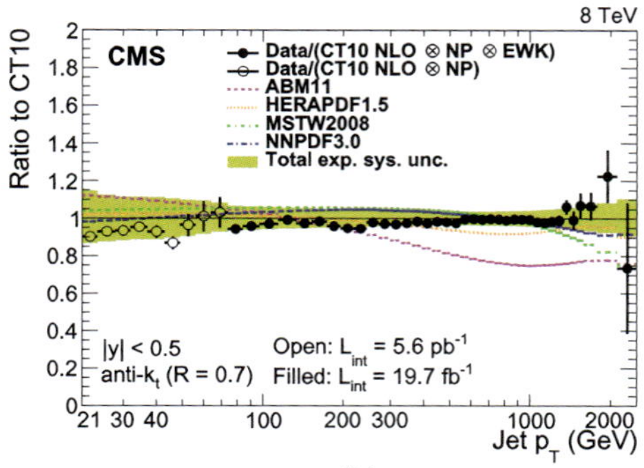

(b)

Fig. 9.3 Double-differential inclusive jet cross-section measurement at $\sqrt{s} = 8\,\text{TeV}$ from Ref. [Khachatryan *et al.* (2016c)]. (a) gives an overview of all data points compared to the NLO QCD prediction. (b) shows the ratio of the data at central rapidity to NLO predictions with different PDF sets.

cross section probed. The cross section at low p_T is gigantic in comparison to the luminosity the LHC is able to deliver and measurements are typically performed using small data sets taken at low instantaneous luminosity.

The measurements at higher $p_T > 100\,\text{GeV}$, where the effects of non-perturbative corrections are progressively less important, may provide information on the gluon PDF at larger x and are routinely used in PDF fits. The recent completion of calculation of the NNLO corrections [Currie *et al.* (2017d,c)] is a milestone to put the interpretation of the data on a more solid footing and promises a reduction of the theoretical uncertainties to a level less than the PDF and the measurement uncertainties. However, contrary to the common expectation, a significant uncertainty from scale variations remains at NNLO. In addition, the choice of the scale as either the leading jet p_T or the individual jet p_T has a significant impact on the visual agreement with the data, where the latter choice even leads to a worse agreement with the data than the NLO prediction. Both of these features can be appreciated in Fig. 9.4 in comparison to a recent measurement. Future studies of the theory calculation and definition of the experimental measurement may be able to resolve this issue in the interpretation of the data.

An interesting extension to inclusive hadronic jets are measurements of jets produced in the fragmentation of heavy quarks. The production of charm and beauty hadrons has been performed down to very low transverse momentum and in the forward region by the LHCb collaboration [Aaij *et al.* (2016c)]. This data is sensitive to the gluon density at very low $x \sim 10^{-6}$, a region not probed by any other collider data so far. However, the theoretical uncertainties on the NLO predictions are large, up to $\sim$30%, and the interpretation relies on the cancellation of these large uncertainties in suitably defined ratios [Zenaiev *et al.* (2015); Gauld and Rojo (2017)].

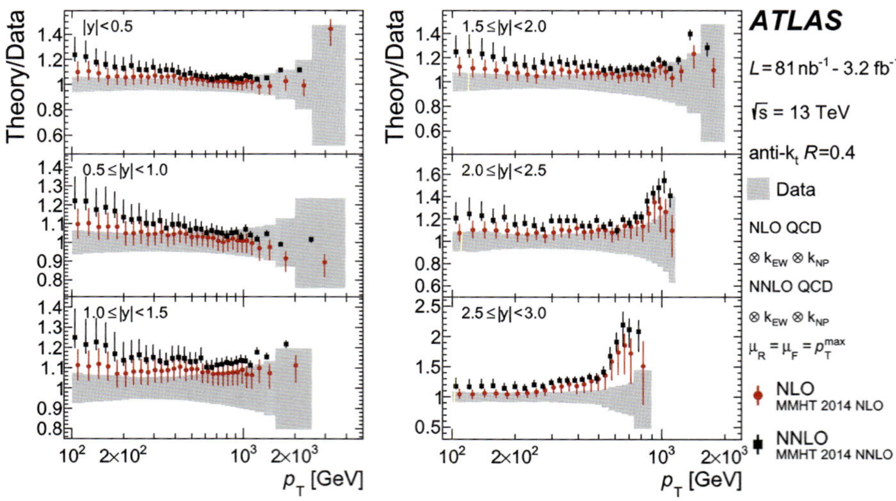

Fig. 9.4 Double-differential inclusive jet cross-section measurement at $\sqrt{s} = 13\,\text{TeV}$ from ref. [Aaboud *et al.* (2017b)] compared to NLO and NNLO QCD predictions with the scale set by the leading jet p_T.

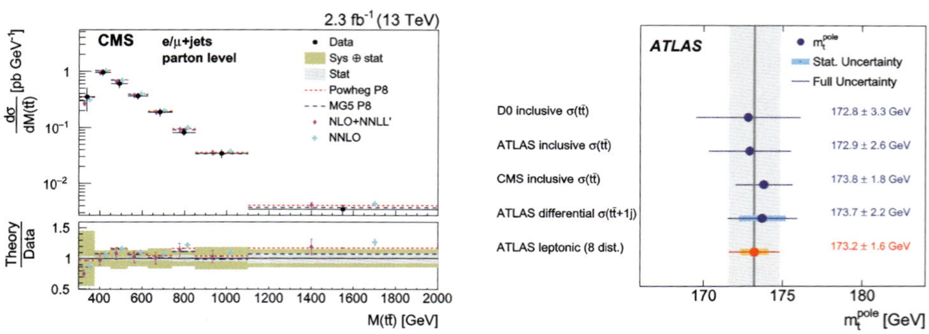

(a) (b)

Fig. 9.5 Example of recent measurements with top-quarks: (a) shows a measurement of the differential cross section for $t\bar{t}$ production at $\sqrt{s} = 13\,\text{TeV}$ compared to a selection of predictions by MC programs and fixed order calculations [Khachatryan *et al.* (2017b)]. (b) shows a recent compilation of top-quark mass extractions using cross-section data [Aaboud *et al.* (2017c)].

The LHC has moved the study of top-quark pair production through the strong interaction to a new level compared to the Tevatron, where the top-quark was discovered and many studies were pioneered. The remainder of this chapter will not be able to do justice to this active field, but just give a glimpse of the more recent results. Predictions for top-quark pair production are available up to NNLO QCD differentially in the stable top-quark kinematics [Czakon *et al.* (2016b)] and these are generally seen to describe the data well, as shown in Fig. 9.5(a) for a recent measurement differential in the $t\bar{t}$-system mass. The measurement of the top-quark mass has reached a precision of about 0.5 GeV [Khachatryan *et al.* (2016e); ATLAS Collaboration (2017c)]. At this level it is likely that measurements of the "MC mass" differ from the (theoretically desired) "pole mass" for example due to non-perturbative and colour-reconnection effects. Methods exploiting the total [Khachatryan *et al.* (2016d)] or differential cross-section measurements [Aaboud *et al.* (2017c)] are better controlled with respect to non-perturbative effects and have recently reached uncertainties of well below 2 GeV as shown in Fig. 9.5(b), typically limited by the uncertainties on the PDFs and missing higher order corrections.

Finally, the production of jets may be used to probe the strong coupling α_S, which is obviously "a crucial ingredient of perturbative QCD" [Altarelli (2013b)] and in fact the least well known of the fundamental couplings with a precision of about 1% at a scale of the Z-boson mass. Unfortunately it is not possible to extract $\alpha_S(m_Z)$ from LHC jet measurements with competitive precision. Instead, the data may be used to probe α_S at very large scales of up to 2 TeV [Khachatryan *et al.* (2017a)]. Determinations from the $t\bar{t}$ cross section have been shown to reach an improved precision compared to inclusive jets [Chatrchyan *et al.* (2014); Klijnsma *et al.* (2017)]. The uncertainty in α_S is important for several LHC predictions, such as Higgs-boson production. The size of the uncertainty in α_S is a topic of debate and Guido Altarelli certainly had an opinion on this issue, advocating caution to the estimates of theory uncertainties [Altarelli (2013b)].

9.6 Di-lepton production in the Drell–Yan process

Di-lepton production in hadron collisions through the Drell–Yan (DY) process [Drell and Yan (1970)] has played a fundamental role in establishing the SM. It is here considered to include both the neutral current $q\bar{q} \to Z/\gamma^* \to \ell^+\ell^-$ as well as the charged current $q\bar{q}' \to W^\pm \to \ell^\pm \overset{(-)}{\nu}$ reactions. Of critical importance was the discovery of the heavy intermediate vector bosons W and Z by the UA1 and UA2 experiments [Arnison *et al.* (1983a); Banner *et al.* (1983); Arnison *et al.* (1983b); Bagnaia *et al.* (1983)] in DY reactions. Beyond these discoveries, the DY process was a test-bed to develop the procedure to systematically calculate and improve cross sections in hadron collisions through the factorisation theorem and an expansion in orders of α_S. In fact, this process remains one of the few examples in hadron-hadron collisions, where factorisation has been rigorously proven to hold [Collins *et al.* (1984); Bodwin (1985)].

Each major step forward in the theory calculation typically took a decade. The first step beyond the initial leading order calculation came through major contributions of the "Rome group" with Guido Altarelli [Altarelli (2012)], who completed "one of the first calculations of NLO corrections in QCD" in 1978/79 [Altarelli *et al.* (1978a, 1979a); Kubar–Andre and Paige (1979); Kubar *et al.* (1980)]. A critical part was to treat the PDFs and their evolution in a consistent way in DIS and DY reactions. The NLO corrections "turned out to be surprisingly large" [Altarelli (2012)],

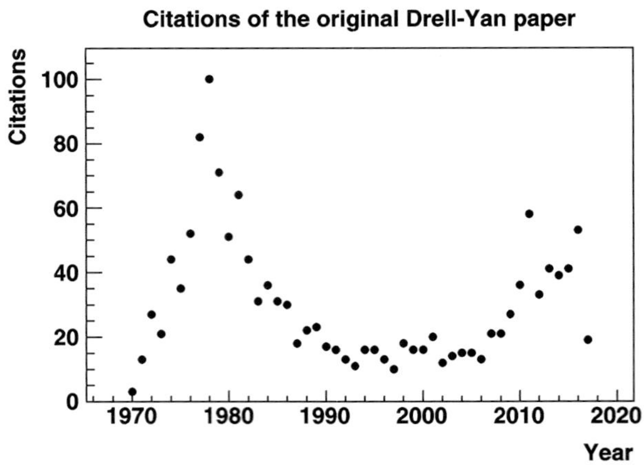

Fig. 9.6 History of citations of the original paper on the Drell–Yan process [Drell and Yan (1970)] taken from `inspire.net`.

necessitating calculations beyond NLO. The NNLO corrections were completed in the early 1990s [Hamberg *et al.* (1991); van Neerven and Zijlstra (1992); Rijken and van Neerven (1995)] and the fully differential NNLO predictions became available until about 2010 [Anastasiou *et al.* (2004); Melnikov and Petriello (2006); Catani *et al.* (2009); Gavin *et al.* (2011)].

Figure 9.6 shows the number of citations of the original paper by Drell and Yan as a function of time. This reflects the initial interest in the process to establish the SM in the 1980s, as well as the more recent "resurrection" that is closely linked to new analyses of this process at the LHC. This renewed interest has diverse reasons, which includes the study of PDFs, precision measurements of SM parameters as the W-boson mass and the effective weak mixing angle $\sin^2\theta_W$, and the background to searches beyond the SM as for example particle candidates for dark matter [Lindert *et al.* (2017)]. The measurements are facilitated by large samples of tens of millions of events reconstructed with small backgrounds and well understood detector performance for the final-state leptons.

Measurements of differential cross sections and distributions for Z/γ^* and $W^\pm$ production have been performed by the ATLAS, CMS and LHCb collaboration in pp collisions at $\sqrt{s} = 2.76$–$13\,\text{TeV}$ and as a function of many kinematic variables. The classical variables to constrain PDFs are the di-lepton rapidity and mass, $y_{\ell\ell}$ and $m_{\ell\ell}$ for Z/γ^* and the lepton rapidity η_ℓ for $W^\pm$ production with some recent examples given by refs. [Aaboud *et al.* (2017h); Aad *et al.* (2016a); Khachatryan *et al.* (2015d, 2016b); Aaij *et al.* (2016a,b)]. Figure 9.7 shows two examples of such measurements compared to state-of-the-art cross-section calculations with different PDF sets. The measurements have reached an accuracy of a percent or better, which is below the uncertainties of a typical PDF set and often also below the uncertainty

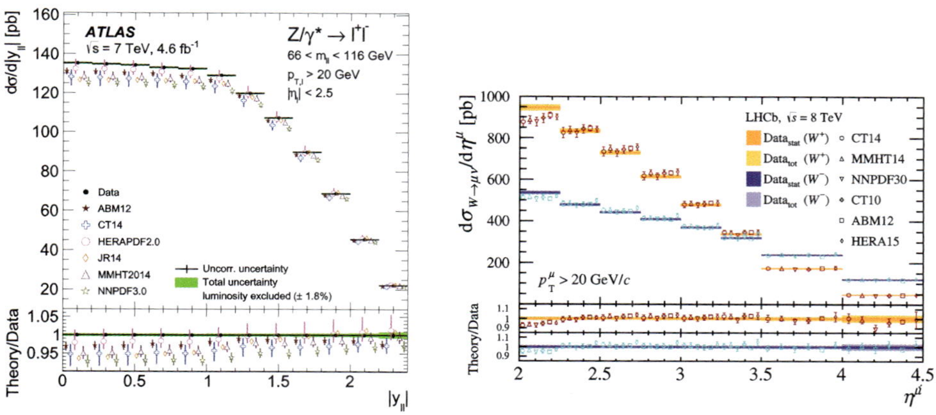

(a) (b)

Fig. 9.7 Two examples of recent Drell–Yan measurements at the LHC compared to NNLO QCD predictions with a variety of different PDF sets. (a) shows a measurement of the differential Z/γ^* cross section as a function of the di-lepton rapidity $y_{\ell\ell}$ at $\sqrt{s} = 7\,\text{TeV}$ [Aaboud *et al.* (2017h)] (b) shows a measurement of W^+ and W^- production as a function of the lepton pseudo-rapidity η_ℓ [Aaij *et al.* (2016a)].

of the NNLO theory calculation. The fact that these precise $W^\pm$ and Z/γ^* cross-section data generally agree with the predictions, is another testament to the success of our understanding of perturbative QCD.

Such data can now be included in PDF analyses based on either the HERA DIS data set [Abramowicz *et al.* (2015)] or global PDF fits [Harland–Lang *et al.* (2015a); Dulat *et al.* (2016); Alekhin *et al.* (2017); Ball *et al.* (2017b)] to improve our knowledge on the proton structure in, broadly speaking, three areas. Firstly, the LHCb data gives unique information on PDFs at very low Bjorken $x \sim 10^{-4}$ or very large $x \sim 0.5$, a range where the information of the HERA data set is limited and information at high x from fixed-target data may not be fully reliable. Second, as illustrated in Fig. 9.8(a), data on $W^\pm$-boson production helps to improve the information on low and intermediate x valence quarks distributions. Finally, Z production at the LHC at central rapidity has a significant contribution from strange quarks, which are only constrained from charm production in neutrino-nucleon DIS on heavy targets at somewhat larger x. As demonstrated in Fig. 9.8(b), the LHC data has here a unique impact and strongly hints at a larger strange-quark density than previously inferred from the neutrino data.

Another key variable is the transverse momentum of the boson, which poses an interesting theoretical challenge and has implications for many studies at the LHC. The distribution is more easily measured in the neutral current process as the di-lepton $p_{T,\ell\ell}$, see e.g. refs. [Aad *et al.* (2016b); Khachatryan *et al.* (2015c, 2017c)]. The spectrum can be roughly separated into two regimes at a transverse momentum of above and below half the boson mass. The higher-$p_{T,\ell\ell}$ region may

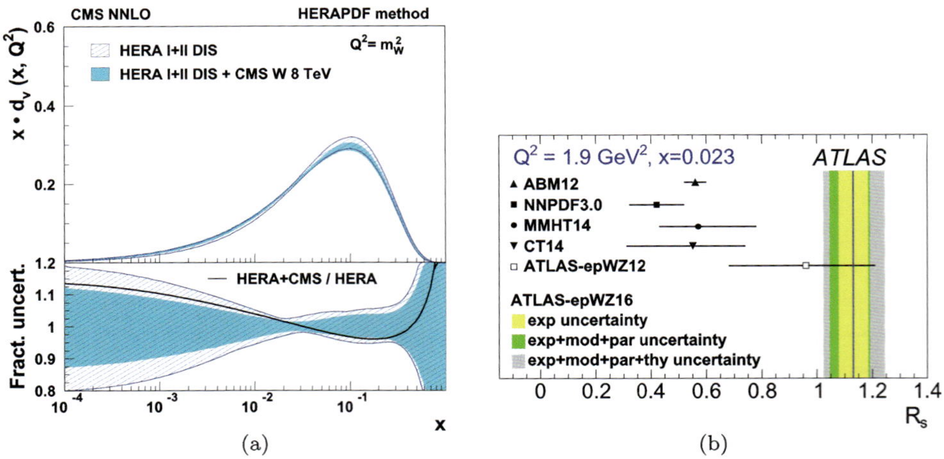

Fig. 9.8 Two examples showing the impact of LHC $W^{\pm}$ and Z/γ^* production data on the knowledge on the proton PDFs. (a) shows a study differential $W^{\pm}$ data where a PDF fit based on HERA DIS data is included, comparing the uncertainties on the d-valence density with and without the $W^{\pm}$ data [Khachatryan *et al.* (2016b)]. (b) shows the ratio $R_s = (s+\bar{s})/(\bar{u}+\bar{d})$ as a measure of the size strange-quark distribution compared to the up- and down-quark sea. The result from a fit with LHC Z/γ^* and $W^{\pm}$ data labelled as "ATLAS-epWZ16" is compared to the prediction of other PDF fits [Aaboud *et al.* (2017h)].

be described by a fixed order calculation, while an all-order resummation calculation is required to describe the lower-$p_{T,\ell\ell}$ domain that is dominated by multiple soft-gluon emission and possibly also influenced by non-perturbative effects. Figure 9.9 shows the high $p_{T,\ell\ell} > 20$ GeV part of a measurement differential in both the di-lepton transverse momentum and mass compared to a very recent NNLO fixed order calculation for the $pp \to Z + \text{jet}$ process at order α_S^3 [Gehrmann-De Ridder *et al.* (2016b); Boughezal *et al.* (2016)]. The correction to the α_S^2 calculation is seen to be significant, although within the uncertainty estimate in the range $p_{T,\ell\ell} \gtrsim m_{\ell\ell}$. In parts of the phase space, especially near the Z-boson resonance, the data is more precise even than the α_S^3 prediction. There remains a tension between the data and the prediction that may be related to the PDFs and can possibly be improved by including the data in a PDF analysis as discussed in ref. [Boughezal *et al.* (2017)].

The large data sets of Z/γ^* and $W^{\pm}$ can be used for precision measurements of the effective weak mixing angle $\sin^2\theta_W$ and the W-boson mass, respectively. These measurements are interesting to pin down the fit of precision electroweak observables [Baak *et al.* (2014b); de Blas *et al.* (2016)] and search indirectly for deviations from the SM [Heinemeyer *et al.* (2013)]. The $\sin^2\theta_W$ measurement exploits the change of the "Forward-Backward Asymmetry" in the $q\bar{q} \to Z/\gamma^* \to \ell^-\ell^+$ process in the vicinity of the Z resonance. The W-boson mass can be extracted through minuscule changes in the shapes of kinematic variables such as the lepton p_T in W events.

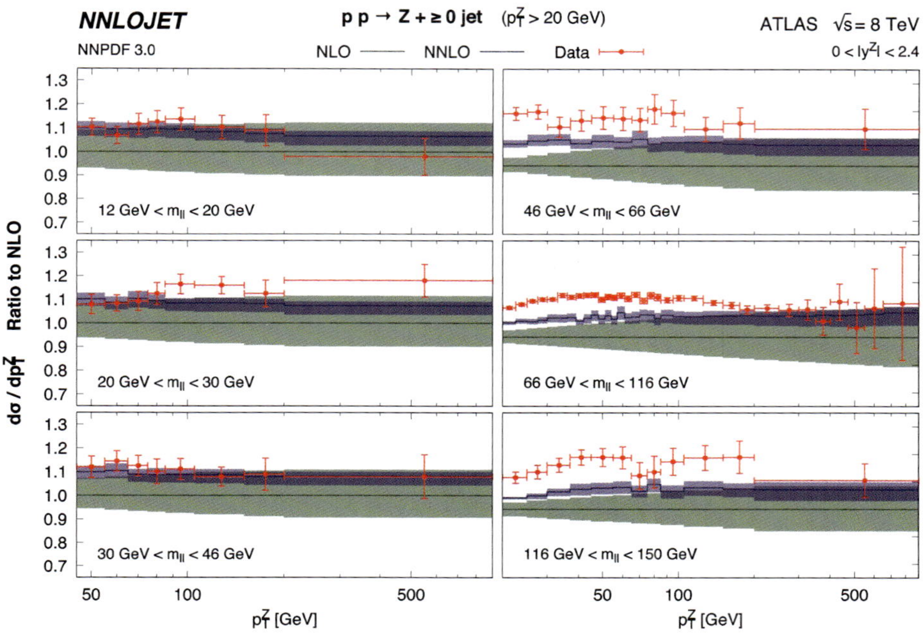

Fig. 9.9 Double-differential di-lepton cross-section measurement as a function of mass and transverse momentum at $\sqrt{s} = 8\,\text{TeV}$ compared to a fixed order NNLO calculation [Gehrmann-De Ridder *et al.* (2016c)].

These electroweak measurements require an extreme experimental precision, for example the lepton momentum calibration has to be understood to the 10^{-4} level. Even more critical is the control of uncertainties related to the modelling of the sensitive distributions with respect to QCD effects related to PDFs, heavy-flavour initiated processes and the transverse-momentum spectrum of the bosons. The LHC environment poses such a challenge, that Guido Altarelli as many of his peers was skeptical about the feasibility of such an enterprise, stating that "... anyone [who] wants to measure the W mass at the LHC with high precision, better he commits suicide ..." [Altarelli (1998)].

The achievements of the first competitive measurements of the W-boson mass [Aaboud *et al.* (2018)] and $\sin^2\theta_W$ [CMS Collaboration (2017d)] in spite of the major effort are therefore all the more remarkable. Figure 9.10(a) gives an overview of $\sin^2\theta_W$ from e^+e^- colliders, the Tevatron and the LHC. The most recent CMS measurement with the $\sqrt{s} = 8\,\text{TeV}$ data set reaches a precision similar to the Tevatron experiments and "just" a factor of two worse than the LEP and SLD results. In Fig. 9.10(b) an overview of measurements of the W-boson mass is shown including the recent ATLAS determination using the $\sqrt{s} = 7\,\text{TeV}$ data set, which has reached the same uncertainty as the previously best measurement by CDF.

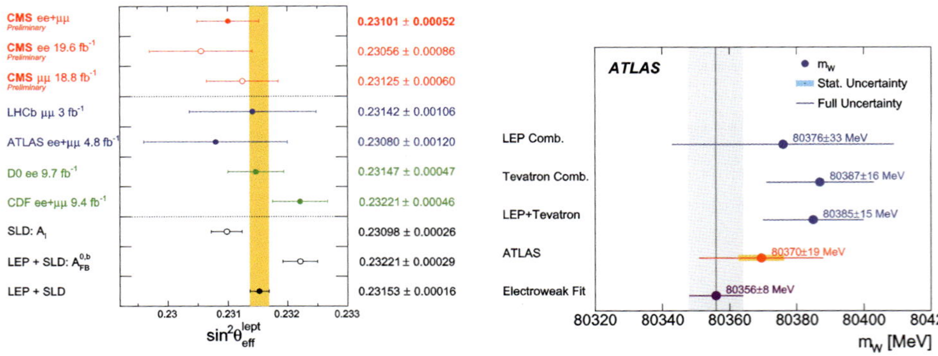

(a) (b)

Fig. 9.10 Overview of measurements of precision electroweak parameters at different colliders for (a) the effective weak mixing angle $\sin^2\theta_W$ [CMS Collaboration (2017d)] and (b) the mass of the W boson [Aaboud *et al.* (2018)].

9.7 Multiboson production and electroweak processes

In addition to the study of the simpler and abundant processes, the LHC allows also the measurements of rarer and more complex final states. In this section we will discuss processes that go beyond the leading order in the electroweak couplings, where either two or three bosons $V = W, Z, \gamma$ are produced or two of these bosons "fuse" (vector boson fusion, VBF) or "scatter" (vector boson scattering, VBS). The diagrams involved contain vertices sensitive to the triple and quartic couplings of the electroweak bosons, which are non-trivial in the SM due to the non-Abelian $U(1)_Y \times SU(2)_L$ structure, which is "unbroken in the interaction vertices, ... [but] is instead badly broken in the masses" [Altarelli (2011b)] by the Higgs mechanism. In the SM the interplay of various diagrams, such as those containing γWW and ZWW vertices, guarantees a result that does not violate unitarity. Another well known example is the scattering of longitudinally polarised massive vector bosons, e.g. $W_L W_L \to W_L W_L$, where unitarity is violated already at about 1 TeV without the contribution from the Higgs boson [Veltman (1977); Lee *et al.* (1977a,b)]. Any deviation from the SM predictions would imply physics with new forces or particles.[2]

In Fig. 9.11(a) an overview of measurements of di-boson cross sections is given. All possible combinations of VV processes have been measured, with the majority of measurements considering the decays with (charged) leptons of the massive W

[2]Similarly, it is interesting to study the electroweak properties of the top-quark, which may have a special role in the process of electroweak symmetry breaking due to its large mass. Such studies can be performed e.g. through electroweak production of single top-quarks or the production of top-quarks in association with electroweak bosons $t\bar{t}V$ or tV, but a detailed discussion is beyond the scope of this section.

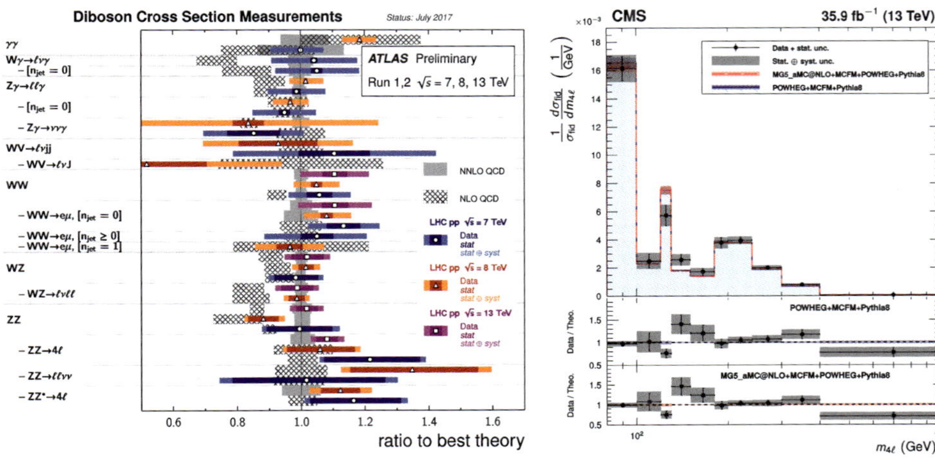

(a) (b)

Fig. 9.11 (a) A summary of di-boson cross-section measurements by the ATLAS collaboration [ATLAS Collaboration (2017a)]. Measurements are compared to SM theory predictions calculated at NLO (hatched) and NNLO QCD (solid, where available). (b) The four-lepton invariant mass spectrum, $m_{4\ell}$, compared to the SM prediction including the processes $Z \to 4\ell$, $H \to 4\ell$ and $ZZ^{(*)} \to 4\ell$ [Sirunyan *et al.* (2017d)].

or Z-bosons. The precision has reached a remarkable level of about 5% for several of the combinations. An interesting feature, visible even on this coarse overview, are the large differences between the theory calculations at NLO and NNLO QCD, where the NNLO predictions are in much better agreement with the data. Such features were noticed already in the early analyses of the $W\gamma$ [Aad *et al.* (2013)], and later explained by large NNLO corrections [Grazzini *et al.* (2015a)]. A similar feature is also noticeable in the WZ final state [Grazzini *et al.* (2017)], which shares with the $W\gamma$ process the feature of a "radiation-zero" suppressed LO prediction with large higher order QCD corrections.

For neutral combinations such as $\gamma\gamma$, WW and ZZ, the higher order QCD corrections have also been found to be large, mainly because of the gluon-initiated contribution $gg \to VV$ through a box diagram, which appear first as part of the NNLO (α_S^2) calculation and receive large α_S^3 corrections [Caola *et al.* (2015a)]. The final state with four charged leptons, 4ℓ ($\ell = e, \mu$), is particularly instructive, as the excellent lepton identification capabilities of the LHC detectors allow a full reconstruction without significant backgrounds in a wide range of four-lepton invariant masses from about $m_{4\ell} = 80$–$1000\,\text{GeV}$ [Aad *et al.* (2016c); Sirunyan *et al.* (2017d)], although the rate is low because of the small branching of the Z boson to two charged leptons. As illustrated in Fig. 9.11(b), this distribution contains a host of interesting features: a resonance at the mass of the Z boson from the radiative decay $Z \to 4\ell$, a resonance at the mass of the Higgs boson from its golden decay $H \to ZZ^* \to 4\ell$, a threshold at twice the Z-boson mass for the production of two on-shell Z bosons, and finally a negative interference of gluon-induced ZZ-production diagrams with and without the Higgs boson, important especially above $m_{4\ell} > 2m_{\rm top}$, which is also sensitive to the width of the Higgs boson [Caola and Melnikov (2013)].

In contrast to the measurements of di-boson final states, which is entering the precision phase, the searches for tri-boson processes will require more luminosity, with the exception of the $W\gamma\gamma$ and $Z\gamma\gamma$ final states, for which evidence and observations have been reported, respectively [Aad *et al.* (2015b, 2016e); Sirunyan *et al.* (2017c)].

The electroweak production of one or two electroweak bosons in the VBF and VBS processes, respectively, tests the self-interactions of three or four electroweak bosons. Experimentally these are challenging to measure due to small cross sections and large backgrounds from processes mediated by the strong interaction, which are typically separated from the signal through "tagging" the signature by a rapidity gap and a high-mass dijet system [Bjorken (1993)] as illustrated for the case of electroweak Zjj production in Fig. 9.12(a). An overview of the status of the field is shown in Fig. 9.12(b). At this point in time, the electroweak production of W and Z bosons has been observed with high significance and studied in detail [Khachatryan *et al.* (2015b, 2016a); Aad *et al.* (2014a); Aaboud *et al.* (2017g,d)]. For VBS processes, most analyses have not yet reached the observation stage. A significant exception is the VBS production of same-sign $W^{\pm}W^{\pm}jj$, the golden VBS process due to low strong-interaction backgrounds, where recently an observation could be reported [Sirunyan *et al.* (2017e)].

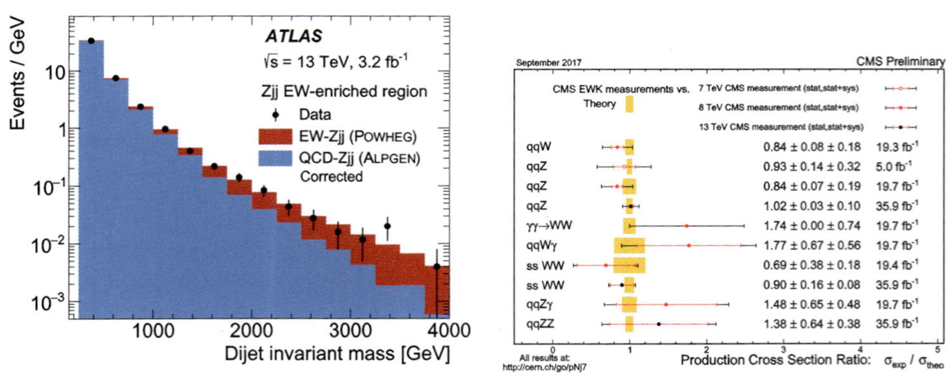

(a) (b)

Fig. 9.12 (a) The dijet invariant mass spectrum in Zjj events, where little hadronic activity is required in the rapidity region between the two tag-jets. The electroweak VBF process is clearly separated from the strong-interaction background at large m_{jj}. [Aaboud *et al.* (2017d)] (b) A summary for the electroweak production of single or di-bosons by the CMS collaboration [CMS Collaboration (2017a)].

9.8 Status of the studies of the Higgs boson

The Higgs boson was the last part of the Standard Model to be discovered. Until then, the existence of the Higgs sector was seen as "a mere conjecture" by many, including Guido Altarelli [Altarelli (2011b)]. Thanks to the well-working LHC and its experiments, the discovery of "a" Higgs boson with a mass near 125 GeV could be announced already in July 2012, delivering on the promise that "the LHC [was] designed to solve the Higgs problem" [Altarelli (2011b)]. Since then, the properties have been studied further using the data collected in Run 1 and now anew using the Run 2 data. Already the Run 1 data allowed a precise measurement of the mass to $m_H = 125.09 \pm 0.24\,\text{GeV}$ [Aad *et al.* (2015a)], fixing the last free parameter of the SM. Also the rates for the main production and decay rates could be confirmed to be as predicted by the SM within a typical experimental precision of 20–50%. The overall agreement of Higgs production and decay with the SM expectation may be summarised with the signal strength of $\mu = 1.09 \pm 0.11$ [Aad *et al.* (2016d)], well compatible with the SM expectation of 1. The quantum numbers were confirmed to be $J^{CP} = 0^{++}$ [Khachatryan *et al.* (2015a); Aad *et al.* (2015c)] with high probability. As shown in Fig. 9.13(a), the Run 1 data established a reasonable agreement with the expectation, that the masses of fermions and bosons are indeed given by Yukawa couplings proportional to the particle masses, although also here the experimental precision was typically just a few 10%.

As commented by Guido Altarelli, the good agreement of the measurements with the minimal version of the Standard Model and the Englert–Brout–Higgs mechanism and at the same time an absence of other evidence for physics beyond the

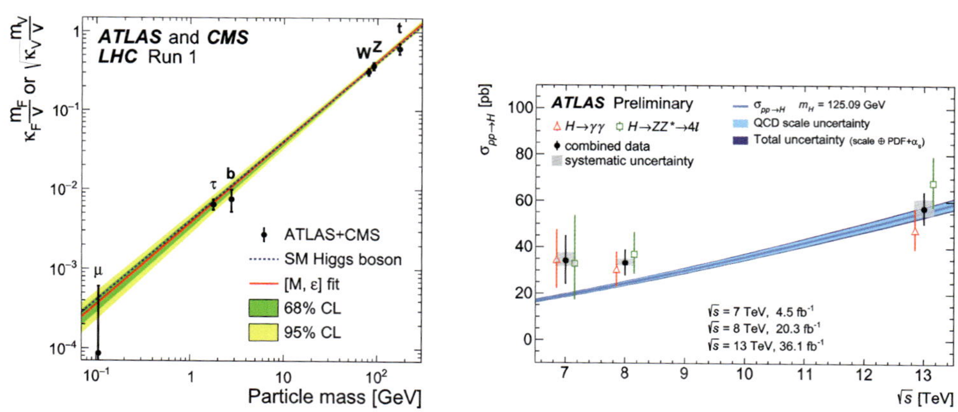

(a) (b)

Fig. 9.13 (a) Measurements of the coupling of the Higgs boson to fermions and vector bosons as a function of the particle mass compared to the SM expectation. [Aad *et al.* (2016d)] (b) Combination of Higgs-boson production cross sections measured in the $H \to 4\ell$ and $H \to \gamma\gamma$ decay channels at centre-of-mass energies of $\sqrt{s} = 7, 8, 13\,\text{TeV}$ compared to the SM prediction [ATLAS Collaboration (2017b)].

Standard Model, poses new theoretical challenges, as the Higgs boson is "simple yet so unnatural" [Altarelli (2013a)]. One of the paths to elucidate the mechanism of electroweak symmetry breaking further are precision measurements of the properties of the Higgs boson. With the availability of the Run 2 data, Higgs measurements have entered now a new phase. Exploiting the rarer but well reconstructed decays $H \to ZZ^* \to 4\ell$ and $H \to \gamma\gamma$ of just one experiment, the total production cross section is already now measured at a precision of about 10% as shown in Fig. 9.13(b), which is similar to the full Run 1 combination of all ATLAS and CMS data.

In addition to more data, the precise measurements of Higgs-boson properties depend critically on a further improved understanding of the QCD aspects, both for the signal and background processes. A recent highlight was the completion of the N^{3}LO calculation for the dominant gluon-fusion process [Anastasiou *et al.* (2015)]. While this significantly reduces the theoretical uncertainty in this particular calculation, uncertainties from the gluon PDF, the strong coupling α_S, as well as the missing N^{3}LO corrections to the PDF evolution and DIS, DY and other relevant cross sections will be a challenge for the next years.

9.9 Conclusion

Since the year 2010 the LHC and the large experimental collaborations have successfully delivered a detailed picture of the fundamental constituents of matter and interactions at the electroweak scale and beyond into the TeV range. As promised, the LHC solved the "Higgs problem": the last missing piece of the Standard Model was discovered and the work is ongoing to measure the properties of the newly discovered particle. The Standard Model as we know it today is often discussed as something that is "unnatural", just a "low-energy effective theory" and that "just describes 5% of the energy content of the universe", yet it is "excessively" successful. Guido Altarelli was hopeful that there could be a fundamental break through out of this "puzzling situation". Meanwhile we should also celebrate the achievements of the SM which was built in decades of systematic work of theorists and experimentalists and which is able to describe the diverse LHC data.

The work ahead for the LHC community is certainly a challenge, but also full of exciting possibilities. The abundant and rich data provided by the LHC collisions can and will be analysed in diverse ways to deepen our understanding of the strong and electroweak interactions. In many measurements and searches we enter the "precision regime" as we accumulate large data sets and continue to improve the understanding of our detectors. The methodology to understand the proton structure and calculate predictions in QCD have advanced to reach the percent level, which is truly remarkable given the inherent complexity.

Guido Altarelli was often skeptical of what concerned the theoretical accuracy of QCD predictions which we experimentalists rely on so critically and which we love to challenge. It would be interesting to learn Guido's reaction to achievements such as the first measurement of the W-boson mass at the LHC, which indeed requires a control of many QCD effects at unprecedented level and skepticism is in order. It is vital that we stay alert to the limitations of the theory.

Acknowledgements

I would like to thank Stefano Forte, Aharon Levy and Giovanni Ridolfi for taking the initiative for this book in memory and honour of Guido Altarelli and for inviting me to contribute.

Chapter 10

A journey through small-x resummation

Fabrizio Caola
Institute for Particle Physics Phenomenology
Durham University, Durham DH1 3LE, UK

10.1 Introduction

The study of QCD in the high-energy regime

$$\Lambda_{\mathrm{QCD}}^2 \ll Q^2 \ll s, \quad \text{or} \quad x \equiv \frac{Q^2}{s} \ll 1, \tag{10.1}$$

where Q is the typical scale of the process under consideration and s is the centre-of-mass energy, is a complex and fascinating one. On the one hand, this region is perturbative since $Q^2 \gg \Lambda_{\mathrm{QCD}}^2$. On the other hand, subtle all-order effects must be taken into account for its proper description. Analyses of this region shed more light on many interesting aspects of QCD, from more formal ones — like for instance a quantum field theory approach to Regge theory (see e.g. [Gribov *et al.* (1983); Lipatov (1997)]) — to more phenomenological ones — like for example the onset of parton saturation in proton and heavy–ion collisions (see e.g. [Mueller (2001)]) or the interplay between the Dokshitzer–Gribov–Lipatov–Altarelli–Parisi (DGLAP) [Altarelli and Parisi (1977); Gribov and Lipatov (1972a); Dokshitzer (1977)] and Balitsky–Fadin–Kuraev–Lipatov (BFKL) [Lipatov (1976); Fadin *et al.* (1975); Kuraev *et al.* (1977); Balitsky and Lipatov (1978)] evolution and its implications for the physics of parton distribution functions (PDFs), see e.g. [Forte *et al.* (2009)].

As in many other cases, Guido's contributions to this field have been substantial. Over a period spanning nearly a decade, together with S. Forte and R. D. Ball he clarified the role of BFKL dynamics in high-energy QCD evolution and studied in depth its phenomenological implications. The body of work on perturbative QCD evolution at small x by Altarelli, Ball and Forte (ABF) [Ball and Forte (1999); Altarelli *et al.* (2000a, 2001, 2002, 2003, 2006, 2008)] now belongs to the "classics" on which a new generation of QCD theorists, including myself, were trained. Apart from pushing forward our knowledge of theoretical QCD, the results obtained by ABF, alongside similar results from another group [Salam (1998); Ciafaloni and Colferai (1999); Ciafaloni *et al.* (1999, 2000, 2002, 2003a,b, 2004)], also led to a better understanding of many peculiar features of small-x HERA data. For example, a naive high-energy QCD analysis seems to suggest that the gluon distribution function G should grow fast, $G \approx x^{-0.5}$ at small x. This is the so called *hard pomeron* behavior. Such a steep rise was not seen in HERA data. Guido's work showed that the analysis leading to the hard pomeron prediction misses crucial physical effects like for example momentum conservation constraints or running coupling corrections. Once all of this is properly taken into account, the gluon small-x growth is softer, $G \approx x^{-0.2}$, fully compatible with HERA data.

Guido's contributions laid the ground work for a proper inclusion of small-x effects into PDF fits. Such an endeavour can be quite important to properly describe small-x HERA data. Indeed, there seems to be a tension between the HERA legacy data set [Abramowicz *et al.* (2015)] in the small-x small-Q^2 region and standard fixed order DGLAP evolution. The availability of small-x resummed parton distributions would allow for a deeper scrutiny of this region, and could actually establish the onset of BFKL dynamics in *ep* HERA data. Although fully resummed parton distributions are still not available, a first important step in this direction has been made recently in ref. [Ball *et al.* (2017a)], where the authors performed PDFs fits taking partially into account small-x effects, computed within the ABF framework. Although the reader is referred to the original work [Ball *et al.* (2017a)] for more details, in Fig. 10.1 we report results for the longitudinal DIS structure function F_L, which is sensitive to small-x physics.

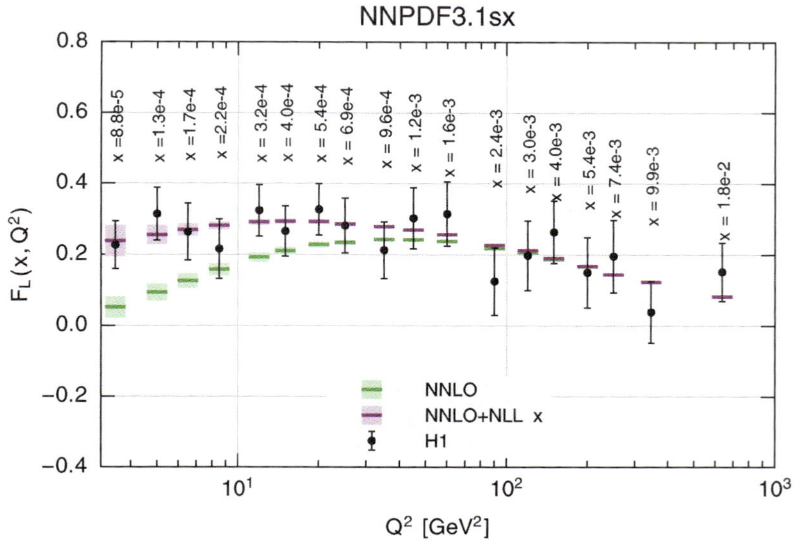

Fig. 10.1 The longitudinal DIS structure function F_L at small x measured by H1 (black dots), compared against theoretical predictions without (green, "NNLO") and with (purple, "NNLO+LLx") small-x resummation effects. Figure taken from ref. [Ball *et al.* (2017a)]. See Ref. [Ball *et al.* (2017a)] and text for details.

This figure shows that the inclusion of resummation effects seems to improve the data/theory comparison. In [Ball *et al.* (2017a)], this qualitative statement is quantitatively substantiated by detailed studies on the quality of the PDFs fits with and without small-x resummation. In general, the tension between HERA data and theoretical predictions in the small-x small-Q^2 region is alleviated if all-order small-x effects are taken into account. In the resummed case, it is also possible to include "dangerous" low-x low-Q^2 data — which lead to unstable (fixed order) fits and hence are typically discarded — in the fit, without significantly decreasing its quality. Studies such as ref. [Ball *et al.* (2017a)] may then not only give us information on perturbative small-x effects, but also shed some interesting light on the onset of genuinely non-perturbative phenomena at low scales.

Moving from HERA to the LHC, the role of high-energy resummation changes. In a perhaps funny twist of fate, the growth of the gluon distribution at small x — at least partially driven by high-energy QCD effects — is also the reason why small-x resummation does not play a central role at the LHC.[1] The argument is simple: the hadronic cross section σ is obtained from its partonic counterpart $\hat{\sigma}$ through a

[1]We note however that there are LHC analyses which are sensitive to the very low value of $x \sim 10^{-5}$–10^{-6} and thus would require a proper resummation. Examples are LHCb measurements of low mass forward Drell–Yan production [LHCb Collaboration (2012)], and of D-meson production at small p_T [Gauld *et al.* (2015)].

convolution integral with parton distribution functions. Schematically,

$$\sigma(\tau) = \int_\tau^1 \frac{dx}{x} \mathcal{L}_{ij}\left(\frac{\tau}{x}\right) \hat{\sigma}_{ij}(x), \tag{10.2}$$

where $\tau = Q^2/s$ and $\mathcal{L}_{ij}$ are the parton luminosities. If the latter are peaked at small x, then the integral is dominated by the large-x region and $\hat{\sigma}_{ij}$ is never probed at very high energy. This argument can be made sharper if one introduces a *Mellin transform* to decouple the convolution integral Eq. (10.2):

$$f(N) = \int_0^1 dx\, x^{N-1} f(x), \quad f(x) = \int_{c-i\infty}^{c+i\infty} \frac{dN}{2\pi i}\, x^{-N} f(N). \tag{10.3}$$

The Mellin transform maps the $x \to 0$ region to the $N \to 0$ one, and it maps convolution products to ordinary ones:

$$\sigma(N) = \mathcal{L}_{ij}(N)\, \hat{\sigma}_{ij}(N). \tag{10.4}$$

Since parton luminosities are steeply falling functions, a reliable estimate for the hadronic cross section can be obtained from results in N space by evaluating the Mellin inversion integral using the saddle point technique: to the first approximation the physical hadronic cross section is determined by the product $\hat{\sigma}_{ij}(N){\cdot}\mathcal{L}_{ij}(N)$ evaluated at a single point $N = N^*$, the saddle of the Mellin inversion integral [Bonvini *et al.* (2012)]. Because of the rapid fall of parton luminosities, in many phenomenologically interesting cases, N^* turns out to be far from the high-energy $N \sim 0$ region. For example, for Higgs production at the LHC, $N^* \sim 1$ [Bonvini *et al.* (2012)].[2] The dominant high-energy contributions to the partonic cross section are poles in $N = 0$, $\hat{\sigma} \sim \alpha_s^k/N^k$. For $N \sim 1$, they are captured by fixed order perturbation theory and don't require resummation.

This does not imply however that a good understanding of small-x physics is irrelevant for the LHC. Indeed, in the recent past it became clearer and clearer that the study of QCD in the high-energy regime can provide useful information on the general structure of higher order perturbative corrections and lead to insight that is quite hard to obtain otherwise. An interesting example is provided by Higgs investigations. As is well known, the Higgs main production mechanism at the LHC — gluon fusion — receives very large $\mathcal{O}(1)$ QCD radiative corrections that must be properly taken into account for any reliable investigation. In the Standard Model, higher order calculations require prohibitively complicated multi-loop amplitudes involving the propagation of virtual heavy quarks, which makes any computation extremely challenging. To overcome this issue, typically one works in

[2] Note that in ref. [Bonvini *et al.* (2012)] the Mellin transform is defined such that the $x \to 0$ region is mapped into the $N \to 1$ one.

the limit of very large top mass, where the Higgs couples directly to gluons through an effective vertex and the aforementioned problem disappears. In many relevant cases this approximation — referred to as Higgs Effective Theory (HEFT) — is fully justified. Still, there are cases like boosted Higgs where the HEFT approximation fails dramatically. Furthermore, even when it holds the accuracy goals for Higgs predictions make it desirable to obtain at least some control on heavy quark effects. The small-x region is very sensitive to them. Indeed, in the regime $s \gg Q^2 \sim m_H^2$ the Higgs process is probed by highly energetic gluons, that always resolve the heavy quark loops. High-energy predictions can then provide important insight on the structure of higher order heavy quark effects in the Higgs cross section and differential distributions.

This example illustrates how high-energy resummation, first developed almost 50 years ago to understand multi-Regge processes in QCD and subsequently applied for example to describe the QCD evolution of PDFs and to study scaling violations at HERA, is still today an interesting part of the LHC phenomenology toolbox. The aim of this contribution is to give the reader a taste of this fascinating subject, of its foundations and of some of its most recent applications. Some comments are in order. First, the topic is vast and has deep ramifications in many disparate areas of high-energy theory and phenomenology. This contribution is not intended to be an exhaustive review of the subject, but rather a quick journey focusing on some very peculiar aspects, with an eye on LHC applications. As such, many interesting developments, like for example on the high-energy structure of the $\mathcal{N} = 4$ sYM theory, or on the physics of parton saturation, won't be covered here. Similarly, references are not meant to be comprehensive, but only to provide extra resources for the reader to learn more on the topics presented here. Second, I make no claim of originality: the literature on small-x physics is copious, and most of the discussion presented here is inspired by classic works on the subject. Still, I hope that on the one hand the expert reader will find at least some of the standard material presented under a different light, and on the other hand that the non-initiate's curiosity will be piqued enough for him/her to try and dig deeper into this fascinating subject.

The remainder of this contribution is organised as follows. In Section 10.2 we describe the origin of leading high-energy effects, and study them in the simplest possible process, namely multi-gluon scattering. In Section 10.3 we apply the intuition gained in Section 10.2 to establish a high-energy factorisation theorem, and in Section 10.4 we show how this factorisation theorem allows to resum all the leading high-energy effects, provided that proper small-x evolution kernels are known. In Section 10.5, we discuss some of the problems that one encounters in the determination of these kernels, and sketch their solution. In Section 10.6, we illustrate some recent LHC application of the small-x technology developed in the previous Sections by studying heavy quark effects in Higgs physics. Finally, we conclude in Section 10.7.

10.2 Multi-Regge kinematics, multiple gluon emission and t-channel ladders

Whenever in quantum field theory there is a hierarchy of scales, on very general grounds we expect radiative corrections to generate large logarithms of their ratios. The high-energy regime, characterised by

$$s \gg Q^2 \gg \Lambda_{\mathrm{QCD}}^2, \tag{10.5}$$

is no exception: large logs $\sim \alpha_s^k \ln^m x$ with $x = Q^2/s$ appear in the cross section and differential distributions, and for $x \ll 1$ they spoil the convergence of the perturbative series and must be resummed. Naively, one may expect double logarithmic corrections $\sigma \sim \alpha_s^k \ln^{2k-1} x$ to appear. However, for inclusive enough quantities there are large cancellations between real and virtual contributions, and as a consequence only single logarithmic terms $\sigma \sim \alpha_s^k \ln^k x$ survive, at least in non pathological theories [Lipatov (1997, 1976); Fadin *et al.* (1975); Kuraev *et al.* (1977); Balitsky and Lipatov (1978); Ciafaloni (1988); Catani *et al.* (1990b,c, 1991d)].[3]

To study the origin of these large logarithmic corrections, we note that they are generated by both phase space and loop integrations. However, since it is possible to write loop integrals as phase space ones (and the other way around), here we focus on real emission only. We consider the generic hadro-production process $f(p)+f(n) \to V$, where V is a generic hard final state with invariant mass m_V (e.g. $V = H, Z, t\bar{t}\ldots$) and $f(k)$ is a generic (massless) parton with momentum k. We focus on the related real emission process $f(p) + f(n) \to V + f(k)$. To single out the leading logarithmic (LL) region of the phase space, it is convenient to introduce a Sudakov parametrisation of momenta. Working in the partonic centre-of-mass frame, one writes $p = \sqrt{s}/2\,(1,0,0,1)$ and $n = \sqrt{s}/2\,(1,0,0,-1)$ and parametrises k as

$$k = \alpha p + \beta n + k_\perp, \qquad s\alpha\beta = {k_\perp}^2. \tag{10.6}$$

In this equation $\alpha, \beta \in (0,1)$, $k_\perp$ is a two-dimensional vector spanning the space transverse to p and n, $k_\perp = (0, k_\perp, 0)$, and the second relation is the on-shell condition $k^2 = 0$. Choosing α and $k_\perp$ as independent variables, the phase space volume for k can be written as

$$\frac{d^3k}{2E(2\pi)^3} = \frac{s}{16\pi^3}\frac{d\alpha}{\alpha} d^2k_\perp. \tag{10.7}$$

The α integration gives rise to logarithmically enhanced terms in the $\alpha \sim 1$ and $\alpha \sim 0$ regions. The $\alpha \to 0$ region corresponds to soft emission, and is relevant for example for threshold resummation. The $\alpha \to 1$ region on the other hand

---

[3] A famous counterexample is Higgs production in the HEFT approximation. Here the appearance of double logs is a manifestation of the breakdown of the effective field theory approach at high enough energy [Hautmann (2002)].

corresponds to the small-x regime. Indeed, momentum conservation implies

$$m_V^2 = (1-\alpha)(1-\beta)s - {k_\perp}^2. \tag{10.8}$$

Since typically $x = m_V^2/s$ and $k_\perp/s \ll 1$ at large s, from Eq. (10.8) it follows that $x \to 0$ in the $\alpha \to 1$ limit. This simple analysis shows that the small-x regime is then dominated by the emission of highly energetic partons, that carry away most of the available energy of the process.

Generalisation to the multi-particle emission process $f(p)+f(n) \to V + f(k_1) + f(k_2) + \cdots + f(k_n)$ is simple. In this case, it is convenient to write the Sudakov parametrisation as

$$\begin{aligned}
k_1 &= (1-z_1)p + y_1 n + k_{\perp 1},\\
k_2 &= (z_1 - z_2)p + (y_2 - y_1)n + k_{\perp 2} - k_{\perp 1},\\
k_3 &= (z_2 - z_1)p + (y_3 - y_2)n + k_{\perp 3} - k_{\perp 2},\\
&\cdots\\
k_n &= (z_{n-1} - z_n)p + (y_n - y_{n-1})n + k_{\perp n} - k_{\perp n-1},\\
p_V &= z_n p + (1-y_n)q - k_{\perp n},
\end{aligned} \tag{10.9}$$

where we used momentum conservation to determine p_V. The n-particle measure becomes

$$\begin{aligned}
d\phi_n = \prod_{i=1}^{n} \frac{d^3k_i}{(2\pi)^3 2E_i} &= \left(\frac{s}{16\pi^3}\right)^n \\
&\times \prod_{i=1}^{n} dz_i dy_i \delta[s(z_{i-1}-z_i)(y_i - y_{i-1}) - (k_{\perp i} - k_{\perp i-1})^2], \quad z_0 = y_0 = 1.
\end{aligned} \tag{10.10}$$

The same argument as for single parton emission shows that the small-x region corresponds to the $z_n \to 0$ limit. Furthermore, by inspecting the phase space Eq. (10.10) it is easy to see that the LL-x contribution comes from the strongly ordered region[4]

$$z_1 \gg z_2 \gg \cdots \gg z_n, \quad y_1 \ll y_2 \ll \cdots \ll y_n. \tag{10.11}$$

Apart from this, no other particular ordering is required. In particular, although $k_{\perp i} \ll s$, there is no hierarchy in the transverse space

$$\frac{{k_{\perp 1}}^2}{s} \approx \frac{{k_{\perp 2}}^2}{s} \approx \cdots \approx \frac{{k_{\perp n}}^2}{s} \ll 1. \tag{10.12}$$

[4]Note that, because of the on-shell constraints, z and y are not independent variables. Typically, it is convenient to choose $z(y)$ as an independent variable when considering emissions originating from the $p(n)$ leg. Since the interference between emissions from two different legs is suppressed at high energy, this choice is unambiguous.

This analysis shows that the LL-x contribution stems from the emission of fewer and fewer energetic particles: most of the (longitudinal) momentum available is carried away by k_1, then k_2 carries away most of the remaining momentum and so on so forth. The kinematic configuration just described is usually referred to as "multi-Regge kinematics".

To study the dynamical implications of this configuration, we now consider the simplest possible process $V = 0$, i.e. $2 \to n-2$ scattering in pure QCD. More precisely, we start by analysing multi-gluon scattering and work in the all-outgoing notation, $0 \to g_1^{h_1} + g_2^{h_2} + \cdots + g_n^{h_n}$. Furthermore, we choose the helicity configuration where all but two gluons' helicities are the same, say $h_k = 1$ for $k \neq i, j$ and $h_i = h_j = -1$. It is well known that the (colour ordered) amplitude in this case is given by the MHV formula

$$\mathcal{A}(1,2,\ldots,n) = \frac{\langle ij\rangle^4}{\langle 12\rangle\langle 23\rangle\langle 34\rangle\ldots\langle (n-1)n\rangle}, \tag{10.13}$$

where $\langle ij\rangle$ denotes the usual spinor product, see e.g. [Dixon (1996)] for details. To make contact with our previous discussion, we now cross two gluons in the initial state by identifying $p = -p_1$, $n = -p_n$ and $k_i = p_{i+1}$. Defining as usual $s_{ij} = 2p_i \cdot p_j$, in the multi-Regge limit we then have $s = s_{1n} \approx s_{1n-1} \approx s_{2n} \approx s_{2n-1}$ much larger than all the other invariants s_{ij}. Since $\langle ij\rangle \sim \sqrt{s_{ij}}$, the MHV formula then shows that there is a leading high-energy contribution only for the choices $\{i,j\} = \{(1,2),(n,n-1)\}$. The physical interpretation of this selection rule is clear: as discussed above, the multi-Regge region is characterised by the emission of increasingly soft gluons. But soft gluons are emitted eikonally, and cannot change the helicity of a hard line. In this process, there is a large momentum flow along the p light-cone component from p to p_2, and a large momentum flow along the n component from n to p_{n-1}. Hence the helicity must be conserved along these two lines (or, equivalently, it must be flipped in the all-outgoing notation). The leading high-energy terms exposed by the MHV formula are exactly the ones which respect this helicity conservation rule.

In the high-energy limit, these leading configurations are all identical up to an (irrelevant) overall phase. For definiteness, we consider the case $\{1^-, n^-\}$, for which the amplitude reads

$$\mathcal{A}(1^-,2^+,3^+,\ldots,n^-) = \frac{s}{\mathfrak{K}_1\mathfrak{K}_2\mathfrak{K}_3\ldots\mathfrak{K}_{n-2}}\left[1+\mathcal{O}\left(\frac{1}{\sqrt{s}}\right)\right]. \tag{10.14}$$

In Eq. (10.14), we have introduced the complex notation $\mathfrak{K}_j = k_{j,x} + ik_{j,y}$. By introducing effective "t-channel" propagator momenta $q_i = \sum_{l=0}^{i} k_l$, $\mathfrak{q} = q_x + iq_y$, Eq. (10.14) can be recast in the following form

$$\mathcal{A}(1^-,2^+,\ldots,n^-) = s\frac{1}{|\mathfrak{q}_1|^2}\left(\frac{\mathfrak{q}_1^*\mathfrak{q}_2}{\mathfrak{K}_2}\right)\frac{1}{|\mathfrak{q}_2|^2}\left(\frac{\mathfrak{q}_2^*\mathfrak{q}_3}{\mathfrak{K}_3}\right)\frac{1}{|\mathfrak{q}_3|^2}\times\cdots\times\frac{1}{|\mathfrak{q}_{n-3}|^2}\frac{\mathfrak{K}_{n-2}^*}{\mathfrak{K}_{n-2}}. \tag{10.15}$$

This is particularly suggestive since it seems to indicate that effectively in the high-energy limit the amplitude can be thought of as iterated t-channel emissions with an effective coupling $\mathfrak{q}_i^* \mathfrak{q}_{i+1}/\mathfrak{K}_{i+1}$. Note that this is *not* equivalent to the QCD three-gluon vertex (at least in a generic gauge). It is however simple to see that it is equivalent to the following non-local *Lipatov vertex*

$$V^\mu(q_i, q_{i+1}; k_l) = (q_i + q_{i+1})_\perp^\mu + \left(\frac{s_{1l}}{s} + 2\frac{q_{i+1}^2}{s_{nl}}\right) p_n^\mu + \left(\frac{s_{nl}}{s} + 2\frac{q_i^2}{s_{1l}}\right) p_1^\mu. \quad (10.16)$$

Although here we derived this result only in the MHV case, it is possible to show [Lipatov (1997); Bartels (1979)] that the form Eq. (10.15) is generic and holds for the emission of gluons of arbitrary helicity along the ladder. The different helicity configurations are taken into account by complex conjugation of the relevant effective vertex $\mathfrak{q}_i^* \mathfrak{q}_{i+1}/\mathfrak{K}_{i+1}$, or equivalently by contracting the Lipatov vertex Eq. (10.16) with the appropriate polarisation vector. Hence, in the small-x limit a *generic* n-gluon amplitude can be effectively thought of as multiple t-channel emissions, with interaction described by the vertex Eq. (10.16) — see Fig. 10.2. We stress that this result is highly non-trivial, and it stems from the sum of many different Feynman diagrams. In particular, it originates from the interplay of iterated emissions both in the t *and* in the s and u channels. Even less trivial, similar simplifications also occur at the loop level (see e.g. refs. [Lipatov (1997); Bartels (1979)]). This leads to an important property of high-energy QCD: in a physical gauge, amplitudes factorise in the t-channel. This feature is at the basis of the BFKL theory and of the k_T factorisation theorem, that we will illustrate in the next section.

Before doing so, we now consider, however, amplitudes involving $q\bar{q}$ pairs. In this case, it is easy to understand that a leading high-energy contribution is only present if either the $q\bar{q}$ pair correspond to p_1 and p_2, or if it corresponds to p_n and p_{n-1}. In other words, either $f(p)\,(f(n))$ is a quark, and then the first(last) rung of the ladder is a quark, or the amplitude is suppressed in the high-energy regime. The reason is simple: along the $p(n)$ light-cone component, $k_1(k_{n-2})$ is hard and all the other k_i are soft. But it is well known that in QCD leading soft singularities only come

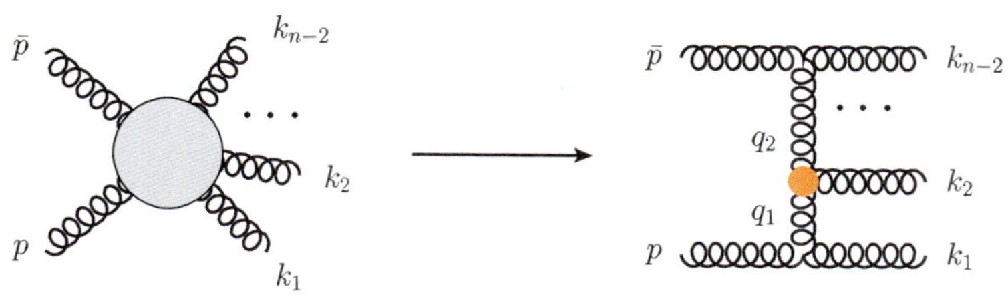

Fig. 10.2 The structure of a $gg \to (n-2)g$ amplitude in the high-energy limit. On the right, the (orange) blob represents the non-local Lipatov vertex Eq. (10.16). See text for details.

from gluon emission. Hence, we can only have a leading high-energy contribution if the hard leg is a quark emitting another hard quark and a soft gluon, that proceeds into the ladder. If on the other hand the soft particle is a quark, there can be no leading small-x singularity. This simple argument also suggests that in the small-x limit quark- and gluon-channels should behave very similarly, as they only differ by the first rung of a ladder. Indeed, it turns out that their contribution to the cross section is identical up to a trivial overall colour factor:

$$\sigma_{qg}^{\text{small}-x} = \frac{C_F}{C_A}\,\sigma_{gg}^{\text{small}-x}, \qquad \sigma_{qq'}^{\text{small}-x} = \left(\frac{C_F}{C_A}\right)^2 \sigma_{gg}^{\text{small}-x}, \tag{10.17}$$

where q and q' stand for both identical and/or different quarks/anti-quarks.

10.3 k_T factorisation, off-shell cross section and unintegrated gluon distribution function

In the previous section, we illustrated how the amplitude for multi-gluon emission in the multi-Regge kinematics can be recast in a t-channel ladder structure. As we mentioned, this feature generalises and allows in the LL-x approximation to write (differential) cross sections for generic processes in a factorised form

$$d\sigma_{\text{small}-x} \sim H \otimes L, \tag{10.18}$$

where H represents a process-dependent "hard part" and L a universal ladder part describing all-order gluon emission [Catani *et al.* (1990a, 1991b,a, 1993)].

We now illustrate in details how this factorisation works for the photo-production process[5]

$$\gamma^*(n) + g(p) \to \mathcal{V} + X. \tag{10.19}$$

The key point is that also in this case the leading high-energy behaviour comes from t-channel ladder-type contributions. More precisely, in a physical gauge $n \cdot A = 0$ only cut diagrams that are at least 2-gluon reducible in the t-channel give rise to small-x enhanced terms [Lipatov (1976); Fadin *et al.* (1975); Kuraev *et al.* (1977); Balitsky and Lipatov (1978); Ciafaloni (1988); Catani *et al.* (1990b,c, 1991d)], see Fig. 10.3. This allows the dimensionless cross section for this process to be written in the factorised form[6]

$$\sigma = \int \frac{Q^2}{2s} H^{\mu\nu}(n, p_L, p_{\mathcal{F}}, \mu_R\mu_F, \alpha_s)\, L_{\mu\nu}(p_L, p, \mu_R, \mu_F, \alpha_s)[dp_L], \tag{10.20}$$

[5]Generalisation to hadroproduction is straightforward once the photo-production case is understood, see e.g. [Catani *et al.* (1991a)].

[6]In this section, we only deal with partonic cross sections. To simplify the notation, we then write $\hat{\sigma}_{ij} \to \sigma$.

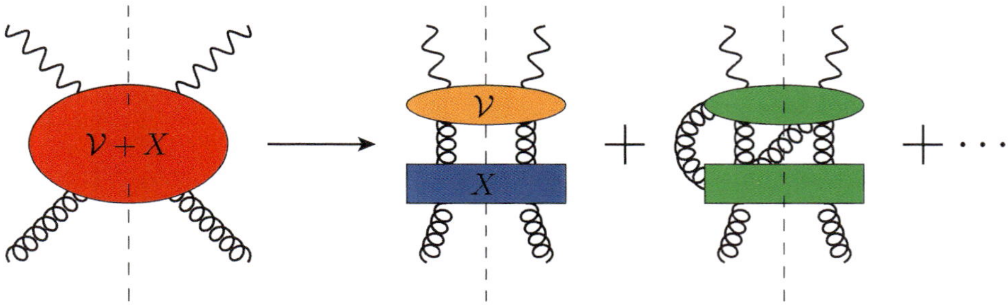

Fig. 10.3 Schematic depiction of the cross section for the process $\gamma + g \to \mathcal{V} + X$. The sum of cut diagrams (left-hand side) can be split into a two-gluon reducible part where the process-dependent part $\mathcal{V}$ and the multi-particle emission X are factorised (right-hand side, first term) and a two-gluon irreducible part where t- and s-channel gluon exchanges entangle $\mathcal{V}$ and X. Only the two-gluon reducible part contributes in the high-energy limit.

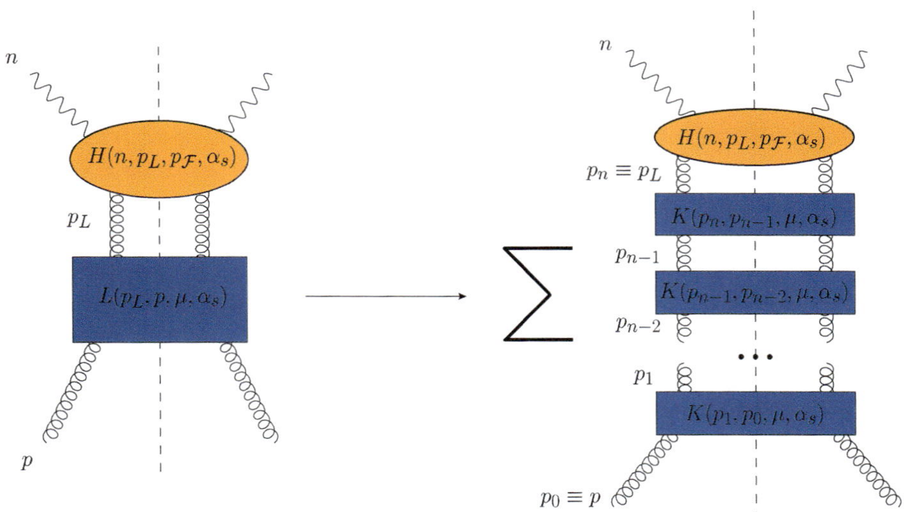

Fig. 10.4 Left: factorisation of the cross section in the high-energy limit in terms of process-dependent hard function H and generalised ladder L. Right: further expansion of the ladder part in 2PI kernels. See text for details.

where Q^2 is a typical scale of the process, H is a process-dependent "hard" part which is two-gluon irreducible (2GI), represented by the orange blob in Fig. 10.3, and L is a universal "ladder" part that describes multiple QCD emission along the ladder, represented by the blue rectangle in Fig. 10.3, see refs. [Catani *et al.* (1990a, 1991b,a)]. In Eq. (10.20) $p_{\mathcal{F}}$ is a collective set of momenta describing the final state $\mathcal{V}$, p_L is the ladder momentum connecting the hard and ladder part, and $[dp_L]$ is the (cut) loop integral measure, see Fig. 10.4-left. In the Sudakov decomposition described in the previous section, $p_L = zp - k_\perp + \frac{{k_\perp}^2}{s(1-z)}n$ and then

$$[dp_L] = \frac{dz}{2(1-z)} d^2k_\perp, \tag{10.21}$$

where here z corresponds to what was called z_n in Section 10.2. Note that in the small-x region

$$z \ll 1, \qquad \frac{{k_\perp}^2}{s} \ll 1, \tag{10.22}$$

p_L is essentially transverse, $p_L^2 \approx -{k_\perp}^2 \equiv -k_T^2$.

Before describing the structure of Eq. (10.20) in more detail, we note that its physical interpretation is straightforward for single gluon emission, $X = g$. Indeed, in this case the "ladder" part is just a single gluon exchanged in the t-channel, while the "hard" part is the reduced cut diagram for the original process, $\gamma^* + g^* \to \mathcal{V}$, initiated by an off-shell gluon with momentum p_L. The loop integral measure $[dp_L]$ is the phase space measure for the emitted gluon. We will see that a simple generalisation of this picture persists to all orders. We first need however to simplify Eq. (10.20) a little bit further.

In general, after real and virtual contributions are combined, the unrenormalised partonic cross section σ contains both UV and initial-state collinear divergences. The former leads to the running of the coupling α_s and to a dependence on the renormalisation scale μ_R in the cross section. These effects however are next-to-leading-log (NLL) at small x, so we neglect them here. Initial state collinear singularities on the other hand cannot be neglected. To further simplify our derivation, we now assume that the hard part is not only 2GI but also two-particle irreducible (2PI), which ensures that it is free of collinear singularities and hence it does not depend on the factorisation scale μ_F.[7] Under these assumptions, the hard function then only depends on the incoming momenta p_L and n, on the $\mathcal{V}$ momenta $p_{\mathcal{F}}$ and on α_s, while the ladder part only depends on z, the incoming momenta p and p_L, the factorisation scale $\mu_F \equiv \mu$ and α_s.

To proceed further and simplify the Lorentz structure, one decomposes $H^{\mu\nu}$ and $L^{\mu\nu}$ in terms of dimensionless scalar form factors. In a physical gauge, one can write

$$\begin{aligned} H^{\mu\nu} &= \left(-g^{\mu\nu} + \frac{p_L^\mu p_L^\nu}{p_L^2}\right) H_\perp + p_L^2 \left(\frac{p_L^\mu}{p_L^2} - \frac{n^\mu}{n\cdot p_L}\right)\left(\frac{p_L^\nu}{p_L^2} - \frac{n^\nu}{n\cdot p_L}\right) H_{||}, \\ L^{\mu\nu} &= \frac{1}{p_L^2}\left(-g^{\mu\nu} + \frac{p_L^\mu p_L^\nu}{p_L^2}\right) L_\perp + \left(\frac{p_L^\mu}{p_L^2} - \frac{n^\mu}{n\cdot p_L}\right)\left(\frac{p_L^\nu}{p_L^2} - \frac{n^\nu}{n\cdot p_L}\right) L_{||}. \end{aligned} \tag{10.23}$$

Note that we have explicitly extracted the ladder propagator factor p_L^2 from the definition of $L_{\perp,||}$. The scalar form factors $H_{\perp,||}$ and $L_{\perp,||}$ are then dimensionless

[7] Generalisation to the two-particle reducible case is tedious but simple, see e.g. [Marzani and Ball (2009)] for a thorough discussion.

by construction. In the small-x kinematics Eq. (10.22), one can write

$$H_{\perp,||} = H_{\perp,||}\left(\frac{Q^2}{n\cdot p_L}, \frac{-p_L^2}{Q^2}, \Omega_{\mathcal{F}}, \alpha_s\right) = H_{\perp,||}\left(\frac{Q^2}{zs}, \frac{k_T^2}{Q^2}, \Omega_{\mathcal{F}}, \alpha_s\right)(1+\mathcal{O}(z)),$$
$$L_{\perp,||} = L_{\perp,||}\left(z, \frac{-p_L^2}{p\cdot p_L}, \frac{\mu^2}{-p_L^2}, \alpha_s\right) = L_{\perp,||}\left(z, \frac{\mu^2}{k_T^2}, \alpha_s\right), \qquad (10.24)$$

where $\Omega_{\mathcal{F}}$ denotes a set of dimensionless variables that characterise the final state $\mathcal{F}$ and in the second equation we used $p_L^2/(p\cdot p_L) = 1/2$ constant.

By performing the Lorentz algebra, using the fact that the longitudinal and transverse form factors have the same small-x behaviour [Catani *et al.* (1991a)], and systematically neglecting all terms that are sub-leading at high energy, one obtains for the cross section in the high-energy limit

$$\sigma_{\text{small}-x} = \int \frac{Q^2}{2s}\left[-\frac{2}{z^2}H_{||}\left(\frac{Q^2}{zs}, \frac{k_T^2}{Q^2}, \Omega_{\mathcal{F}}, \alpha_s\right) L_{||}\left(z, \frac{\mu^2}{k_T^2}, \alpha_s\right)\right]\frac{d^2k_\perp}{k_T^2}dz$$
$$= \int \left[-\frac{Q^2}{2sz}H_{||}\left(\frac{Q^2}{zs}, \frac{k_T^2}{Q^2}, \Omega_{\mathcal{F}}, \alpha_s\right)\right]\left[2\pi L_{||}\left(z, \frac{\mu^2}{k_T^2}, \alpha_s\right)\right]\frac{dz}{z}\frac{dk_T^2}{k_T^2}\frac{d\theta}{2\pi}. \qquad (10.25)$$

This factorised form for $\sigma_{\text{small}-x}$ encodes the leading high-energy behaviour to all orders. Both terms in the square bracket have a clear physical interpretation, that we now discuss.

We start from the term involving H. It is easy to see that at small x the longitudinal form factor $H_{||}$ can be extracted from $H^{\mu\nu}$ through the following projector

$$\mathcal{P}^{\mu\nu} = \frac{k_\perp^\mu k_\perp^\nu}{k_T^2}, \qquad \mathcal{P}_{\mu\nu}H^{\mu\nu} = -H_{||}(1+\mathcal{O}(z)). \qquad (10.26)$$

The term in the first square bracket in Eq. (10.25) then defines an *impact factor* C

$$C\left(\frac{Q^2}{zs}, \frac{k_T^2}{Q^2}, \Omega_{\mathcal{F}}, \alpha_s\right) = \int \frac{d\theta}{2\pi}\frac{Q^2}{2zs}\left[\mathcal{P}_{\mu\nu}H^{\mu\nu}\right], \qquad (10.27)$$

which can be thought of as the *off-shell cross section* for the process

$$\gamma^*(n) + g^*(q) \to \mathcal{V}, \qquad (10.28)$$

with $q = zp + k_\perp$, $q^2 = -k_T^2$ (note that $q \approx p_L$ in the small-x region). Indeed, as we already mentioned, $H^{\mu\nu}$ contains all the cut diagrams for the process Eq. (10.28). Furthermore, the term $Q^2/(zs)$ provides the proper flux factor (note the extra z term), and the action of the projector $\mathcal{P}$ can be interpreted as the average over polarisations of the initial state off-shell gluon. Indeed, it is easy to show that

$$\int \frac{d\theta}{2\pi}\mathcal{P}^{\mu\nu} = \frac{1}{2}\left(-g^{\mu\nu} + \frac{q^\mu n^\nu + n^\mu q^\nu}{q\cdot n}\right), \qquad (10.29)$$

and that in the on-shell $k_T^2 \to 0$ limit

$$\left\langle C\left(\frac{Q^2}{zs}, \frac{k_T^2}{Q^2}, \alpha_s\right)\right\rangle_\theta = \sigma_{\text{on-shell}}(\gamma^*, g(zp) \to \mathcal{F}), \tag{10.30}$$

i.e. the azimuthally averaged impact factor C correctly reproduces the on-shell partonic cross section for the process $\gamma^* + g \to \mathcal{F}$ in the $k_T \to 0$ limit.

We now consider the second bracket in Eq. (10.25), namely the term involving L. As we said before, at LO this represents just the t-channel exchange of a single gluon. Beyond that, it contains multiple gluon emissions both in the t- and s-channels. It can be interpreted [Gribov *et al.* (1983); Catani *et al.* (1990a, 1991b,a); Ball and Forte (1997); Lipatov (1997)] as a generalised (perturbative) unintegrated gluon distribution $\mathcal{G}$, which describes all-order gluon emission and that depends non-trivially on the transverse momentum k_T. It satisfies evolution equations both in Q^2 (DGLAP) and in x (BFKL). In terms of the off-shell cross section C and of the unintegrated distribution $\mathcal{G}$, Eq. (10.25) can be written as

$$\sigma_{\text{small}-x} = \int \frac{dk_T^2}{k_T^2}\frac{dz}{z} C\left(\frac{Q^2}{zs}, \frac{k_T^2}{Q^2}, \alpha_s\right) \mathcal{G}\left(z, \frac{k_T^2}{\mu^2}, \alpha_s\right). \tag{10.31}$$

This is the famous k_T factorisation formula [Catani *et al.* (1990a, 1991b,a)]. It states that leading small-x effects for a generic process $\mathcal{V}$ can be obtained by the convolution of the cross section for the process of interest, calculated with incoming off-shell gluons, with a universal gluon distribution function encoding all-order gluon emission in the high-energy regime.

10.4 From factorisation to resummation, the generalised ladder expansion and duality

The k_T factorisation formula Eq. (10.31) allows the all-order resummation of LL-x contributions to the cross section to be performed. In this section, we briefly sketch how this is done, i.e. how Eq. (10.31) connects to the actual small-x resummation. We begin by introducing the two dimensionless variables

$$x = \frac{Q^2}{s}, \quad \xi = \frac{k_T^2}{Q^2}, \tag{10.32}$$

in terms of which Eq. (10.31) becomes

$$\sigma_{\text{small}-x}(x, Q^2/\mu^2) = \int \frac{d\xi}{\xi}\frac{dz}{z} C\left(\frac{x}{z}, \xi, \alpha_s\right) \mathcal{G}\left(z, \frac{Q^2\xi}{\mu^2}, \alpha_s\right). \tag{10.33}$$

From now on, we set $\mu = Q^2$ to simplify the notation. Consequently, we drop the explicit μ dependence from our formulae. The complete μ dependence of the cross section can be trivially recovered from renormalisation group arguments. To deal with Eq. (10.33), it is convenient to disentangle the z convolution integral through a N-Mellin transform, see Eq. (10.3). In N-Mellin space, Eq. (10.33) becomes

$$\sigma_{\text{small}-x}(N) = \int \frac{d\xi}{\xi} C(N, \xi, \alpha_s) \mathcal{G}(N, \xi, \alpha_s). \tag{10.34}$$

As we mentioned in the introduction, $\ln x$ enhanced terms are mapped by the Mellin transform into poles in $N = 0$. In Eq. (10.34) there are two sources of N dependence: the off-shell impact factor C and the unintegrated gluon distribution $\mathcal{G}$. In non pathological theories, the 2GI requirement on the hard coefficient function H implies that C is regular in $N = 0$, $C(N, \xi, \alpha_s) = C(0, \xi, \alpha_s) + \mathcal{O}(N)$. All the leading small-$x$ enhanced terms are then encoded in the universal unintegrated gluon distribution $\mathcal{G}$. To extract them, one should find an explicit representation for $\mathcal{G}$.

To achieve this, one notes that $\mathcal{G}$ satisfies a BFKL x-evolution equation (see e.g. the discussion in ref. [Catani *et al.* (1991a)]). It reads

$$\frac{d}{d\ln(1/z)} \mathcal{G}(z, \xi, \alpha_s) = \int \frac{d\xi'}{\xi'} \chi(\alpha_s, \xi') \mathcal{G}\left(z, \frac{\xi}{\xi'}, \alpha_s\right), \tag{10.35}$$

where χ is the BFKL kernel. It is convenient to disentangle the convolution integral in Eq. (10.35) with a Mellin transform w.r.t. ξ,

$$F(M) = \int_0^\infty d\xi \xi^{M-1} F(\xi). \tag{10.36}$$

In the double (N, M) space it is then straightforward to solve Eq. (10.35):

$$\mathcal{G}(N, M, \alpha_s) = \frac{\mathcal{G}_0(M, \alpha_s)}{N - \chi(\alpha_s, M)}, \tag{10.37}$$

with $\mathcal{G}_0(M, \alpha_s)$ a boundary condition. The ξ space result is then obtained by inverting the M-Mellin transform. Up to subleading power corrections, one obtains

$$\mathcal{G}(N, \xi, \alpha_s) = \int_{c-i\infty}^{c+i\infty} \frac{dM}{2\pi i} e^{M \ln \xi} \mathcal{G}(N, M, \alpha_s) = \mathcal{N}(M^*, \alpha_s)\, \xi^{M^*}, \tag{10.38}$$

where $\mathcal{N}(M^*, \alpha_s) = -\mathcal{G}_0(M^*, \alpha_s)/\chi'(\alpha_s, M^*)$ and $M^* = M^*(N)$ is the position of the pole in Eq. (10.37). It is implicitly defined by

$$N = \chi(\alpha_s, M^*). \tag{10.39}$$

Plugging the result Eq. (10.38) for $\mathcal{G}(N, \xi, \alpha_s)$ in the k_T factorisation formula Eq. (10.34), one obtains

$$\sigma_{\text{small}-x} = \mathcal{N}(M^*) \int d\xi \xi^{M^*-1} C(N, \xi, \alpha_s). \tag{10.40}$$

The all-order small-x behaviour is encoded in the function $M^*(N)$. To explicitly see this, we expand the BFKL kernel in power of α_s:

$$\chi(\alpha_s, M) = \alpha_s \chi_0(M) + \alpha_s^2 \chi_1(M) + \mathcal{O}(\alpha_s^3). \tag{10.41}$$

Equation (10.39) then becomes

$$\chi_0(M^*(\alpha_s, N)) + \alpha_s \chi_1(M^*(\alpha_s, N)) = \frac{N}{\alpha_s}. \tag{10.42}$$

This implies that M^* admits the following perturbative expansion

$$\begin{aligned} M^*(\alpha_s, N) &= M_0^*\left(\frac{\alpha_s}{N}\right) + \alpha_s M_1^*\left(\frac{\alpha_s}{N}\right) + \cdots \\ &= \sum_{k=1}^{\infty} g_{0,k}\left(\frac{\alpha_s}{N}\right)^k + \alpha_s \sum_{k=1}^{\infty} g_{1,k}\left(\frac{\alpha_s}{N}\right)^k + \cdots \end{aligned} \tag{10.43}$$

Performing an inverse Mellin transform, this becomes

$$\int_{c-i\infty}^{c+i\infty} \frac{dN}{2\pi i} e^{N \ln \frac{1}{x}} M^*(\alpha_s, N) = \sum_{k=1}^{\infty} \frac{\alpha_s^k \ln^{k-1}(1/x)}{(k-1)!}\left[g_{0,k} + \alpha_s g_{1,k} + \mathcal{O}(\alpha_s^2)\right], \tag{10.44}$$

i.e. $M_0^*(M_1^*)$ contains the small-x leading(next-to-leading)-logarithmic towers. M_i^* are determined by (see Eq. (10.42))

$$\chi_0\left[M_0^*\left(\frac{\alpha_s}{N}\right)\right] = \frac{N}{\alpha_s}, \quad M_1^*\left(\frac{\alpha_s}{N}\right) = -\frac{\chi_1(M_0^*(\alpha_s/N))}{\chi_0'(M_0^*(\alpha_s/N))}. \tag{10.45}$$

It is then clear from the above discussion that leading(next-to-leading) small-x terms are determined by the fixed-order leading (next-to-leading) BFKL kernel.

Note that interestingly enough the expression for the N-space k_T factorisation formula Eq. (10.40) has the structure of a ξ-Mellin transform. Pragmatically then to include all-order (leading) high-energy effects for a generic process $\gamma^* + g \to \mathcal{V} + X$ one should

(1) Compute the reduced cross section $C(x, \xi, \alpha_s)$ for the process $\gamma^* + g^* \to \mathcal{V}$, initiated by an off-shell (transverse) gluon with momentum $q = zp + k_\perp$, $q^2 = -k_T^2$. Perform the polarisation average with the projector Eq. (10.26).
(2) Perform a double (N, M) Mellin transform to obtain $C(N, M, \alpha_s)$.
(3) Expand $C(N, M, \alpha_s)$ around $M = 0$ and substitute $M \to M^*(\alpha_s, N)$.
(4) Perform an inverse N-Mellin transform to go back in x-space, where all the $1/N^k$ poles are mapped into logarithms of x.

Although this procedure is at least in principle straightforward, its physical origin is not immediately transparent. In the remainder of this section, we show that the "Mellin" structure of Eq. (10.40) actually has a very clear physics interpretation [Caola *et al.* (2011)].

We start by analysing in more detail the factorised expression Eq. (10.25). Its structure bears a striking formal resemblance with the traditional QCD collinear factorisation formula, which separates short distance partonic cross sections from all-order collinear emission. To explore further the connections between the two factorisations, we first make the collinear singularities present in the ladder kernel L explicit. As is well known [Ellis *et al.* (1979)], these come from two-particle reducible sub-graphs. We then write the full kernel L as an infinite sum of 2PI basic kernels K, see Fig. 10.4. For the reasons explained in Section 10.2, at LL-x we only need to consider gluonic ladders. The 2PI kernels K are free of collinear singularities, which are then only generated by the integrations over momenta of the lines connecting the different rungs of the ladder.

The kernel K contains all-order real and virtual (gluon) emissions, that generate the non-trivial small-x dynamics. Before discussing it in full generality, it is instructive to consider a simpler case, namely small-x resummation in the double logarithmic LL-x *and* LL-Q^2 approximation. This regime is characterised not only by strong ordering of longitudinal momenta (see Eq. (10.11)), but also of transverse ones

$$k_{\perp 1}^{\,2} \ll k_{\perp 2}^{\,2} \ll \cdots \ll k_{\perp n}^{\,2} \ll s. \tag{10.46}$$

This implies that only strongly ordered t-channel gluon emission contributes here. In other words, in the LL-x LL-Q^2 approximation the kernel K describes the emission of a *single* (strongly ordered) gluon along the ladder. We now consider the contribution of a single insertion of K to the factorised expression formula Eq. (10.25). It reads[8]

$$\sigma_1(x,\mu^2/Q^2,\alpha_s) = \int \frac{dz}{z}\int \frac{d\xi}{\xi} C\left(\frac{x}{z},\xi,\alpha_s\right)\left[2\pi K\left(z,\frac{\mu^2}{Q^2\xi},\alpha_s\right)\right], \tag{10.47}$$

or in N-Mellin space

$$\sigma_1(N,\mu^2/Q^2,\alpha_s) = \int \frac{d\xi}{\xi} C\left(N,\xi,\alpha_s\right)\left[\frac{2\pi}{N} K\left(N,\frac{\mu^2}{Q^2\xi},\alpha_s\right)\right]. \tag{10.48}$$

Here the subscript "1" indicates that we are only considering one insertion of the kernel K. In the LL-Q^2 approximation, K is linear in α_s. The ξ integral gives rise to a collinear $\xi \sim 0$ singularity, that must be regulated. We use dimensional regularization, to obtain[9]

$$\sigma_1(N,\mu^2/Q^2,\alpha_s;\epsilon) = \int \frac{d\xi}{\xi^{1+\epsilon}} C\left(N,\xi,\alpha_s;\epsilon\right)\left[\frac{2\pi}{N} K\left(N,\left(\frac{\mu^2}{Q^2}\right)^{\epsilon}\alpha_s;N^{\epsilon},\epsilon\right)\right]. \tag{10.49}$$

[8]We now reintroduce the explicit dependence on μ, to discuss the factorisation of collinear singularities.

[9]It is possible to show that the structure of k_T factorisation is unchanged in d-dimension. The only difference w.r.t. the $d = 4$ case is that the "phase space" $d\xi$ integration measure is continued $d\xi \to d\xi\ \xi^{-\epsilon}$. Intuitively, the reason is the following: as discussed in Section 10.2, the dominant LL-x contribution comes from soft gluon emission, which does not change the spin of hard lines. Hence, it is determined by a single helicity configuration, independently of the number of dimensions. See the discussion in [Catani *et al.* (1993)] for further details.

In this formula, we have indicated after a semicolon, terms that vanish in the $\epsilon \to 0$ limit. Writing as usual the $1/\xi^{1+\epsilon}$ factor in terms of plus distributions,

$$\frac{1}{\xi^{1+\epsilon}} = -\frac{1}{\epsilon}\delta(\xi) + \sum_{i=0}^{\infty} \left[\frac{\ln^i \xi}{\xi}\right]_+ \frac{(-\epsilon)^i}{i!}, \tag{10.50}$$

and using the on-shell limit Eq. (10.30), we immediately get

$$\sigma_1(N, \mu^2/Q^2, \alpha_s; \epsilon) = -\frac{1}{\epsilon}\sigma_{\text{on-shell}} \times \left[\frac{2\pi}{N}K(N, \alpha_s)\right] + \text{finite}. \tag{10.51}$$

Equation (10.51) shows that $2\pi K(N, \alpha_s)/N$ is the residue of the $\mathcal{O}(\alpha_s)$ collinear pole. In the LL-Q^2 approximation, this is just the leading order DGLAP anomalous dimension. By equating it to its known expression for it in the small-x limit, one gets

$$\frac{2\pi}{N}K^1(\alpha_s) = \frac{\alpha_s C_A}{\pi}\gamma_0(N) = \bar{\alpha}_s\gamma_0(N), \qquad \gamma_0(N) = \frac{1}{N}, \tag{10.52}$$

where we defined $\bar{\alpha}_s = \alpha_s C_A/\pi$. In terms of the anomalous dimension γ_0, the one-gluon cross section Eq. (10.49) becomes

$$\sigma_1(N, \mu^2/Q^2, \alpha_s; \epsilon) = \int \frac{d\xi}{\xi^{1+\epsilon}} C\left(N, \xi, \alpha_s; \epsilon\right)\bar{\alpha}_s \left(\frac{\mu^2}{Q^2}\right)^{\epsilon}\gamma_0(N). \tag{10.53}$$

The generalisation to n kernel insertions (i.e. to n gluon emissions) is straightforward

$$\begin{aligned}\sigma_n(N, \mu^2/Q^2, \alpha_s; \epsilon) = &\left[\bar{\alpha}_s \left(\frac{\mu^2}{Q^2}\right)^{\epsilon}\gamma_0(N)\right]\int_0^{\infty}\frac{d\xi_n}{\xi_n^{1+\epsilon}}C\left(N, \xi, \alpha_s; \epsilon\right)\\ &\times \int_0^{\xi_n}\left[\bar{\alpha}_s \left(\frac{\mu^2}{Q^2}\right)^{\epsilon}\gamma_0(N)\right]\frac{d\xi_{n-1}}{\xi_{n-1}^{1+\epsilon}}\\ &\times \cdots \times \int_0^{\xi_2}\left[\bar{\alpha}_s \left(\frac{\mu^2}{Q^2}\right)^{\epsilon}\gamma_0(N)\right]\frac{d\xi_1}{\xi_1^{1+\epsilon}},\end{aligned} \tag{10.54}$$

where we have restored the integral limits for clarity. Note the nested integration regions, typical of the strongly ordered LL-Q^2 regime considered here. Equation (10.54) contains multiple collinear poles in ϵ. The $\overline{\text{MS}}$ procedure to remove them is well-understood [Ellis *et al.* (1979)]: schematically, the poles are subtracted iteratively by requiring that Eq. (10.54) is finite after each ξ_i integration. After performing $n-1$ subtractions, the collinear $(n-1)$-regulated cross section becomes [Caola *et al.* (2011)]

$$\begin{aligned}\sigma_n(N, \mu^2/Q^2, \alpha_s; \epsilon) = &\left[\bar{\alpha}_s \left(\frac{\mu^2}{Q^2}\right)^{\epsilon}\gamma_0(N)\right]\int_0^{\infty}\frac{d\xi_n}{\xi_n^{1+\epsilon}}C\left(N, \xi, \alpha_s; \epsilon\right)\\ &\times \frac{1}{(n-1)!}\frac{1}{\epsilon^{n-1}}\left[\bar{\alpha}_s\gamma_0(N)\left((1 - \left(\frac{\mu^2}{Q^2\xi_n}\right)^{\epsilon}\right)\right]^{n-1}.\end{aligned} \tag{10.55}$$

The total cross section $\sigma_{\text{small}-x}$ is then just the sum of the individual n-kernel contributions σ_n, i.e.

$$\sigma_{\text{small}-x} = \sum_{i=1}^{\infty} \sigma_i = \left[\bar{\alpha}_s \left(\frac{\mu^2}{Q^2}\right)^{\epsilon} \gamma_0(N)\right] \times \int_0^{\infty} \frac{d\xi}{\xi^{1+\epsilon}} C(N,\xi,\alpha_s;\epsilon) e^{\bar{\alpha}_s \gamma_0(N) \frac{1}{\epsilon}\left[1-\left(\frac{\mu^2}{Q^2 \xi}\right)^{\epsilon}\right]}. \tag{10.56}$$

This is finite in the $\epsilon \to 0$ limit:

$$\sigma_{\text{small}-x}(N, \mu^2/Q^2, \alpha_s(\mu^2)) = \bar{\alpha}_s(\mu^2)\gamma_0(N) \int_0^{\infty} d\xi \; \xi^{\bar{\alpha}_s(\mu^2)\gamma_0(N)-1} e^{\bar{\alpha}_s(\mu^2)\gamma_0(N) \ln \frac{Q^2}{\mu^2}} C(N,\xi,\alpha_s(\mu^2)). \tag{10.57}$$

Setting as before $\mu^2 = Q^2$, this becomes

$$\sigma_{\text{small}-x}(N, \alpha_s(Q^2)) = \bar{\alpha}_s(Q^2)\gamma_0(N) \int_0^{\infty} d\xi \; \xi^{\bar{\alpha}_s(Q^2)\gamma_0(N)-1} C(N,\xi,\alpha_s(Q^2)). \tag{10.58}$$

The structure of Eq. (10.58) is the same as Eq. (10.40), but now its physical interpretation is clear: the "Mellin" structure is just the finite remainder of the all-order exponentiation of t-channel multi-kernel insertion, and M^* is just the anomalous dimension for the DGLAP Q^2 evolution in the small-x limit.

Although we derived this result in the LL-Q^2 approximation, it is easy to realise that the same structure also holds in the generic case. The only modification is that the LO DGLAP kernel γ_0 should be replaced by a suitable all-order small-x kernel γ, see [Caola *et al.* (2011)] for details. As for the M^* case, γ admits an expansion in α_s at fixed α_s/N, see Eq. (10.43)

$$\gamma(\alpha_s, N) = \gamma_s\left(\frac{\alpha_s}{N}\right) + \alpha_s \gamma_{ss}\left(\frac{\alpha_s}{N}\right) + \mathcal{O}(\alpha_s^{k+2}/N^k). \tag{10.59}$$

Contrary to the LL-Q^2 case, the full kernel γ has a non-trivial ϵ dependence, which makes the $\overline{\text{MS}}$ subtraction procedure more cumbersome. Nevertheless, it is possible to carry it out along the same lines discussed above. The only difference is that in the general case an overall scheme-dependent prefactor R appears, a leftover of the collinear subtraction procedure (see e.g. [Catani *et al.* (1993)] and the discussion in [Caola *et al.* (2011)]). Relaxing the LL-Q^2 requirement, the LL-x resummed cross section then reads

$$\sigma_{\text{small}-x}(N, \alpha_s) = R_{\overline{\text{MS}}}\left[\gamma_s\left(\frac{\alpha_s}{N}\right)\right] \gamma_s\left(\frac{\alpha_s}{N}\right) \int_0^{\infty} d\xi \; \xi^{\gamma_s\left(\frac{\alpha_s}{N}\right)-1} C(N,\xi,\alpha_s), \tag{10.60}$$

where the all-order scheme-dependent normalisation $R_{\overline{\text{MS}}}$ can be fully determined (in the LL-x approximation) from the knowledge of the LO BFKL kernel, see ref. [Catani *et al.* (1993)] for details.

We note that the derivation of Eq. (10.60) discussed above presents some advantages. First, it elucidates the physics underlying the structure of Eq. (10.40). Second, it explicitly exposes the close connection between k_T and collinear factorisation. This allows one to easily extend the techniques routinely used in the collinear factorisation framework to the k_T factorisation case. As an example, using the ideas sketched above it is quite simple to generalise the k_T factorisation framework — originally developed for total cross sections — to more differential observables like e.g. rapidity [Caola *et al.* (2011)] and transverse momentum distributions [Forte and Muselli (2016); Muselli (2017)].[10]

Although the formal structure of Eq. (10.60) is clear, its physical content and predictive power rely on the knowledge of the all-order small-x anomalous dimension γ (and $R_{\overline{\rm MS}}$). The comparison of Eqs. (10.40) and (10.60) implies that one should interpret M^* as the anomalous dimension γ describing the Q^2 evolution encoded in the all-order kernel K. In this context, the M^* implicit definition Eq. (10.39) has a simple physical interpretation, that we now illustrate. In order to do this, it is convenient to first consider the integrated version of $\mathcal{G}$

$$G(x, Q^2) \equiv \int dk^2 \mathcal{G}(x, k^2, \alpha_s)\theta(Q^2 - k^2), \tag{10.61}$$

which can be interpreted as a (perturbative) gluon distribution function.[11] G satisfies a DGLAP Q^2-evolution equation, with kernel γ

$$Q^2 \frac{d}{dQ^2} G(N, Q^2) = \gamma(\alpha_s, N) G(N, Q^2). \tag{10.62}$$

Its solution is

$$G(N, Q^2) = G_0(N) e^{\gamma(N,\alpha_s)\ln\frac{Q^2}{Q_0^2}}. \tag{10.63}$$

From the definition Eq. (10.61), it is also clear that G satisfies the same BFKL x-evolution equation of $\mathcal{G}$. Its solution can be written as (c.f. Eqs. (10.37), (10.38))

$$G(N, Q^2) = \tilde{G}_0(M^*) e^{M^* \ln\frac{Q^2}{Q_0^2}}, \tag{10.64}$$

with $\tilde{G}_0(M^*)$ a boundary condition and M^* implicitly defined by $N = \chi(\alpha_s, M^*)$. Provided that the boundaries $G_0(N)$ and $\tilde{G}_0(M^*)$ are compatible, Eqs. (10.63) and (10.64) lead to the *duality relation* [Jaroszewicz (1982); Ball and Forte (1999)] $M^* = \gamma_S(\alpha_s, N)$, or

$$\chi(\alpha_s, \gamma(\alpha_s, N)) = N. \tag{10.65}$$

As we saw earlier, this implicit equation allows one to extract the desired all-order small-x anomalous dimension γ from a fixed-order determination of the BFKL

---

[10] Generalisation to observables that can resolve the internal ladder kinematics — like for example many interesting jet dynamics observables — is still problematic though.

[11] To follow the same notation of most of the literature, we omit the explicit dependence on α_s in G. Note however that all the analyses here apply to the perturbative gluon distribution function. As such, any higher twist effect is systematically neglected.

kernel χ. This analysis show that the implicit definition of M^* in Eq. (10.39) can be interpreted as a mutual consistency requirement on the x- and Q^2-evolution equations that the gluon distribution G satisfies.

10.5 Evolution kernels beyond LO: the ABF approach

In this section, we explore in more details the DGLAP and BFKL kernels introduced in Section 10.4, which control the small-x behaviour of the gluon distribution G.

We start by studying the BFKL equation at LO

$$\frac{d}{d\ln(1/x)}G(x,M) = \alpha_s\chi_0(M)G(x,M). \tag{10.66}$$

The explicit form of the LO kernel reads

$$\chi_0(M) = \frac{C_A}{\pi}\left[2\psi(1) - \psi(M) - \psi(1-M)\right], \tag{10.67}$$

where $\psi(x)$ is the digamma function, $\psi(x) = \Gamma'(x)/\Gamma(x)$. The solution of Eq. (10.66) is given by the inverse Mellin transform

$$G(x,Q^2) = \int_{c-i\infty}^{c+i\infty}\frac{dM}{2\pi i}G_0(M)e^{\alpha_s\chi_0(M)\ln\frac{1}{x}+M\ln\frac{Q^2}{Q_0^2}}. \tag{10.68}$$

If both x is small and Q^2 is large, the leading asymptotic for $G(x,Q^2)$ can be obtained by evaluating Eq. (10.68) in the saddle point approximation. In the limit $x \to 0$ at fixed Q^2, it is immediate to see that the saddle is determined by the minimum of the LO kernel Eq. (10.67), which leads to the following asymptotic behaviour

$$G(x,Q^2) \propto \exp\left(\alpha_s\chi_0(1/2)\ln\frac{1}{x}\right) = x^{-\lambda_0}, \qquad \lambda_0 = \alpha_s\frac{C_A}{\pi}4\ln 2. \tag{10.69}$$

This is the famous *pomeron solution*, which predicts a steep rise of the gluon distribution at small x. This region has been probed at HERA, whose kinematics coverage includes data with $x < 10^{-3}$ and $Q \lesssim 10$ GeV. At such scales, one can take $\alpha_s \approx 0.2$ and thus $\lambda_0 \approx 0.5$. As mentioned in the introduction, the strong rise $G \propto x^{-0.5}$ was not observed at HERA.

The aforementioned analysis is however based on the LO kernel only. It is natural to ask whether higher-order terms can lead to a softer growth for G. The NLO corrections to BFKL evolution were computed long ago [Fadin and Lipatov (1998); Camici and Ciafaloni (1997); Ciafaloni and Camici (1998)]. LO and NLO results for the kernel χ are shown in Fig. 10.5. It is apparent that corrections are very large and completely change the qualitative features of the kernel. Given this result, it is far from obvious that in the high-energy regime one can trust the perturbative expansion at all.

These issues motivated a series of detailed studies by two different groups — Altarelli, Ball, Forte [Ball and Forte (1999); Altarelli *et al.* (2000a, 2001, 2002, 2003, 2006, 2008)] and Ciafaloni, Colferai, Salam, Stasto (CCSS) [Salam (1998);

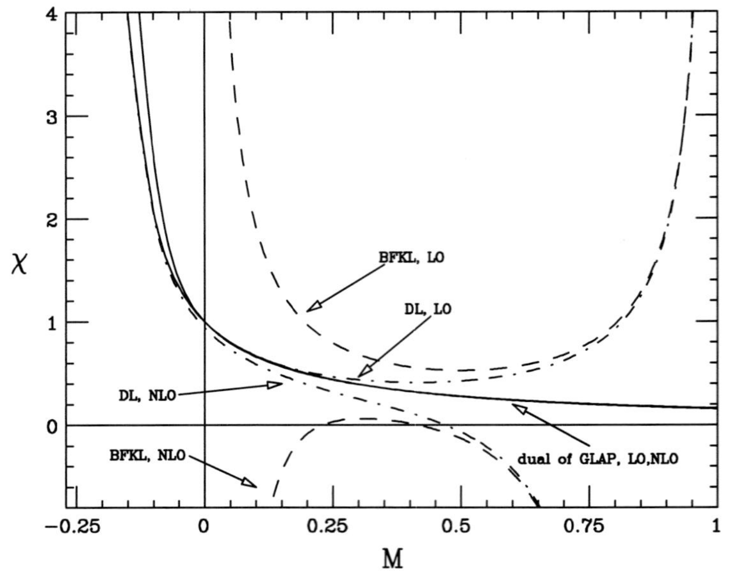

Fig. 10.5 Different approximations to the BFKL kernel, with $\alpha_s = 0.2$. Plot taken from ref. [Altarelli *et al.* (2000b)]. See text for details.

Ciafaloni and Colferai (1999); Ciafaloni *et al.* (1999, 2000, 2002, 2003a,b, 2004)] — to ascertain whether it is possible to cure the perturbative instabilities. These investigations shed more light on the all-order structure of high-energy evolution kernels, and showed that although a naive perturbative expansion is problematic at small x, it is still possible to use perturbation theory to obtain robust and reliable results. In the following, we will briefly illustrate the main ingredients for constructing perturbatively stable evolution kernels. We refer the reader to the original literature for a more comprehensive treatment of the many technical details.

The starting point is the duality relation Eq. (10.65), which allows, as explained in Section 10.4, the resummation of all-order small-x effects in the anomalous dimension,

$$\gamma(\alpha_s, N) = \gamma_s\left(\frac{\alpha_s}{N}\right) + \alpha_s\gamma_{ss}\left(\frac{\alpha_s}{N}\right) + \mathcal{O}\left(\alpha_s^{k+2}/N^k\right). \tag{10.70}$$

Equation (10.70) properly captures leading (γ_s) and next-to-leading (γ_{ss}) $\ln x$ effects, but it does not lead to a proper resummation of $\ln Q^2$ terms that are not enhanced at small x. This situation can be easily improved by matching Eq. (10.70) against the fixed order DGLAP kernels. More precisely, starting from Eq. (10.70) and from the standard perturbative expansion of the DGLAP anomalous dimension

$$\gamma(\alpha_s, N) = \alpha_s\gamma_0(N) + \alpha_s^2\gamma_1(N) + \mathcal{O}(\alpha_s^3), \tag{10.71}$$

one can construct a matched "double leading" (DL) anomalous dimension as a double expansion in α_s and α_s/N:

$$\begin{aligned}\gamma_{\rm DL}(\alpha_s,N) = &\left[\alpha_s\gamma_0(N)+\gamma_s\left(\frac{\alpha_s}{N}\right)-\text{d.c.}\right] \\ &+\left[\alpha_s^2\gamma_1(N)+\alpha_s\gamma_{ss}\left(\frac{\alpha_s}{N}\right)-\text{d.c.}\right]+\mathcal{O}\left(\alpha_s^3,\alpha_s^{k+2}/N^k\right), \qquad (10.72)\end{aligned}$$

where "d.c" stands for the double counting terms that must be subtracted. The double leading kernel Eq. (10.72) contains both all-order small-x (γ_s, γ_{ss}) *and* large-Q^2 (γ_0, γ_1) information: small-x terms are resummed explicitly in Eq. (10.72), large-Q^2 ones are resummed by solving the DGLAP evolution equation.

Along the same lines, one can construct a double leading BFKL kernel

$$\begin{aligned}\chi_{\rm DL}(\alpha_s,M) = &\left[\alpha_s\chi_0(M)+\chi_s\left(\frac{\alpha_s}{M}\right)-\text{d.c.}\right] \\ &+\left[\alpha_s^2\chi_1(M)+\alpha_s\chi_{ss}\left(\frac{\alpha_s}{M}\right)-\text{d.c.}\right]+\mathcal{O}(\alpha_s^3,\alpha_s^{k+2}/M^k), \qquad (10.73)\end{aligned}$$

where now $\chi_s(\chi_{ss})$ are determined by duality from the LO(NLO) DGLAP anomalous dimension and explicitly resum leading(next-to-leading) $\ln Q^2$ terms, while the fixed order kernels χ_0 and χ_1 contain the small-x information. The physical content of $\chi_{\rm DL}$ is the same of Eq. (10.72), although now the role of x and Q^2 is exchanged: in Eq. (10.73) the logarithms of Q^2 are resummed explicitly, while the logarithms of x are resummed by solving the BFKL evolution equation.[12]

The duality relation Eq. (10.65) not only allows a simultaneous resummation of $\ln Q^2$ and $\ln 1/x$ terms, it also provides important information on the all-order structure of the evolution kernels. Indeed, it allows one to translate physical properties of the DGLAP kernel to the BFKL one, and the other way around. To illustrate this point, we note that within the DGLAP framework there is an important constraint coming from momentum conservation considerations: the DGLAP anomalous dimension γ vanishes exactly for $N = 1$. Indeed, in the OPE formalism the $N = 1$ operator is the stress-energy tensor, which is protected from quantum corrections. Duality then implies

$$\chi(\alpha_s,0) = \chi(\alpha_s,\gamma(\alpha_s,1)) = 1, \qquad (10.74)$$

to all orders. This relation explains the instability of the BFKL kernel: since in the LO approximation $\chi_0(M) \sim 1/M$ for $M \sim 0$, the perturbative expansion *must* be badly behaved. Interestingly, this issue is less severe for the double leading kernel.

[12]Incidentally, this discussion shows that — as far as the gluon distribution G is concerned — the DGLAP and BFKL equations are completely equivalent, provided that a suitable evolution kernel is chosen. Starting from DGLAP evolution, one can use the duality relation Eq. (10.65) to obtain a *leading-twist*, *forward* BFKL equation for G at any given order in perturbation theory, see e.g. the discussion in [Marzani *et al.* (2007)]. This is in sharp contrast with the traditional analysis of BFKL evolution, which is more general but becomes problematic beyond NLO, see e.g. [Lipatov (1997)].

Indeed, in this case it is straightforward to check that the momentum conservation constraint $\chi(\alpha_s, 0) = 1$ is automatically satisfied by the LO double leading kernel. At NLO, there is still a (small) violation which can be however easily removed by adding a suitable (subleading) term to $\chi_{\rm DL}$. This results in a much better perturbative expansion, as shown in Fig. 10.5: the DL kernel is very stable in the small-M region $M \lesssim 0.3 - 0.4$.

For larger M, there are still residual instabilities, which originate from the $M = 1$ singularity of the fixed order BFKL kernel. To cure this problem, one can again use symmetry arguments. At LO, the BFKL kernel Eq. (10.67) is symmetric under the exchange $M \leftrightarrow 1 - M$. Beyond LO, this symmetry is broken, but in a controlled way [Ciafaloni *et al.* (1999)]. One can then use this feature to further improve the double leading expansion. Schematically, one first switches off the symmetry breaking terms, then symmetrises the kernel and finally restores the symmetry breaking effects. This procedure completely removes the large M instabilities: the resulting kernel is free from singularities for $\Re(M) > -1$, and its perturbative expansion is stable, see Fig. 10.6-left. In this figure, we show the final results from both the ABF and the CCSS groups. The agreement between the two predictions is clearly very good. Since the two groups implement the ideas sketched above in quite a different way, this agreement is non-trivial and it provides an important validation for the resummation program.

So far, we mostly focused on the BFKL kernel χ. However, using the duality relation Eq. (10.65) it is immediate to express the results discussed so far in terms of the DGLAP anomalous dimension γ. As an illustration of the resulting features for the kernel γ, we now briefly study its singularity structure. More precisely, we discuss the right-most N-space singularity of $\gamma(\alpha_s, N)$, which determines the leading small-x behavior of the gluon distribution G (see the Mellin inversion formula Eq. (10.3)). At LO and NLO, the leading singularity is a simple pole in $N = 0$. At the resummed level, the situation is different. Indeed, Fig. 10.6-left shows that

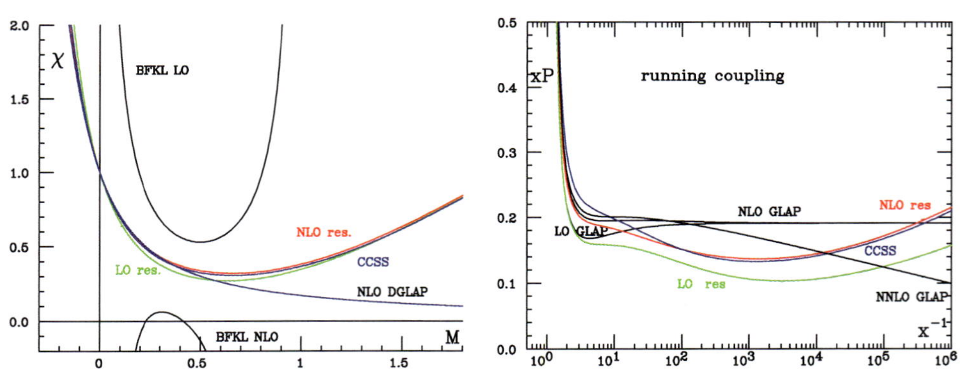

Fig. 10.6 The final results for the resummed kernels. Left: BFKL kernel. Right: DGLAP splitting function. Plots taken from ref. [Forte *et al.* (2009)]. See text for details.

the resummed BFKL kernel has a minimum. One would then conclude that the leading singularity of the DGLAP anomalous dimension is a branch cut, originating from the inversion of the BFKL kernel near its minimum. It turns out however that this is only true for fixed α_s; running coupling effects change this picture substantially.

Since it is easy to see that order-by-order running coupling effects are subleading at small x, one might naively imagine that they would only lead to small corrections to the fixed-order behaviour. The situation is however more delicate. Indeed, it turns out that an order-by-order treatment of running coupling effects is not enough to properly describe the small-x region and they must be resummed to all orders to obtain reliable results. Such a resummation is highly non-trivial. Here we refrain from explaining the details, and only state the final results [Altarelli *et al.* (2002, 2006); Ciafaloni *et al.* (1999, 2002, 2003b)]. The resummation procedure removes the branch cut singularity of the anomalous dimension and replaces it with a simple pole in $N = N_0 > 0$. The position N_0 of the pole is fully determined by the behaviour of the BFKL kernel near its minimum. It is interesting to study the phenomenological implications of this behaviour at HERA. As we mentioned, the relevant kinematic region is $x < 10^{-3}$, $Q \lesssim 10$ GeV. Here one can use $\alpha_s \approx 0.2$, to obtain $N_0 \approx 0.2$. This implies that the gluon distribution at small-x grows like $G(x, Q^2) \approx x^{-0.2}$, a much tamer behaviour with respect to the "hard pomeron" one. As already mentioned in the introduction, this softer growth is fully compatible with small-x HERA data.

The analysis above makes it clear that once all the relevant physical input is properly taken into account, one is left with stable kernels for small-x evolution, which can be reliably obtained through (properly resummed) perturbation theory. It is interesting to compare resummed and fixed-order results. In Fig. 10.6-right, we show this comparison for the DGLAP splitting function (i.e. the inverse Mellin transform of the anomalous dimension γ). As for the BFKL kernel, see Fig. 10.6-left, also here the ABF and CCSS results agree very well.[13] Compared to the (N)NLO result, the resummed splitting function exhibits a characteristic "dip" in the small-x HERA region, $10^{-2} \lesssim x \lesssim 10^{-4}$. Nevertheless, the resummed and fixed-order results start deviate substantially only for very small values of $x \sim 10^{-6}$. This is in sharp contrast with a "naive" small-x analysis, which predicts strong violations of fixed-order DGLAP evolution, as shown at the beginning of this section.

The agreement of fixed-order and resummed evolution in the small-x HERA region explains the phenomenological success of the former. This is of crucial importance for PDFs extractions, which are routinely performed within a fixed-order DGLAP framework. If high-energy effects would have been as large as naively expected, fixed-order DGLAP could have been inapplicable in a significant fraction

[13] ABF and CCSS results only slightly differ in the transition region $x \sim 0.1$. This is most likely due to a different treatment of spurious (subleading) running coupling effects, which in the ABF approach are removed through a matching to the fixed-order result.

of the HERA kinematics. This would have led to a smaller data set available for (fixed-order) PDFs extraction, and hence to less precise PDFs. Thanks to the (accidentally) small impact of all-order high-energy corrections, only a small fraction of very small-x and Q^2 HERA data has to be excluded from PDFs fits.

Of course, the optimal strategy would be to incorporate all-order small-x effects in the PDFs fitting framework. Thanks to the results sketched in this section, this endeavour is possible. For complete small-x phenomenological studies, the analysis presented here must however be extended. First, beyond the leading logarithmic accuracy, one should extend the framework to include quark effects. This does not pose particular conceptual challenges, although it is quite involved. It was performed in ref. [Altarelli *et al.* (2008)]. Second, resummed evolution alone is not enough: resummed coefficient functions are also required. They can be obtained within the k_T factorisation approach described in Sections 10.3 and 10.4. Although conceptually straightforward, actual computations are non-trivial.

As mentioned in the introduction, a first important step towards resummed phenomenology was performed recently [Ball *et al.* (2017a)]. In ref. [Ball *et al.* (2017a)], the authors used resummed evolution and DIS coefficient functions as a baseline for a small-x resummed fit. The resulting PDFs lead to a better data/theory agreement for small-x sensitive observables, see e.g. Fig. 10.1, and allowed the kinematic coverage of the fit to be extended to small-x data, which are notoriously difficult to incorporate in fixed-order PDFs determinations. After almost 10 years from the completion of the small-x resummation program for the evolution kernels, ref. [Ball *et al.* (2017a)] shows that a fully resummed fit could finally be within reach in the near future.

10.6 Small-x techniques at the LHC: Higgs physics beyond the heavy top approximation

In the introduction, we mentioned that the direct phenomenological impact of small-x resummation at the LHC is somewhat less important than for example at HERA: only a few selected measurements probe the very small-x region where resummation is crucial. Nevertheless, a control on small-x effects is far from useless for LHC investigations. Indeed, it can provide important information on the structure of QCD radiative corrections for many interesting cases. This information can be used for example to construct approximate higher order predictions for important processes, like e.g. Higgs boson [Ball *et al.* (2013a)] or top quark [Muselli *et al.* (2015)] production. In general, these approximate results may provide an extra handle on the residual perturbative uncertainty of higher order results, as well as elucidate the pattern of radiative corrections. However, they become particularly useful in cases where the exact computations are prohibitively complicated. In this section, we illustrate this point using examples relevant for Higgs-boson investigations.

We start by discussing the Higgs-boson total production cross section in gluon-fusion, which suffers from very large QCD radiative corrections. At LO, the cross section is localised at $x = m_H^2/s = 1$. At higher orders, the bulk of the contribution to the cross section still comes from the $x \approx 1$ region, but extra gluon emission can populate the $x < 1$ region. If $x \approx 1$, the emitted gluons are relatively soft and effectively do not resolve the internal structure of the (top) quark loop that mediates Higgs production. As a consequence, one can integrate out the top quark and work in the HEFT approximation, which simplifies computations dramatically. In the HEFT the total Higgs production cross section in gluon-fusion is known to an impressive next-to-next-to-next-to-leading (N^3LO) order [Anastasiou *et al.* (2015)], and has an estimated residual perturbative uncertainty of just a few percent [de Florian *et al.* (2016b)]. Given this level of accuracy, it is important to investigate the impact of quark mass effects, beyond the HEFT approach. Although a comparison of the full and HEFT results at NLO confirms the robustness of the latter, extra information at higher order is very valuable.

At NNLO, a full calculation retaining exact quark mass effects is not available. Still, small-x analyses can provide important information on them. The reason is simple: small-x gluons carry large momentum, and as such they always resolve the top loop. As a consequence, the HEFT approximation dramatically fails at small-x. Indeed, in the high-energy region the HEFT result exhibits an unphysical *spurious* double logarithmic behavior, not present in the full Standard Model (SM) case. More specifically, at small-x the Higgs partonic cross section behaves as

$$\hat{\sigma}_{\rm HEFT} \sim \sum_{k=1}^{\infty} \alpha_s^k \ln^{2k-1} \frac{\hat{s}}{m_H^2}, \qquad \hat{\sigma}_{\rm SM} \sim \sum_{k=1}^{\infty} \alpha_s^k \ln^{k-1} \frac{\hat{s}}{m_H^2}. \tag{10.75}$$

Although the cross section is dominated by the large $x \approx 1$ region, the spurious double logarithmic behaviour Eq. (10.75) can easily induce $\mathcal{O}(1\%)$ effects. At the current level of theoretical accuracy, such a shift is relevant and must be under control. To obtain a reliable prediction over the whole x range, one can combine higher order HEFT predictions at large x with the correct SM small-x behavior, determined with high-energy techniques. As explained in Section 10.4, computing the leading small-x contribution to the Higgs-boson cross section basically amounts to performing a one-loop computation with incoming off-shell gluons — see Fig. 10.7. Although this is non-trivial, it is much simpler than the full SM NNLO calculation. In ref. [Marzani *et al.* (2008)], the leading small-x term was obtained numerically, and subsequently matched to the HEFT calculation in [Harlander *et al.* (2010); Pak *et al.* (2011)]. These studies showed that top quark effects can indeed affect the cross section at the few percent level. To further constrain the cross section, it would be interesting to match HEFT results to more refined high-energy predictions. Unfortunately, this is not possible with current technology, which only allows the computation of the leading small-x contribution.

Apart from the total cross section, there are other interesting Higgs differential observables where virtual heavy quark effects play an important role. Probably the most interesting one is the Higgs transverse momentum distribution $d\sigma/dp_T$. For this observable, clearly the HEFT is inadequate in boosted configurations: at large

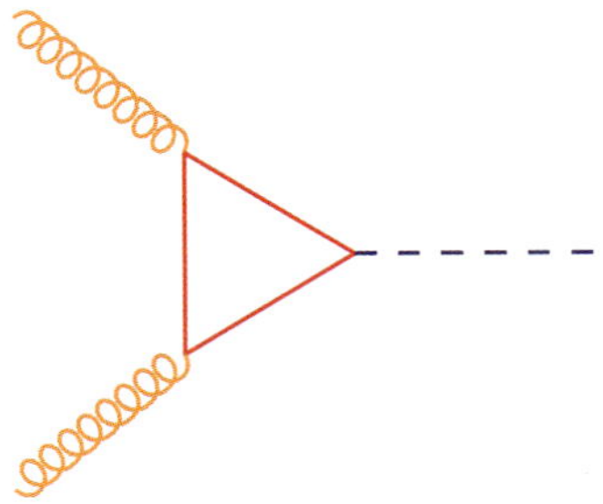

Fig. 10.7 The diagram contributing to the leading small-x contribution to the Higgs cross section. Massive quarks circulate in the loops, external gluons are off-shell and carry a non-vanishing transverse momentum. See text for details.

p_T, the Higgs must recoil against energetic partons, and they resolve the top loop. At moderate transverse momentum $p_T < m_t$, one would instead imagine that the HEFT is a good approximation. Although by and large this is indeed the case, there is one caveat. Indeed, in the region $m_{b,c} \lesssim p_T \lesssim m_H$ the bottom and charm quark contribution to the Higgs transverse momentum spectrum are enhanced by (power suppressed) Sudakov-like double logarithms $m_{b,c}^2/m_H^2 \ln^2(p_t^2/m_{b,c}^2)$ [Mantler and Wiesemann (2013); Grazzini and Sargsyan (2013); Banfi *et al.* (2014); Bagnaschi *et al.* (2016)]. In the Standard Model, these terms affect the Higgs p_T distribution at the few percent level, but they can be much larger in non-standard scenarios. As such, they provide important information on the structure of the Higgs couplings to b and c quarks, which is quite hard to constrain at the LHC. Clearly, these effects are not captured by the HEFT approximation, where one integrates out the top quark and completely neglects contributions from lighter quarks.

Because of the complexity of the amplitudes involved, the Higgs transverse momentum is currently only known at LO if the exact dependence on the top and bottom masses is retained. Small-x techniques can then provide at least partial information on radiative corrections. The crucial point in obtaining the leading high-energy contribution to the transverse momentum distribution is the realisation [Forte and Muselli (2016)] that at LL-x the Higgs p_T is entirely originated from the off-shellness of the incoming k_T-factorised gluons. The high-energy calculation for the transverse momentum distribution then closely follows the analogous one for the total cross section, only in the former case the Higgs p_T is kept fixed, which constrains the transverse momentum of the off-shell gluons. As for the total cross section, the dominant high-energy contribution to the transverse momentum distribution is then determined by the off-shell triangle diagram in Fig. 10.7. This is in sharp contrast with the full SM computation, which also involves complicated two-loop double box diagrams.

We now discuss to which extent the knowledge of the leading high-energy contribution can constrain quark mass effects in the Higgs transverse momentum spectrum. We study first the intermediate p_T region. Unfortunately, in this case a small-x analysis alone is not enough to constrain even the leading $\ln(p_T^2/m_{b,c}^2)$ terms at higher orders [Caola *et al.* (2016b)]. Indeed, it is possible to show that these (power suppressed) logarithmically enhanced effects are of both collinear and soft origin, see e.g. [Banfi *et al.* (2014)]. It is easy to realise, following the analysis in Section 10.3, that high-energy gluons only probe the collinear region. Still, the knowledge of the LL-x contribution gives access to part of the *subleading* $\ln(p_T^2/m_{b,c}^2)$ terms in the p_T distribution, the ones that come with an extra small-x enhancement. This is particularly useful because the leading $m_{b,c}$ effects can be extracted using different techniques [Melnikov and Penin (2016); Liu and Penin (2017)]. A combination of the small-x [Caola *et al.* (2016b)] and soft [Melnikov and Penin (2016); Liu and Penin (2017)] results then allows the prediction of both the leading *and* part of the subleading $\ln(p_T^2/m_{b,c}^2)$ corrections to the spectrum.

The full two-loop amplitude needed to compute NLO corrections to the Higgs p_T spectrum was recently computed in the $m_{b,c} \ll p_{t,H} \ll m_H$ limit [Melnikov *et al.* (2016, 2017)]. A perfect agreement with the small-x prediction for the subleading logarithmic $\ln(p_T^2/m_{b,c}^2)$ term was found. The high-energy result then provides an important check for the (very complicated) two-loop calculation. Also, it would be interesting to investigate to which extent the combination of the soft [Melnikov and Penin (2016); Liu and Penin (2017)] and small-x [Caola *et al.* (2016b)] predictions reproduces the bulk of the full result [Melnikov *et al.* (2016, 2017)]. Finally, we note that the small-x analysis also sheds more light on the structure of the corrections to the physical p_T spectrum. Indeed, at high energy there are delicate cancellations between real and virtual contributions to physical observables. Individually, virtual and real contributions contain double logarithmic high-energy terms, but only single logarithmic ones survive in (inclusive enough) observables. In the case at hand, a small-x analysis shows that some of the $\ln(p_T^2/m_{b,c}^2)$ terms present in the two-loop amplitude would exactly cancel against similar terms coming from real radiation, and thus don't contribute to the physical spectrum. This is hard to see in the traditional approach, where real emission contributions are integrated numerically.

We now discuss the boosted Higgs region. As we mentioned, here the HEFT approximation dramatically fails. Small-x analyses can provide interesting information in the region $m_H \ll p_t \ll s$. In practice, this can provide an estimate of QCD corrections in the region 500 GeV $\lesssim p_t \lesssim$ 1 TeV [Caola *et al.* (2016b)]. Similarly to the total cross-section case, also here the failure of the HEFT approximation translates into a spurious scaling behaviour of the physical distribution. Indeed, in the boosted region the Higgs transverse momentum spectrum scales as $d\sigma^{\mathrm{LL}x-\mathrm{LO}}/dp_t^2 \sim 1/(p_t^2)^a$, with $a = 2$ in the full SM and $a = 1$ in the HEFT case. Contrary to the examples discussed before, in the boosted region the top quark mass effects induce $\mathcal{O}(1)$ corrections to the HEFT approximation: the virtual top loop is almost *always* probed by high-energy partons. As a consequence, the HEFT result is about one order of magnitude larger than the correct one at $p_t \sim 1$ TeV. This feature is however already present at LO. It is unclear whether — on top of

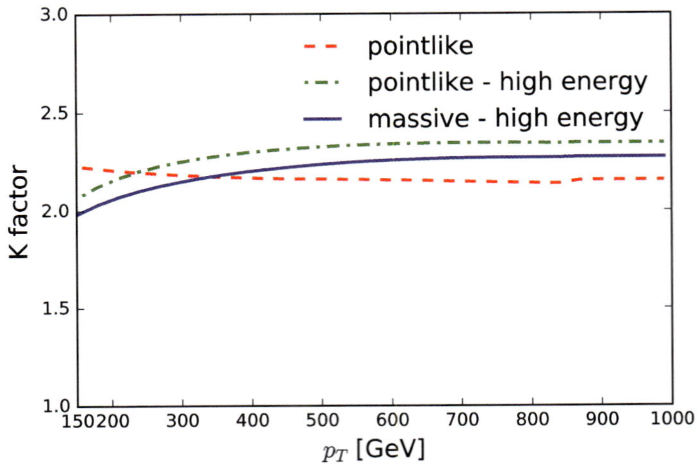

Fig. 10.8 The NLO K-factor for the Higgs transverse momentum distribution at the 13 TeV LHC. The full NLO result in the HEFT ("pointlike" in the key) approximation (red, dashed) is compared to the high-energy result both in the HEFT approximation (green, dot-dashed) and in the full SM (blue, solid, "massive" in the key). Plot taken from ref. [Caola *et al.* (2016b)]. See text for details.

the different LO scaling — QCD corrections behave in a significantly different way in the HEFT and SM case or not. It is then interesting to investigate the impact of finite top quark mass effects on the shape of QCD corrections.

To this extent, in Fig. 10.8 we show the NLO K-factor for the Higgs transverse momentum distribution at the 13 TeV LHC, in different approximations. More precisely, we consider

$$K_{\rm th} = \frac{d\sigma^{\rm th,NLO}/dp_T^2}{d\sigma^{\rm th,LO}/dp_T^2}, \tag{10.76}$$

where "th" indicates either the full HEFT result or the small-x approximation in both the HEFT and full SM cases. In the HEFT case, where one has full control on the NLO corrections, Fig. 10.8 shows that the small-x approximation slightly overshoots the size of corrections but gives a qualitative behaviour in agreement with the full result. It is then reasonable to expect that also beyond the HEFT approximation the small-x result would provide a fair approximation of the exact result. Figure 10.8 then suggests that the NLO HEFT and exact K-factors are quite similar, despite the corresponding absolute NLO predictions being very different.

This result is quite interesting, but it does not come as a surprise. Indeed, in the lack of the full NLO computation, other approximations for the boosted region were developed, see e.g. refs. [Buschmann *et al.* (2015); Frederix *et al.* (2016b); Greiner *et al.* (2017); Neumann and Williams (2017)]. Although they are based on completely different formalisms, they all point to the same conclusion: even if the actual spectrum differs dramatically in the HEFT and full SM cases, the shape of *higher order corrections* is relatively similar in the two cases. The fact that many different approaches lead to similar conclusions can be taken as an indication of the robustness of it.[14] Finally, we would like to mention that very recently NLO results with top mass dependence have become available [Kudashkin *et al.* (2017); Lindert *et al.* (2018); Jones *et al.* (2018)]. These results seem to support the conclusion obtained here: the NLO K-factor in the HEFT approximation and in the full SM are very similar (see e.g. [Caola (2017)] for preliminary result).

The examples shown in this section illustrate how a good control on high-energy effects can provide important physical information on the structure of higher order QCD corrections, even in cases where an all-order high-energy resummation is not mandatory. Unfortunately, current small-x techniques only allow computation of the *leading* high-energy term. Extending the framework to allow for a *systematic expansion* in the small-x region would be very useful: in combination with information coming from other sources, it could significantly constrain the structure of higher order corrections in many important cases. For example, in the Higgs case, combining systematic high-energy and soft expansions could reliably determine higher order (QCD) corrections to both the total cross section and differential distributions to very high accuracy.

Ideally, one would like to extend the high-energy formalism both to arbitrary logarithmic accuracy *and* at the subleading power level. Although this endeavour may seem hopeless, here we stress that an extension of the full resummation machinery is actually not needed for many interesting LHC applications. Indeed, as mentioned before more often than not one only needs to have control on the first few terms of the perturbative expansion, and not on its all-order structure. This dramatically simplifies the task, and in our opinion makes it worth investigating it further.

10.7 Conclusions

In this contribution, we tried to give the reader a glimpse of the vast and interesting field of small-x resummation. As we mentioned in the introduction, many important topics have not been covered here, but we hope to have given the reader at least an idea of the foundations of the high-energy resummation framework, of its implication for the physics of parton distribution functions and their evolution, and of some of the more modern applications of small-x techniques for LHC physics.

Guido's contributions to this field have been very important, especially in elucidating the many subtle physical effects that invalidate a naive approach to small-x resummation, and how to properly take them into account. His work was fundamental in extending the validity of the standard "Altarelli–Parisi" evolution to the

[14]We note however that, despite using quite different formalism, all the approaches described here more or less tacitly rely on the assumption that in the boosted region higher order corrections are dominated by the emission of extra real partons.

deep small-x region. Despite the fact that the bulk of the analyses of the small-x evolution kernels pioneered by Guido have been completed almost ten years ago, we now finally have all the ingredients in place to actually apply the formalism to parton distribution function fits. This, combined with the high quality of the HERA legacy data, has the potential of shedding new light on the structure of all-order QCD evolution.

At the LHC, a key role of small-x resummation is to provide important information on the structure of higher order QCD corrections, especially in cases where exact calculations are prohibitively complicated, like for example in Higgs physics. Currently, the main limitation for these analyses is that the existing high-energy framework only gives access to the leading small-x term. While this still provides interesting information, it would be extremely useful to extend the current technology to cope with both higher logarithmic and subleading power corrections. Although this endeavour is very far from trivial, it may be possible to obtain results for the first few terms in the perturbative expansion. This would already be enough for many interesting LHC cases. Recent years have seen an enormous progress in the development of technical tools for higher order QCD calculation, and it is very possible that some of these new ideas could be combined with insight originating from all-order small-x investigations, to tackle this problem. We hope that this contribution will pique the interest of a wider audience, and stimulate them to work on these theoretically interesting and phenomenologically relevant issues.

Acknowledgements

I am grateful to S. Forte, A. Levy and G. Ridolfi for inviting me to contribute to this volume. I am indebted to S. Forte, S. Marzani, C. Muselli and G. Diana for countless interesting discussions on the topics presented here. Finally, I would like to thank K. Melnikov for many stimulating discussions on heavy quark effects in Higgs physics.

Chapter 11

The proton spin, semi-inclusive processes, and measurements at a future Electron Ion Collider

Richard D. Ball[1] and Abhay Deshpande[2]

[1] *The Higgs Centre for Theoretical Physics, University of Edinburgh, JCMB, KB, Mayfield Rd, Edinburgh EH9 3JZ, Scotland*

[2] *Center for Frontiers in Nuclear Science, Stony Brook University, Stony Brook, NY 11794-3800 & Brookhaven National Laboratory, Upton, NY 11973-5000, USA*

Abhay Deshpande (left) and Richard D. Ball (right)

We discuss spin physics, Guido's contribution to it, and what we still have to learn. We set out in particular a programme for incorporating constraints from semi-inclusive data into global fits of polarised PDFs, and discuss the need for the EIC to increase the precision and kinematic coverage of current measurements.

11.1 Introduction

Guido's interest in spin physics goes right back to the start of the partonic approach to perturbative QCD [Altarelli and Parisi (1977); Altarelli *et al.* (1978a, 1979a,b)]: already, in the Altarelli–Parisi paper the structure function $g_1(x, Q^2)$, measurable in inclusive longitudinally polarised deep inelastic scattering, is expressed in terms of polarised parton densities (pPDFs) $\Delta q(x, Q^2)$, $\Delta\bar{q}(x, Q^2)$, $\Delta g(x, Q^2)$, and the corresponding polarised splitting functions are computed in parton language. The results were consistent with the LO polarised anomalous dimensions calculated one year previously in the OPE formalism [Ahmed and Ross (1976)]. Section 4.6 of what became known as Guido's 'Bible of QCD' [Altarelli (1982b)] is devoted to the theory of polarised parton densities, and remains recommended reading.

The EMC experiment [Ashman *et al.* (1988)], using a longitudinally polarised muon beam and a stationary target that contained polarised protons, was the first experiment to explore $g_1(x, Q^2)$ down to momentum fractions x as low as 0.01. When extrapolated over unmeasured $x < 0.01$ and combined with the couplings in leptonic hyperon decays and the assumption of SU(3) flavour symmetry, this led to the famous conclusion that the quark and anti-quark spins constitute only a small fraction of the proton spin. In addition, with these assumptions, the polarisation of the strange quark sea in the polarised proton was found to be large and negative.

The resolution of this 'spin crisis' by Altarelli and Ross [Altarelli and Ross (1988)] is a beautiful example of a deep theoretical insight that is of direct relevance to experiment. The nonsinglet first moments of the quark distributions $q^+ \equiv q + \bar{q}$, the axial charges $a_3 = \Delta u^+ - \Delta d^+$ and $a_8 = \Delta u^+ + \Delta d^+ - 2\Delta s^+$, correspond in the OPE to matrix elements of conserved currents, and thus there exist factorisation schemes (such as $\overline{\rm MS}$) in which they are independent of Q^2, and can thus be identified with axial decay constants measured in hyperon decays, interpreted through the quark model. The singlet first moment a_0 is different however, since the conservation of the singlet axial current is violated by the axial anomaly, and a_0 thus evolves in $\overline{\rm MS}$ (at two loops). This scale dependence makes the identification of the axial singlet charge $a_0(Q^2)$, extracted from the EMC data, with the total helicity of the quarks problematic, since quark model predictions rely on conserved currents and are thus scale independent.

However there exist alternative factorisation schemes in which the axial anomaly is not renormalised [Adler and Bardeen (1969)], and in these schemes

$$a_0(Q^2) = \Delta\Sigma - n_f \tfrac{\alpha_s(Q^2)}{2\pi} \Delta g(Q^2), \tag{11.1}$$

where $\Delta\Sigma$ is the singlet first moment of the polarised quark distributions, and $\Delta g(Q^2)$ is the first moment of the polarised gluon distribution. $\Delta\Sigma$ is independent of Q^2, to all orders, and it may thus be identified with $\Delta u^+ + \Delta d^+ + \Delta s^+$, and interpreted as the total helicity of the quarks [Altarelli and Ross (1988)]. Moreover, although in Eq. (11.1) the second term appears to be of $\mathcal{O}(\alpha_s)$, and thus small, because $a_0(Q^2)$ only evolves at two loops, Δg grows as $1/\alpha_s$ at large Q^2, and thus the difference between a_0 and $\Delta\Sigma$ can be of order one even at large scales. It follows that if the first moment of polarised gluon distribution $\Delta g(Q^2)$ is large and positive, it can compensate $\Delta\Sigma$, reduce a_0, and thus explain the EMC measurement.

For a number of years this resolution of the 'spin crisis' was rather controversial, with many people either confirming or contesting the arguments in [Altarelli and Ross (1988); Altarelli and Stirling (1989); Altarelli and Lampe (1990)], often with more heat than light (see for example [Altarelli (1990); Ball (1995); Forte (1995)], and refs. therein). This was in part because the issues are rather subtle and the result surprising, but also perhaps because the mathematical structure of chiral anomalies had only recently been unravelled (see ref. [Ball (1989)] for a contemporary review). A complete NLO calculation [Mertig and van Neerven (1996)] of the $\overline{\text{MS}}$ polarised splitting functions settled most of the theoretical issues regarding the scheme dependence, and in particular clarified the relation with the Adler–Bardeen condition [Ball *et al.* (1996b)]. Some theoretical studies were made of the small-x behaviour of the pPDFs [Berera (1992); Ball *et al.* (1995)], in an attempt to improve the small-x extrapolation when computing first moments. However, the central questions remained unanswered: is Δg large and positive, or does Δs violate expectations from SU(3)? Since pPDFs are by nature non perturbative quantities, these questions can only really be answered by experiment.

11.2 Current polarised DIS and p-p data

Significant progress has been made since the EMC observations on the proton's spin composition. One main focus has been on measurements with longitudinally polarised lepton beams scattering off longitudinally polarised nucleons in stationary targets. Inclusive data have been obtained in experiments at CERN [Adeva *et al.* (1998); Alexakhin *et al.* (2007); Alekseev *et al.* (2010b)], DESY [Ackerstaff *et al.* (1997); Airapetian *et al.* (2007)], Jefferson Laboratory [Zheng *et al.* (2004); Dharmawardane *et al.* (2006)], and SLAC [Anthony *et al.* (1996); Abe *et al.* (1998, 1997); Anthony *et al.* (1999, 2000)] in scattering off targets with polarised protons and neutrons. The kinematic reach and precision of the data on $g_1(x, Q^2)$ so far is similar to that of the unpolarised structure function $F_2(x, Q^2)$ just prior to the experimental programme at the HERA electron-proton collider.

Figure 11.1 provides a survey of the regions in x and Q^2 covered by the world polarised-DIS data, which is roughly $0.004 < x < 0.8$ for $Q^2 > 1\,\text{GeV}^2$. For a representative value of $x \simeq 0.03$, the $g_1(x, Q^2)$ data are in the range $1\,\text{GeV}^2 < Q^2 < 10\,\text{GeV}^2$. This is to be compared to $1\,\text{GeV}^2 < Q^2 < 2000\,\text{GeV}^2$ for the unpolarised data on $F_2(x, Q^2)$ at the same x. The figure also shows the vast expansion in x, Q^2 reach that an Electron Ion Collider (EIC) would provide, as will be discussed below. Over the past 15 years, an additional powerful line

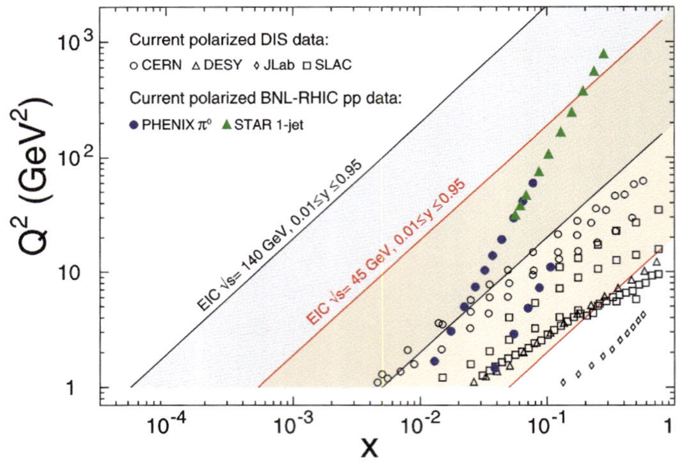

Fig. 11.1 Kinematic reach of the current data from fixed target electron and muon scattering experiments around the world (SLAC, JLab and CERN). Also shown for comparison are from RHIC at a scale $Q^2 = p_T^2$, where p_T is the observed jet (pion) transevrse momentum and an x value that is representative for the measurement scale. The x ranges probed at different scales are wide and have considerable overlap. Areas covering the $x - Q^2$ reach of the future Electron Ion Collider are also shown for the minimum and maximum $\sqrt{s}$ of 45 and 140 GeV.

of experimental study of nucleon spin structure has emerged: *semi-inclusive* deep-inelastic scattering (SIDIS). In these measurements, a charged or identified hadron is observed in addition to the scattered lepton. The relevant spin structure functions then depend on fragmentation functions [Altarelli *et al.* (1979b)] which depend on the momentum fraction that is transferred from the outgoing quark or anti-quark to the observed hadron. The non perturbative fragmentation functions are at present determined primarily from precision data on hadron production in e^+e^- annihilation, but data from the B-factories and the LHC are now helping to further improve their determination (see [Albino (2010); Metz and Vossen (2016)] for recent reviews). Insights from the semi-inclusive measurements are complementary to those from the inclusive measurements. Specifically, they make it possible to delineate the quark and anti-quark spin contributions by flavour, since Δq and $\Delta\bar{q}$ appear with different weights in the semi-inclusive (SI) cross-section. A large body of SI pPDFs data sensitive to nucleon helicity structure has been collected by the experiments at CERN and DESY in the last 20 years.

A further milestone in the study of the nucleon was the advent of RHIC, the world's first polarised proton+proton collider. In the context of the exploration of nucleon spin structure, the RHIC spin programme is a logical continuation. Very much in the spirit of the unpolarised hadron colliders in the 1980's, RHIC entered the scene to provide complementary information on the nucleon that is not readily available in fixed-target lepton scattering. The measurement of the spin-dependent gluon distribution $\Delta g(x, Q^2)$ in the proton is a major focus and strength of RHIC.

Here the main tools are spin asymmetries in the production of inclusive pions [Adler *et al.* (2004); Adare *et al.* (2007, 2009); Abelev *et al.* (2009); Adare *et al.* (2014)] and jets [Abelev *et al.* (2006, 2008); Adamczyk *et al.* (2015)] at large transverse momentum perpendicular to the beam axis, which sets the hard scale Q in these reactions. Their reach in x and Q^2 is also indicated in Fig. 11.1. Unlike DIS, the processes used at RHIC do not probe the partons locally in x, but rather sample over a region in x. RHIC also provides complementary information on $\Delta u, \Delta\bar{u}, \Delta d, \Delta\bar{d}$ for $0.05 < x < 0.5$ [Adare *et al.* (2011); Aggarwal *et al.* (2011); Adamczyk *et al.* (2014); Gal (2014)], with a beautiful technique that exploits the violation of parity and does not rely on knowledge of fragmentation. The carriers of the charged-current weak interactions, the W bosons, naturally select left-handed quarks and right-handed anti-quarks, and their production in p+p collisions at RHIC and calculable leptonic decay hence provide an elegant probe of nucleon helicity structure.

11.3 Nucleon spin structure and polarised PDF fits

11.3.1 *Global fits of polarised PDFs*

The earliest attempts [Altarelli and Stirling (1989); Altarelli *et al.* (1994b); Altarelli and Ridolfi (1995); Gehrmann and Stirling (1995); Ball *et al.* (1995)] to determine pPDFs from the data on inclusive polarised DIS relied on LO perturbation theory, since that was all that was available at the time. Various issues, in particular the Q^2 dependence of the asymmetries, and the importance of the small-x extrapolation region, were better understood, and significant improvements in the quality of the data meant that first attempts could be made to determine Δg from scaling violations. Following the publication of the NLO splitting functions [Mertig and van Neerven (1996)], these fits were upgraded to full NLO perturbative QCD [Ball *et al.* (1996b); Gehrmann and Stirling (1996); Altarelli *et al.* (1997a, 1998a)], and incorporated all the most recent inclusive pDIS data from SMC and E142, E143 and E154 on both proton and deuteron targets. Final results for the first moments were however still inconclusive: the suppression of the axial singlet charge a_0 was confirmed, SU(3) breaking seemed small, and the data showed a mild preference for a positive Δg, but the uncertainties were too large to allow more definite conclusions.

By this time, the difficulties were becoming clear. Reliable PDF determination requires reliable uncertainties, and for this it was necessary to develop new statistical techniques. Parametrisation bias was a major issue: this is particularly the case for pPDFs since they can cross over from positive to negative, and thus have quite a complicated shape. On the experimental side, inclusive DIS data are insufficient to determine Δq and $\Delta\bar{q}$ separately, and since strangeness is important, the flavour separation provided by proton and deuteron targets is insufficient. The narrow range in Q^2 of the experimental data made extraction of the gluon only from scaling violations very challenging, while the limited range in x created substantial uncertainty in the extrapolation required to determine first moments.

These issues have been thrown into sharp relief by the development of more sophisticated techniques for determining global PDFs (see [Ball *et al.* (2015, 2017b); Thorne *et al.* (2015); Dulat *et al.* (2016); Ball (2016)] and refs. therein), with in particular more reliable error estimation through the use of Hessians [Pumplin *et al.* (2001)], Lagrange multipliers [Stump *et al.* (2001)], Monte Carlo replicas [Giele and Keller (1998)] and unbiased parametrisations using neural networks [Forte *et al.* (2002)]. Applying the NNPDF methodology, developed for unpolarised PDFs, to the determination of pPDFs from inclusive DIS, showed just how large the uncertainties actually are [Ball *et al.* (2013b); Nocera (2016)]. Even when recent data from COMPASS and JLAB are included [Leader *et al.* (2015); Jimenez–Delgado *et al.* (2014); Sato *et al.* (2016b)], realistically determined uncertainties on Δs^+ and Δg are so large that they are both still consistent with zero. It thus becomes essential to attempt to incorporate other types of data into the determination of pPDFs.

11.3.2 *Role of SIDIS and fragmentation functions*

Semi-inclusive data are useful for separating quark from anti-quark distributions, and giving information on strangeness. Coefficient functions for polarised semi-inclusive DIS (pSIDIS) and single particle hadroproduction [Jager *et al.* (2003)] are known at NLO, and there is now a wealth of data from SMC, HERMES and COMPASS, and from PHENIX and STAR. Inclusion of these data in the global fits of pPDFs was first accomplished by DSSV [de Florian *et al.* (2008, 2009)] and LSS [Leader *et al.* (2010)]. The DSSV fits seemed to indicate that Δs is positive at large x, becoming negative at small x, and likewise that the Δg is small, with a first moment consistent with zero. While demonstrating that semi-inclusive data might have a significant part to play in the determination of pPDFs, their impact was inevitably limited by the large uncertainties in the fragmentation functions (FFs). Indeed, it is for this reason that SI data are rarely included in global determinations of unpolarised PDFs.

The FFs used by DSSV [de Florian *et al.* (2007a,b)] for $\pi^\pm$, $K^\pm$ and $p/\bar{p}$ were determined from semi-inclusive annihilation data (SIA) and unpolarised SIDIS. Although the uncertainties were large, they were not well quantified, and this inevitably led to questions about the reliability of the inclusion of pSIDIS in the determination of pPDFs. In recent years however the situation has improved, with a new generation of FFs [de Florian *et al.* (2015, 2017); Sato *et al.* (2016a); Bertone *et al.* (2017); Nocera (2018)] with more reliable uncertainties. It is thus now possible to employ polarised SI data in a global fit of pPDFs, propagating the uncertainties from the FFs into the pPDFs. A general scheme for combining, consistently, fits to PDFs, FFs and pPDFs, making the best use of available unpolarised and polarised SI data, will be set out in the next section.

11.3.3 *RHIC spin data and polarised gluon and sea quark PDFs*

Perhaps the most significant development in spin physics in recent years has been the advent of Drell–Yan, W/Z production, and inclusive π^0 and jet data from longitudinally polarised proton-proton collisions at RHIC. All these processes can be computed at NLO [Gehrmann (1997, 1998); Gluck *et al.* (2001); Jager *et al.* (2004); de Florian and Vogelsang (2010)], and as in unpolarised PDF determination they allow us to separate quark and anti-quark distributions, and give a more direct determination of the gluon than is available in inclusive DIS. In recent years, truly global determinations of pPDFs have thus become possible [de Florian *et al.* (2014); Nocera *et al.* (2014); Nocera (2017)], incorporating data on inclusive π^0, jets and $W^{\pm}$ production from STAR and PHENIX alongside the inclusive data from pDIS. The pPDFs determined in this way are shown in Fig. 11.2, and some results for first moments are collected in Tables 11.1 and 11.2. Now at last some evidence for a positive polarised gluon distribution emerges, at least in the measured region.

The combination of the large body of inclusive deep inelastic scattering data off targets containing polarised protons and neutrons has established that the up quarks and anti-quarks combine to have net polarisation along the proton spin, whereas the down quarks and anti-quarks combine to carry negative polarisation. The Δu^+ and Δd^+ distributions are very well constrained by now at medium to large x (see Fig. 11.2).

The light sea quark and anti-quark distributions still carry large uncertainties, even though there are some constraints by the semi-inclusive data and, most recently, from measurements of spin-dependence in leptonic W decay in $\sqrt{s} = 500\,\text{GeV}$ polarised proton+proton collisions at RHIC [Adamczyk *et al.* (2014); Gal (2014)]. RHIC probes the Δu, Δd, $\Delta\bar{u}$ and $\Delta\bar{d}$ densities for $0.05 < x < 0.5$ at a scale set by the W-mass. The sea shows hints of not being SU(2)-flavour symmetric: the $\Delta\bar{u}$ distribution has a tendency to be mainly positive, while the $\Delta\bar{d}$ anti-quarks carry opposite polarisation, see Table 11.1. This pattern has been predicted at least qualitatively by a number of models of the nucleon. Better constraints on Δu, Δd, $\Delta\bar{u}$ and $\Delta\bar{d}$ are anticipated [de Florian and Vogelsang (2010)] from additional RHIC measurements with higher integrated luminosity. Furthermore the large luminosities and high resolution available at the Jefferson Laboratory after an upgrade to 12 GeV electron beam energy will extend the kinematic reach of the existing JLab inclusive and semi-inclusive deep inelastic scattering data to twice smaller x as well as to larger x than have thus far been measured.

Strange quarks appear to be deeply involved in nucleon spin structure. As we mentioned earlier, from the inclusive deep inelastic data, along with SU(3) flavour symmetry considerations, one finds a negative value for the integrated strange helicity distribution. Strange quarks and anti-quarks would thus be polarised opposite to the nucleon. This would need to be viewed as part of the reason why the total quark and anti-quark spin contribution $\Delta\Sigma$ is so much smaller than expected in simple models. A variety of mechanisms for SU(3) flavour symmetry breaking have

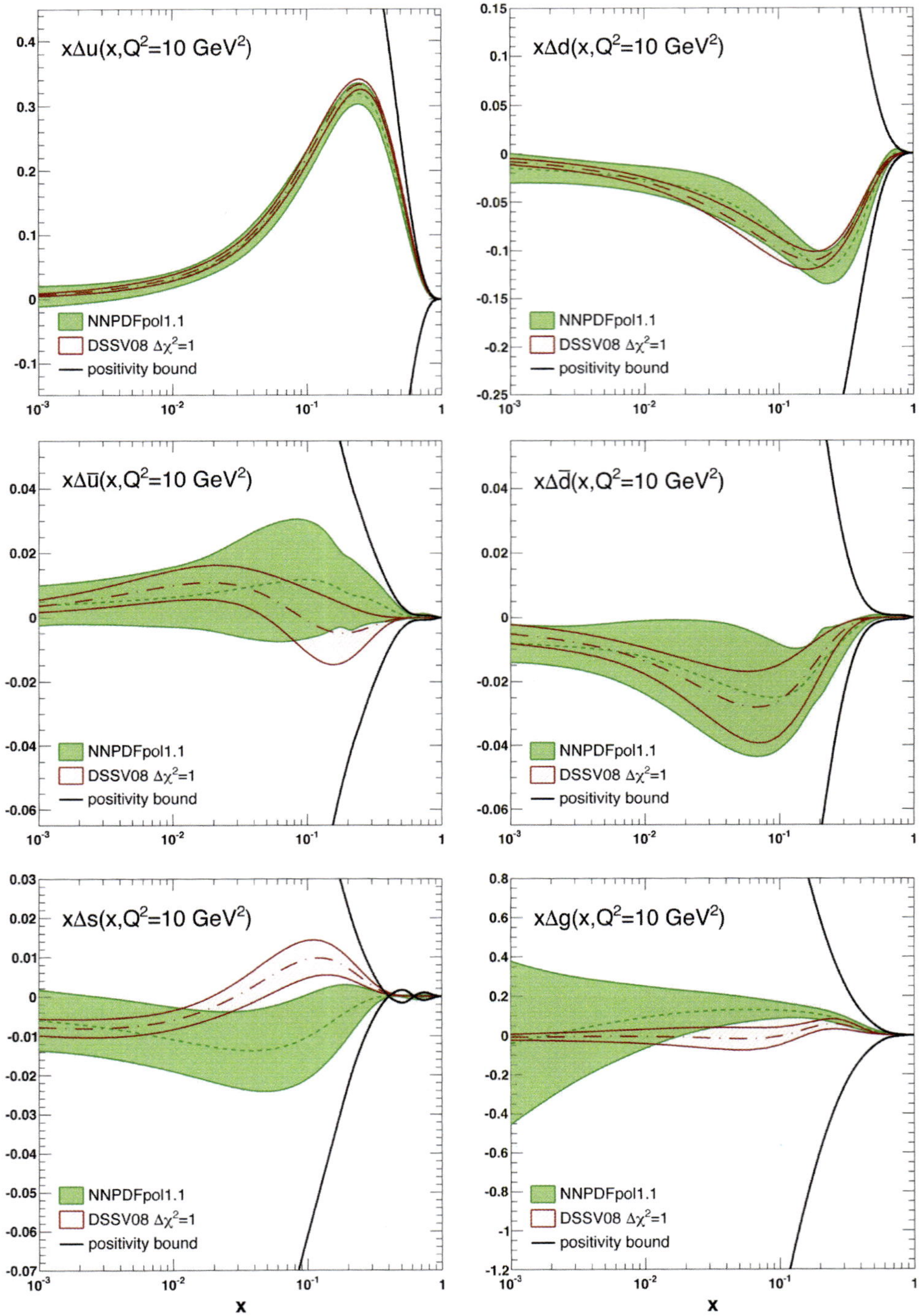

Fig. 11.2 The NNPDFpol1.1 parton set [Nocera (2017)] compared to DSSV08 [de Florian *et al.* (2009)] at $Q^2 = 10$ GeV^2. Both are presented in the $\overline{\text{MS}}$ scheme.

Table 11.1 Full and truncated first moments of the polarised quark distributions at $Q^2 = 10$ GeV^2, for NNPDFpol1.1 [Nocera (2017)], NNPDFpol1.0 [Nocera *et al.* (2014)] (based on inclusive DIS data only) and DSSV08 [de Florian *et al.* (2009)] (which included SIDIS data). The uncertainties shown are one-sigma for NNPDF and Lagrange multiplier with $\Delta\chi^2/\chi^2 = 2\%$ for DSSV. The number in parenthesis for DSSV08 is the contribution that should be added to the truncated moment in order to obtain the full moment. All results are in $\overline{\text{MS}}$ scheme.

	$\langle\Delta f\rangle^{[0,1]}$	$\langle\Delta f\rangle^{[10^{-3},1]}$	
Δf	NNPDFpol1.1	NNPDFpol1.1	DSSV08
Δu^+	$+0.79 \pm 0.07$	$+0.76 \pm 0.04$	$+0.793^{+0.028}_{-0.034}$ (+0.020)
Δd^+	-0.47 ± 0.07	-0.41 ± 0.04	$-0.416^{+0.035}_{-0.025}$ (−0.042)
$\Delta\bar{u}$	$+0.06 \pm 0.06$	$+0.04 \pm 0.05$	$+0.028^{+0.059}_{-0.059}$ (+0.008)
$\Delta\bar{d}$	-0.11 ± 0.06	-0.09 ± 0.05	$-0.089^{+0.090}_{-0.080}$ (−0.026)
Δs	-0.07 ± 0.05	-0.05 ± 0.04	$-0.006^{+0.028}_{-0.031}$ (−0.051)
a_0	$+0.18 \pm 0.21$	$+0.25 \pm 0.10$	$+0.366^{+0.042}_{-0.062}$ (+0.124)

Table 11.2 Same as Table 11.1 but for the $\overline{\text{MS}}$ gluon. Results are also shown for the truncated moment in the range of the RHIC jet data, for NNPDFpol1.1 [Nocera (2017)], DSSV08 [de Florian *et al.* (2009)], and DSSV++ [de Florian *et al.* (2014)].

	$\langle\Delta g\rangle^{[0,1]}$	$\langle\Delta g\rangle^{[10^{-3},1]}$	$\langle\Delta g\rangle^{[0.05,0.2]}$
NNPDFpol1.1	$+0.03 \pm 3.24$	$+0.49 \pm 0.75$	$+0.17 \pm 0.06$
DSSV08	—	$0.01^{+0.70}_{-0.31}$ (+0.10)	$0.01^{+0.13}_{-0.16}$
DSSV++	—	—	$0.10^{+0.06}_{-0.07}$

been discussed over the years [Savage and Walden (1997); Zhu *et al.* (2002); Cabibbo *et al.* (2003); Ratcliffe (2004); Sasaki and Yamazaki (2009)]. The SI measurements with identified kaons [Alekseev *et al.* (2010a); Airapetian *et al.* (2005)] are hence of particular interest since they potentially yield the most direct measurements of the polarisation of strange quarks and anti-quarks, albeit with considerable dependence on the kaon fragmentation functions [Leader *et al.* (2011)]. No evidence for sizable $\Delta s(x, Q^2)$ or $\Delta\bar{s}(x, Q^2)$ has so far been found in polarised SI measurements with fixed targets (see Table 11.1). As a consequence, Δs would need to obtain its negative integral purely from the contribution from the thus far unmeasured small-x region. This situation demonstrates rather clearly the need for simultaneous measurements of the kaon production cross sections and their spin-dependence in SI deep inelastic scattering at smaller x.

Constraints on the spin-dependent gluon distribution, Δg, predominantly come from RHIC, with some information also entering from scaling violations of the deep inelastic structure function $g_1(x, Q^2)$. The production cross sections for inclusive hadrons and jets at RHIC receive contributions from gluon-gluon and quark-gluon scattering and probe $\Delta g(x, Q^2)$ over the range $0.02 < x < 0.4$. Note that the x

is not explicitly resolved in measurements of inclusive pion and jet probes. Initial results from RHIC saw small double spin asymmetries for inclusive jets and hadrons. As a result, the older DSSV fits [de Florian *et al.* (2008, 2009)] concluded that there were no indications of a sizable contribution of gluon spins to the proton spin. The latest much more precise RHIC data for the double spin asymmetry in π^0 and jet production [Adamczyk *et al.* (2015)] provide, for the first time, evidence of a non-vanishing polarisation of gluons in the nucleon in the RHIC kinematic regime [de Florian *et al.* (2014); Nocera (2017)], see Table 11.2.

The limited x-range and unresolved small-x dependence still preclude definitive conclusions on the total gluon spin contribution to the proton spin, Δ_g, although it appears likely now that gluons are an important player. Final results from RHIC at $\sqrt{s} = 200\,\mathrm{GeV}$ will enhance the sensitivity primarily at large x, and measurements of correlated probes are anticipated to yield insights in x-dependence. Forthcoming measurements at $\sqrt{s} = 500\,\mathrm{GeV}$ are expected to extend the small-x reach to $2 \div 3$ times smaller values, and modest further gains may be possible with new instruments at larger pseudorapidity.

While the data from RHIC have had a tremendous impact, particularly as they improve in precision, the essential difficulty in determining the first moments $\Delta\Sigma$ and Δg in Eq. (11.1) remains: the kinematic reach of current experiments is still insufficient to reliably extrapolate from the measured region to small x. We now have tantalising indications that both Δs and Δg may become negative at small x, with obvious implications for their first moments, so this is a key issue. Theoretically, the perturbative expansion of the polarised splitting functions becomes increasingly unstable at small x [Ball *et al.* (1995)], with double logarithms of the form $\frac{\alpha_s^n}{x} \log^{2n-2} \frac{1}{x}$ for $n = 1, 2, \ldots$ (see [Ermolaev *et al.* (2010)] and references therein). Although leading log resummations exist, it is unlikely that they could increase the precision of the pPDFs extrapolated to small-x sufficiently to be able to determine a reliable first moment. Thus all our hopes rest on improved measurements, at a new machine such as the EIC. We discuss this possibility, and what it might achieve, in the last section of this article.

11.4 Determining polarised PDFs from SI processes

In this section we will consider the incorporation of SI processes in determinations of PDFs and pPDFs, through the consistent determination of FFs. Guido was always rather skeptical of the use of SI data in PDF fits, for the simple reason that he remained unconvinced that the uncertainties due to the FFs were properly accounted for. Here we will attempt to convince him, belatedly perhaps, that we now have the tools to do this properly. In particular, we will show how uncertainties can be propagated consistently through the fits, in such a way that the final results are all mutually self-consistent. The discussion will be described in the NNPDF framework, using a replica representation [Giele and Keller (1998)] of PDFs, pPDFs and FFs, but is easily adapted to Hessian representations. We begin with a general discussion of theoretical uncertainties in PDF fits, and then use the results to discuss SI processes. A general fitting strategy is outlined at the end.

11.4.1 *Theoretical uncertainties in PDF fits*

We denote the n data in a particular process by a vector y, and the corresponding theoretical predictions by a vector $t[f]$, where f is a generic ensemble of PDF replicas. Then by Bayes' Theorem

$$P(t|yf)P(y|f) = P(y|tf)P(t|f), \tag{11.2}$$

so that if we integrate over all possible predictions,

$$P(y|f) = \int d^n t\, P(y|tf)P(t|f). \tag{11.3}$$

Now if we assume Gaussian uncertainties for the data, $P(y|tf) \sim \exp(-\frac{1}{2}\chi^2)$, where the tilde means 'up to a normalisation factor', and

$$\chi^2(y, t, \sigma) = (y - t)^T \sigma^{-1}(y - t) \tag{11.4}$$

is the usual χ^2 function for the inclusive data, T denoting the transpose vector, and summation over the n data points being implicit. The n by n matrix σ is the complete covariance matrix for the data y, including both statistical and systematic uncertainties. Likewise we also assume that the theoretical uncertainties are Gaussian,

$$P(t|f) \sim \exp(-\tfrac{1}{2}(t - t[f])^T s^{-1}(t - t[f])), \tag{11.5}$$

where again, summation over the data points is implicit, and s is the covariance matrix for the theoretical uncertainties. Substituting these two Gaussians into Eq. (11.3), the integrand is Gaussian, and we find after some algebra that again $P(y|f) \sim \exp(-\frac{1}{2}\chi^2[f])$, but now with

$$\chi^2[f] = \chi^2(y, t[f], \sigma + s) = (y - t[f])^T(\sigma + s)^{-1}(y - t[f]). \tag{11.6}$$

Thus to include Gaussian theoretical uncertainties in a PDF fit, all we need to do is take our estimate s for the covariance matrix of the theoretical uncertainties in the space of the data, and add it to the experimental covariance matrix σ when computing the χ^2 function to be minimised in the fit. This result is with hindsight rather obvious: there is no distinction mathematically between an experimental and a theoretical uncertainty, so both should be included together on the same footing. In the limit when theoretical uncertainties are ignored, $s \to 0$, Eq. (11.5) becomes a δ-function, and Eq. (11.6) reduces immediately to the usual expression Eq. (11.4) in which the only uncertainties are experimental.

It remains to estimate the theoretical covariance matrix s. A number of sources of theoretical uncertainties might be envisaged as commonly arising in a PDF fit: (i) statistical theoretical uncertainties from Monte Carlo generators. These are perhaps the easiest to treat, since the matrix s is diagonal, and the uncertainties

can either be read off from the MC generator itself, or else estimated by studying the point-to-point fluctuations of the predictions [Ball *et al.* (2017b); Harland–Lang *et al.* (2017)];
(ii) systematic theoretical uncertainties from, for example, missing higher order perturbative corrections. Here the covariance matrix might be estimated by varying renormalisation and/or factorisation scales: then

$$s_{ij} = \langle (t[f;\mu] - t[f;\mu_0])_i (t[f;\mu] - t[f;\mu_0])_j \rangle, \tag{11.7}$$

where μ_0 is the scale adopted in the central prediction (for example Q in DIS), and the angled brackets denote averaging over a given range of μ according to some prescription. Of course this estimate is subject to all the usual caveats of scale variation, and in particular will depend explicitly on the range of scale adopted, and assumptions regarding the independence (or otherwise) of scales in different processes;
(iii) systematic theoretical uncertainties from, for example nuclear corrections. Here one might adopt a procedure similar to (ii), but now averaging over different models of the corrections. Again the result will be only qualitative, but should at least capture the correlations of the corrections between different values of x and Q. Alternatively one might use fits to nuclear data, particularly when these are quite well determined (for example using data from LHC collisions of heavy nuclei);
(iv) in determinations of polarised PDFs, there is a further source of theoretical uncertainty, namely that due from the extraction of the polarised cross sections from the asymmetries, which requires knowledge of the corresponding unpolarised cross section. Here the covariance matrix, to be added to the experimental covariance matrix for the extracted polarised cross sections, could be computed by averaging an expression similar to Eq. (11.7) over replicas of the unpolarised PDFs. Uncertainties in positivity constraints [Altarelli *et al.* (1998c)] could be handled in the same way.

In the next section, we will consider a fifth example, the systematic theoretical uncertainties in semi-inclusive processes from fragmentation functions.

11.4.2 *Uncertainties in PDFs and pPDFs due to uncertainties in FFs*

Now consider semi-inclusive processes, in which the theoretical prediction $t[f,d]$ depends on both PDFs f and FFs d. One way to proceed would be to attempt to determine f and d simultaneously through a minimisation of $\chi^2[f] + \tilde{\chi}^2[f,d]$, where

$$\tilde{\chi}^2[f,d] \equiv \chi^2(\tilde{y}, \tilde{t}[f,d], \tilde{\sigma}), \tag{11.8}$$

with the χ^2 function given by Eq. (11.4), and we denote all SI quantities by adding a tilde. For simplicity, from now on the usual sources of theoretical uncertainty will be absorbed into the experimental uncertainties, so that we can focus on the FFs.

There are a number of difficulties with this approach:
(a) theoretical: in perturbative QCD, both PDFs and FFs are factorised from the hard cross section, and are thus universal, in the sense that they are independent

of any particular process [Altarelli *et al.* (1978a)]. This suggests that they must be mutually independent, and indeed this is the way they are always used. (Note that generalised crossing, while being useful to relate perturbative cross sections, is not expected to hold non-perturbatively, and thus cannot be used to relate PDFs and FFs [Altarelli *et al.* (1979b)].) Determining PDFs and FFs simultaneously by minimising Eq. (11.8) would however induce non-universal correlations between them. However, simply ignoring these correlations would affect the reliability of the residual PDF and FF uncertainties;
(b) statistical: both PDFs and FFs enter the cross section multiplicatively. This means that an uncertainty in the FFs induces a multiplicative uncertainty in the PDFs (and vice versa). Accordingly, the value of the FFs rescales uncertainties in the PDFs, meaning that lower values of the FFs will be preferred. This is a rather subtle version of the d'Agostini bias [D'Agostini (1994)], and seems unavoidable in a joint minimisation;
(c) computational: besides the technical complications of a joint minimisation, $\tilde{t}[f,d]$ is computationally intensive, because of the extra convolutions: while inclusive DIS or SIA have only one convolution per iteration per data point, SIDIS has two (the same as an inclusive hadronic process), while a SI hadronic process has three. This would probably prohibit the use of SI hadronic processes in a fit using lookup tables, as the tables would be very large, and the fitting correspondingly slow.

The solution to all of these problems is to fit PDFs and FFs independently, but consistently, by treating the uncertainties in one or the other as theoretical uncertainties in the fit. Consider for example adding SI processes to a global PDF fit using only inclusive processes, with FFs already determined using SIA. Then for a given fixed PDF the theoretical uncertainty on $\tilde{t}[f,d]$ due to the FFs is

$$\tilde{s}^d_{ij}[f] = \frac{1}{N}\sum_d (\tilde{t}[f,d]) - \tilde{t}[f,d_0])_i(\tilde{t}[f,d]) - \tilde{t}[f,d_0])_j, \tag{11.9}$$

where the sum is over the N FF replicas, and d_0 is the central value of the FF, i.e. replica zero (or, in a Hessian approach, a similar expression summing over Hessian eigenvectors). We then minimise $\chi^2[f] + \tilde{\chi}^2[f,d_0]$ to determine PDF replicas f, where the inclusive χ^2 is given by Eq. (11.4) (or Eq. (11.6) if theoretical uncertainties are also included), and the semi-inclusive contribution

$$\tilde{\chi}^2[f,d_0] = \tilde{\chi}^2(\tilde{y}, \tilde{t}[f,d_0], \tilde{\sigma} + \tilde{s}^d[f_0]), \tag{11.10}$$

the FF uncertainty being treated as a theoretical uncertainty. Note that employing $\tilde{s}^d[f]$ in $\tilde{\chi}^2$ would lead to a d'Agostini bias, so instead we use $\tilde{s}^d[f_0]$, just as in the unbiased treatment [Ball *et al.* (2010)] of multiplicative experimental uncertainties, where $\sigma[f]$ is replaced by $\sigma[f_0]$. The value of f_0 can then be iterated to self-consistency in the usual way.

Now this procedure is evidently symmetrical: we can use the same SI data to instead improve the FFs, by computing the PDF uncertainties

$$\tilde{s}^f_{ij}[d] = \tfrac{1}{N}\sum_f (\tilde{t}[f,d]) - \tilde{t}[f_0,d])_i(\tilde{t}[f,d]) - \tilde{t}[f_0,d])_j, \tag{11.11}$$

and determining the FFs through replicas determined by minimising $\chi^2[d]+\tilde{\chi}^2[f_0,d]$, where now $\chi^2[d]$ is the χ^2 for SIA, given by a formula analogous to Eq. (11.4), and the SI data contribute

$$\tilde{\chi}^2[f_0,d] = \chi^2(\tilde{y},\tilde{t}[f_0,d],\tilde{\sigma}+\tilde{s}^f[d_0]), \tag{11.12}$$

with the PDF uncertainties now being treated as theoretical. Indeed it probably makes more sense to start this way around, since as noted previously SIA is not sufficient to determine the charge separation of the FFs, while the PDFs are already well known from the global fit to inclusive data.

Having obtained a first estimate of FFs in this way, we can of course iterate back and forth, until both FFs and PDFs are determined self-consistently: starting from $(f^{(0)},d^{(0)})$ determined independently from the global inclusive PDF and SIA fits respectively, we can use the SI data to improve the FFs, then use these FF to update the PDFs, then the updated PDFs to improve the FFs: $(f^{(0)},d^{(0)}) \to (f^{(0)},d^{(1)}) \to (f^{(1)},d^{(1)}) \to \cdots$, updating the covariance matrices as we go. It is clear that since after the intial step the further effect of the SI data will be small, this process will converge very rapidly. Moreover the resulting PDF and FF replicas will be statistically independent, as required for universality, there is no d'Agostini bias, and since at each stage only one or the other is being fitted, the convolution with the unfitted distributions can be precomputed before each fit. Consequently the joint determination should take little longer than conventional PDF and FF fits, and most importantly, needs no new technology.

Finally, once we have determined the FFs in this way, we can use them to improve the global determination of the polarised PDFs, by again minimising $\chi^2[\Delta f] + \tilde{\chi}^2[\Delta f, d_0]$, but now with

$$\chi^2[\Delta f] = \chi^2(y, t[\Delta f], \sigma) \tag{11.13}$$

the χ^2 for the inclusive polarised data, $t[\Delta f]$) being the usual polarised predictions for inclusive processes, and

$$\tilde{\chi}^2[\Delta f, d_0] = \chi^2(\tilde{y},\tilde{t}[\Delta f,d_0],\tilde{\sigma}+\tilde{s}^d[f_0]), \tag{11.14}$$

the contribution from the polarised SI data. Again the procedure might be iterated, now between polarised PDFs and fragmentation functions using

$$\tilde{\chi}^2[\Delta f_0, d] = \chi^2(\tilde{y},\tilde{t}[\Delta f_0,d],\tilde{\sigma}+\tilde{s}^{\Delta f}[d_0]), \tag{11.15}$$

but this seems unlikely to lead to significant further improvement in the FFs when one remembers that polarised SI data are rather harder to obtain than unpolarised, and thus inevitably have larger uncertainties.

Note that when determining pPDFs from SI data, there is a significant advantage in fitting the asymmetries rather than extracted polarised cross sections or structure functions: in the asymmetries not only do many experimental systematics cancel, but also much of the dependence on the FFs will also cancel. The contribution to the covariance matrix from FF uncertainties (the polarised equivalent of Eq. (11.9)) will thus be significantly smaller for the SI asymmetries than for the polarised cross sections. Of course it will also be necessary to add a contribution to the experimental covariance matrix from the uncertainties in the unpolarised PDFs, using Eq. (11.11), but this should be smaller still.

11.4.3 *A strategy for the use of SI data to determine pPDFs*

Putting all this together, we might use the following strategy to improve the determination of pPDFs using SI data. Ideally this should be done using a common methodological framework for determining PDFs, FFs and pPDFs, to ensure consistency of the uncertainties passed between them. Fortunately such a framework now exists: we can use as a baseline global PDFs (determined from a wide range of inclusive DIS and hadronic data) from NNPDF3.1 [Ball *et al.* (2017b)], FFs (determined from SIA data) from NNFF1.0 [Bertone *et al.* (2017)], and global pPDFs (determined from polarised inclusive DIS and hadronic data) from NNPDFpol1.1 [Nocera *et al.* (2014)], all determined using NLO pQCD. Alternatively, we might start with the DSSV FFs [de Florian *et al.* (2015, 2017)], and pPDFs [de Florian *et al.* (2014)]. We then proceed in two steps:

- We add unpolarised SI data (both SIDIS and hadronic data), with charge identification of $\pi^{\pm}$, $K^{\pm}$, and $p/\bar{p}$, to upgrade the NNFF1.0 FFs determined from SIA, and in particular separate out quark and anti-quark FFs. Once this is done, we could in turn use the same data to improve the NNPDF3.1 PDFs: here we expect changes to be small, except possibly in the strange sector, where $K^{\pm}$ SI processes could significantly improve the determination of s and $\bar{s}$. Iteration to consistency should then be very rapid.
- taking the FFs from the first step, we can now use polarised SI data, again with charge identification of $\pi^{\pm}$, $K^{\pm}$, and $p/\bar{p}$, to upgrade the NNPDFpol1.1 pPDFs, improving the separation of quark and anti-quark, and again with significant improvements expected in the polarised strange quark distributions. Again, once this is done, the polarised SI data can be used to improve the FFs, though here the impact should be relatively small. Consequently, iteration to consistency should be even more rapid.

To carry out this sequence of fits, we would need to prepare FF tables for both unpolarised and polarised SI predictions, with either the (p)PDF or FF fixed to the appropriate central values, thus $\tilde{t}[f_0, d]]$, $\tilde{t}[f, d_0]$, $\tilde{t}[\Delta f, d_0]$, $\tilde{t}[\Delta f_0, d]$, with the corresponding covariance matrices $\tilde{s}^d[f_0]$, $\tilde{s}^f[d_0]$. However once this is done, the actual fitting should be no more challenging than in any other parton fit.

11.5 Experimental prospects at the EIC

Due to its high luminosity and extended kinematic reach (see Fig. 11.1), the EIC [Deshpande *et al.* (2005); Accardi *et al.* (2016b); Aschenauer *et al.* (2014)] offers the possibility of finally measuring the first moments of the parton helicities with sufficient accuracy to finally resolve the proton spin puzzle. Various studies have been performed [Aschenauer *et al.* (2012, 2013); Ball *et al.* (2014)] by adding simulations of EIC data to DSSV and NNPDFpol fits of pPDFs. Figure 11.3 shows the results from the DSSV studies for the sea quark and gluon helicity distributions. For comparison, the present uncertainty bands are also displayed. As one can see, an impressive reduction in the width of the bands is expected from EIC data, in particular, towards lower values of x. Evidently, extractions of Δg from scaling violations, and of the light-flavour helicity distributions Δu, Δd and their anti-quark distributions from semi-inclusive scattering will be possible with exquisite precision. With dedicated studies of kaon production, the strange and anti-strange distributions would also be accessible.

The projected reduction in the uncertainties of the gluon helicity Δg and the total quark helicity $\Delta\Sigma$ (in $\overline{\text{MS}}$ scheme, so this is actually $a_0(Q^2)$), comparing

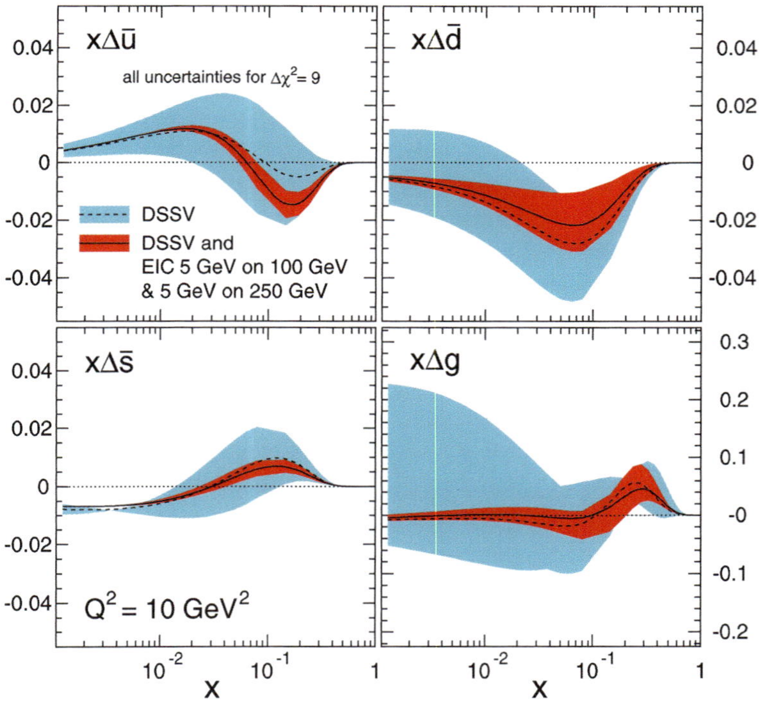

Fig. 11.3 The uncertainty bands on polarised parton distributions in $\overline{\text{MS}}$ scheme, in the DSSV analysis (blue bands) and with EIC pseudo-data (red bands) using projected DIS and SIDIS EIC data expected in the future.

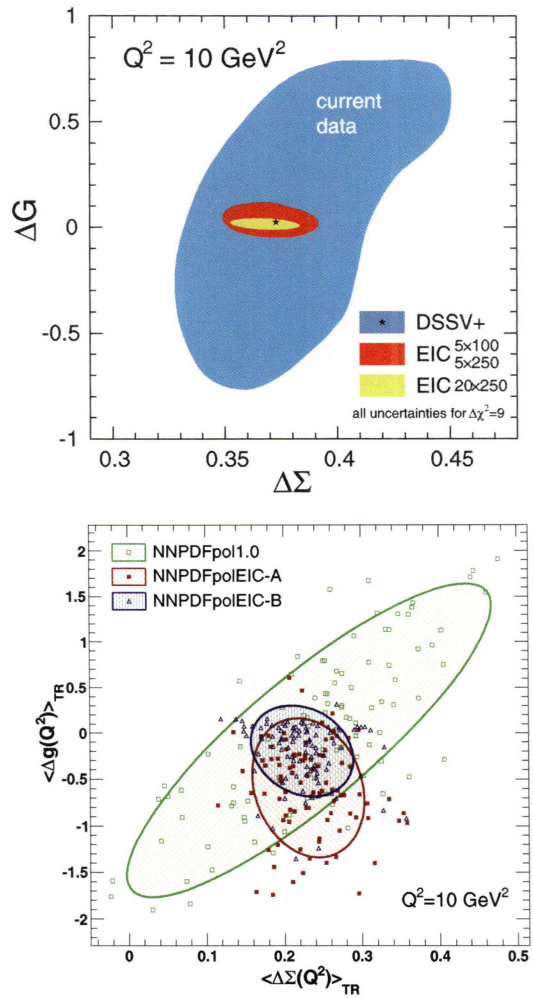

Fig. 11.4 The projected reduction in the uncertainties of the gluon helicity Δg and the total quark helicity $\Delta\Sigma$, comparing the current situation with what might be achieved by the EIC with operation at different beam energy combinations. The first moments are evaluated in the measured region $[0.001, 1]$. The upper plot shows results from the DSSV analysis [Aschenauer *et al.* (2012)], current situation in blue, and projections in red and yellow. The uncertainties are determined with $\Delta\chi^2 = 9$. The lower plot shows similar results obtained by NNPDF [Ball *et al.* (2014)], current situation in green, and projections in red and blue. The uncertainties are now genuine 1-σ uncertainties obtained by averaging over replicas (also shown). Note the different scales on the two plots.

the current situation with what might be achieved by the EIC with operation at different beam energy combinations, is shown in Fig. 11.4. Results are shown both for the DSSV analysis [Aschenauer *et al.* (2012)] and from an independent analysis using NNPDF methodology [Ball *et al.* (2014)]. Although the results of the NNPDF analysis are more conservative, as might be expected, the general conclusions are the same: the anticipated kinematic range and precision of EIC data will give unprecedented insight into the size of the contributions Δg and $\Delta\Sigma$.

Besides polarised proton beams, the EIC design envisions beams of polarised deuterons or helium-3. The neutron's $g_1(x, Q^2)$ can thus be determined, potentially with a precision that is comparable to the data on $g_1(x, Q^2)$ of the proton. The difference in the moments of proton and neutron $g_1(x, Q^2)$ allows for a test of the fundamental sum rule by Bjorken [Bjorken (1969)]. The data from polarised fixed target experiments have so far verified the sum rule to no better than 20% of its value [Nocera *et al.* (2014)]. The extended kinematic range and improved precision of EIC data will offer a more stringent test of this sum rule, as well as its corrections, to an accuracy that is expected to be driven mostly by advances in hadron beam polarimetry.

An additional, and unique, avenue for delineating the flavour structure of the quark and anti-quark spin contribution to the proton spin at the EIC is electroweak deep inelastic scattering. At high Q^2, the deep inelastic process also proceeds significantly via exchange of Z and $W^\pm$ bosons. This gives rise to novel structure functions that are sensitive to different combinations of the proton's helicity distributions. For instance, in the case of charged-current interactions through W^-, the inclusive structure functions contribute

$$g_1^{W^-} = \Delta u + \Delta\bar{d} + \Delta c + \Delta\bar{s}, \quad g_5^{W^-} = -\Delta u + \Delta\bar{d} - \Delta c + \Delta\bar{s}, \tag{11.16}$$

where Δc denotes the proton's polarised charm-quark distribution. The analysis of these structure functions does not rely on knowledge of fragmentation. Studies show that both charged-current [Aschenauer *et al.* (2014)] and neutral-current [Zhao *et al.* (2017)] interactions would be observable at the EIC, even with relatively modest integrated luminosities. To fully exploit the potential of the EIC for such measurements, positron beams are required, albeit not necessarily polarised. Besides the new insights into nucleon structure this would provide, studies of spin-dependent electroweak scattering at short distances with an EIC would be beautiful physics in itself, much in the line of past and ongoing electroweak measurements at HERA, Jefferson Laboratory, RHIC, and the LHC.

11.6 EIC realisation and status

The US nuclear scientific community meets approximately every five years to prepare a long range plan that is effective for 10 subsequent years. In their last such planning exercise held in 2015, the community recommended construction of a high-energy, high-luminosity polarised electron-proton and electron-nucleus collider, now called the Electron Ion Collier (EIC), as the highest-priority new construction facility. The primary aim of this machine that convinced the broader community to stand behind the recommendation included the role of gluons and sea quarks in building the nucleons and nuclei, and the partonic dynamics that underlies the interactions in QCD. Understanding the proton's spin structure, the principal aspect of the present article, is only one of the three major thrusts of the scientific goals of this future machine. To achieve the scientific programme outlined in the science case the machine's designs should be able to deliver:

- highly polarised ($\sim$ 70%) electron and nucleon beams;
- ion beams from deuteron to the heaviest nuclei (uranium or lead);
- variable centre-of-mass energies from $\sim 40-\sim 100$ GeV, upgradable to $\sim$150 GeV;
- high collision luminosity $\sim 10^{33-34}$ cm^{-2}s^{-1};
- possibilities of having more than one interaction region.

Currently there are two possible technical designs being considered by the DOE, based on existing infrastructure investments. One proposes to add an electron beam facility to the RHIC complex at BNL to realise the EIC as eRHIC. A linac based design, ambitious in its R&D requirements, has been worked out, and a more conventional ring-ring design is now being finalised as an option to reduce the uncertainties associated with the challenges in the linac-ring design. The conceptual design of eRHIC is shown in Fig. 11.5. The other uses the recently upgraded CEBAF at JLab as an electron beam injector into a green field electron and hadron/nuclear beam complex to be built adjacent to CEBAF, in a novel bow-tie (or 8) shape and plans to achieve the abovementioned collision parameters. The conceptual layout of JLEIC is shown in Fig. 11.6. Both designs are expected to push the limits of our knowledge and control of accelerator parameters way beyond the state-of-the art,

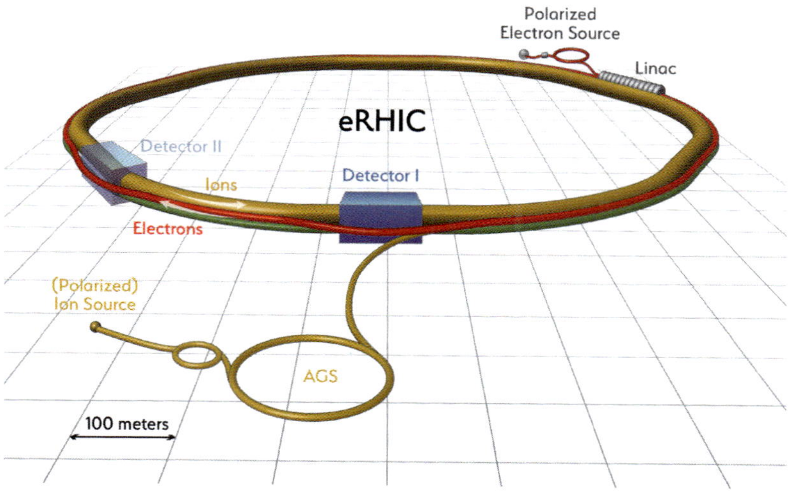

Fig. 11.5 The layout of eRHIC: The blue and yellow rings of RHIC exist, to which the electron beam facility would be added (shown in red) to realise electron ion collisions in two experimental areas as shown.

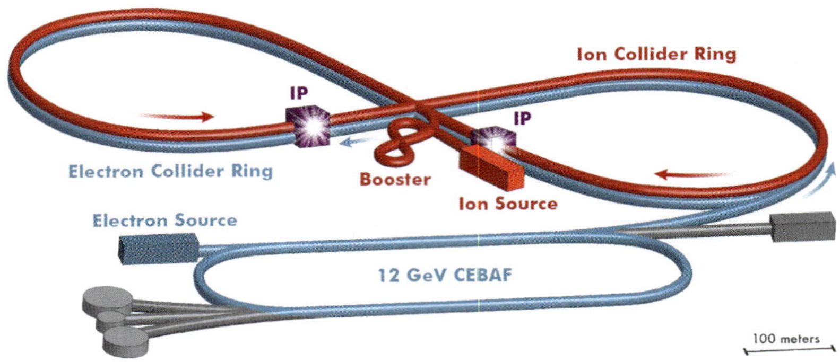

Fig. 11.6 The layout of JLEIC: The 12 GeV upgraded CEBAF in blue exists, to which the bow-tie (or 8) shaped complex with red (ion) and electron (blue) rings would be added, to realise collisions in two experimental areas as shown.

and hence are expected to be critical for future development and advancement in accelerator science.

The physics-driven requirements on the EIC accelerator parameters and extreme demands on the kinematic coverage for measurements make integration of the detector into the accelerator a particularly challenging and yet absolutely essential feature of the EIC. Lessons learned from HERA have been considered while designing the EIC interaction region. Driven by the demand for high precision on particle detection and identification of final state particles in both e+p and e+A programmes, modern particle detector systems will be critical at the EIC. In order to keep the detector costs manageable, R&D efforts are under way on various novel ideas for compact (fibre sampling and crystal) calorimetry, tracking (NaI coated GEMs, GEM size and geometries), particle identification (compact DIRC, dual radiator RICH and novel TPC) and high radiation tolerance for electronics.

Motivated by the prospects of realisation of EIC in the US, a worldwide group of interested scientists have come together to form an EIC Users Group (eicug.org). Current participants, not including students, include $\sim$750 scientists, from $\sim$160 institutions and $\sim$29 countries around the world. It is expected that two interaction regions will provide for two independent detector systems built and operated by a worldwide user community forming at least two large experimental collaborations.

Considering the historical trends of US funding over the last 10 years, and assuming the same for the next 10 years, the Long Range Planning exercise suggests that the construction of the EIC could begin as early as 2022. Depending on which design is adopted (eRHIC or the JLEIC), one could then expect first collisions between 2027 and 2030. Currently the US National Academy of Science (NAS) is holding its own independent review of the EIC Science and is expected to give its verdict by Spring 2018. After a positive evaluation of its science, the community expects a Critical Decisions Zero (CD0) status for the US EIC to be granted by the DOE.

11.7 Conclusions

In this article we focused on one aspect of QCD and the possible physics at a future EIC, that of the helicity contributions from the quarks and gluons to the nucleon's spin, due to Guido's work and interest in this area. While there are now indications that gluons may carry a non-zero fraction of the proton's helicity, the situation is still inconclusive due to large uncertainties that come from unmeasured low-x regions. We have outlined two strategies for improving this situation in coming years: the systematic inclusion of polarised semi-inclusive data, to increase precision in the measured region, and improving the kinematic reach by polarised beam collisions at the future EIC. We still have a long way to go before we fully understand the partonic content and the dynamical origin of the proton spin in terms of the proton's constituents, however, we now know exactly what is needed to get us there.

Acknowledgements

RBD: I would like to take this opportunity to thank Guido, in absentia, for inspiration and support over many years. I first met Guido at Erice in 1983, and the impression he made on me then never left me. Many people's memories of Guido revolve around the slightly menacing question, calmly and simply expressed in his inimitable style, that identified precisely the weak point in any argument. However this was only one side of Guido; on a more personal level, he could also be both constructive and supportive. At the first scientific meeting of the Higgs Centre in Edinburgh in 2013, I was discussing with Guido some of the political difficulties we faced. His response was characteristically cryptic: "the start is a most crucial step". Looking back now at his career, Guido made many crucial first steps: in QCD, Altarelli–Parisi, Altarelli–Ellis–Martinelli, Altarelli–Ross have each laid the foundations for an area of research that remains active to this day. It is generally much harder to start something than it is to end it, but at present there is no end in sight to the legacy of Guido Altarelli.

AD: I met Guido first in 1996 when he first presented new data on g_1^d from the SMC collaboration's NA47 experiment at the CERN theory seminar. Thereafter I remember fondly the discussions with Guido and Vernon Hughes about the need for a polarised electron-proton/deuteron collider. The idea of polarised HERA was very attractive then [Ball *et al.* (1996a)], but unfortunately never happened due to the technical challenges of realising polarised protons. However, after I mentioned the new idea of a polarised EIC (which used to be called eRHIC then) in the US, Guido strongly encouraged me to pursue it. Again his attitude was to encourage the crucial first step. Guido was very supportive of the RHIC spin programme as a proof of principle for polarising protons at high energy, making way for a future

EIC. He would doubtless be very pleased that the EIC project, with high-energy polarised proton, light ion and electron beams has evolved to the point that we can be optimistic that it will actually get built. I would like to thank Guido for the enthusiasm that he showed for these new projects, and the support he gave to a young postdoctoral fellow like me.

Chapter 12

Are neutrino masses modular forms?

Ferruccio Feruglio

Dipartimento di Fisica e Astronomia 'G. Galilei', Università di Padova
INFN, Sezione di Padova, Via Marzolo 8, I-35131 Padua, Italy

We explore a new class of supersymmetric models for lepton masses and mixing angles where the role of flavour symmetry is played by modular invariance. The building blocks are modular forms of level N and matter supermultiplets, both transforming in representations of a finite discrete group Γ_N. *In the simplest version of these models, Yukawa couplings are just modular forms and the only source of flavour symmetry breaking is the vacuum expectation value of a single complex field, the modulus. In the special case where modular forms are constant functions the whole construction collapses to a supersymmetric flavour model invariant under* Γ_N, *the case treated so far in the literature. The framework has a number of appealing features. Flavon fields other than the modulus might not be needed. Neutrino masses and mixing angles are simultaneously constrained by the modular symmetry. As long as supersymmetry is exact, modular invariance determines all higher-dimensional operators in the superpotential. We discuss the general framework and we provide complete examples of the new construction. The most economical model predicts neutrino mass ratios, lepton mixing angles, Dirac and Majorana phases uniquely in terms of the modulus vacuum expectation value, with all the parameters except one within the experimentally allowed range. As a byproduct of the general formalism we extend the notion of non-linearly realised symmetries to the discrete case.*

Guido literally dragged me into the world of neutrinos during the summer of 1998, when I was visiting CERN. He did it with his contagious enthusiasm and energy, starting a long collaboration in which scientific passion got mixed up with friendship. I owe him a big debt of gratitude and I consider myself very lucky to have had the chance of being close to him for a long time. Without meeting Guido my life would have been much more poor, both on the professional and on the human sides. I remember him with deep affection.

12.1 Introduction

A considerable interest in discrete flavour symmetries [Altarelli and Feruglio (2010); Ishimori *et al.* (2010); Hernandez and Smirnov (2012); King and Luhn (2013); King *et al.* (2014); King (2017); Hagedorn (2017)] has been fostered by early models of quark masses and mixing angles [Pakvasa and Sugawara (1978); Wilczek and Zee (1977)] and, more recently, by the discovery of neutrino oscillations. Early data were well-compatible with a highly symmetric lepton mixing pattern, the tri-bimaximal one [Harrison *et al.* (2002)], which could be derived from small non-Abelian discrete symmetry groups such as A_4 [Ma and Rajasekaran (2001); Babu *et al.* (2003); Altarelli and Feruglio (2005)]. Other discrete groups like S_4 and A_5 produced interesting alternative mixing patterns, which could be adopted as zeroth-order approximation to the data. Today this approach is facing several difficulties. The formidable recent experimental progress has sharpened the neutrino oscillation parameters, revealing many details that require a precise description, such as the non-vanishing value of the reactor angle, the deviation of the atmospheric angle from the maximal value and a non-trivial Dirac CP-violating phase. Inclusion of these features in a realistic model based on discrete symmetries requires departure from minimality. Large corrections to the zeroth-order approximation can be introduced at the price of spoiling predictability, due to the ignorance about the non-negligible higher order contributions. Alternatively, groups of large dimensionality can be invoked to correctly fit the data [de Adelhart Toorop *et al.* (2011, 2012); Holthausen *et al.* (2013a); King *et al.* (2013); Hagedorn *et al.* (2014); Fonseca and Grimus (2014)]. Discrete flavour symmetries can also be combined with CP invariance in predictive models [Feruglio *et al.* (2013); Holthausen *et al.* (2013b)]. Apart from the loss of minimality, there are several drawbacks in this programme. The breaking of flavour symmetries typically relies on a generous set of scalar multiplets, the so-called flavons, and the Yukawa interactions generally include non-renormalisable operators with flavon insertions. Higher-dimensional operators with multiple flavon insertions come with unknown coefficients that affect the model predictions. Moreover the flavon energy density has to be cleverly designed to get the correct vacuum alignment. The approach is mainly focused on lepton mixing angles while neutrino masses are reproduced by tuning the available parameters. Finally, it is not straightforward to extend the construction to the quark sector that seems not to like discrete symmetries. In view of these disadvantages, anarchy [Hall *et al.* (2000); Haba and Murayama (2001); de Gouvea and Murayama (2003); Espinosa (2003); de Gouvea and Murayama (2015)] and its generalisations have gained considerable momentum. Anarchy in the neutrino sector can arise in a variety of different frameworks providing a common description to both quark and lepton mass/mixing parameters, also in the context of grand unified theories [Feruglio (2015)]. However in the anarchy paradigm the observed lepton mixing angles are regarded as environmental quantities [Schellekens (2013)] and cannot be accurately predicted. For their intrinsic nature, models based on anarchy essentially escape experimental tests aiming at an accuracy that matches the experimental precision.

In this wavering between order and anarchy we feel encouraged to investigate new directions. The aim of the present work is to explore a new class of models generalising the current approach based on discrete symmetry groups. These models are required to be invariant under transformations of the modular group, acting on the complex modulus τ ($\mathtt{Im}(\tau) > 0$) as linear fractional transformations:

$$\tau \to \frac{a\tau + b}{c\tau + d}, \qquad (a, b, c, d \text{ integers}, \qquad ad - bc = 1).$$

In a supersymmetric theory these transformations naturally induce transformations of the matter multiplets according to representations of Γ_N, the so-called finite modular groups. Moreover there are holomorphic combinations of the modulus τ, the modular forms of level N, that also transform in representations of Γ_N. We can thus exploit group theoretical properties of Γ_N to build supersymmetric, modular-invariant models of fermion masses, where matter fields and modular forms are used as building blocks. A particular case of modular forms is that of constant functions. In this case the whole construction collapses to the case that has been considered in the literature so far when dealing with discrete flavour symmetries. On the other hand a new, presently unexplored, class of models arises when modular forms carry a non-trivial dependence on τ. For small N the finite modular groups are permutation groups and the new models are generalisations of well known existing models.

In this paper we illustrate how the new construction can be realised, by discussing the general formalism and by illustrating it with simple examples. In the minimal examples, see Section 12.3.1.4, there are no flavons and the only source of flavour symmetry breaking is the vacuum expectation value (VEV) of the modulus τ. A truly remarkable feature of these examples is that all higher-dimensional operators in the superpotential are unambiguously determined in the limit of unbroken supersymmetry. The possible corrections to the theory predictions can only come from the Kahler potential or from supersymmetry breaking contributions. Moreover the modular symmetry constrains lepton mixing angles and neutrino masses at the same time. In the most economical examples, discussed in Sections 12.3.1.2 and 12.3.1.3, neutrino mass ratios, lepton mixing angles, Dirac and Majorana phases do not depend on any Lagrangian parameter, but only on the modulus VEV. In the non-trivial task of reproducing five known observables in terms of a single complex parameter, the model of Section 12.3.1.2 scores relatively well: only $\sin^2\theta_{13}$, predicted to be about 0.045, is out of the allowed experimental range. The model predicts inverted neutrino mass ordering and determines completely the Majorana phases. The model of Section 12.3.1.3 predicts normal neutrino mass ordering and is almost equally well-performing.

An unexpected byproduct of the new framework is a new concept of non-linearly realised symmetry, applied to discrete groups. This is a built-in feature of modular forms of level N but, in the second part of this work, we will analyse it in a more general framework. In the context of relativistic quantum field theories, non-linear realisations of a continuous symmetry are synonymous of symmetry breaking and of the systematic low-energy expansion describing the massless modes predicted by the Goldstone theorem. The applications of these concepts are ubiquitous, ranging from the description of fundamental particle interactions to the theory of phase transitions. Extension of this formalism to the case of discrete symmetries may look unmanageable, especially when thinking to the Callan–Coleman–Wess–Zumino construction [Coleman *et al.* (1969); Callan *et al.* (1969)], which is tightly connected to the Goldstone theorem. Moving from the properties of level-3 modular forms, we show how non-linear realisations of discrete symmetries can be defined and how they are related to the symmetry breaking pattern of a discrete group. In the continuous case non-linear realisations can be defined also by imposing invariant constraints on the field space, such as for instance $\varphi^T\varphi = M^2$, the multiplet φ transforming in the fundamental of $SO(N)$. In the discrete case this approach benefits from an important generalisation. While a (connected, semisimple) Lie group has only the trivial one-dimensional representation, a discrete group can possess a number of non-invariant singlets χ_i. Therefore the invariant conditions $\chi_i = 0$ have no counterpart in (connected, semisimple) Lie groups. If the product of a set of multiplets φ contains one or more non-invariant singlets $\chi_i(\varphi)$, the conditions $\chi_i(\varphi) = 0$ may define a non-linear realisation of the discrete group. This is precisely what happens with modular forms of level 3. Non-linear realisations of a discrete symmetry do not define low-energy effective theories. They should rather be viewed as consistent truncations of some ultraviolet completion. As we will see, in general, the requirement $\chi_i(\varphi) = 0$ leaves no non-trivial residual symmetry. We identify the cases in which a non-trivial residual symmetry group survives and we discuss the relation of such events with the properties of the so-called orbit space, the space spanned by the invariants of the group. This last discussion is only very tentative, but we hope that in the future it may be better clarified, offering further possibilities in model building.

12.2 Modular forms and modular-invariant theories

In this section we briefly recap definitions and properties of modular forms. We also review the framework of modular invariant supersymmetric theories. The material of this section is quite known, but will be presented here from a different perspective, emphasising the role of finite discrete symmetries in modular invariant theories. Modular invariance has a long history both in string and in field theories. Target space modular invariance has been investigated soon after the discovery that the spectrum of a closed string, when compactified on a circle of radius R, is invariant under the duality transformation $R \to 1/2R$, which corresponds to the linear fractional transformation $\tau \to -1/\tau$ restricted to the imaginary part of τ ($\tau = 2iR^2$). Modular invariance controls orbifold compactifications of the heterotic string and requires that the couplings among twisted states are modular forms [Hamidi and Vafa (1987); Dixon *et al.* (1987); Lauer *et al.* (1989, 1991); Erler *et al.* (1992)]. This holds in particular for realistic Yukawa couplings [Ibanez (1986); Casas *et al.* (1993); Lebedev (2001); Kobayashi and Lebedev (2003)]. In orientifold compactifications of Type II strings the Yukawa couplings are functions with specific modular properties [Cremades *et al.* (2003); Blumenhagen *et al.* (2005); Abel and Goodsell (2007); Blumenhagen *et al.* (2007); Marchesano (2007); Antoniadis *et al.* (2009); Kobayashi *et al.* (2017)]. A similar feature is found also in magnetised extra dimensions [Cremades *et al.* (2004); Abe *et al.* (2009)]. Modular invariant supersymmetric field theories have been analysed in the late 80s [Ferrara *et al.* (1989a,b)], both for global and local supersymmetry. Modular invariance in field theory constructions has been invoked while addressing several aspects of the flavour problem in model building [Brax and Chemtob (1995); Binetruy and Dudas (1995); Dudas *et al.* (1996); Dudas (1996); Leontaris and Tracas (1998); Dent (2001a,b)]. Duality and modular invariance have been suggested as underlying properties of the quantum Hall effect [Lutken and Ross (1992); Cappelli and Zemba (1997); Dolan (1999b,a); Burgess and Dolan (2001, 2002, 2007); Lutken and Ross (2007, 2011)].

12.2.1 *Modular forms of level N*

We consider the series of groups $\Gamma(N)$ ($N = 1, 2, 3, \ldots$) defined by:

$$\Gamma(N) = \left\{ \begin{pmatrix} a & b \\ c & d \end{pmatrix} \in SL(2,Z), \begin{pmatrix} a & b \\ c & d \end{pmatrix} = \begin{pmatrix} 1 & 0 \\ 0 & 1 \end{pmatrix} \pmod{N} \right\}, \tag{12.1}$$

where $\Gamma \equiv SL(2,Z)$ is the group of two by two matrices with integer entries and determinant equal to one, also called homogeneous modular group. We have $\Gamma = \Gamma(1)$, and the groups $\Gamma(N)$ $(N \geq 2)$ are infinite normal subgroups of Γ, called principal congruence subgroups. The group $\Gamma(N)$ acts on the complex variable τ, varying in the upper-half complex plane $\mathcal{H} = \mathtt{Im}(\tau) > 0$, as the linear fractional transformation

$$\gamma\tau = \frac{a\tau + b}{c\tau + d}. \tag{12.2}$$

The quotient space $\mathcal{H}/\Gamma(N)$ can be compactified by adding special points called cusps. They coincide with $i\infty$ or with rational real numbers. While it is very convenient to deal with two by two matrices, the groups $\overline{\Gamma}(N)$ of linear fractional transformations are slightly different from the groups $\Gamma(N)$. Since γ and $-\gamma$ induce the same linear fractional transformation, for $N = 1$ and $N = 2$, the transformations (12.2) are in a one-to-one correspondence with the elements of the group $\overline{\Gamma} \equiv \Gamma/\{\pm\mathbb{1}\}$ and $\overline{\Gamma}(2) \equiv \Gamma(2)/\{\pm\mathbb{1}\}$, respectively. $\overline{\Gamma}$ is called inhomogeneous modular group, or simply modular group. When $N > 2$ the element $-\mathbb{1}$ does not belong to $\Gamma(N)$ and we have $\overline{\Gamma}(N) \equiv \Gamma(N)$. The group $\overline{\Gamma}$ is generated by two elements S and T satisfying:

$$S^2 = \mathbb{1}, \quad (ST)^3 = \mathbb{1}. \tag{12.3}$$

They can be represented by the $SL(2,Z)$ matrices:

$$S = \begin{pmatrix} 0 & 1 \\ -1 & 0 \end{pmatrix} \qquad T = \begin{pmatrix} 1 & 1 \\ 0 & 1 \end{pmatrix}, \tag{12.4}$$

corresponding to the transformations

$$S : \tau \to -\frac{1}{\tau}, \qquad T : \tau \to \tau + 1. \tag{12.5}$$

The quotient groups $\Gamma_N \equiv \overline{\Gamma}/\overline{\Gamma}(N)$ are called finite modular groups. Some of their properties can be found in ref. [de Adelhart Toorop *et al.* (2012)].

Modular forms $f(\tau)$ of weight $2k$ and level N are holomorphic functions of the complex variable τ with well defined transformation properties under the group $\Gamma(N)$ [Gunning (1962)]:

$$f(\gamma\tau) = (c\tau + d)^{2k} f(\tau) \qquad \gamma = \begin{pmatrix} a & b \\ c & d \end{pmatrix} \in \Gamma(N), \tag{12.6}$$

where $k \geq 0$ is an integer.[1] The function $f(\tau)$ is required to be holomorphic in $\mathcal{H}$ and at all the cusps. In the special case when it vanishes at all the cusps, $f(\tau)$ is called a cusp form. For $\Gamma(N)$ $(N \geq 2)$, N is the level of the group and T^N belongs to $\Gamma(N)$. It follows that $f(\tau + N) = f(\tau)$ and we have the Fourier expansion

[1]Following [Gunning (1962)], here we only consider modular forms of even weight.

(or simply q-expansion):

$$f(\tau) = \sum_{i=0}^{\infty} a_n q_N^n \qquad q_N = e^{\frac{i2\pi\tau}{N}}. \tag{12.7}$$

For $k < 0$ there are no modular forms. If $k = 0$ the only possible modular form is a constant. Modular forms of weight $2k$ and level N form a linear space $\mathcal{M}_{2k}(\Gamma(N))$ of finite dimension $d_{2k}(\Gamma(N))$. The dimensions $d_{2k}(\Gamma(N))$ for few levels N are shown in Table 12.1, derived from ref. [Gunning (1962)]. General formulas are reported in Appendix A. The product of two modular forms of level N and weights $2k$, $2k'$ is a modular form of level N and weight $2(k+k')$ and the set $\mathcal{M}(\Gamma(N))$ of all modular forms of level N is a ring

$$\mathcal{M}(\Gamma(N)) = \bigoplus_{k=0}^{\infty} \mathcal{M}_{2k}(\Gamma(N)), \tag{12.8}$$

generated by few elements. For instance $\mathcal{M}(\Gamma)$ is generated by two modular forms $E_4(\tau)$ and $E_6(\tau)$ of weight 4 and 6 respectively, so that each modular form in $\mathcal{M}_{2k}(\Gamma)$ can be written as a polynomial $\sum_{ij} c_{ij}\ E_4(\tau)^{n_i} E_6(\tau)^{n_j}$, with powers satisfying $2k = 4n_i + 6n_j$.

Central to the construction illustrated in the next sections is the following result, explicitly proved in Appendix B. Modular forms of weight $2k$ and level $N \geq 2$ are invariant, up to the factor $(c\tau + d)^{2k}$, under $\Gamma(N)$ but they transform under the quotient group $\Gamma_N \equiv \overline{\Gamma}/\overline{\Gamma}(N)$. As shown in Appendix B, it is possible to

Table 12.1 Some properties of modular forms: g is the genus of the space $\mathcal{H}/\Gamma(N)$ after compactification, $d_{2k}(\Gamma(N))$ the dimension of the linear space $\mathcal{M}_{2k}(\Gamma(N))$, μ_N is the dimension of the quotient group $\Gamma_N \equiv \overline{\Gamma}/\overline{\Gamma}(N)$, which, for $N \leq 5$, is isomorphic to a permutation group.

N	g	$d_{2k}(\Gamma(N))$	μ_N	Γ_N
2	0	$k+1$	6	S_3
3	0	$2k+1$	12	A_4
4	0	$4k+1$	24	S_4
5	0	$10k+1$	60	A_5
6	1	$12k$	72	
7	3	$28k-2$	168	

choose a basis in $\mathcal{M}_{2k}(\Gamma(N))$ such that this transformation is described by a unitary representation ρ of Γ_N:

$$f_i(\gamma\tau) = (c\tau + d)^k \rho(\gamma)_{ij} f_j(\tau), \tag{12.9}$$

where

$$\gamma = \begin{pmatrix} a & b \\ c & d \end{pmatrix} \tag{12.10}$$

stands for a representative element of Γ_N. In practice, it is sufficient to determine the representation ρ for the two elements S and T that generate the entire modular group and, for $N \geq 2$, are not contained in $\overline{\Gamma}(N)$. While the groups $\overline{\Gamma}(N)$ are infinite, the quotient groups Γ_N are finite. In Table 12.1 we show few groups Γ_N and their dimensions.

12.2.2 *Modular-invariant supersymmetric theories*

Here we briefly review the formalism [Ferrara *et al.* (1989a,b)], starting from the case of $N = 1$ global supersymmetry, where the action takes the general form[2]:

$$\mathcal{S} = \int d^4x d^2\theta d^2\bar{\theta}\; K(\Phi, \bar{\Phi}) + \int d^4x d^2\theta\; w(\Phi) + h.c. \tag{12.11}$$

where $K(\Phi, \bar{\Phi})$, the Kahler potential, is a real gauge-invariant function of the chiral superfields Φ and their conjugates and $w(\Phi)$, the superpotential, is a holomorphic gauge-invariant function of the chiral superfields Φ. With $\Phi = (\tau, \varphi)$ we denote collectively all chiral supermultiplets of the theory, including the modulus[3] τ plus a number of additional chiral supermultiplets φ separated into sectors $\varphi^{(I)}$. In general each sector I contains several chiral supermultiplets $\varphi^{(I)}_{i_I}$ but, to keep our notation more concise, we will not write the additional index i_I. We ask for invariance under transformations of the modular group $\overline{\Gamma}$, under which the supermultiplets $\varphi^{(I)}$ of each sector I are assumed to transform in a representation $\rho^{(I)}$ of a quotient group Γ_N, kept fixed in the construction, with a weight $-k_I$:

$$\begin{cases} \tau \to \dfrac{a\tau + b}{c\tau + d} \\ \varphi^{(I)} \to (c\tau + d)^{-k_I} \rho^{(I)}(\gamma) \varphi^{(I)} \end{cases}. \tag{12.12}$$

The supermultiplets $\varphi^{(I)}$ are not modular forms and there are no restrictions on the possible value of k_I, a priori. The invariance of the action $\mathcal{S}$ under (12.12) requires the invariance of the superpotential $w(\Phi)$ and the invariance of the Kahler potential

[2] We focus on Yukawa interactions and we turn off the gauge interactions.

[3] In the literature it is common to make use of the field $T = -i\tau$. The modulus τ describes a dimensionless chiral supermultiplet, depending on both space-time and Grassmann coordinates. By introducing a fundamental scale Λ, a chiral supermultiplet $\sigma = \Lambda\tau$ with mass dimension one can be defined. In the limit of unbroken supersymmetry the results presented in this paper depend on σ only through the τ combination.

up to a Kahler transformation:

$$\begin{cases} w(\Phi) \to w(\Phi) \\ K(\Phi,\bar{\Phi}) \to K(\Phi,\bar{\Phi}) + f(\Phi) + f(\overline{\Phi}) \end{cases}. \tag{12.13}$$

The requirement of invariance of the Kahler potential can be easily satisfied. An example of Kahler potential invariant under (12.12) up to Kahler transformations is

$$K(\Phi,\bar{\Phi}) = -h\log(-i\tau + i\bar{\tau}) + \sum_I (-i\tau + i\bar{\tau})^{-k_I} |\varphi^{(I)}|^2, \tag{12.14}$$

where h is a positive constant. Since $(\tau - \bar{\tau}) \to |c\tau + d|^{-2}(\tau - \bar{\tau})$, we have:

$$K(\Phi,\bar{\Phi}) \to K(\Phi,\bar{\Phi}) + h\log(c\tau + d) + h\log(c\bar{\tau} + d). \tag{12.15}$$

Assuming that only the modulus τ acquires a vacuum expectation value (VEV), this Kahler potential gives rise to kinetic terms for the scalar components of the supermultiplets τ and $\varphi^{(I)}$ of the kind

$$\frac{h}{\langle -i\tau + i\bar{\tau}\rangle^2}\partial_\mu\bar{\tau}\partial^\mu\tau + \sum_I \frac{\partial_\mu\overline{\varphi}^{(I)}\partial^\mu\varphi^{(I)}}{\langle -i\tau + i\bar{\tau}\rangle^{k_I}}. \tag{12.16}$$

On the contrary, the invariance of the superpotential $w(\Phi)$ under the modular group provides a strong restriction on the theory. Consider the expansion of $w(\Phi)$ in power series of the supermultiplets $\varphi^{(I)}$:

$$w(\Phi) = \sum_n Y_{I_1\ldots I_n}(\tau)\ \varphi^{(I_1)}\ldots\varphi^{(I_n)}. \tag{12.17}$$

For the n-th order term to be modular invariant the functions $Y_{I_1\ldots I_n}(\tau)$ should be modular forms of weight $k_Y(n)$ transforming in the representation ρ of Γ_N:

$$Y_{I_1\ldots I_n}(\gamma\tau) = (c\tau + d)^{k_Y(n)}\rho(\gamma)\ Y_{I_1\ldots I_n}(\tau), \tag{12.18}$$

with $k_Y(n)$ and ρ such that:

1. The weight $k_Y(n)$ should compensate the overall weight of the product $\varphi^{(I_1)}\ldots\varphi^{(I_n)}$:

$$k_Y(n) = k_{I_1} + \cdots + k_{I_n}. \tag{12.19}$$

2. The product $\rho \times \rho^{I_1} \times \cdots \times \rho^{I_n}$ contains an invariant singlet.

When we have $k_I = 0$ in all sectors I of the theory, we get $k_Y(n) = 0$ for all n and the functions $Y_{I_1\ldots I_n}(\tau)$ are τ-independent constants since, as we have seen, there are no non-trivial modular forms of weight zero. The product $\rho^{I_1} \times \cdots \times \rho^{I_n}$ should contain an invariant singlet and $Y_{I_1\ldots I_n}$ is an invariant tensor under the group Γ_N. This particular limit reproduces the well-studied case of a supersymmetric theory invariant under the discrete symmetry Γ_N. Some of the fields $\varphi^{(I)}$ can be gauge singlets, *i.e.* flavons, whose vacuum expectation values break Γ_N in the appropriate direction. This scheme applies to most of the models of fermion masses based on discrete symmetries that have been proposed so far. However this is just a particular case of the more general setup considered here. When $k_I \neq 0$ in some sector I, the Yukawa couplings $Y_{I_1 \ldots I_n}(\tau)$ should carry a non-trivial τ dependence. The whole modular group acts on the field space and $Y_{I_1 \ldots I_n}(\tau)$ are strictly constrained by the relatively small number of possible modular forms that match the above requirements.

This setup can be easily extended to the case of $N = 1$ local supersymmetry where Kahler potential and superpotential are not independent functions since the theory depends on the combination

$$\mathcal{G}(\Phi, \bar{\Phi}) = K(\Phi, \bar{\Phi}) + \log w(\Phi) + \log w(\bar{\Phi}). \tag{12.20}$$

The modular invariance of the theory can be realised in two ways. Either $K(\Phi, \bar{\Phi})$ and $w(\Phi)$ are separately modular invariant or the transformation of $K(\Phi, \bar{\Phi})$ under the modular group is compensated by that of $w(\Phi)$. An example of this second possibility is given by the Kahler potential of Eq. (12.14), with the superpotential $w(\Phi)$ transforming as

$$w(\Phi) \to e^{i\alpha(\gamma)}(c\tau + d)^{-h} w(\Phi). \tag{12.21}$$

In the expansion (12.17) the Yukawa couplings $Y_{I_1 \ldots I_n}(\tau)$ should now transform as

$$Y_{I_1 \ldots I_n}(\gamma\tau) = e^{i\alpha(\gamma)}(c\tau + d)^{k_Y(n)} \rho(\gamma)\ Y_{I_1 \ldots I_n}(\tau), \tag{12.22}$$

with $k_Y(n) = k_{I_1} + \cdots + k_{I_n} - h$ and the representation ρ subject to the requirement 2. When we have $k_{I_1} + \cdots + k_{I_n} = h$, we get $k_Y(n) = 0$ and the functions $Y_{I_1 \ldots I_n}(\tau)$ are τ-independent constants. This occurs for supermultiplets belonging to the untwisted sector in the orbifold compactification of the heterotic string.

The above framework is known and we have reported it here for completeness. However there are two aspects that we have emphasised in this presentation. First of all, the close relationship between supersymmetric modular invariant theories and finite discrete symmetries. As stressed above the whole class of models invariant under discrete groups of the type Γ_N can be regarded as a particular case of the wider class of modular invariant theories. The second aspect is that this formulation is particularly suitable to a bottom-up approach. Namely, by explicitly constructing the whole ring $\mathcal{M}(\Gamma(N))$ of modular forms of level N, one can systematically explore all possible Yukawa couplings $Y_{I_1 \ldots I_n}(\tau)$ which can occur in the superpotential $w(\Phi)$. As we shall see in a specific example, the limited number of generators of $\mathcal{M}(\Gamma(N))$ severely constrain the candidate couplings and lead to a new class of predictive models of lepton masses.

12.3 Models of lepton masses

In this section we apply the formalism outlined above to models of lepton masses and mixings. Here we give the general rules to build modular invariant $N = 1$ supersymmetric models with matter supermultiplets transforming according to representations of a finite modular group Γ_N, with a fixed integer N. The relevant matter

Table 12.2 Chiral supermultiplets, transformation properties and weights. Matter supermultiplets E_i^c, N_i^c and L_i come in three copies and their representations ρ_I $(I = E, N, L)$ can be reducible. If this is the case, different generations can have different weight. φ denotes the flavons, a set of gauge-invariant chiral supermultiplets that can contribute to the spontaneous breaking of Γ_N through a non-vanishing VEV.

	E^c	N^c	L	H_d	H_u	φ
$SU(2)_L \times U(1)_Y$	$(1, +1)$	$(1, 0)$	$(2, -1/2)$	$(2, -1/2)$	$(2, +1/2)$	$(1, 0)$
Γ_N	ρ_E	ρ_N	ρ_L	ρ_d	ρ_u	ρ_φ
k_I	k_{E_i}	k_{N_i}	k_{L_i}	k_d	k_u	k_{φ_i}

fields and their transformation properties are collected in Table 12.2. The field content includes a chiral supermultiplet for the modulus τ, transforming as in (12.12) under modular transformations and singlet under gauge transformations. There is a minimal class of models where the only source of breaking of the modular symmetry is the VEV of τ. In this class of models we do not introduce any flavon multiplet. The role of flavons is replaced by the modular forms $Y(\tau)$ of level N. In a more general context, we can allow for the existence of a set of gauge-singlet flavons φ, which transform non-trivially under Γ_N and contribute to the breaking of the discrete symmetry through their VEVs. Without losing generality the flavon fields can be assumed dimensionless, as the modulus τ. As a preliminary step, we need to identify the modular forms $Y(\tau)$ and their transformation properties under the modular group, i.e. their representations ρ under Γ_N. We assume a Kahler potential of the form (12.14). For the superpotential, we distinguish two cases.

1. When neutrino masses originate from a type I see-saw mechanism, the superpotential in the lepton sector reads:

$$w = \alpha\ (E^c H_d L\ f_E(Y, \varphi))_1 + g(N^c H_u L\ f_N(Y, \varphi))_1 + \Lambda(N^c N^c f_M(Y, \varphi))_1, \tag{12.23}$$

   where $f_I(Y, \varphi)$ $(I = E, N, M)$ denotes the most general combination of modular forms $Y(\tau)$ and flavon fields φ such that the total weight of each term in (12.23) is zero. Moreover $(\ldots)_1$ stands for the invariant singlet of Γ_N. Several independent invariant singlets can occur. In this case a sum over all contributions with arbitrary coefficients is understood and in w there are more parameters than those explicitly indicated.

2. When neutrino masses directly originate from the Weinberg operator, the superpotential in the lepton sector reads:

$$w = \alpha\ (E^c H_d L\ f_E(Y, \varphi))_1 + \frac{1}{\Lambda}(H_u H_u\ LL\ f_W(Y, \varphi))_1. \tag{12.24}$$

The supermultiplet τ and the modular forms $Y(\tau)$ do not carry lepton number. In the above superpotential the lepton number of the matter multiplets is assigned in the usual way and the total lepton number is effectively broken either by the VEV of some flavon field or by the parameter Λ, that can be regarded as a spurion. We will not address here the vacuum alignment problem. We will not attempt to build the most general supersymmetric and modular invariant scalar potential for τ and φ. To explore the ability of this class of models in reproducing the existing data, both τ and φ will be treated as spurions whose values will be varied when scanning the parameter space of the model. In principle, for each value of τ the kinetic terms of the chiral multiplets in Table 12.2 have to be rescaled to match their canonical form. In practice, in the concrete models we will explore, this effect can be absorbed into the unknown parameters of the superpotential. By scanning the parameter space of the model (α, g, Λ) and the VEVs $(\langle\tau\rangle, \langle\varphi\rangle)$ we can compute lepton masses, mixing angles and phases and compare them to the data.

12.3.1 *Models based on* $\mathbf{\Gamma_3}$

To assess the viability of the above formalism, we build in this section models of lepton masses based on modular forms of level 3. The group Γ_3 is isomorphic to A_4 — see Table 12.1 — and we will be able to compare the present construction with a class of models widely discussed in the literature, especially in connection to the tribimaximal mixing. Though the explicit example will refer to the case $N = 3$, it is pretty clear from the previous section that there is no conceptual obstacle to develop a similar construction for any N. A comprehensive analysis of all models that can be constructed along these lines is beyond the scope of the present work. The examples discussed here, admittedly not fully realistic, are meant to illustrate how this new type of construction can be brought to completion and compared with the data.

12.3.1.1 *Modular forms of level 3*

We focus on modular forms of level 3, satisfying

$$f(\gamma\tau) = (c\tau + d)^{2k} f(\tau) \qquad \gamma = \begin{pmatrix} a & b \\ c & d \end{pmatrix} \in \Gamma(3), \tag{12.25}$$

with

$$\Gamma(3) = \{\begin{pmatrix} a & b \\ c & d \end{pmatrix} \in SL(2, Z), \begin{pmatrix} a & b \\ c & d \end{pmatrix} = \begin{pmatrix} 1 & 0 \\ 0 & 1 \end{pmatrix} \quad (\text{mod } 3)\}. \tag{12.26}$$

The quotient space $\mathcal{H}/\Gamma(3)$ can be described by a fundamental domain $\mathcal{F}$ for $\Gamma(3)$, that is a connected region of $\mathcal{H}$ such that each point of $\mathcal{H}$ can be mapped into $\mathcal{F}$ by a $\Gamma(3)$ transformation, but no two points in the interior of $\mathcal{F}$ are related under $\Gamma(3)$. The space $\mathcal{H}/\Gamma(3)$ is simply $\mathcal{F}$ with certain boundary points identified. A fundamental domain for $\Gamma(3)$ is shown in Fig. 12.1. $\mathcal{H}/\Gamma(3)$ can be made compact by

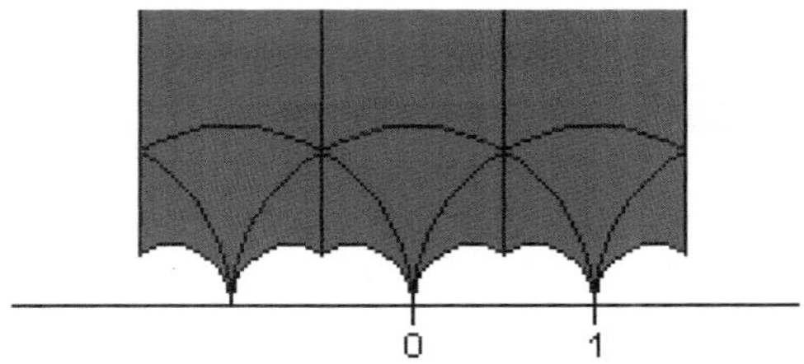

Fig. 12.1 Fundamental domain for $\Gamma(3)$.

adding the points $i\infty$, -1, 0 and $+1$, which are the cusps. The compactified space $\overline{\mathcal{H}/\Gamma(3)}$ has genus zero and can be thought of as a tetrahedron whose vertices are the cusps. Indeed the cusps are related by transformations of $\Gamma_3 = \bar{\Gamma}/\overline{\Gamma}(3)$, which is the symmetry group of a regular tetrahedron, given the isomorphism between Γ_3 and A_4. The group A_4 is generated by the elements S and T, satisfying the relations:

$$S^2 = T^3 = (ST)^3 = \mathbb{1}. \tag{12.27}$$

Basic properties of A_4 are summarised in Appendix C. Modular forms of weight $2k$ and level 3 transform according to unitary representations of A_4. In the final part of this work we will see that they provide a non-linear representation of A_4. Modular forms of level 3 can be constructed starting from those of lower weight, $k = 1$. From Table 12.1 we see that there are three linearly independent such forms, which we call $Y_i(\tau)$. Three linearly independent weight 2 and level-3 forms are constructed in Appendix C. They read:

$$\begin{aligned}
Y_1(\tau) &= \frac{i}{2\pi}\left[\frac{\eta'\left(\frac{\tau}{3}\right)}{\eta\left(\frac{\tau}{3}\right)} + \frac{\eta'\left(\frac{\tau+1}{3}\right)}{\eta\left(\frac{\tau+1}{3}\right)} + \frac{\eta'\left(\frac{\tau+2}{3}\right)}{\eta\left(\frac{\tau+2}{3}\right)} - \frac{27\eta'(3\tau)}{\eta(3\tau)}\right] \\
Y_2(\tau) &= \frac{-i}{\pi}\left[\frac{\eta'\left(\frac{\tau}{3}\right)}{\eta\left(\frac{\tau}{3}\right)} + \omega^2\,\frac{\eta'\left(\frac{\tau+1}{3}\right)}{\eta\left(\frac{\tau+1}{3}\right)} + \omega\,\frac{\eta'\left(\frac{\tau+2}{3}\right)}{\eta\left(\frac{\tau+2}{3}\right)}\right] \\
Y_2(\tau) &= \frac{-i}{\pi}\left[\frac{\eta'\left(\frac{\tau}{3}\right)}{\eta\left(\frac{\tau}{3}\right)} + \omega\,\frac{\eta'\left(\frac{\tau+1}{3}\right)}{\eta\left(\frac{\tau+1}{3}\right)} + \omega^2\,\frac{\eta'\left(\frac{\tau+2}{3}\right)}{\eta\left(\frac{\tau+2}{3}\right)}\right]
\end{aligned} \tag{12.28}$$

where $\eta(\tau)$ is the Dedekind eta-function, defined in the upper complex plane:

$$\eta(\tau) = q^{1/24}\prod_{n=1}^{\infty}(1-q^n), \qquad q \equiv e^{i2\pi\tau}. \tag{12.29}$$

They transform in the three-dimensional representation of A_4. In a vector notation where $Y^T = (Y_1, Y_2, Y_3)$ we have

$$Y(-1/\tau) = \tau^2\,\rho(S)Y(\tau), \qquad Y(\tau+1) = \rho(T)Y(\tau),$$

with unitary matrices $\rho(S)$ and $\rho(T)$

$$\rho(S)=\frac{1}{3}\begin{pmatrix}-1 & 2 & 2\\ 2 & -1 & 2\\ 2 & 2 & -1\end{pmatrix},\quad \rho(T)=\begin{pmatrix}1 & 0 & 0\\ 0 & \omega & 0\\ 0 & 0 & \omega^2\end{pmatrix},\quad \omega=-\frac{1}{2}+\frac{\sqrt{3}}{2}i.$$

The q-expansion of $Y_i(\tau)$ reads:

$$\begin{aligned}Y_1(\tau)&=1+12q+36q^2+12q^3+\ldots,\\ Y_2(\tau)&=-6q^{1/3}(1+7q+8q^2+\ldots),\\ Y_3(\tau)&=-18q^{2/3}(1+2q+5q^2+\ldots).\end{aligned}$$

From the q-expansion we see that the functions $Y_i(\tau)$ are regular at the cusps. Moreover $Y_i(\tau)$ satisfy the constraint:

$$Y_2^2+2Y_1Y_3=0. \tag{12.30}$$

As discussed explicitly in Appendix D, the constraint (12.30) is essential to recover the correct dimension of the linear space $\mathcal{M}_{2k}(\Gamma(3))$. On the one side from Table 12.1 we see that this space has dimension $2k+1$. On the other hand the number of independent homogeneous polynomial $Y_{i_1}Y_{i_2}\cdots Y_{i_k}$ of degree k that we can form with Y_i is $(k+1)(k+2)/2$. These polynomials are modular forms of weight $2k$ and, to match the correct dimension, $k(k-1)/2$ among them should vanish. Indeed this happens as a consequence of Eq. (12.30). Therefore the ring $\mathcal{M}(\Gamma(3))$ is generated by the modular forms $Y_i(\tau)$ $(i=1,2,3)$.

There are two special sets of VEV for τ that preserve a subgroup of the modular group (and of Γ_3). The subgroup generated by T is preserved by $\langle\tau\rangle=i\infty$.[4] This induces the well known VEV for the triplet Y_i:

$$(Y_1,Y_2,Y_3)|_{\tau=i\infty}=(1,0,0). \tag{12.31}$$

The subgroup generated by S is preserved by $\langle\tau\rangle=i$, which results into:

$$(Y_1,Y_2,Y_3)|_{\tau=i}=Y_1(i)(1,1-\sqrt{3},-2+\sqrt{3}). \tag{12.32}$$

This VEV pattern is completely different from $(1,1,1)$, the eigenvector of the matrix S corresponding to the eigenvalue $+1$. Indeed we should satisfy:

$$Y(-1/\tau)|_{\tau=i}=-\rho(S)Y(\tau)|_{\tau=i}. \tag{12.33}$$

For this reason $Y(\tau)|_{\tau=i}$ must be an eigenvector of the matrix S with eigenvalue -1. There are two such eigenvectors and the one in (12.32) satisfies the relation

[4]This can be seen more clearly by going from $\mathcal{H}$ to the unit disk $|w|<1$, via $w=(\tau-1)/(\tau+1)$. The transformation T acts on w as $w\to(1+w)/(3-w)$, having a unique fixed point at $w=1$, belonging to the boundary of the unit disk and corresponding to $\tau=i\infty$.

(12.30). Notice that the configuration $(Y_1, Y_2, Y_3) = (1, 1, 1)$, widely used in model building, cannot be realised since it violates the constraint (12.30).

Of course, values of $\langle\tau\rangle$ related to i by a modular transformation, have little groups isomorphic to Z_2, the subgroup generated by S. A similar consideration applies to the vacuum configurations related to $\langle\tau\rangle = i\infty$ by modular transformations.

12.3.1.2 *Example 1: neutrino masses from the Weinberg operator*

We consider an example where the neutrino masses and mixing angles originate directly from the Weinberg operator. In this model we have no contribution to the mixing from the charged lepton sector. This can be achieved by choosing the field content of Table 12.3. In particular we need a flavon φ_T transforming as a triplet of Γ_3 and developing a VEV of the type:

$$\langle\varphi_T\rangle = (u, 0, 0). \tag{12.34}$$

Such a VEV breaks Γ_3 down to the Z_3 subgroup generated by T. We choose the weights k_{E_i}, k_L, k_d and k_φ such that $k_{E_i} + k_L + k_d + k_\varphi = 0$. Moreover, to forbid a dependence of the charged lepton masses on $Y(\tau)$ (and a dependence of the Weinberg operator on φ_T), we take, for instance, $k_\varphi = -3$. The superpotential for the charged lepton sector reads:

$$w_e = \alpha\ E_1^c H_d (L\ \varphi_T)_1 + \beta\ E_2^c H_d (L\ \varphi_T)_{1'} + \gamma\ E_3^c H_d (L\ \varphi_T)_{1''}. \tag{12.35}$$

The VEV of Eq. (12.34) leads to a diagonal mass matrix for the charged leptons:

$$m_e = \mathtt{diag}(\alpha, \beta, \gamma) u\ v_d. \tag{12.36}$$

The charged lepton masses can be reproduced by adjusting the parameters α, β and γ, with an ambiguity related to the freedom of permuting the eigenvalues. As a result, the lepton mixing matrix U_{PMNS} is determined up to a permutation of the rows. Finally, by choosing $k_L = +1$ and $k_u = 0$, we uniquely determine the form of

Table 12.3 Chiral supermultiplets, transformation properties and weights.

	(E_1^c, E_2^c, E_3^c)	L	H_d	H_u	φ_T
$SU(2)_L \times U(1)_Y$	$(1, +1)$	$(2, -1/2)$	$(2, -1/2)$	$(2, +1/2)$	$(1, 0)$
$\Gamma_3 \equiv A_4$	$(1, 1'', 1')$	3	1	1	3
k_I	$(k_{E_1}, k_{E_2}, k_{E_3})$	k_L	k_d	k_u	k_φ

the Weinberg operator:

$$w_\nu = \frac{1}{\Lambda}(H_u H_u \; LL \; Y)_1. \tag{12.37}$$

The superpotential $w = w_e + w_\nu$ depends on the four parameters $\alpha, \beta, \gamma, \Lambda$. The charged lepton masses m_e, m_μ and m_τ are in a one-to-one correspondence with α, β and γ, which can be taken to be real without losing generality. The neutrino mass matrix is given by:

$$m_\nu = \begin{pmatrix} 2Y_1 & -Y_3 & -Y_2 \\ -Y_3 & 2Y_2 & -Y_1 \\ -Y_2 & -Y_1 & 2Y_3 \end{pmatrix} \frac{v_u^2}{\Lambda}. \tag{12.38}$$

We see that the fourth parameter, Λ, controls the absolute scale of neutrino masses. A remarkable feature of this model is that neutrino mass ratios, lepton mixing angles, Dirac and Majorana phases are completely determined by the modulus τ. We have eight dimensionless physical quantities that do not depend on any coupling constant. Assuming the VEV of Eq. (12.34) for the flavon φ_T, they all uniquely depend on τ. Four of these parameters, $r = \Delta m^2_{sol}/|\Delta m^2_{atm}|$, $\sin^2\theta_{12}$, $\sin^2\theta_{13}$ and $\sin^2\theta_{23}$, have been measured with good precision. A fifth one, δ_{CP}, starts to be constrained by the present data, see Table 12.4. It is a significant challenge to reproduce all of them by varying a single complex parameter. By scanning a portion of the upper complex plane where τ varies, we found that the agreement between predictions and data is optimised by the choice:

$$\tau = 0.0111 + 0.9946i, \tag{12.39}$$

giving rise to an inverted neutrino mass ordering. Assuming that the charged lepton sector induces a permutation between the second and the third rows, the neutrino

Table 12.4 Values of observables and their 1σ errors used to optimise the model parameters, through a χ^2 scan. Oscillation parameters are from ref. [Capozzi *et al.* (2017)] and ratios of charged lepton masses from ref. [Ross and Serna (2008)]. We use $|\Delta m^2_{atm}| = |m_3^2 - (m_1^2 + m_2^2)/2|$ where m_i are the neutrino masses. The ratios $\frac{m_e}{m_\mu}$ and $\frac{m_\mu}{m_\tau}$ are evaluated at the scale 2×10^{16} GeV. For $\frac{m_\mu}{m_\tau}$ the average between the values obtained with $\tan\beta = 10$ and $\tan\beta = 38$ has been used. There is a sizable difference between the allowed 1σ ranges of $\sin^2\theta_{23}$ for the cases of normal and inverted ordering. For simplicity we have adapted the ranges quoted in ref. [Capozzi *et al.* (2017)] and we use a unique range for the two cases. The value of $\frac{\delta_{CP}}{\pi}$ has not been used in the scan.

	$r = \frac{\Delta m^2_{sol}}{\lvert\Delta m^2_{atm}\rvert}$	$\sin^2\theta_{12}$	$\sin^2\theta_{13}$	$\sin^2\theta_{23}$	$\frac{\delta_{CP}}{\pi}$	$\frac{m_e}{m_\mu}$	$\frac{m_\mu}{m_\tau}$
best value	0.0292	0.297	0.0215	0.5	1.4	0.0048	0.0565
1σ error	0.0008	0.017	0.0007	0.1	0.2	0.0002	0.0045

mass/mixing parameters are predicted to be[5]

$$\begin{aligned}
&\frac{\Delta m^2_{sol}}{|\Delta m^2_{atm}|} = 0.0292 \\
&\sin^2\theta_{12} = 0.295 \qquad \sin^2\theta_{13} = 0.0447 \quad \sin^2\theta_{23} = 0.651 \\
&\frac{\delta_{CP}}{\pi} = 1.55 \qquad \frac{\alpha_{21}}{\pi} = 0.22 \qquad \frac{\alpha_{31}}{\pi} = 1.80.
\end{aligned} \tag{12.40}$$

Also, the absolute scale of neutrino masses is determined, since v_u^2/Λ can be fixed by requiring that the individual square mass differences $|\Delta m^2_{atm}|$ and Δm^2_{sol} are reproduced. We find the central values:

$$m_1 = 4.998 \times 10^{-2}\ eV \quad m_2 = 5.071 \times 10^{-2}\ eV \quad m_3 = 7.338 \times 10^{-4}\ eV. \tag{12.41}$$

Some comments are in order.

- The value of τ that minimises the χ^2 is intriguingly close to the self-dual point $\tau = i$ where S is unbroken. It is easy to verify that the self-dual point leads to an inverted-ordering spectrum with $\Delta m^2_{sol} = 0$ and relative Majorana phase π between the first two neutrino levels. For $\tau = i$ the CP symmetry is unbroken and the non-trivial phases in Eq. (12.40) are entirely generated by the departure of τ from i.[6]
- Three observables, $r = \Delta m^2_{sol}/|\Delta m^2_{atm}|$, $\sin^2\theta_{12}$ and δ_{CP} are within the 1σ experimental range. A fourth one, $\sin^2\theta_{23}$, is very close to the 3σ range [Capozzi *et al.* (2017)] for the inverted ordering.
- The minimum χ^2 is completely dominated by $\sin^2\theta_{13}$, which is many σs away from the allowed range. Nevertheless θ_{13} is predicted to be the smallest angle, in qualitative agreement with the data.
- The observed value of δ_{CP} was not included in the data set used to minimise the χ^2, but it turns out to be in very good agreement with the current determination.
- The two Majorana phases are predicted and the parameter $|m_{ee}|$ of neutrinoless double beta decay is completely determined.
- Had we assumed no permutation between the second and third rows of the mixing matrix, we would have obtained $\sin^2\theta_{23} = 0.349$ and $\delta/\pi = -1.55$, with no change for the other predictions.
- Notice that modular symmetry and supersymmetry completely determine the superpotential in the lepton sector in terms of the free parameters α, β, γ and Λ.

---

[5]Dirac and Majorana phases are in the PDG convention.

[6]When $\tau = i$ the mixing matrix U is real and, from Eq. (12.32), we find $\sin^2\theta_{12} = (11 - 6\sqrt{3})/26 \approx 0.02$, $\sin^2\theta_{23} = (16 - 4\sqrt{3})/26 \approx 0.35$ and $|U_{e3}| = (3 - \sqrt{3})/6 \approx 0.21$.

There are no extra contributions we can add to w that cannot be absorbed in a redefinition of those parameters. In particular the couplings of the modulus τ to the matter supermultiplets are fixed to any order in the τ power expansion and all higher-dimensional operators involving τ are known as a function of a finite set of parameters. This remarkable feature has no counterpart in the known models based on discrete symmetries where, in general, to fix all higher-dimensional operators involving the relevant flavon fields, an infinite number of parameters is needed. As a consequence, corrections to the above predictions can only come from a few sources. One is the vacuum alignment. In this example we have kept fixed the VEV of φ_T. Additional, small contributions to the mixing can be expected if we relax this assumption. Other sources of corrections can be either supersymmetry breaking contributions or corrections to the Kahler potential. A candidate modification of the Kahler potential (12.14) able to induce corrections to the above results is an additive contribution depending explicitly on both the matter supermultiplets and on the modular forms $Y_i(\tau)$. The problem of accounting for these additional terms will be analysed elsewhere. The properties discussed here are not specific to the example under examination, but rather general features of the set-up proposed here. A small correction to the model, coming for instance from the charged lepton sector, might bring the model to agreement with θ_{13}, perhaps without spoiling the predictions for the remaining observables. A more quantitative analysis will be pursued elsewhere. We consider the above zeroth-order approximation as an excellent starting point.

12.3.1.3 *Example 2: neutrino masses from the see-saw mechanism*

In this example neutrino masses and mixing angles come from the see-saw mechanism. The field content is the one of Table 12.5 and includes a triplet of chiral supermultiplets N^c describing right-handed neutrinos. The charged lepton sector is exactly the same as in the previous example. The relevant superpotential w_e is the one of Eq. (12.35). As before the weights k_{E_i}, k_L, k_d and k_φ satisfy $k_{E_i} + k_L + k_d + k_\varphi = 0$ and we choose $k_\varphi = -3$. We assume that the flavon φ_T, transforming as a triplet of Γ_3, develops the VEV in Eq. (12.34), which guarantees

Table 12.5 Chiral supermultiplets, transformation properties and weights.

	(E_1^c, E_2^c, E_3^c)	N^c	L	H_d	H_u	φ_T
$SU(2)_L \times U(1)_Y$	$(1, +1)$	$(1, 0)$	$(2, -1/2)$	$(2, -1/2)$	$(2, +1/2)$	$(1, 0)$
$\Gamma_3 \equiv A_4$	$(1, 1'', 1')$	3	3	1	1	3
k_I	$(k_{E_1}, k_{E_2}, k_{E_3})$	k_N	k_L	k_d	k_u	k_φ

a diagonal mass matrix for the charged leptons, as in Eq. (12.36). There is no contribution to the mixing from w_e and the charged lepton masses are in a one-to-one correspondence with the parameters α, β and γ. Also in this case there is an ambiguity related to the freedom of permuting the eigenvalues of m_e and the lepton mixing matrix U_{PMNS} is determined up to a permutation of the rows. By choosing the weights $k_N + k_u + k_L = 0$ and $k_N = +1$, the superpotential of the neutrino sector is given by:

$$w_\nu = g\,(N^c H_u L)_1 + \Lambda (N^c N^c Y)_1. \tag{12.42}$$

With the above choice of weights this is the only gauge-invariant holomorphic polynomial singlet under Γ_3 that also satisfies the constraint of Eq. (12.19). Using a vector notation ($E^{cT} = (E_1^c, E_2^c, E_3^c), \ldots$), we can write the superpotential $w = w_e + w_\nu$ as:

$$\begin{aligned}
w &= E^{cT}\mathcal{Y}_e H_d L + N^{cT}\mathcal{Y}_\nu H_u L + N^{cT}\mathcal{M}_R N^c \\
\mathcal{Y}_e &= \begin{pmatrix} \alpha\ \varphi_{T1} & \alpha\ \varphi_{T3} & \alpha\ \varphi_{T2} \\ \beta\ \varphi_{T2} & \beta\ \varphi_{T1} & \beta\ \varphi_{T3} \\ \gamma\ \varphi_{T3} & \gamma\ \varphi_{T2} & \gamma\ \varphi_{T1} \end{pmatrix} \\
\mathcal{Y}_\nu &= g \begin{pmatrix} 1 & 0 & 0 \\ 0 & 0 & 1 \\ 0 & 1 & 0 \end{pmatrix} \\
\mathcal{M}_R &= \begin{pmatrix} 2Y_1 & -Y_3 & -Y_2 \\ -Y_3 & 2Y_2 & -Y_1 \\ -Y_2 & -Y_1 & 2Y_3 \end{pmatrix} \Lambda.
\end{aligned} \tag{12.43}$$

The light neutrino mass matrix m_ν derived from the see-saw and the charged lepton mass matrix m_e read

$$\begin{aligned}
m_\nu &= -\mathcal{Y}_\nu^T \mathcal{M}_R^{-1} \mathcal{Y}_\nu \langle H_u \rangle^2 \\
m_e &= \mathcal{Y}_e \langle H_d \rangle.
\end{aligned} \tag{12.44}$$

The number of independent low-energy parameters is the same as in the previous example, since g only enters in the combination g^2/Λ.[7] The best agreement between data and predictions is obtained with the choice:

$$\tau = -0.195 + 1.0636i, \tag{12.45}$$

giving rise to a normal neutrino mass ordering. Assuming that the charged lepton sector induces a permutation between the second and the third rows, the neutrino

[7] The phases of g and Λ are unobservable.

mass/mixing parameters are predicted to be:

$$\begin{aligned}
&\frac{\Delta m^2_{sol}}{|\Delta m^2_{atm}|} = 0.0280 \\
&\sin^2\theta_{12} = 0.291 \qquad \sin^2\theta_{13} = 0.0486 \quad \sin^2\theta_{23} = 0.331 \\
&\frac{\delta_{CP}}{\pi} = 1.47 \qquad \frac{\alpha_{21}}{\pi} = 1.83 \qquad \frac{\alpha_{31}}{\pi} = 1.26.
\end{aligned} \tag{12.46}$$

Also the absolute scale of neutrino masses is determined, since $g^2 v_u^2/\Lambda$ can be fixed by requiring that the individual square mass differences $|\Delta m^2_{atm}|$ and Δm^2_{sol} are reproduced. We find the central values:

$$m_1 = 1.096 \times 10^{-2}\ eV \quad m_2 = 1.387 \times 10^{-2}\ eV \quad m_3 = 5.231 \times 10^{-2}\ eV. \tag{12.47}$$

Most of the comments on the previous model apply to the present example as well. The distinctive feature of the model is the normal ordering of neutrino masses. Beyond the prediction of θ_{13}, which deviates by many standard deviations from the experimental range, also the value of $\sin^2\theta_{23}$ is slightly below the 3σ allowed range for normal ordering. Nevertheless θ_{13} is the smallest angle and θ_{23} the largest one, in qualitative agreement with the data. Corrections coming either from the Kahler potential or from the VEV of the flavon φ_T might restore the full agreement between data and predictions.

12.3.1.4 *Example 3: no flavons*

The following example can be regarded as minimal, in that it makes no use of any flavon field other than the modulus τ. We chose for lepton and Higgs supermultiplets the transformations of Table 12.6. We can exploit the fact that the whole ring $\mathcal{M}(\Gamma(3))$ is generated by the A_4 triplet $Y_i(\tau)$ $(i = 1, 2, 3)$, to cast the general superpotential of Eq. (12.23) in the form:

$$\begin{aligned}
w &= w_e + w_\nu \\
w_e &= \alpha\ E_1^c H_d (L\, Y^{a_1})_1 + \beta\ E_2^c H_d (L\, Y^{a_2})_{1'} + \gamma\ E_3^c H_d (L\, Y^{a_3})_{1''}, \\
w_\nu &= g(N^c H_u L\, Y^b)_1 + \Lambda(N^c N^c Y^c)_1
\end{aligned} \tag{12.48}$$

Table 12.6 Chiral supermultiplets, transformation properties and weights.

	(E_1^c, E_2^c, E_3^c)	N^c	L	H_d	H_u
$SU(2)_L \times U(1)_Y$	$(1, +1)$	$(1, 0)$	$(2, -1/2)$	$(2, -1/2)$	$(2, +1/2)$
$\Gamma_3 \equiv A_4$	$(1, 1'', 1')$	3	3	1	1
k_I	$(k_{E_1}, k_{E_2}, k_{E_3})$	k_N	k_L	k_d	k_u

where Y^a denotes a insertions of the basic modular forms $Y_i(\tau)$ and $(\ldots)_r$ $(r = 1, 1', 1'', 3)$ stands for the r representation of A_4. It can occur that there are several independent combinations of the type $(\ldots)_r$. In this case a sum over all contributions with arbitrary coefficients is understood. The invariance of w under modular transformations implies the following relations for the weights:

$$\begin{cases} 2a_i = k_{E_i} + k_d + k_L & (i = 1, 2, 3) \\ 2b = k_N + k_u + k_L \\ 2c = 2k_N \end{cases} . \tag{12.49}$$

To get canonical kinetic terms we have to rescale the supermultiplets of the theory. The effect of such rescaling can be absorbed into the arbitrary coefficients α, β, γ, g and Λ. We now proceed by selecting one example with specific choices of the weights k_I. By choosing $k_{u,d} = 0$ and $k_{E_i} = k_N = k_L = 1$, we have $a_i = b = c = 1$ and all couplings are linear in Y_i:

$$\begin{aligned} w_e &= \alpha\ E_1^c H_d (L\ Y)_1 + \beta\ E_2^c H_d (L\ Y)_{1'} + \gamma\ E_3^c H_d (L\ Y)_{1''} \\ w_\nu &= g(N^c H_u L\ Y)_1 + \Lambda (N^c N^c Y)_1. \end{aligned} \tag{12.50}$$

Using a vector notation $(E^{cT} = (E_1^c, E_2^c, E_3^c), \ldots)$, we can write:

$$\begin{aligned} w &= E^{cT} \mathcal{Y}_e H_d L + N^{cT} \mathcal{Y}_\nu H_u L + N^{cT} \mathcal{M}_R N^c \\ \mathcal{Y}_e &= \begin{pmatrix} \alpha Y_1 & \alpha Y_3 & \alpha Y_2 \\ \beta Y_2 & \beta Y_1 & \beta Y_3 \\ \gamma Y_3 & \gamma Y_2 & \gamma Y_1 \end{pmatrix} \\ \mathcal{Y}_\nu &= \begin{pmatrix} 2g_1 Y_1 & (-g_1 + g_2) Y_3 & (-g_1 - g_2) Y_2 \\ (-g_1 - g_2) Y_3 & 2g_1 Y_2 & (-g_1 + g_2) Y_1 \\ (-g_1 + g_2) Y_2 & (-g_1 - g_2) Y_1 & 2g_1 Y_3 \end{pmatrix} \\ \mathcal{M}_R &= \begin{pmatrix} 2Y_1 & -Y_3 & -Y_2 \\ -Y_3 & 2Y_2 & -Y_1 \\ -Y_2 & -Y_1 & 2Y_3 \end{pmatrix} \Lambda. \end{aligned} \tag{12.51}$$

Notice that the contribution $g(N^c H_u L\ Y)_1$ depends on two parameters g_1 and g_2 corresponding to the two independent invariant singlets that we can form out of N^c, L and Y. The light neutrino mass matrix m_ν derived from the see-saw and the charged lepton mass matrix m_e read

$$\begin{aligned} m_\nu &= -\mathcal{Y}_\nu^T \mathcal{M}_R^{-1} \mathcal{Y}_\nu \langle H_u \rangle^2 \\ m_e &= \mathcal{Y}_e \langle H_d \rangle. \end{aligned} \tag{12.52}$$

From the matrices m_e and m_ν we can compute lepton masses, mixing angles and phases in terms of the effective parameters: α, β, γ, $g_{1,2}$ and τ. We focus on mass ratios and, without loss of generality, we can set the spurion Λ to one. Moreover, by exploiting field redefinitions, α, β, γ and g_1 can be taken real. Notice that the light neutrino mass matrix m_ν depends on a single complex parameter g_2/g_1, up to an overall factor that does not affect mass ratios and mixing angles. Mass ratios among charged leptons depend on α/γ and β/γ. By scanning the parameter space of the model we could identify the point $\alpha/\gamma = 15.3$, $\beta/\gamma = 0.054$, $g_2/g_1 = 0.029i$, $\tau = 0.008 + 0.98i$ which better approaches the data, collected in Table 12.4, though this does not provide a realistic set of observables. We find that neutrino masses have inverted ordering. Mass ratios are in good agreement with the data:

$$\frac{\Delta m^2_{sol}}{|\Delta m^2_{atm}|} = 0.0292, \qquad \frac{m_e}{m_\mu} = 0.0048, \qquad \frac{m_\mu}{m_\tau} = 0.0567. \tag{12.53}$$

The mixing angles are:

$$\sin^2\theta_{12} = 0.459, \qquad \sin^2\theta_{13} = 0.001, \qquad \sin^2\theta_{23} = 0.749. \tag{12.54}$$

An appealing feature of this pattern is that it predicts large solar and atmospheric mixing angles and a small reactor angle in qualitative agreement with the observations. Nevertheless, given the present experimental accuracy this set of angles is clearly excluded by many standard deviations. For completeness we also list the Dirac and Majorana phases in the PDG convention:

$$\frac{\delta_{CP}}{\pi} = 1.25, \qquad \frac{\alpha_{21}}{\pi} = 1.04, \qquad \frac{\alpha_{31}}{\pi} = 1.02. \tag{12.55}$$

The observed value of δ_{CP} was not included in the data used to search for the best region of the parameter space and it turns out to be in good agreement with the current determination. As in the previous examples, modular symmetry and supersymmetry completely determine the superpotential in the lepton sector in terms of the free parameters α, β, γ, $g_{1,2}$ and Λ. In this case the model does not include flavons and corrections to the above predictions can only come from two sources — either from supersymmetry breaking contributions or from corrections to the Kahler potential.

12.3.2 *Symmetry realisations*

Consider the minimal theory defined by Eqs. (12.11), (12.14) and (12.17) in the absence of flavon multiplets φ. It is reminiscent of a non-linear sigma model and it is interesting to better understand how the underlying symmetries are effectively realised. If we set the superpotential $w(\Phi)$ to zero, the action becomes invariant under the full $SL(2, R)$ continuous group. Indeed the action

$$\mathcal{S} = \int d^4x d^2\theta d^2\bar{\theta}\; K(\Phi, \bar{\Phi}) \tag{12.56}$$

where

$$K(\Phi, \bar{\Phi}) = -h \log(-i\tau + i\bar{\tau}) + \sum_I (-i\tau + i\bar{\tau})^{-k_I} |\varphi^{(I)}|^2, \tag{12.57}$$

is invariant under $SL(2,R)$, with no restriction on the real parameters a, b, c, d other than $ad-bc=1$. Starting from a generic τ ($\mathtt{Im}\tau>0$), it is always possible to reach i by a transformation of $SL(2,R)$ and we can choose $\tau=i$ as representative of the vacuum configuration. Such configuration has an $SO(2)$ invariance, since

$$\frac{\cos\alpha\ i+\sin\alpha}{-\sin\alpha\ i+\cos\alpha}=i. \tag{12.58}$$

Hence, the modulus τ parametrises the coset space $SL(2,R)/SO(2)$. In this limit $SL(2,R)$ is non-linearly realised and the theory describes the spontaneous breaking of $SL(2,R)$ down to the subgroup $SO(2)$, τ describing the corresponding massless Goldstone bosons.

In the full theory, which includes a non-vanishing superpotential, the continuous symmetry $SL(2,R)$ is explicitly broken down to $SL(2,Z)$. Here the group Γ_N comes into play. Level-N modular forms and matter supermultiplets transform in representations of the groups Γ_N. Modular forms depend on the modulus τ and, in the examples considered so far, the symmetry Γ_N is always in the broken phase, as in the case of non-linearly realised continuous symmetries. The analogy can be pushed further. In the continuous case non-linear realisations can be defined by imposing invariant constraints on the field space of the theory, such as for instance $\varphi^T\varphi=M^2$ ($M^2>0$), the multiplet φ transforming in the fundamental of $SO(N)$. This constraint induces the breaking of $SO(N)$ down to $SO(N-1)$. Consider the case of a supersymmetric modular-invariant theory with matter multiplets in representations of Γ_3, as in Section 12.3.1. We have seen that the three modular forms of level 3 and weight 2 obey the relation:

$$(YY)_{1''}=Y_2^2+2Y_1Y_3=0. \tag{12.59}$$

The left-hand side is covariant, since $(YY)_{1''}$ transforms with a phase factor under A_4. Thus the constraint itself is left invariant by A_4 transformations and induces a restriction on the field space similar to that induced by $\varphi^T\varphi=M^2$ in the case of $SO(N)$. For generic values of $Y(\tau)$ satisfying the relation (12.59), the symmetry Γ_3 is completely broken. Only for the special values related to $\tau=i$ or $\tau=i\infty$ by a Γ_3 transformation, a subgroup of Γ_3 is left unbroken. Therefore, in some sense, Γ_3 is non-linearly realised in the models considered here.

Of course there are several differences with the continuous case. In the discrete case there are no Goldstone bosons and the modulus τ is expected to be massive. Non-linear covariant constraints of the type (12.59) do not define low-energy effective theories in the discrete case. These theories should rather be viewed as consistent truncations of some ultraviolet completion. In general, as in the case of the models under discussions, covariant constraints leave no non-trivial residual symmetry. We can nevertheless identify the cases in which a non-trivial residual symmetry group survives, by discussing the properties of the so-called orbit space, the space spanned by the invariants of the group. We do this in the remaining part of this work.

12.4 Non-linear realisations of discrete symmetries

Consider a Lagrangian $\mathcal{L}(\varphi, \psi)$ invariant under the action of a discrete group G_d, depending on a set of scalar fields φ responsible for the spontaneous breaking of G_d, plus additional fields ψ that do not play a role in the symmetry breaking. As in the case of a continuous symmetry, we can restrict the field space $\mathcal{M}$ to which φ belong by means of constraints invariant under the action of G_d. We classify these constraints into two types:

1. Constraints of the type I, I standing for *invariant*, are relations such as:
$$I_i(\varphi) = 0 \qquad (i = 1, \dots, n) \qquad [\texttt{type I}] \tag{12.60}$$
where $I_i(\varphi)$ are invariants under G_d.
2. Constraints of the type C, C standing for *covariant*, are defined by the conditions:
$$C_i(\varphi) = 0 \qquad (i = 1, \dots, m) \qquad [\texttt{type C}] \tag{12.61}$$
where $C_i(\varphi)$ are covariant, but not invariant, combinations of φ. Here we focus on those transforming with a phase factor under G_d
$$C_i(\varphi') = e^{i\alpha_i} C_i(\varphi), \tag{12.62}$$
where α_i depend on the G_d transformation.

We will consider polynomial invariants, that is $I_i(\varphi)$ are polynomials in the components of the multiplet φ. The ring of invariant polynomials is infinite, but it is generated by a finite number of invariants $\gamma_\alpha(\varphi)$, which means that any invariant polynomial can be written as a polynomial in γ_α. The invariants γ_α might be related by a number of algebraic relations, or syzygies, $\mathcal{Z}_s(\gamma) = 0$. Also in the case of C constraints we will consider only polynomial expressions. Constraints of type I are those usually considered to define non-linear realisations of continuous symmetries. In the discrete case there are other possibilities to restrict the field space $\mathcal{M}$ in an invariant way. While a (connected, semisimple) Lie group has only the trivial one-dimensional representation, a discrete group can possess a number of non-invariant one-dimensional representations, or singlets, χ_i. Therefore the invariant conditions $\chi_i = 0$ have no counterpart in (connected, semisimple) Lie groups and give rise to constraints of type C. Notice that both conditions of type I and type C are left invariant by transformations of G_d.

In this section we would like to analyse the pattern of symmetry breaking induced by the above constraints. To this purpose it is useful to inspect the orbits of the group, i.e. the set of points in the field space $\mathcal{M}$ that are related by group transformations. Each point of a given orbit has the same little group, up to a conjugation. The union of orbits having isomorphic little groups forms a stratum. The full field space $\mathcal{M}$ is partitioned into several strata. For instance the origin of $\mathcal{M}$ is the stratum of type G_d, since for $\varphi = 0$ the symmetry is unbroken. Most of the field space $\mathcal{M}$ is made of orbits having trivial little group, i.e. the symmetry G_d is completely broken. This subset of $\mathcal{M}$ is called principal stratum. To analyse the symmetry breaking pattern induced by a constraint, it is sufficient to determine the orbit type of the strata affected by the constraint. An even more effective picture is obtained by moving to the orbit space $\mathcal{M}_I$, spanned by the invariants γ_α of the theory. A whole orbit of $\mathcal{M}$ is mapped into a single point of $\mathcal{M}_I$. The crucial property of $\mathcal{M}_I$ is that while $\mathcal{M}$ has no boundaries, $\mathcal{M}_I$ has boundaries that describe the possible breaking chains of the group. The tools that allow one to characterise the orbit space $\mathcal{M}_I$ are the Jacobian matrix [Cabibbo and Maiani (1970)]

$$J \equiv \frac{\partial \gamma}{\partial \varphi}, \tag{12.63}$$

and the so-called $\mathcal{P}$-matrix [Talamini (2006)]

$$\mathcal{P} = JJ^T. \tag{12.64}$$

The manifold $\mathcal{M}_I$ is identified by the requirement that the matrix $\mathcal{P}$ is positive semidefinite, resulting in a set of inequalities involving the invariants γ_α. The boundaries of $\mathcal{M}_I$ can be found by studying the rank of J. In the interior of $\mathcal{M}_I$ the matrix J has maximum rank r_{max}. This region corresponds to the principal stratum where G_d is completely broken. On the boundaries $\mathcal{P}$ has some vanishing eigenvalue and the rank of J is reduced. If the dimension of $\mathcal{M}_I$ is d, in general we have $(d-1)$-dimensional boundaries where $\mathrm{rank}(J) = r_{max} - 1$. They correspond to strata of type G'_d, with $G'_d \subset G_d$. These boundaries meet along $(d-2)$-dimensional spaces, where $\mathrm{rank}(J) = r_{max} - 2$. They correspond to strata of type G''_d, with $G'_d \subset G''_d \subset G_d$. And so on, until the 1-dimensional boundaries meet in a point corresponding to the stratum of type G_d. Thus the symmetry breaking induced by a constraint can be immediately visualised by identifying the region of $\mathcal{M}_I$ involved by the constraint. Here we provide some examples, by discussing the case of the groups S_3 and A_4.

12.4.1 *Covariant constraints in S_3*

S_3 is the group of permutations of three objects. It is also the symmetry group of an equilateral triangle, generated by a reflection a with respect to an axis crossing a vertex and orthogonal to the opposite side and a $2\pi/3$ rotation ab around the centre. In terms of the two generators a and b its six elements are given by

$$S_3 = \{e, a, b, ab, ba, bab\}. \tag{12.65}$$

The elements (ab, ba) are of order three and the elements (a, b, bab) are of order two. They fall into three distinct conjugacy classes $C_1 = \{e\}$, $C_2 = \{ab, ba\}$, $C_3 = \{a, b, bab\}$ and the three irreducible representations of the group are an invariant singlet 1, a singlet $1'$ and a doublet 2. A real basis for the generators a and b is given by:

$$\begin{aligned}
&1: \; a=1 && b=1\\
&1': \; a=-1 && b=-1\\
&2: \; a=\begin{pmatrix}1 & 0\\ 0 & -1\end{pmatrix} && b=\begin{pmatrix}-\frac{1}{2} & -\frac{\sqrt{3}}{2}\\ -\frac{\sqrt{3}}{2} & \frac{1}{2}\end{pmatrix}.
\end{aligned} \tag{12.66}$$

Given two doublets $\varphi = (\varphi_1, \varphi_2)$ and $\psi = (\psi_1, \psi_2)$, we have

$$\begin{aligned}
(\varphi\psi)_1 &= \varphi_1\psi_1 + \varphi_2\psi_2\\
(\varphi\psi)_{1'} &= \varphi_1\psi_2 - \varphi_2\psi_1\\
(\varphi\psi)_2 &= \begin{pmatrix}-\varphi_1\psi_1 + \varphi_2\psi_2\\ \varphi_1\psi_2 + \varphi_2\psi_1\end{pmatrix}.
\end{aligned} \tag{12.67}$$

We consider a theory invariant under S_3 and depending on a single real doublet $\varphi(x)$ in the scalar sector. The ring of invariant polynomials is generated by the two basic invariants[8]:

$$\gamma_1 = (\varphi\varphi)_1, \qquad \gamma_2 = (\varphi\varphi\varphi)_1. \tag{12.68}$$

In our real basis these read:

$$\gamma_1 = \varphi_1^2 + \varphi_2^2, \qquad \gamma_2 = \varphi_1^3 - 3\varphi_1\varphi_2^2. \tag{12.69}$$

The invariants $\gamma_{1,2}$ are independent: there are no algebraic relations among them. Examples of possible constraints of type I are

$$\varphi_1^2 + \varphi_2^2 = M_1^2, \tag{12.70}$$

$$\varphi_1^3 - 3\varphi_1\varphi_2^2 = M_2^3. \tag{12.71}$$

We notice that the constraint (12.70) is invariant under the full orthogonal group $O(2)$ and is an analogue to the constraint defining a non-linear sigma model G/H, $G = O(2)$ and $H = \{e\}$. In general, imposing these constraints breaks completely the group S_3, in the sense that a generic pair (φ_1, φ_2) satisfying (12.70), (12.71) or both is not left invariant by any subgroup of S_3. These constraints might still be useful in the model building, since they limit the possible set of vacuum configuration of the theory.

The group S_3 has a unique one-dimensional non-invariant representation, the singlet $1'$. One can verify that the most general combination of (φ_1, φ_2) transforming as a $1'$ can be written as

$$I(\varphi)\chi(\varphi), \tag{12.72}$$

[8]The Molien function of S_3 in the doublet representation is $M(x) = 1/(1-x^2)(1-x^3)$. It follows that there are two independent invariants, one quadratic and one cubic [Ramond (2010)].

where $I(\varphi)$ is an S_3 invariant while $\chi(\varphi)$ is given by

$$\chi(\varphi) = (\varphi\varphi\varphi)_{1'} = 3\varphi_1^2\varphi_2 - \varphi_2^3, \tag{12.73}$$

where in the last equality we display the explicit expression in the real basis. Thus the unique non-trivial example of type C constraint for S_3 with a real doublet φ is

$$3\varphi_1^2\varphi_2 - \varphi_2^3 = 0. \tag{12.74}$$

This has no analogue in $O(2)$. It defines a domain $\mathcal{D}$, contained in the full field space $\mathcal{M} \equiv R^2$, with the property of being invariant under the action of S_3. It is satisfied by points belonging to the orbits

$$\mathcal{O}_\xi = \Big((\xi, 0), (-\xi/2, -\sqrt{3}/2\xi), (-\xi/2, +\sqrt{3}/2\xi)\Big) \qquad -\infty \leq \xi \leq +\infty. \tag{12.75}$$

For $\xi \neq 0$ these orbits have little group isomorphic to Z_2 while for $\xi = 0$ the little group coincides with S_3. When ξ varies from $-\infty$ to $+\infty$, $\mathcal{O}_\xi$ describes a domain $\mathcal{D}$ consisting of three straight lines that are permuted under the action of S_3. The domain $\mathcal{D}$ is the union of the stratum of type Z_2 and the stratum of type S_3. The orbits $\mathcal{O}_\xi$ are not generic ones. Generic orbits of S_3 consist of six points in the field space $\mathcal{M}$, see Figure 12.2. In generic orbits the symmetry S_3 is completely broken. The full field space $\mathcal{M}$ is partitioned into three strata: the origin, the stratum of type Z_2 and the principal stratum, having trivial little group, see Figure 12.2. Restricting the field space of φ through the covariant constraint (12.74) amounts

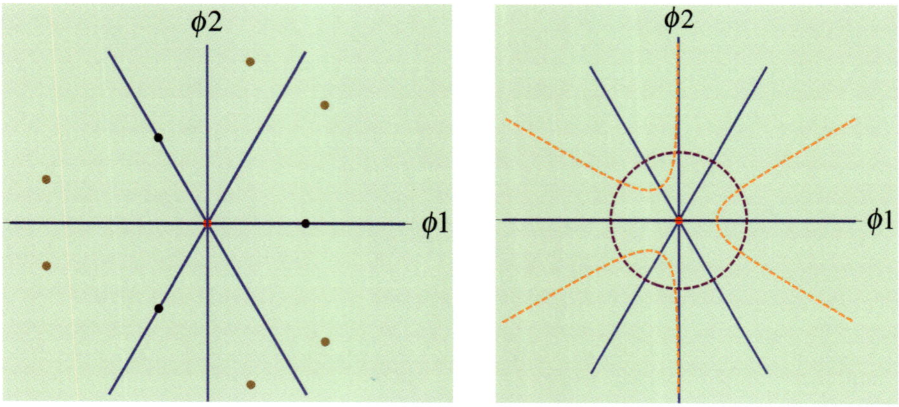

Fig. 12.2 **Left plot**: the field space $\mathcal{M}$, partitioned in strata of different types. The origin (red dot) has little group S_3. The set made of the three straight lines is the stratum of type Z_2. For example, an orbit in this stratum is given by the three black points. The rest of the plain is the stratum of type $\{e\}$. This stratum is called principal stratum. An orbit belonging to this stratum is displayed by the six brown points. **Right plot**: constraints of type I, Eq. (12.70) (dashed, purple) and Eq. (12.71) (dashed, orange), represented in the field space $\mathcal{M}$. In generic points along these lines no S_3-subgroup is preserved.

to describing the breaking of S_3 down to Z_2 but for the particular case where $\varphi = 0$, which leaves S_3 unbroken. The condition (12.74) is invariant under S_3 transformations and fields $\varphi = (\varphi_1, \varphi_2)$ obeying such conditions do not abandon the domain $\mathcal{D}$ when undergoing an S_3 transformation. We can say that the discrete symmetry S_3 is non-linearly realised, reflecting the fact that the domain $\mathcal{D}$ cannot be globally parametrised in terms of a single real field variable. The Lagrangian

$$\mathcal{L} = \frac{1}{2}\partial_\mu\varphi_1\partial^\mu\varphi_1 + \frac{1}{2}\partial_\mu\varphi_2\partial^\mu\varphi_2 - V(\gamma_1,\gamma_2) + \delta\mathcal{L}(\varphi,\psi) \quad \varphi \in \mathcal{D} \tag{12.76}$$

where $\delta\mathcal{L}(\varphi,\psi)$ represents S_3-invariant terms depending on possible additional fields ψ, describes an S_3-invariant theory, where S_3 is unbroken or spontaneously broken down to Z_2. At variance with non-linear realisations of a continuous group, the scalar field ξ that locally parametrises the domain $\mathcal{D}$ is not a Goldstone boson. Moreover it is expected to have a non-vanishing mass, since it parametrises a continuous direction linking different group orbits, albeit of the same orbit type. In general, we cannot interpret the theory restricted to $\mathcal{D}$ as a low-energy effective description obtained by integrating out the heavy degrees of freedom included in $\varphi = (\varphi_1, \varphi_2)$, unless appropriate assumptions on the parameters of the scalar potential $V(\gamma_1,\gamma_2)$ are made. Thus the theory described by the Lagrangian (12.76) should be viewed as a consistent truncation of the full theory, rather than a low-energy approximation.

The orbit space $\mathcal{M}_I$ is spanned by (γ_1,γ_2). A whole orbit of $\mathcal{M}$ is mapped into a single point of $\mathcal{M}_I$, see Figure 12.3. The Jacobian matrix J and the

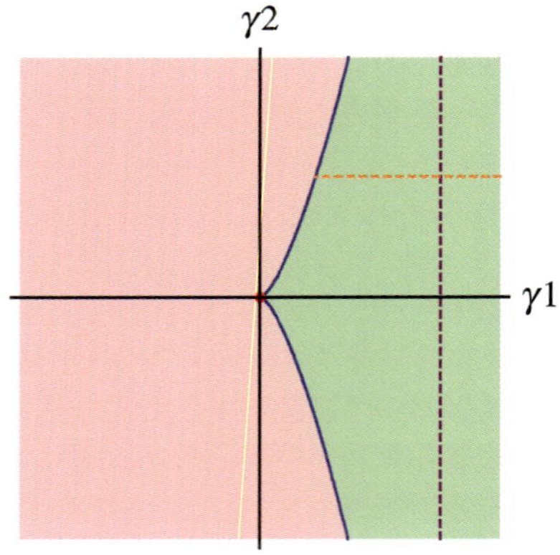

Fig. 12.3 The space of invariants $\mathcal{M}_I$. Each orbit of the field space $\mathcal{M}$ is mapped into a single point of $\mathcal{M}_I$. The origin of $\mathcal{M}$ goes into the origin of $\mathcal{M}_I$ (red point). The stratum of type Z_1, composed of the three straight lines of Fig. 12.2, is mapped into the boundary of $\mathcal{M}_I$ represented by a blue line. The principal stratum is mapped into the region to the right of the blue line. The region to the left of the blue line is not allowed (see text). The vertical (horizontal) dashed line represents the constraint of Eq. (12.70) (Eq. (12.71)).

$\mathcal{P}$ matrix read

$$J \equiv \frac{\partial \gamma}{\partial \varphi} = \begin{pmatrix} 2\varphi_1 & 2\varphi_2 \\ 3\varphi_1^2 - 3\varphi_2^2 & -6\varphi_1\varphi_2 \end{pmatrix}, \quad \mathcal{P} = JJ^T = \begin{pmatrix} 4\gamma_1 & 6\gamma_2 \\ 6\gamma_2 & 9\gamma_1^2 \end{pmatrix}. \tag{12.77}$$

The manifold $\mathcal{M}_I$ is identified by requiring that the matrix $\mathcal{P}$ is positive semidefinite, which defines the region:

$$\gamma_1 \geq 0, \qquad |\gamma_2| \leq \sqrt{\gamma_1^3}. \tag{12.78}$$

We have:

$$\begin{aligned} &\text{rank}(J) = 1 \leftrightarrow 3\varphi_1^2\varphi_2 - \varphi_2^3 = 0 \quad (\varphi \neq 0), \\ &\text{rank}(J) = 0 \leftrightarrow \varphi = 0. \end{aligned} \tag{12.79}$$

The 2-dimensional orbit space $\mathcal{M}_I$ is shown in Fig. 12.3. The conditions $3\varphi_1^2\varphi_2 - \varphi_2^3 = 0$ $(\varphi \neq 0))$ and $\texttt{rank}(J) = 1$ are equivalent and identify the 1-dimensional boundary, corresponding to the stratum of type Z_2. The rank of J vanishes for $\varphi = 0$, where $\gamma_1 = \gamma_2 = 0$. Thus the origin of $\mathcal{M}_I$ represents the stratum of type S_3. In the interior of $\mathcal{M}_I$, the principal stratum, we have $\texttt{rank}(J) = 2$ and S_3 is completely broken.

Even when we do not make use of constraints to restrict the field space of the theory, the above conditions can still be very useful to identify extrema of the scalar potential $V(\gamma_1, \gamma_2)$ in the unrestricted field space $\mathcal{M}$. Indeed the boundary of $\mathcal{M}_I$ provides an important tool to characterise the extrema of V [Cabibbo and Maiani (1970)]. The extremum condition reads:

$$\delta V = \sum_i \frac{\partial V}{\partial \gamma_i} \frac{\partial \gamma_i}{\partial \varphi_\alpha} \delta\varphi_\alpha = \sum_i \frac{\partial V}{\partial \gamma_i} J_{i\alpha}\, \delta\varphi_\alpha = 0. \tag{12.80}$$

One can prove that extrema of V with respect to the points of the boundary of $\mathcal{M}_I$ are also extrema of V. Moreover extrema on the boundary of $\mathcal{M}_I$ are more natural than extrema in the interior of $\mathcal{M}_I$, in the sense that they require the vanishing of a smaller number of derivatives of V, since on the boundary the rank of the Jacobian matrix J is reduced. Therefore the condition $\chi(\varphi) = 3\varphi_1^2\varphi_2 - \varphi_2^3 = 0$ parametrises possible natural extrema of the scalar potential. Additional conditions are in general required to guarantee the existence of an extremum on the boundary of $\mathcal{M}_I$ [Michel and Radicati (1970); Cabibbo and Maiani (1970)]. Invariants of continuous groups relevant to particle physics have been extensively used in the literature [Jenkins and Manohar (2009); Degee *et al.* (2013); Espinosa *et al.* (2013); Alonso *et al.* (2013); Fong and Nardi (2014)].

The previous example shows a first relation between constraints of type C and the breaking of a discrete symmetry. The covariant constraint $3\varphi_1^2\varphi_2 - \varphi_2^3 = 0$ enforces the breaking of S_3 down to a Z_2 subgroup. This is not a general result, but rather a special outcome related to the fact that with a single real doublet there is only one independent constraint of type C that can be constructed. To analyse a more general case we move to the group A_4.

12.4.2 *Covariant constraints in A_4*

We replicate the previous analysis in the case of A_4, isomorphic to Γ_3 and of direct interest to modular forms of level 3. The basic properties of A_4 are summarised in Appendix C. We consider a theory invariant under A_4, whose scalar sector depends on a single real triplet $\varphi(x)$. As we did in the example of the S_3 group, we restrict the field space $\mathcal{M} = R^3$ by means of constraints. Also in this case we can formulate two types of constraints. Constraints of type I are of the form $I_i(\varphi) = 0$ as in Eq. (12.60), where $I_i(\varphi)$ are polynomial invariants. With a single real triplet the ring of invariant polynomials is generated by the four invariants γ_i $(i = 1, \ldots, 4)$ listed in Table 12.7. These invariants are not algebraically independent. They satisfy the relation, or Syzygy [Ramond (2010)]:

$$\mathcal{Z}(\gamma) = 4\gamma_4^2 - 2\gamma_3^3 + 108\gamma_2^4 + \gamma_1^6 + 36\gamma_3\gamma_2^2\gamma_1 - 20\gamma_2^2\gamma_1^3 + 5\gamma_3^2\gamma_1^2 - 4\gamma_3\gamma_1^4 = 0. \tag{12.81}$$

As we discussed in the case of S_3, imposing a generic constraint $I_i(\varphi) = 0$ of type I breaks completely the A_4 symmetry. We turn to constraints of type C, which we formulate by using the non-invariant singlets χ_i $(i = 1, \ldots, 4)$ shown in Table 12.8. By working in the real basis, we start by examining the constraint:

$$\chi_1 = \varphi_1^2 + \omega^2\varphi_2^2 + \omega\varphi_3^2 = 0. \tag{12.82}$$

By taking the real and the imaginary parts of Eq. (12.82) we get

$$\varphi_1^2 = \varphi_2^2 = \varphi_3^2, \tag{12.83}$$

Table 12.7 Invariants that generate the ring of invariant polynomials in A_4 and depend on the real triplet φ. They are not algebraically independent. See the text.

		Real basis	Complex basis
γ_1	$(\varphi^2)_1$	$\varphi_1^2 + \varphi_2^2 + \varphi_3^2$	$\varphi_1^2 + 2\varphi_2\varphi_3$
γ_2	$\frac{1}{6}(\varphi^3)_1$	$\varphi_1\varphi_2\varphi_3$	$\frac{1}{3\sqrt{3}}\left(\varphi_1^3 + \varphi_2^3 + \varphi_3^3 - 3\varphi_1\varphi_2\varphi_3\right)$
γ_3	$\frac{1}{3}\left[(\varphi^2)_1^2 + 2(\varphi^2)_{1'}(\varphi^2)_{1''}\right]$	$\varphi_1^4 + \varphi_2^4 + \varphi_3^4$	$\frac{1}{3}\left[\varphi_1^4 + 12\varphi_1^2\varphi_2\varphi_3 + 4\varphi_1(\varphi_2^3 + \varphi_3^3) + 6\varphi_2^2\varphi_3^2\right]$
γ_4	$-\frac{i}{3\sqrt{3}}\left[(\varphi^2)_{1'}^3 - (\varphi^2)_{1''}^3\right]$	$(\varphi_1^2 - \varphi_2^2)(\varphi_2^2 - \varphi_3^2)(\varphi_3^2 - \varphi_1^2)$	$-\frac{i}{3\sqrt{3}}\left[\varphi_3^6 - \varphi_2^6 + 6\varphi_1\varphi_2\varphi_3(\varphi_3^3 - \varphi_2^3) - 8\varphi_1^3(\varphi_3^3 - \varphi_2^3)\right]$

Table 12.8 Some covariant combinations of the real triplet φ. The product (φ^4) contains several non-invariant singlets and $(\varphi^4)_{1',1''}$ represent two possible combinations.

		Real basis	Complex basis
χ_1	$(\varphi^2)_{1'}$	$\varphi_1^2+\omega^2\varphi_2^2+\omega\varphi_3^2$	$\varphi_3^2+2\varphi_1\varphi_2$
χ_2	$(\varphi^2)_{1''}$	$\varphi_1^2+\omega\varphi_2^2+\omega^2\varphi_3^2$	$\varphi_2^2+2\varphi_1\varphi_3$
χ_3	$(\varphi^4)_{1'}$	$\varphi_2^2\varphi_3^2+\omega^2\varphi_3^2\varphi_1^2+\omega\varphi_1^2\varphi_2^2$	$\frac{1}{3}(\varphi_2^4-2\varphi_2(\varphi_1^3+\varphi_3^3)+3\varphi_1^2\varphi_3^2)$
χ_4	$(\varphi^4)_{1''}$	$\varphi_2^2\varphi_3^2+\omega\varphi_3^2\varphi_1^2+\omega^2\varphi_1^2\varphi_2^2$	$\frac{1}{3}(\varphi_3^4-2\varphi_3(\varphi_1^3+\varphi_2^3)+3\varphi_1^2\varphi_2^2)$

which implies also $\chi_2=0$. The constraint is satisfied by points belonging to the orbits

$$\mathcal{O}_{2\xi}\equiv((\xi,\xi,\xi),(-\xi,\xi,-\xi),(-\xi,-\xi,\xi),(\xi,-\xi,-\xi))\qquad-\infty\le\xi\le+\infty. \tag{12.84}$$

For $\xi\neq0$ these orbits have little group isomorphic to Z_3 while for $\xi=0$ the little group coincides with A_4. When ξ varies from $-\infty$ to $+\infty$, $\mathcal{O}_{2\xi}$ describes a domain $\mathcal{D}_2$ consisting of four straight lines that are permuted under the action of A_4. The domain $\mathcal{D}_2$ is the union of the stratum of type Z_3 and the stratum of type A_4. As a whole, the set of four lines is left invariant under A_4. To complete the partition of the field space $\mathcal{M}$ into strata, we analyse the constraint[9]

$$\chi_3=\varphi_2^2\varphi_3^2+\omega^2\varphi_3^2\varphi_1^2+\omega\varphi_1^2\varphi_2^2=0. \tag{12.85}$$

This constraint is solved by all the points in $\mathcal{D}_2$ and also by points belonging to the orbits

$$\mathcal{O}_{4\xi}\equiv((\xi,0,0),(-\xi,0,0),(0,\xi,0),(0,-\xi,0),(0,0,\xi),(0,0,-\xi))\qquad0\le\xi\le+\infty\}. \tag{12.86}$$

For $\xi\neq0$ these orbits have little group isomorphic to Z_2 while for $\xi=0$ the little group coincides with A_4. When ξ varies from $-\infty$ to $+\infty$, $\mathcal{O}_{4\xi}$ describes a domain $\mathcal{D}_4$ consisting of three straight lines that are permuted under the action of A_4. Excluding the origin, these lines form a stratum of type Z_2. Thus the two covariant constraints (12.82) and (12.85) provide a partition of the field space $\mathcal{M}$ into strata. The domains $\mathcal{D}_2$ and $\mathcal{D}_4$ describe, for $\xi\neq0$, strata of type Z_3 and Z_2 respectively. The origin is the stratum of type A_4 and the rest is the principal stratum.

[9]The conditions $\chi_3=0$ and $\chi_4=0$ have the same solutions, for a real triplet.

Moving to the space of invariants $\mathcal{M}_I$, it is easy to check that both the strata $\mathcal{D}_2$ and $\mathcal{D}_4$ are mapped into boundaries of $\mathcal{M}_I$. Indeed along the orbits $\mathcal{O}_{2\xi}$ and $\mathcal{O}_{4\xi}$, the rank of the Jacobian matrix

$$J^T \equiv \left(\frac{\partial \gamma}{\partial \varphi}\right)^T = \begin{pmatrix} 2\varphi_1 & \varphi_2\varphi_3 & 4\varphi_1^3 & 2\varphi_1(\varphi_2^2-\varphi_3^2)(\varphi_2^2-2\varphi_1^2+\varphi_3^2) \\ 2\varphi_2 & \varphi_3\varphi_1 & 4\varphi_2^3 & 2\varphi_2(\varphi_3^2-\varphi_1^2)(\varphi_3^2-2\varphi_2^2+\varphi_1^2) \\ 2\varphi_3 & \varphi_1\varphi_2 & 4\varphi_3^3 & 2\varphi_3(\varphi_1^2-\varphi_2^2)(\varphi_1^2-2\varphi_3^2+\varphi_2^2) \end{pmatrix}, \tag{12.87}$$

is equal to one for $\xi \neq 0$ and zero when $\xi = 0$, while for generic orbits the rank of J is equal to three. The space of invariants $\mathcal{M}_I$ is more complicated than the one discussed in the S_3 case, due to the presence of the Syzygy $\mathcal{Z}(\gamma) = 0$. $\mathcal{M}_I$ lies on the three-dimensional hyper-surface defined by $\mathcal{Z}(\gamma) = 0$ and it is identified by the requirement that the matrix $\mathcal{P} = JJ^T$ is positive semidefinite. There are no two-dimensional boundaries, as it might be expected in a smooth three-dimensional manifold, and $\mathcal{M}_{\mathcal{I}}$ is a sort of hyperconical surface.

Concerning the applications to fermion masses, the models that we can build using the constraints (12.82) and (12.85) are still far from realistic. In most of the known models specific vacuum alignments among several multiplets of a discrete group are required. For instance, in the lepton sector, typically we ask distinct residual symmetries in the neutrino and in the charged lepton sectors. If the model contains just a single real triplet of A_4, the above constraints only allow to break A_4 down to a Z_2 or a Z_3 subgroup. Nevertheless this approach can be easily generalised to models with a larger scalar sector and may help in exploring different paths in model building.

An interesting case occurs when a covariant constraint does not correspond to a non-trivial residual symmetry, as it happens in the class of supersymmetric modular-invariant models built by using modular forms $Y_i(\tau)$ of level 3. We consider an A_4-invariant theory depending on a complex scalar triplet $\varphi = (\varphi_1, \varphi_2, \varphi_3)$ and for convenience we adopt the complex basis for the three-dimensional representation. We restrict to holomorphic constraints, having in mind the case of supersymmetric theories where φ represents a triplet of chiral supermultiplets of $N = 1$ supersymmetry. We can read the relevant invariant and covariant combinations of φ in Tables 12.7 and 12.8, second columns. In particular we focus on the covariant constraint

$$(\varphi\varphi)_{1''} \equiv \varphi_2^2 + 2\varphi_1\varphi_3 = 0. \tag{12.88}$$

This is precisely the algebraic relation met when examining modular forms of level 3 and weight 2. We see that this constraint does not imply any more $(\varphi\varphi)_{1'} = 0$, due to the complex nature of the φ_i components. Therefore imposing (12.88) does not necessarily enforce the breaking of A_4 into Z_3, as in the case of a real triplet. Indeed this is what we found when exploring the models of Section 12.3.1. Generic orbits of the field space satisfying this constraint belong to the principal stratum and are mapped into the interior of the orbit space, where the symmetry is completely broken. There can be special orbits satisfying the constraint and belonging to strata with a non-trivial little group, but these orbits do not represent generic solutions of the constraint. The region defined by the constraint is invariant under A_4 transformations and the theory can be regarded as a non-linear realisation of the discrete symmetry A_4.

12.5 Discussion

We have proposed a new bottom-up approach to the problem of lepton masses and mixing angles based on supersymmetric modular invariant theories. By recalling general properties of such theories we have seen how they naturally involve the finite modular groups Γ_N. When chiral multiplets transform, up to an overall factor, in unitary representations of Γ_N, the Yukawa couplings have to be modular forms of level N. Modular invariance plays the role of flavour symmetry. Flavon fields might not be needed and the flavour symmetry can be entirely broken by the VEV of the modulus. The superpotential of the theory is very restricted and modular invariance constrains both neutrino masses and mixing angles. As long as supersymmetry is exact all higher-dimensional operators in the superpotential are completely determined by modular invariance. Possible sources of corrections are supersymmetry-breaking contributions and modifications of the Kahler potential. When the weights of all matter supermultiplets vanish, the theory collapses to a supersymmeric Γ_N-invariant model, of the type that has been extensively studied in the past. We have provided several complete examples based on Γ_3, by explicitly constructing the ring of modular forms of level 3. In the most economical example, discussed in Section 12.3.1.2, neutrino mass ratios, lepton mixing angles, Dirac and Majorana phases do not depend on any Lagrangian parameter, but only on the modulus VEV. In the difficult challenge of reproducing five known observables in terms of a single complex parameter, this model scores relatively well. Only $\sin^2\theta_{13}$, predicted to be about 0.045, is out of the allowed experimental range, while providing a reasonable zeroth-order approximation. The model predicts inverted neutrino mass ordering. The example presented in Section 12.3.1.3 predicts normal neutrino mass ordering and is almost equally well-performing. The last example, discussed in Section 12.3.1.4, is minimal in the sense that there are no additional fields breaking the flavour symmetry other than the modulus. Lepton mass ratios and the CP-violating phase are in agreement with the observations, but the mixing angles are only qualitatively reproduced. Solar and atmospheric angles are large and the reactor angle is small, but they are out of range by many standard deviations. Nevertheless the example provides an additional proof that this type of constructions can be carried out and that they are worth exploring.

When discussing modular forms of level 3 we came across a new feature, allowing the notion of non-linearly realised symmetry to be extended to the discrete case. While in the continuous case non-linear realisations can be defined by imposing invariant constraints on the field space, in the discrete case we can also consider covariant constraints of the type $\chi_i(\varphi) = 0$, χ_i denoting non-invariant singlets of the group, a possibility that is precluded to (connected, semisimple) Lie groups. These conditions can be used to define non-linear realisations of a discrete group and we have explicitly analysed how they can arise for the groups S_3 and A_4. Non-linear realisations of a discrete symmetry do not define low-energy effective theories. They should rather be viewed as consistent truncations of some ultraviolet completion. In general covariant constraints leave no non-trivial residual symmetry. We identify the cases in which a non-trivial residual symmetry group survives, by discussing the properties of the so-called orbit space, the space spanned by the invariants of the group. The discussion given here about non-linearly realised discrete symmetries is very provisional and has been motivated by a puzzle about the dimension of the linear spaces of modular forms. Nevertheless we hope that in the future it may turn into a useful tool when looking for new possibilities in model building.

Acknowledgements

I am grateful to Stefano Forte, Aharon Levy and Giovanni Ridolfi for giving me the chance to honour the memory of Guido Altarelli with this contribution. I thank Franc Cameron for useful correspondence about modular forms and Gianguido Dall'Agata and Roberto Volpato for comments on the manuscript. This work was supported in part by the MIUR-PRIN project 2015P5SBHT 003 "Search for the Fundamental Laws and Constituents" and by the European Union network FP10 ITN ELUSIVES and INVISIBLES-PLUS (H2020- MSCA- ITN- 2015-674896 and H2020- MSCA- RISE- 2015- 690575).

Appendix A

We list here general formulae for the dimensions $d_{2k}(\Gamma(N))$ of the linear space $\mathcal{M}_{2k}(\Gamma(N))$ of modular forms of level N and weight $2k$ [Gunning (1962)]. The dimensions read

$$d_{2k}(\Gamma) = \begin{cases} [k/6] & k = 1 \pmod 6 \\ [k/6] + 1 & k \neq 1 \pmod 6 \end{cases} \tag{12.89}$$

where $[x]$ is the integer part of x, and

$$\begin{cases} d_{2k}(\Gamma(2)) = k + 1 \\ d_{2k}(\Gamma(N)) = \dfrac{(2k-1)N+6}{24} N^2 \displaystyle\prod_{p|N} (1 - \frac{1}{p^2}) & N > 2 \end{cases} \tag{12.90}$$

where the product is over the prime divisors p of N.

Appendix B

We show that modular forms $f_i(\tau)$ of weight $2k$ and level $N \geq 2$ transform under Γ_N as

$$f_i(\gamma\tau) = (c'\tau + d')^{2k} \rho(\gamma)_{ij} f_j(\tau) \tag{12.91}$$

where

$$\gamma = \begin{pmatrix} a' & b' \\ c' & d' \end{pmatrix} \tag{12.92}$$

is representative of an element in Γ_N and $\rho(\gamma)$ is a unitary representation of Γ_N. It is sufficient to show that (12.91) holds for γ equal to S and T, which generate the entire modular group $\overline{\Gamma}$ and are not elements of $\overline{\Gamma}(N)$ for $N \geq 2$. When γ is equal to S and T Eq. (12.91) becomes

$$\begin{cases} (S\tau)^{2k} f_i(S\tau) = \rho(S)_{ij} f_j(\tau) \\ f_i(T\tau) = \rho(T)_{ij} f_j(\tau) \end{cases}. \tag{12.93}$$

We start by observing that the holomorphic functions $F_{Ti}(\tau) = f_i(T\tau)$ and $F_{Si}(\tau) = (S\tau)^{2k} f_i(S\tau)$ are modular forms of weight $2k$ and level N. Let

$$h = \begin{pmatrix} a & b \\ c & d \end{pmatrix} \tag{12.94}$$

be a generic element of $\Gamma(N)$. Making use of the fact that $\Gamma(N)$ are normal subgroups of Γ, we have:

$$F_{Ti}(h\tau) = f_i(ThT^{-1}T\tau) = (c\tau + d)^{2k} F_{Ti}(\tau)$$

and

$$F_{Si}(h\tau) = (Sh\tau)^{2k} f_i(ShS^{-1}S\tau) = (c\tau + d)^{2k} (S\tau)^{2k} f_i(S\tau) = (c\tau + d)^{2k} F_{Si}(\tau) .$$

Thus $f_i(T\tau)$ and $(S\tau)^{2k} f_i(S\tau)$ are in $\mathcal{M}_{2k}(\Gamma(N))$ and can be written as linear combinations of $f_j(\tau)$, like in Eq. (12.93). The fact that the coefficients $\rho_{ij}(S)$ and $\rho_{ij}(T)$ define representations of Γ_N follows from the algebraic relations satisfied by S and T. For instance, the generators S and T of Γ_3 satisfy the relations

$$S^2 = T^3 = (ST)^3 = \mathbb{1}. \tag{12.95}$$

It follows that

$$\begin{aligned} \rho_{ik}(S)\rho_{kj}(S) &= \rho_{ik}(T)\rho_{km}(T)\rho_{mj}(T) = \delta_{ij} \\ \rho_{ik}(S)\rho_{km}(T)\rho_{mp}(S)\rho_{pq}(T)\rho_{qr}(S)\rho_{rj}(T) &= \delta_{ij}, \end{aligned} \tag{12.96}$$

and $\rho_{ij}(S)$ and $\rho_{ij}(T)$ are linear representations of the generators of Γ_3. The presentations of several groups Γ_N in terms of the elements S and T can be found in ref. [de Adelhart Toorop *et al.* (2012)]. Finally, the unitarity of the representation stems from the finite dimensionality of Γ_N.

Appendix C

A_4 is the group of even permutations of four objects. It is also the symmetry group of a regular tetrahedron. It has 12 elements and two generators, S and T, satisfying

$$S^2 = T^3 = (ST)^3 = 1. \tag{12.97}$$

The 12 elements fall into four conjugacy classes $C_1 = \{e\}$, $C_2 = \{T, ST, TS, STS\}$, $C_3 = \{T^2, ST^2, T^2S, TST\}$, $C_4 = \{S, T^2ST, TST^2\}$. The group has four irreducible representations: an invariant singlet 1, two non-invariant singlets $1', 1''$ and a triplet 3. The elements S and T in the irreducible representations of A_4 are shown in Table 12.9. The one-dimensional representations are determined non-ambiguously by the conditions (12.97), while the three-dimensional representation is determined up to a unitary transformation, representing a change of basis. The last two columns of Table 12.9 display S and T in two convenient bases, the real one and the complex one, respectively. Given two triplets $\varphi = (\varphi_1, \varphi_2, \varphi_3)$ and $\psi = (\psi_1, \psi_2, \psi_3)$, their product decomposes in the sum $1 + 1' + 1'' + 3_S + 3_A$, where $3_{S(A)}$ denotes the symmetric (antisymmetric) combination. We show the result in Table 12.10, both in the real and in the complex bases.

Table 12.9 Elements S and T in the irreducible representations of A_4, $\omega = -1/2 + i\sqrt{3}/2$ denoting a cubic root of unity. For the triplet representation, two possible choices of basis are displayed.

	1	$1'$	$1''$	3 (Real)	3 (Complex)
S	1	1	1	$\begin{pmatrix} 1 & 0 & 0 \\ 0 & -1 & 0 \\ 0 & 0 & -1 \end{pmatrix}$	$\frac{1}{3}\begin{pmatrix} -1 & 2 & 2 \\ 2 & -1 & 2 \\ 2 & 2 & -1 \end{pmatrix}$
T	1	ω	ω^2	$\begin{pmatrix} 0 & 1 & 0 \\ 0 & 0 & 1 \\ 1 & 0 & 0 \end{pmatrix}$	$\begin{pmatrix} 1 & 0 & 0 \\ 0 & \omega & 0 \\ 0 & 0 & \omega^2 \end{pmatrix}$

Table 12.10 Decomposition of the product of two triplets $\varphi = (\varphi_1, \varphi_2, \varphi_3)$ and $\psi = (\psi_1, \psi_2, \psi_3)$ in irreducible representations, both for the real and the complex bases.

	Real basis	Complex basis
$(\varphi\psi)_1$	$\varphi_1\psi_1 + \varphi_2\psi_2 + \varphi_3\psi_3$	$\varphi_1\psi_1 + \varphi_2\psi_3 + \varphi_3\psi_2$
$(\varphi\psi)_{1'}$	$\varphi_1\psi_1 + \omega^2\varphi_2\psi_2 + \omega\varphi_3\psi_3$	$\varphi_3\psi_3 + \varphi_1\psi_2 + \varphi_2\psi_1$
$(\varphi\psi)_{1''}$	$\varphi_1\psi_1 + \omega\varphi_2\psi_2 + \omega^2\varphi_3\psi_3$	$\varphi_2\psi_2 + \varphi_3\psi_1 + \varphi_1\psi_3$
$(\varphi\psi)_{3_S}$	$\begin{pmatrix} \varphi_2\psi_3 + \varphi_3\psi_2 \\ \varphi_3\psi_1 + \varphi_1\psi_3 \\ \varphi_1\psi_2 + \varphi_2\psi_1 \end{pmatrix}$	$\frac{1}{\sqrt{3}}\begin{pmatrix} 2\varphi_1\psi_1 - \varphi_2\psi_3 - \varphi_3\psi_2 \\ 2\varphi_3\psi_3 - \varphi_1\psi_2 - \varphi_2\psi_1 \\ 2\varphi_2\psi_2 - \varphi_3\psi_1 - \varphi_1\psi_3 \end{pmatrix}$
$(\varphi\psi)_{3_A}$	$\begin{pmatrix} \varphi_2\psi_3 - \varphi_3\psi_2 \\ \varphi_3\psi_1 - \varphi_1\psi_3 \\ \varphi_1\psi_2 - \varphi_2\psi_1 \end{pmatrix}$	$\begin{pmatrix} \varphi_2\psi_3 - \varphi_3\psi_2 \\ \varphi_1\psi_2 - \varphi_2\psi_1 \\ \varphi_3\psi_1 - \varphi_1\psi_3 \end{pmatrix}$

Appendix D

We explicitly construct a basis of weight 2 modular forms for $\Gamma(3)$. The dimension of the space $M_2(\Gamma(3))$ of modular forms of weight 2 for $\Gamma(3)$ has dimension 3 and we look for three linearly elements. We start by observing that if $f(\tau)$ transforms as

$$f(\tau) \to e^{i\alpha}\,(c\tau + d)^k\, f(\tau),$$

then

$$\frac{d}{d\tau}\log f(\tau) \to (c\tau + d)^2 \frac{d}{d\tau}\log f(\tau) + k\; c(c\tau + d).$$

The inhomogeneous term can be removed if we combine several $f_i(\tau)$ with weights k_i

$$\frac{d}{d\tau}\sum_i \log f_i(\tau) \to (c\tau + d)^2 \frac{d}{d\tau}\sum_i \log f_i(\tau) + \left(\sum_i k_i\right) c(c\tau + d),$$

provided the sum of the weights vanishes. Consider the Dedekind eta-function $\eta(\tau)$, defined in the upper complex plane:

$$\eta(\tau) = q^{1/24}\prod_{n=1}^{\infty}(1 - q^n) \qquad\qquad q \equiv e^{i2\pi\tau}.$$

It satisfies

$$\eta(-1/\tau) = \sqrt{-i\tau}\;\eta(\tau), \qquad\qquad \eta(\tau + 1) = e^{i\pi/12}\;\eta(\tau).$$

So $\eta(\tau)^{24}$ is a modular form of weight 12 under the full modular group, actually a cusp form. We further observe that the set composed by $\eta(3\tau)$, $\eta(\tau/3)$, $\eta((\tau+1)/3)$

and $\eta((\tau+2)/3)$ is closed under the modular group. Under T we have

$$\begin{aligned}\eta(3\tau) &\to e^{i\pi/4}\,\eta(3\tau)\\ \eta\left(\frac{\tau}{3}\right) &\to \eta\left(\frac{\tau+1}{3}\right)\\ \eta\left(\frac{\tau+1}{3}\right) &\to \eta\left(\frac{\tau+2}{3}\right)\\ \eta\left(\frac{\tau+2}{3}\right) &\to e^{i\pi/12}\,\eta\left(\frac{\tau}{3}\right).\end{aligned}$$

Under S we have

$$\begin{aligned}\eta(3\tau) &\to \sqrt{1/3}\,\sqrt{-i\tau}\,\eta\left(\frac{\tau}{3}\right)\\ \eta\left(\frac{\tau}{3}\right) &\to \sqrt{3}\,\sqrt{-i\tau}\,\eta(3\tau)\\ \eta\left(\frac{\tau+1}{3}\right) &\to e^{-i\pi/12}\,\sqrt{-i\tau}\,\eta\left(\frac{\tau+2}{3}\right)\\ \eta\left(\frac{\tau+2}{3}\right) &\to e^{i\pi/12}\,\sqrt{-i\tau}\,\eta\left(\frac{\tau+1}{3}\right).\end{aligned}$$

A candidate weight 2 form is

$$Y(\alpha,\beta,\gamma,\delta|\tau) = \frac{d}{d\tau}\left[\alpha\log\eta\left(\frac{\tau}{3}\right) + \beta\log\eta\left(\frac{\tau+1}{3}\right) + \gamma\log\eta\left(\frac{\tau+2}{3}\right) + \delta\log\eta(3\tau)\right],$$

with $\alpha+\beta+\gamma+\delta=0$ to eliminate the inhomogeneous term. Under S and T we have

$$Y(\alpha,\beta,\gamma,\delta|\tau) \xrightarrow{S} \tau^2\,Y(\delta,\gamma,\beta,\alpha|\tau), \qquad Y(\alpha,\beta,\gamma,\delta|\tau) \xrightarrow{T} Y(\gamma,\alpha,\beta,\delta|\tau).$$

We search for three independent forms $Y_i(\tau)$ transforming in the three-dimensional representation of A_4 (in a vector notation where $Y^T=(Y_1,Y_2,Y_3)$)

$$Y(-1/\tau) = \tau^2\,\rho(S)Y(\tau), \qquad Y(\tau+1) = \rho(T)Y(\tau),$$

with unitary matrices $\rho(S)$ and $\rho(T)$

$$\rho(S) = \frac{1}{3}\begin{pmatrix}-1 & 2 & 2\\ 2 & -1 & 2\\ 2 & 2 & -1\end{pmatrix}, \quad \rho(T) = \begin{pmatrix}1 & 0 & 0\\ 0 & \omega & 0\\ 0 & 0 & \omega^2\end{pmatrix}, \quad \omega = -\frac{1}{2}+\frac{\sqrt{3}}{2}i.$$

The transformation T requires:

$$Y_1(\tau) = c_1\,Y(1,1,1,-3|\tau), \qquad Y_2(\tau) = c_2\,Y(1,\omega^2,\omega,0|\tau),$$
$$Y_3(\tau) = c_3\,Y(1,\omega,\omega^2,0|\tau),$$

while the transformation S fixes the coefficients c_i up to an overall factor:

$$c_1 = 3\,c, \qquad c_2 = -6\,c, \qquad c_3 = -6\,c.$$

For convenience we choose $c = i/2\pi$, and explicitly we have

$$\begin{aligned}
Y_1(\tau) &= \frac{i}{2\pi}\left[\frac{\eta'\left(\frac{\tau}{3}\right)}{\eta\left(\frac{\tau}{3}\right)} + \frac{\eta'\left(\frac{\tau+1}{3}\right)}{\eta\left(\frac{\tau+1}{3}\right)} + \frac{\eta'\left(\frac{\tau+2}{3}\right)}{\eta\left(\frac{\tau+2}{3}\right)} - \frac{27\eta'(3\tau)}{\eta(3\tau)}\right] \\
Y_2(\tau) &= \frac{-i}{\pi}\left[\frac{\eta'\left(\frac{\tau}{3}\right)}{\eta\left(\frac{\tau}{3}\right)} + \omega^2\frac{\eta'\left(\frac{\tau+1}{3}\right)}{\eta\left(\frac{\tau+1}{3}\right)} + \omega\frac{\eta'\left(\frac{\tau+2}{3}\right)}{\eta\left(\frac{\tau+2}{3}\right)}\right] \\
Y_2(\tau) &= \frac{-i}{\pi}\left[\frac{\eta'\left(\frac{\tau}{3}\right)}{\eta\left(\frac{\tau}{3}\right)} + \omega\frac{\eta'\left(\frac{\tau+1}{3}\right)}{\eta\left(\frac{\tau+1}{3}\right)} + \omega^2\frac{\eta'\left(\frac{\tau+2}{3}\right)}{\eta\left(\frac{\tau+2}{3}\right)}\right].
\end{aligned} \tag{12.98}$$

The q-expansion of $Y_i(\tau)$ reads:

$$\begin{aligned}
Y_1(\tau) &= 1 + 12q + 36q^2 + 12q^3 + \ldots \\
Y_2(\tau) &= -6q^{1/3}(1 + 7q + 8q^2 + \ldots) \\
Y_3(\tau) &= -18q^{2/3}(1 + 2q + 5q^2 + \ldots).
\end{aligned}$$

It agrees, up to an overall factor, with the q expansion derived in ref. [Franc and Mason (2016)], where the functions $Y_i(\tau)$ are expressed in terms of hypergeometric series. From the q-expansion we see that the modular forms $Y_i(\tau)$ satisfy the constraint:

$$(YY)_{1''} \equiv Y_2^2 + 2Y_1Y_3 = 0. \tag{12.99}$$

Notice that this constraint is left invariant by A_4 transformations, since the combination $(YY)_{1''}$ is invariant up to a phase factor. This constraint has a direct consequence on the number of non-vanishing multilinear combinations of Y_i. If we do not require Eq. (12.99), there are $(k+1)(k+2)/2$ independent k-linear combinations of the type

$$Y_{i_1}Y_{i_2}\cdots Y_{i_k} \qquad (i_1, i_2, i_k = 1, 2, 3). \tag{12.100}$$

For instance, for $k = 2$, we have the 6 combinations Y_1Y_1, Y_2Y_1, Y_3Y_1, Y_2Y_2, Y_3Y_2, Y_3Y_3. These combinations can be arranged into irreducible representations of A_4. For $k = 2$ the six independent Y_iY_j decompose as $3 + 1 + 1' + 1''$ under A_4:

$$\begin{aligned}
Y_3^{(4)} &= (Y_1^2 - Y_2Y_3, Y_3^2 - Y_1Y_2, Y_2^2 - Y_1Y_3) \\
Y_1^{(4)} &= Y_1^2 + 2Y_2Y_3 \\
Y_{1'}^{(4)} &= Y_3^2 + 2Y_1Y_2 \\
Y_{1''}^{(4)} &= Y_2^2 + 2Y_1Y_3,
\end{aligned} \tag{12.101}$$

where, in the left-hand side, the lower index denotes the representation and the upper index stands for $2k$. For $k = 3$, we have the 10 independent combinations $Y_iY_jY_l$. It is not difficult to see that they should decompose as $3 + 3 + 3 + 1$

under A_4. They are

$$\begin{aligned}
Y_1^{(6)} &= Y_1^3 + Y_2^3 + Y_3^3 - 3Y_1Y_2Y_3 \\
Y_{3,1}^{(6)} &= (Y_1^3 + 2Y_1Y_2Y_3, Y_1^2Y_2 + 2Y_2^2Y_3, Y_1^2Y_3 + 2Y_3^2Y_2) \\
Y_{3,2}^{(6)} &= (Y_3^3 + 2Y_1Y_2Y_3, Y_3^2Y_1 + 2Y_1^2Y_2, Y_3^2Y_2 + 2Y_2^2Y_1) \\
Y_{3,3}^{(6)} &= (Y_2^3 + 2Y_1Y_2Y_3, Y_2^2Y_3 + 2Y_3^2Y_1, Y_2^2Y_1 + 2Y_1^2Y_3).
\end{aligned}$$

As a consequence of the constraint in Eq. (12.99) we see that $Y_{1''}^{(4)} = 0$ and $Y_{3,3}^{(6)} = 0$. In general, $k(k-1)/2$ among the $Y_{i_1}Y_{i_2}\cdots Y_{i_k}$ are equal to zero and we are left with $2k+1$ non-vanishing combinations. These should still decompose into A_4 representations, since the constraint (12.99) is A_4-invariant. This explains why the linear space of modular forms of level 3 and weight $2k$ has dimension $2k+1$, as it should, and not $(k+1)(k+2)/2$ as naively derived by counting the homogeneous polynomials of degree k in $Y_i(\tau)$. Thus a modular-invariant supersymmetric theory using as building blocks modular forms of level 3, as the one we have explicitly illustrated as a candidate model for lepton masses, can be regarded as a sort of non-linear realisation of the discrete symmetry A_4.

Chapter 13

The dawn of the post-naturalness era

Gian Francesco Giudice
CERN, Theoretical Physics Department,
CH-1211 Geneva 23, Switzerland

In an imaginary conversation with Guido Altarelli, I express my views on the status of particle physics beyond the Standard Model and its future prospects.

13.1 A master and a friend

Honour a king in his own land; honour a wise man everywhere.

— Tibetan proverb [Cüppers and Sørensen (1998)]

Guido Altarelli was an extraordinary theoretical physicist. Not only was Guido one of the heroes of the Standard Model, but he incarnated the very essence of that theory: a perfect synthesis of pure elegance and brilliance. With his unique charisma, he had a great influence on CERN and contributed much in promoting the role of theoretical physics in the life of the laboratory. With the right mixture of vision, authority, and practical common sense, he led the Theory Division from 2000 to 2004. I have always admired Guido for his brilliance, humour, knowledge, leadership, and intellectual integrity. I learned much from his qualities and his example is a precious legacy for me and for all of his colleagues.

Guido had a very pragmatic attitude towards scientific theories. He was not attracted by elaborate mathematical constructions, but wanted to understand the essence behind the formalism and get straight to the concept. In physics, I would define him as a conservative revolutionary or as an optimistic skeptic. One episode illustrates the meaning of this definition.

In October 2003, I went to Fermilab for a workshop on Future Hadron Colliders. One morning Guido, Fabiola Gianotti, and I went for breakfast in a coffee shop in Naperville. While the blueberry muffins were markedly forgettable, our conversation was memorable. Although at the time ATLAS was only an empty cavern, the LHC was already very present in everyone's mind and so Fabiola started to ask what we should expect the LHC to find: supersymmetry, technicolor, extra dimensions, little Higgs? Nothing, said Guido dispassionately, nothing other than the Higgs. Guido had a natural skepticism towards complicated model building or elaborate constructions whose only rationale seemed to be their ability of ignoring what LEP was telling us. But never did his skepticism turn into a bleak view of the future. He kept a sober attitude towards the latest fashions in new-physics theories, but had a sincere interest in them. Prejudices were not hampering his scientific curiosity. His natural optimism was not incompatible with his rational skepticism. This wise balance is a useful lesson to learn for any scientist who experiences the roller coaster of excitements and disappointments that characterise research at the frontier of knowledge. I will not forget our breakfast in Naperville.

Some of the views expressed during our breakfast conversation reverberated in the summary talk that he gave at the Fermilab workshop. With his inimitable theatrical gestures and distinctive Roman accent, he concluded his talk by saying: "I think that the case for fundamental physics (particle physics and cosmology) is strong beyond fashion and that we should defend it proudly and confidently." [Altarelli (2003)] These words still echo in my mind. I shared with Guido this point of view in 2003 and I still passionately subscribe to it today.

Guido had a significant influence on my personal scientific growth. We were not regular collaborators. We coauthored only three scientific papers, but one of them [Altarelli *et al.* (1997b)] was remembered by Graham Ross as "the most *staffed* paper I have ever seen from CERN." Having four staff members among the authors of the same research paper was peculiar for the CERN Theory Division whose traditional hallmark has always been a respectable (and healthy) dose of individualism.

In spite of the limited collaboration, we had regular physics conversations. Sometimes he was coming into my office, asking me about new trends in physics beyond the Standard Model; often I was visiting his office showing him some new results and asking him for advice. I regarded his opinions highly and our conversations contributed greatly to my scientific development. I consider him one of my masters, but I am honoured to say that he treated me like a friend.

My contribution to this volume has the same style as past conversations with Guido. I will expand freely on views about the status of our field and its future prospects, with no intention to be systematic or complete. My only goal is to have one last conversation with Guido, telling him how I see the field evolving, with no presumption of knowing what is really going to happen in the future. (Pragmatic realism in scientific matters is something I learned from Guido.) My point will be that we live in times of great uncertainties — the best moments for scientific revolutions to happen — so most of what I say today is likely to become obsolete very soon. And I sincerely hope so, because I am eagerly waiting for radical new ideas to disrupt our present beliefs. Guido, too, would like that to happen.

13.2 The naturalness era

Naturalness is the seal of genius.
— François-Joachim de Pierre de Bernis [de Pierre Cardinal de Bernis (1741); Harbottle and Dalbiac (1904)]

If one has to summarise in one word what drove the efforts in physics beyond the Standard Model of the last several decades, the answer is *naturalness.* (For a non-technical introduction to the naturalness problem and for my views on the subject, see [Giudice (2008, 2013)].) Like it or not, naturalness has been the leitmotiv that has accompanied and motivated most of the attempts to incorporate the Standard Model into a bigger framework at the weak scale. There are good reasons for that. First, no mechanism for electroweak breaking can be viewed as complete without an explanation of the size of the Fermi scale relative to other scales that enter particle physics. Second, naturalness has shown to be a good guiding principle to infer the cutoff scale of a theory in a variety of cases: from the emergence of QED beyond energies of the order of the electron mass, to the presence of hadronic resonances in the GeV range, or the existence of the charm quark [Giudice (2008)].

This success is related to the fact that naturalness is an inescapable consequence of the ingredients generally used to construct effective field theories. Finally, there is a much more practical reason for the emphasis on naturalness in research beyond the Standard Model. When applied to the Higgs mechanism, naturalness indicates the need for new physics at a scale within the reach of the present generation of collider experiments. This combination of favourable factors has made naturalness the bulls eye of research in physics beyond the Standard Model.

The effort of understanding the role of naturalness in electroweak physics was rewarded with some of the most extraordinary ideas in theoretical physics: ideas with the ambition to redesign the very structure of spacetime and reinvent the pattern of fundamental forces. Technicolor, supersymmetry, extra dimensions, composite Higgs: all these theories proposed a new interpretation of physical reality and opened new avenues in the exploration of the particle world. These avenues were enthusiastically pursued in experimental research and the LHC is the critical step in this journey.

The LHC delivered an integrated luminosity of about 40 fb^{-1} at $\sqrt{s} = 13$ TeV by the end of 2016 and is rapidly accumulating more data as Run 2 unfolds. This result may look like only a small step in a project that aims at reaching a total of 3000 fb^{-1} by the end of the high-luminosity phase (around 2035). But this small step has given us already a wealth of information, radically influencing the directions of research and the way theorists think of the microworld.

The 2012 discovery of the Higgs boson [Aad *et al.* (2012c); Chatrchyan *et al.* (2012)] was a crucial milestone that revealed how the Standard Model completes the spontaneous breaking of the electroweak symmetry at short distances. The simplest possible UV completion, with a single weakly-interacting scalar particle, received striking experimental confirmation. At the same time no evidence was found of dynamics able to cure the sickness of the Higgs with respect to naturalness. By the summer of 2017, the limits on new particles have reached levels of tension with naturalness of more than the percent: in the simplest supersymmetric set-ups (which correspond to the most aggressive bounds) the gluino mass limits are about 2 TeV [D'Onofrio (2017)], while scalar or fermion top partners are constrained up to 1 TeV [D'Onofrio (2017)]. Although more data is needed to cast the final verdict on the role of naturalness in electroweak physics, the present data have already had a decisive role in ruling out many model-building possibilities and in reorienting the thinking of most theorists.

A celebrated implication of weak-scale solutions to the Higgs naturalness problem is known as the *WIMP miracle.* With apparently innocuous assumptions, often automatically satisfied in realistic models, one finds the intriguing coincidence that the same ingredients needed to cure an unnatural Higgs also lead to a population of particles in today's universe with a density compatible with dark matter observations. The result is striking. Two seemingly unrelated problems — the Higgs unnaturalness (rooted in the quantum structure of the particle world at distances of 10^{-20} metres and below) and the nature of dark matter (observed from galactic distances of 10^{20} metres to the largest scales in the universe) — are linked to the same physical agent. This is the dream of every physicist: a unified and conceptually simple explanation of completely different phenomena.

The conceptual success of the WIMP miracle has ignited a robust experimental programme for the search of weak-scale dark matter particles. So far, no evidence for the existence of such particles has emerged and the results translate into strong constraints on models. In many realistic constructions, direct detection experiments rule out the case of weak-scale thermal relics, if the dark-matter particle has spin-independent couplings with ordinary matter in the non-relativistic regime. The case of spin-dependent interactions is more unconstrained, and here indirect detection gives significant (but mostly model dependent) bounds.

13.3 A krisis?

If a problem is fixable, then there is no need to worry. If it's not fixable, then there is no help in worrying.

— His Holiness the 14th Dalai Lama [Hellstrom (2016)]

In his famous essay *The Structure of Scientific Revolutions* [Kuhn (1962)], the science historian Thomas Kuhn identifies a pattern in the development of scientific theories that is common to all revolutions in science. By freely reinterpreting (and simplifying) Kuhn's structure, I can distinguish three phases in the process. The *phase of discovery* is when new conceptual breakthroughs and experimental results lead to the emergence of a new theory that departs from old paradigms. This is followed by a *phase of consolidation*, in which the theory is understood at a much deeper level and confirmed by precise measurements. This process has the effect of transforming the new theory into the established paradigm of *normal science*. Inescapably, this is superseded by a *phase of crisis*, in which the normal theory can no longer address new conceptual questions or explain experimental data. This phase is characterised by the search for new paradigms and marked by periods of confusion and frustration. Finally a *paradigm shift* occurs, which results in a radical departure from normal science, thus activating a new phase of discovery and marking the beginning of a new cycle.

We can recognise this pattern in the development of both particle physics and cosmology. The Standard Model of particle physics went through its phase of discovery from the late 60's to the early 80's with the rise of gauge theories — from the use of spontaneous symmetry breaking in weak interactions (1967) to asymptotic freedom in strong interactions (1973) — and with the experimental discoveries of charm (1974), tau (1975), bottom (1977), W and Z (1983). The phase of consolidation was characterised, on the theory side, by a careful exploration of the

Standard Model at the quantum level and, on the experimental side, by the confirmation of the gauge doctrine as the ruling principle in the world of particles through the LEP programme and by the discoveries of the top (1995) and Higgs (2012).

The discovery of the cosmic microwave background, the quantitative predictions in nucleosynthesis, the emergence of dark matter and dark energy, the understanding of initial conditions and structure formation in terms of an inflationary theory have marked the birth of cosmology as a science and characterised its discovery phase. This was followed by an extensive programme in observational cosmology (precision studies of the cosmic microwave background, large-scale surveys, gravitational lensing, telescopes from radio to gamma rays), which proclaimed ΛCDM as the standard model of cosmology and which can be identified with the phase of consolidation.

There are many indications that, following the recursive pattern of scientific revolutions, we are now witnessing the beginning of the phase of crisis. The lack of new physics in the initial stages of the LHC project is putting into question the logic of naturalness when applied to the Higgs; the absence of a positive detection in dark matter experiments is casting doubts about nature taking advantage of the WIMP miracle. We are not simply confronted with experimental data excluding a model or a class of models. We are confronted with the need to reconsider the guiding principles that have been used for decades to address the most fundamental questions about the physical world. These are symptoms of a phase of crisis.

The Standard Model of particle physics is a superb monument attesting to the inner beauty of nature and the power of human logical deduction. It is astounding how natural phenomena, in all their complexity, can be summarised by a single principle — the *gauge* principle — and described by a compact set of equations. And it is equally astounding how humans have been able to crack this secret. Along this path, the synthesis of general relativity with the physical laws derived in the microworld has led to the ΛCDM model, which can successfully describe a huge array of cosmological observations, the present large-scale structure of the universe and its early history, in terms of a handful of parameters. This is today's consolidated *normal science.*

And yet, this superb monument of knowledge is insufficient to address some fundamental questions. The Standard Model is incapable of shedding light on the dynamics underlying electroweak symmetry breaking or explaining the structure of quarks, leptons, and their mass pattern at a fundamental level. The theory of inflation, in spite of its stunning conceptual successes, could not be linked univocally with a unified theory of particle physics. Moreover, the ubiquitous phenomenon of eternal inflation has changed the perspective on the outcome of an inflationary universe and its properties. We have plausible explanations for the cosmic baryon asymmetry, but we lack any conclusive empirical confirmation. The nature of dark matter is still unknown. The observed value of the cosmological constant is hard to reconcile with the rules of effective field theory, and quantum gravity is still beyond our grasp.

None of these problems are new, and theoreticians have been tackling them for decades. What is changing is the feeling that the paradigm that so successfully led to the Standard Model may not be the right tool to make further progress. There is a widespread sensation that the organising principles based on symmetry and separation of scales, which follow from an effective quantum field theory approach, in spite of their triumphs, must be superseded by new organising principles. Physicists are in search of new conceptual paradigms, which is another symptom of a phase of crisis.

The privilege of experiencing krisis

Today the word 'crisis' has a sinister connotation, suggesting an approaching downfall, a moment of difficulty or danger. This is not the original meaning of the word. 'Crisis' comes from the Greek *krisis*, which means "decisive moment", "turning point", and was especially used in a medical context by Hippocrates and Galen as the end of a disease. This is the meaning I will refer to and, to stress the idea, I will use the spelling krisis.

Krisis means the opportunity to look at a problem with new eyes; it is a moment of change, a discontinuity between past and future. Krisis does not mean a decline of ideas, but the search for a paradigm change.

Each phase of the Kuhnian cycle is a necessary step for a scientific revolution to happen and an exciting moment for the scientists who live through it. The phase of discovery is the moment of eureka. The phase of consolidation is the moment of maturity. The phase of krisis is the moment of opportunity.[1] A young researcher entering the field can find satisfaction in each of these phases: witnessing the success of a new emerging idea during the phase of discovery, or participating in a common targeted effort during the phase of consolidation. But for an ambitious young mind the phase of krisis is a unique opportunity. This is the most complex and intense

[1]In a 1959 speech John F. Kennedy stated: "In the Chinese language, the word 'crisis' is composed of two characters, one representing danger and the other, opportunity." [Kennedy (1959)] Although perfectly befitting in my context, Kennedy's assertion is unfortunately incorrect as it is based on an etymological misinterpretation of the word *wēi-jī* [Zimmer (1959)]. In 1963, Kennedy had another unfortunate encounter with foreign languages, when he misquoted Dante as saying: "The hottest places in hell are reserved for those who, in times of moral crisis, preserve their neutrality" [Kennedy (1963)], as a result of a dubious translation of the verses: "Questo misero modo tegnon l'anime triste di coloro che visser sanza 'nfamia e sanza lodo." [Alighieri (1321)] Once again, notwithstanding the literary fallacy, Kennedy's words serve my purpose. I concur with him in urging all physicists to give up neutrality and take a stand on the issue of naturalness in these times of scientific crisis.

moment of scientific research, when revolutionary and unprejudiced ideas are needed for a real paradigm change.

The phase of krisis in scientific revolutions is epitomised by the birth period of quantum mechanics, when physicists had to knock down the edifice of classical physics and go through a radical paradigm change. Those were "the thirty years that shook physics" (as Gamow called them [Gamow (1966)]) when physicists, in a rebellion against the sacred principles of classical physics, reinvented the map of physical reality. In the words of Abraham Pais: "Today we live in the midst of upheaval and crisis. We do not know where we are going, nor even where we ought to be going. Awareness is spreading that our future cannot be a straight extension of the past or the present." [Pais (1996)] "Progress leads to confusion leads to progress and on and on without respite. Every one of the many major advances created sooner or later, more often sooner, new problems. These confusions, never twice the same, are not to be deplored. Rather, those who participate experience them as a privilege." [Pais (1986)]

It is striking how perfectly Pais' words capture the spirit of the intellectual tension that characterises the phase of krisis and how they apply equally well to the times of quantum mechanics and to the present situation. The lack of sense of direction, the confusion on the right track to follow in our quest beyond the Standard Model are not reasons for commiseration but for sensing the privilege of the opportunity for an upcoming paradigm change. While the discovery of the Higgs boson has completed a chapter in the book of science, it has also crystallised new conceptual problems that cry out for solutions. Repeatedly, in the history of science, when knowledge consolidated in a consistent theory, there have been voices announcing that "there is nothing new to discover." Invariably, they have been proven to be wrong. The chain of "progress leading to confusion leading to progress" continues without respite and ambitious scientists cherish the moments of confusion.

Of course a big difference between the early 20$^{\rm th}$ century and today is that the paradigm change of quantum mechanics was triggered by many unexpected experimental discoveries, while particle physics today is suffering from a lack of expected experimental discoveries. But the difference may not be so decisive. Although unexpected discoveries are much more sensational and exciting, the lack of expected discoveries can be as effective in triggering a paradigm change. A good example is the failure in measuring the relative motion of the Earth with respect to the aether by Michelson and Morley: a negative result with far-reaching effects! The lack of an expected discovery can be a great source of information able to destroy consolidated principles. What really matters for triggering a paradigm change are two ingredients. One is a broad and intense experimental programme at the frontier of knowledge combined with the vitality of a theory community ready to catch in the data the right lead to phrase the correct questions. The other one is the existence of conceptual puzzles regarding fundamental questions. We are lucky to live in an era in which both ingredients are present.

13.4 The post-naturalness era

The ideal is unnatural naturalness, or natural unnaturalness. I mean it is a combination of both.
— Bruce Lee [Lee (1999)]

I will refer to the present period of rethinking the directions in particle physics as the *post-naturalness era.* What are the guiding principles of the post-naturalness era and what are the emerging new paradigms? Of course I have no idea. Today we lack the historical perspective to see through the fog of the present state of healthy confusion. All I can offer are some simple comments (which, most likely, time will show to be wrong) on aspects that could transform the post-naturalness era into a phase of discovery.

Naturalness

Naturalness is a well-defined concept rooted in the logic of effective quantum field theory. If the LHC rules out dynamical solutions to Higgs naturalness at the weak scale, it does not eradicate the problem: a doctor who is unable to find the right diagnosis cannot simply declare the patient healed. Even in post-natural times, the concept of naturalness cannot simply be ignored, although its use may differ from what it has been so far. One way or another, naturalness will still play a role in the post-naturalness era. But, certainly, ruling out dynamical solutions to Higgs naturalness will be one of the most momentous results in particle physics and will have a radical impact on our approach to physics beyond the Standard Model.

There is already ongoing activity on how the concept of naturalness could be reshaped in post-natural times. Here, instead of giving a comprehensive review, I will only comment on a single new trend: the idea that the explanation of Higgs naturalness may not lie behind some still undiscovered symmetry, but within the cosmological evolution of the universe.

The most daring approach of this kind is based on a multiverse populated by eternal inflation [Vilenkin (1983); Linde (1986a,b)], in conjunction with the idea that fundamental parameters may not necessarily be god-given numbers, but dynamical variables that take different values in a landscape of vacuum states [Bousso and Polchinski (2000); Kachru *et al.* (2003); Susskind (2003); Douglas (2003)]. While this theoretical setup is quite reasonable, the tricky part is to understand what the mechanism that singles out our universe is, selecting the peculiar values of the cosmological constant and the Higgs vev that we observe in our neighbourhood.

There are various possible logical approaches to the selection problem. One — the most hated but the most cogent — is based on anthropic arguments that seem to work reasonably well both for the cosmological constant [Weinberg (1987)] and the Higgs vev [Agrawal *et al.* (1998)], at least in a context in which only a limited number of parameters scan. The difficulty of this approach is to identify observables that make the hypothesis empirically falsifiable. Other approaches — still to be explored, to a large extent — are based on *dynamical selection* (evolution towards a preferred vacuum), *statistical selection* (the vast majority of the possible vacua have features similar to our universe), or *censorship* (criteria that exclude vacua with the wrong properties).

For me, the greatest mystery behind the multiverse is the diffidence some physicists feel towards it. After all, the idea is part of the established toolkit of theoretical physics. For instance, take the well-respected PQ solution to the strong CP problem [Peccei and Quinn (1977)]. The starting point is the empirical observation that a SM parameter ($\theta_{\rm QCD}$) happens to be smaller than 10^{-10} in our universe, while theoretical considerations would give it a value of order one, according to the logic of effective field theory and naturalness. The PQ approach is to promote the SM parameter $\theta_{\rm QCD}$ into a dynamical variable (the axion), so that the actual physical value of $\theta_{\rm QCD}$ in our universe is selected by some underlying dynamics. However, the potential for the axion is generated only after the QCD chiral phase transition, i.e. very late (at least from a particle theorist's perspective) in the history of the universe. Before that time, the axion (and so the effective $\theta_{\rm QCD}$) takes randomly different values in different parts of space, which are then blown up by inflation into different patches of the universe. In other words, a multiverse is created. Although $\theta_{\rm QCD}$ will eventually relax to zero everywhere, the different patches of the "multiverse" contain different physical information, because the energy stored in the axion oscillations around its minimum varies from patch to patch, as it depends on the initial condition of the axion. In other words, the axion can be viewed as an incarnation of the multiverse approach.

The multiverse approach to naturalness is often viewed as the antithesis of the symmetry approach. However, in order to formulate a quantitative criterion of naturalness in the symmetry framework (supersymmetry being the prototypical example) one makes statistical comparisons in a "theory space", whose coordinates are theory parameters [Barbieri and Giudice (1988)]. Ironically, defining the degree of naturalness of a fundamental theory requires a description in terms of a form of "multiverse".

The multiverse is a fertile and constructive framework that could help us to reformulate in different terms some of the open questions in fundamental physics. The idea that the parameters of the Higgs potential or the cosmological constant could be dynamical variables offer a variety of novel approaches to the issue of naturalness. It is intriguing that the Higgs naturalness problem can be rephrased from an issue of UV sensitivity to a problem of criticality [Giudice and Rattazzi (2006)]: why is the Higgs vev so close to the critical point for electroweak phase transition? From this perspective, it appears that the SM parameters are magically chosen such that the theory lives at the boundary between the broken and unbroken phases. Surprisingly, the discovery of the Higgs boson at 125 GeV has revealed another potential problem of criticality within the SM [Buttazzo *et al.* (2013)]: why is the Higgs mass so close to the critical point for a new phase transition? If taken seriously, these questions may suggest an underlying dynamics of the Higgs potential parameters that drive them towards critical points in the phase diagram. The Higgs potential could undergo a self-tuning process, with critical boundaries acting as attractors, just as in the mechanism of self-organised criticality [Bak *et al.* (1987)]. While such a mechanism of *dynamical selection* is uncommon in particle physics, it is rather ubiquitous in other sciences and it has been claimed as an explanation of many disparate phenomena, such as the pattern of earthquakes, forest fires, solar flares, epidemics, and even the behaviour of financial markets.

A concrete realisation of a *dynamical selection* mechanism is the relaxion [Graham *et al.* (2015)]. While the Higgs mass parameter is scanned by a slow-rolling field during the evolution of the universe, the electroweak phase transition generates a back-reaction able to stop the evolution near the critical point. An example of *statistical selection* is the framework proposed in ref. [Arvanitaki *et al.* (2017)], where a large number of vacua is generated in the proximity of a special value for electroweak breaking. Only in that region, the cosmological constant can scan finely enough to be within the narrow range of anthropically allowed values.

Much of the popularity of the multiverse was brought about by the discovery of the enormous *landscape* of possible vacua in string theory [Bousso and Polchinski (2000); Kachru *et al.* (2003); Susskind (2003); Douglas (2003)]. This result had a negative impact on string theory because it conveyed the feeling of practical ineffectiveness of the UV theory to determine the physical properties of our universe, hampering the enthusiasm towards an ultimate fundamental theory. More recent research has inverted the trend by focusing on the restrictions imposed by the quantum theory of gravity on low-energy effective theories. A large variety of theories which, from the low-energy point of view, look perfectly healthy are actually incompatible with quantum gravity and hence cannot describe physical reality. Using current jargon, these non-theories are said to live in the *swampland*, which is believed to be vastly larger than the landscape [Vafa (2005); Ooguri and Vafa (2007)]. This result is a revenge of the ideology professing that the shape of nature is moulded in the UV.

A concrete example of criteria that exclude theories superficially viable from a low-energy perspective is the weak gravity conjecture [Arkani-Hamed *et al.* (2007)]. Although neither the weak gravity conjecture nor other swampland criteria have been proved rigorously, no counterexample in string theory has been found so far. This suggests that an important ingredient for theory selection in the post-naturalness era could be *censorship*. UV information may not be sufficient to determine uniquely the final theory, but can effectively prune the tree of possibilities. This line of thought has been pursued in refs. [Ibanez *et al.* (2017)] and [Lüst and

Palti (2017)] where, using different approaches, it has been suggested that the SM with an electroweak breaking scale much above the TeV scale may only belong to the swampland.

A more prosaic way of implementing *censorship* is to invent cosmological criteria that prevent the physical occurrence of sectors with large electroweak breaking scale. One example is Naturalness [Arkani-Hamed *et al.* (2016a)], in which the reheating process acts as a cosmological censor expunging immoral versions of the universe.

Although none of these examples may be a convincing explanation of Higgs naturalness, these models testify to the activity in trying to explain Higgs naturalness as a result of the cosmological evolution. The general approach is an interesting and a novel line of thought.

While new ideas will emerge in the future, what was learned during the naturalness era will certainly not turn out to be useless for the post-naturalness era. The examples I presented focus on the special features of the Higgs naturalness problem. But it is likely that the Higgs problem is only a small aspect of a bigger naturalness problem. The Higgs boson is unlikely to be the only scalar particle we must reckon with. Other fundamental scalar particles may be required by various patterns of symmetry breaking at high scales, by the axion sector, inflation, and the dark sector. Each of these particles would come with its own naturalness problem. While the Higgs naturalness (or at least the little hierarchy problem) may be cured by alternative medicine, it is plausible that a more radical solution, based on symmetry, is present at the fundamental level and deals with the big hierarchy problem. Supersymmetry is the obvious candidate to reappear in the post-naturalness era in a different guise. One interesting example is split supersymmetry [Arkani-Hamed and Dimopoulos (2005); Giudice and Romanino (2004); Arkani-Hamed *et al.* (2005)], which offers many attractive theoretical features, although it lacks the power to deal with the little hierarchy. Even in the context of the relaxion, supersymmetry can help in making the picture more convincing by supplementing UV ingredients lacking in the non-supersymmetric version [Batell *et al.* (2015)].

Cosmological constant

While discussing the role of naturalness in the post-naturalness era, I focused on the Higgs. However, the cosmological constant poses an even bigger problem. The dominant attitude during the naturalness era was to believe that the two problems could be addressed separately. This was probably not dictated by a deep theoretical conviction, but mostly due to pragmatic reasons. While progress with the cosmological constant was difficult, Higgs naturalness seemed an endless source of revolutionary ideas with promises of direct experimental verification.

From a post-natural perspective, it seems that the problem of the cosmological constant can no longer be ignored. By making a field expansion of the SM effective potential, it appears that the problems with the cosmological constant and the Higgs mass have a common origin. The only difference is that the vacuum energy, unlike the Higgs mass, becomes an observable only through its coupling to gravity. But it seems likely that progress during the post-naturalness era will only come by addressing the two problems simultaneously.

The cosmological constant corresponds to a scale of about 2×10^{-3} eV and affects physics in the deep IR, at astronomical and cosmological distances. Modifications of gravity at large distances are attempts to tackle the problem from an IR perspective [de Rham (2014)]. However, the conceptual problem with the cosmological constant comes from quantum effects in the deep UV. Attempts to tackle the UV problem with the traditional methods used for the Higgs seem hopeless because the corresponding UV cutoff is way lower than any energy scale entering those theories.

This confusion among scales is at the basis of the problem. It is a big source of confusion because the systematic approach of effective field theories has taught us how to separate energy scales in successive shells and make sense of the theory at each shell separately. The cosmological constant seems to resist this approach. Naturalness is an offspring of effective field theory and so it is not surprising that the difficulty we are encountering with the effective theory description leads to a problem with naturalness.

A theory that manifests an active interplay between the IR and the UV would not be simply describable by an effective field theory, as it would violate its inner logic. It is not impossible that quantum gravity will exhibit some kind of IR/UV interplay. An indication could come from the classical behaviour of gravity. Consider the head-on collision of two particles at ever increasing energies. Once you pass the threshold for forming a black hole, the more energy you feed in the system the larger the Schwarzschild radius becomes. In other words, higher-energy collisions are less sensitive to short distances, in contrast with our effective theory intuition for a separation between IR and UV.

A radical conclusion that could be derived from these considerations is that we are facing the end of validity of field theory and the solution of the cosmological constant lies beyond our familiar theories. The new framework should incorporate an interplay of physical effects occurring at all scales. Although it is difficult to tell how such a framework would look, one can easily expect that the cosmological constant problem will play a key role in the post-naturalness era.

Simplicity and complexity

The phenomena that we observe in nature are complex. And yet, humans have always had the aspiration to understand nature in terms of an inner simplicity, whether in the form of deities or of physical laws expressed in mathematical language. This idea reverberates also in the world of art. As the sculptor Constantin Brâncuşi said, "Simplicity is complexity resolved." [Gimenez and Gale (2004)]

Just by staring at the world outside the window, it may seem that this programme of searching for an ultimate physical law governing the entire physical reality is a bit simpleminded. *Why would the complexity of natural phenomena be reducible to a simple set of physical laws?* One of the most stupefying results of modern science was to find empirical evidence that this programme — simple-minded as it may seem — is actually marvellously successful. Nature shows the same aspiration as humans to have simple principles behind the manifest complexity of the universe.[2]

Natural phenomena, when observed with poorer resolution, become simpler. This is simple fact. When we describe the Moon's trajectory orbiting around Earth, we can safely approximate both bodies with material points and neglect all the complexity of geography, crystalline structure, chemical composition, and terrestrial biology. This is technically known as the "spherical-cow approximation", which is a basic tool of effective field theory. The observation that the universe is roughly uniform and isotropic at large scales supports the point of view that simplicity dominates the far IR.

As we move from large to small distances, the complexity of nature grows. We see clusters with thousands of galaxies, each containing billions of stars, planets, life. But, surprisingly, once we reach much smaller distances the trend of growing complexity seems to reverse direction. Simplicity emerges again at small distances, in the form of a finite number of building blocks. Chemistry organises matter into molecules, which can be reduced to universal atoms, constructed out of elementary particles. This is a highly non-trivial, and a priori unexpected, result. Simplicity is not relegated to the spherical-cow approximation, but becomes preponderant also at small distances.

When we enter the particle world, simplicity at small distances takes on a different meaning. As we descend to shorter distances, we discover that physical reality is described not by fewer building blocks, but more. Heavy quarks, heavy gauge bosons and the Higgs are found only when we explore smaller distances. If supersymmetry were correct, many more particles would be hidden at even smaller distances. We don't know if the SM particles are truly fundamental or will be ultimately replaced by new building blocks. But we have learned that simplicity in the UV does not manifest itself in terms of fewer degrees of freedom, but in terms of more powerful organising principles. The full structure of the fundamental laws becomes apparent only when we observe nature at the smallest possible distance. This is the reason why fundamental physics today is focusing so much on particle physics and high-energy collider experiments. It is there that we expect to find the answers to our most fundamental questions.

It is not difficult to imagine designing a universe based on simple laws. The problem is that, most likely, such a universe would also exhibit only simple phenomena and would be unable to produce the complexity that characterises life or consciousness. The original question of the reductionist programme then takes a more technical form: *how can nature get emerging complexity out of simple principles?*

[2]Probably it is the other way around.

We do not yet have a full picture of how nature overcomes the dichotomy between simple fundamental laws and complex emergent phenomena. But particle physics has made huge progress in this direction and the key words are *gauge theory*. Gauge theory is the essential concept out of which the Standard Model is built: a concept that has all the features of a fundamental principle of nature. It is elegant (based on symmetry considerations), robust (no continuous deformations of the theory are generally allowed), and predictive (given the field content, all processes are described by a single coupling constant). In short, it has all the requirements for a physicist to see simplicity in it.

The magic about gauge theory lies in the richness of its structure and its ability to produce, out of a simple conceptual principle, a great variety of different manifestations. Long-range forces, short-range forces, confinement, dynamical symmetry breaking are all phenomena described by the same principle. The vacuum structure of gauge theory is unbelievably rich, with θ-vacua, instantons, chiral and gluon condensates, all being expressions of the same theory. The phase diagram at finite temperature and density exhibits a variety of new phenomena and states of matter. In short, gauge theory is an exquisite tool to make complexity out of simplicity.

Is gauge theory the final answer to overcome the simplicity/complexity dichotomy? In spite of the brilliant success of gauge theory, it is likely that post-natural physicists will look elsewhere for new answers. We cannot be sure that the trend of organising principles hiding in the UV will continue beyond the boundaries of present knowledge. The difficulties with the cosmological constant I was alluding to before may already be a sign of the need for a paradigm change and results from the LHC may reinforce this indication. Losing the belief that nature exhibits its fundamental laws in the UV will be a radical change of perspective.

The multiverse already offers new interpretations for merging simplicity with complexity, as it is a perfect example of how nature can have simple physical laws and generate complexity out of the vacuum structure. Many new approaches are starting to be pursued. Gauge theories, as well as Lorentz invariance and the structure of spacetime, may be emerging concepts. Techniques from quantum information lead to surprising new insights on the AdS/CFT correspondence and the theory of quantum gravity, opening the door to completely new interpretations of physical reality, as summarised by the slogan "It from Qubit." [Simons Foundation, Mathematical and Physical Sciences (1996)] This is physics for the post-naturalness era.

Symmetry

Symmetry has been a powerful engine that fuelled the rise of the Standard Model and guided most of the attempts to go beyond the Standard Model during the naturalness era. Its instrumental role in physics is undeniable and its influence will be everlasting. But now lend me your ears: I come to bury symmetry, not to praise it.

It is believed that no global symmetries exist in a quantum theory of gravity [Kallosh *et al.* (1995); Banks and Seiberg (2011)]. This suggests that global symmetries are not ingredients of the fundamental theory. Global symmetries can only be approximate and accidental. In other words, they are emergent properties at low energy. A simple example is the rotational symmetry that emerges in the spherical-cow approximation. Examples more pertinent to particle physics are baryon and lepton number in the SM, custodial symmetry of the Higgs model in the limit of vanishing hypercharge and quark mass difference, flavour symmetry in the limit of vanishing Yukawa couplings, isospin symmetry in nuclear forces, chiral symmetry in the pion Lagrangian, and many others. These examples show that global symmetry is a useful concept in the IR. It is useful as a classifier in low-energy theories, both in linearly realised versions (establishing selection rules for forbidden local operators) and non-linear versions (characterising the structure of the Lagrangian of light scalar particles). However, in spite of their practical usefulness in the IR, they are probably inconsequential in the UV, where the truly fundamental theory is expected to live.

The same "folk theorem" that rules out continuous global symmetries at the fundamental level applies to discrete (non-gauge) symmetries as well. The Standard Model vouches for this assertion since discrete symmetries like C, P, and T can be explained as the result of Lorentz invariance and the structure of the interactions. Whenever possible, the Standard Model breaks maximally C and CP, both in weak and Yukawa interactions. Surprisingly, this is not true for strong interactions, where the topological structure of the gauge theory allows for a large violation of CP, which is not observed in nature. This anomalous behaviour is taken seriously by most physicists, who believe it cannot be just a fluke of strong interactions, but the indication for some new dynamics yet to be discovered (e.g. the axion).

We are left with gauge symmetry. The problem with gauge symmetry is that it is not a symmetry in the sense of quantum mechanics. A symmetry is the invariance of the Hamiltonian under transformations of quantum states, which are elements of a Hilbert space. Gauge symmetry is not a symmetry because the corresponding transformation does not change the quantum states. Gauge symmetry acts trivially on the Hilbert space and does not relate physically distinct states. A gauge transformation is like a book by James Joyce: it seems that something is going on, but nothing really happens.

Gauge symmetry is the statement that certain degrees of freedom do not exist in the theory. This is why gauge symmetry corresponds only to a redundancy of the theory description. The non-symmetry nature of gauge symmetry explains why gauge symmetry, unlike global symmetry, cannot be broken by adding local operators to the action: gauge symmetry is exact at all scales. The only way to "break" gauge symmetry is adding to the theory the missing degrees of freedom, but this operation is not simply a deformation of the theory (as is the case of adding local operators to an action with global symmetry) but corresponds to considering an altogether different theory. The non-symmetry nature of gauge symmetry also explains trivially the physical content of the Higgs theorem. For a spontaneously-broken global symmetry, an infinite number of vacuum states are related by the symmetry transformation. This leads to the massless modes dictated by the Goldstone theorem. In a spontaneously-broken gauge symmetry, there is a single physical vacuum and thus there are no massless Goldstones. Gauge symmetry does not provide an exception to the Goldstone theorem, simply because there is no symmetry to start with.

For gauge symmetry, the word 'symmetry' is a misnomer, much as 'broken' is a misnomer for spontaneously broken symmetry. But as long as the physical meaning is clear, any terminology is acceptable in human language. The important aspect is that the mathematical language of gauge symmetry (both in the linear and non-linear versions) is extremely powerful in physics and permeates the Standard Model, general relativity, and many systems in condensed matter. As the redundancy of degrees of freedom is mathematically described by the same group theory used for quantum symmetries, the use of the word 'symmetry' seems particularly forgivable.

Does this necessarily make gauge symmetry a fundamental element in the UV? The property of gauge symmetry of being — by construction — valid at all energy scales may naively suggest that gauge symmetry must be an ingredient of any UV theory from which the Standard Model and general relativity are derived. On the contrary, many examples have been constructed – from duality to condensed-matter systems – where gauge symmetry is not fundamental, but only an emergent property of the effective theory [Witten (2017)]. Gauge symmetry could emerge in the IR, without being present in the UV theory. If this is the case, gauge symmetry is not the key that will unlock the mysteries of nature at the most fundamental level. The concept of symmetry has given much to particle physics, but it could be that it is running out of fuel and that, in the post-naturalness era, new concepts will replace symmetry as guiding principles.

Dark matter

The WIMP miracle is an attractive and fairly generic consequence of the naturalness approach to the Higgs. But, no matter how appealing it is, the WIMP paradigm will have to be replaced, if future experiments give no evidence in support of the idea.

How is one to proceed without guidance from the WIMP miracle? An often-used strategy is adding a single elementary particle for the dark matter and, possibly, another particle to mediate a force between the dark and visible sectors. This approach is justified by criteria of "minimality" or "simplicity". While I agree that, given our ignorance about the dark sector, this is the first thing to try, I also think that the approach, rather than "minimal", looks awfully shortsighted.

We have ample evidence that nature follows a grand scheme. We recognise in the Standard Model features of this grand scheme. Adding one or two particles only for the reason of generating dark matter is not something that nature would do, if she indeed has a grand scheme in mind. It is much more plausible that the dark matter is only the tip of the iceberg of a sector that serves a structural purpose. This is indeed the logic behind traditional WIMP models. Take the prototypical example of supersymmetry: the WIMP does not come alone, but it is one out of many new particles whose purpose is to stabilise the Higgs mass against quantum corrections.

Once we accept that dark matter is unlikely to manifest itself in "minimal" ways, the tree of possibilities grows rapidly. Post-naturalness dark matter could manifest itself in innumerably different manners. Why should dark matter be a single particle charged under a single force, while matter in the universe comes in a variety of states and chemical compounds?

Both theoretical activity and experimental searches for dark matter are adapting to this growing number of unconventional and multi-component forms of dark matter. Tension with the predictions of collisionless dark matter at small scales [Weinberg *et al.* (2014)] can be relaxed if dark matter is self-interacting [Wandelt *et al.* (2000)]. Moreover, dark matter could be much lighter than that expected for a WIMP. The extreme case is the proposal of fuzzy dark matter, made of a particle as light as 10^{-22} eV [Hui *et al.* (2017)]. The light dark-matter window, for masses below the GeV range, is not well covered by conventional searches and several new clever techniques have been proposed to explore this region [Battaglieri *et al.* (2017)]. One of the most appealing aspects of these new ideas is that they are employing techniques borrowed from other sectors of science, from atomic physics to condensed matter. This cross-fertilisation between different research fields is especially needed during a phase of krisis.

In unconventional models, one cannot exclude the possibility that dark matter is present in the universe in different forms, from dust to compact objects of astronomical size. The detection of gravitational waves [Abbott *et al.* (2016)] has opened a new field of exploration, providing a new tool for probing the existence of compact objects made of exotic matter [Giudice *et al.* (2016); Cardoso *et al.* (2016)]. These measurements have also revitalised [Bird *et al.* (2016); Sasaki *et al.* (2016); Clesse and Garca-Bellido (2017); Kashlinsky (2016)] the hypothesis that dark matter is not made by new particles, but by a primordial population of black holes [Zel'dovich *et al.* (1967); Hawking (1971)].

The field of dark matter is quickly changing and, both on the theoretical and experimental sides, there is growing interest in alternatives to traditional WIMPs. It is likely that the new trend in dark matter will influence the paradigm search during the post-naturalness era and, conversely, that post-natural research will influence dark matter modelling and experimental strategies.

The role of experiment

Physics is a natural science and experimental observation is the essence of the scientific method upon which physics is based. A phase of krisis cannot mature into a paradigm change without the support of an intense, vigorous, and broad experimental programme. A reason for my optimism that the present phase of krisis will soon blossom into an era of new knowledge is to be found in the exceptional experimental prospects. In every area of fundamental physics, we are witnessing an explosion of innovative projects. A wealth of new experimental results will soon be available, bringing the promise of the spark that will ignite the new paradigm change.

Observational cosmology is moving fast: the B-mode of the CMB is under siege and new missions probing the CMB are already on the horizon; space and ground surveys will make a leap in measurements of the large scale structure and the properties of dark energy. Direct dark matter experiments are expected to reach the neutrino floor and many new techniques are being proposed to explore the light-mass window. Cosmic ray observations can constrain or give a map of the distribution and properties of dark matter particles. Progress in axion and axion-like experiments is advancing rapidly and the hypothesis of dark matter axions will be conclusively tested. Gravitational wave explorations have only started and a new generation of interferometers — on earth and in space — will reach stunning precision with sensitivity to a large range of frequencies, thus opening new and still unimagined ways of probing the universe. The neutrino programme is advancing on various fronts: exploration of the neutrino mass in $\beta\beta_0$ decay, measurements of neutrino mixing angles and CP-violating phases, ordering of the mass hierarchy, existence of light sterile neutrinos, and neutrino cosmic background. Future long- and short-baseline experiments also have the potential of exploring many exotic phenomena beyond the world of neutrinos. B-factories are starting a new campaign of exploration in flavour processes. Experiments on the muon $g-2$, on rare muon processes, and particle EDMs are formidable probes of short-distance physics. Beam dump experiments will explore the dark side of light and feebly-interacting particles. Many table-top experiments probing macroscopic forces and making precision tests of gravity could furnish unexpected surprises.

CERN is contributing to this landscape of experimental activity by promoting studies for possible small-scale experiments that will enrich its scientific programme [PBCWG (2016–2017)]. Of course, the main focus of today's CERN programme is the LHC: data taking is in full swing and the operation of the ten-year high-luminosity phase will start in 2026. As of today, the most tantalising LHC news comes from B physics, in the form of combined anomalies in various decay processes [Blanke (2017)]. The surprising aspect of these results is that they imply violation of lepton universality, which is not the first place at the high-energy frontier where one would have expected new physics to come up. The exciting aspect of these results is that they involve the flavour sector, which remains mysterious from the theory point of view, as no significant progress has ever been made in computing quark and lepton masses and mixing angles or understanding the dynamical origin of their pattern. The craziness of these results is promising. Given our failure in cracking the flavour problem, it is unlikely that its solution is simple and that it manifests itself in simple ways. Only crazy results in flavour physics have a chance to be true.

While LHC operations are underway, there is already feverish activity around the world in planning the new generation of high-energy collider experiments. The physics community will soon be called upon to make strategic choices about the future of experiments in collider physics. What are the goals of the high-energy physics programme in the post-naturalness era?

One sure priority is the research programme in Higgs physics. The discovery of the Higgs boson [Aad *et al.* (2012c)] opened the door to the understanding of the mechanism of electroweak breaking. This mechanism plays a crucial role in nature, giving rise to a fundamental scale that rules not only the microscopic world but many physical properties of our universe, from the size of atoms to the timescale of the processes that make the sun shine. Missing the opportunity to study in depth the mechanism of electroweak breaking would be like giving up the exploration of a new continent in the planet of knowledge.

Although a fundamental milestone, the discovery of the Higgs cannot be regarded as the final resolution to the enigma of electroweak breaking. If anything, its discovery has rendered unavoidable the need to address some of the open problems in particle physics. Indeed, every single Higgs interaction introduces its own puzzle. The Yukawa couplings express the flavour problem, i.e. our inability to compute quark and lepton masses and mixing angles. The Higgs quartic coupling raises issues with the stability of the SM electroweak vacuum. The Higgs quadratic term incarnates the naturalness problem. The constant term in the Higgs potential is an expression of the cosmological constant problem.

The root of these problems is that the Higgs boson introduces 14 new forces (without taking into account other forces associated with neutrinos) besides the 3 known fundamental forces of the Standard Model. Unlike strong, weak, and electromagnetic forces, these new 14 forces are not gauge-like. Not surprisingly, they do not share the properties of elegance, simplicity, robustness, and predictivity that characterise the gauge forces of the Standard Model. Accustomed with the conceptual perfection of the gauge sector of the Standard Model, it is difficult for a theorist to believe that a structure so arbitrary and provisional as the Higgs sector could be the final word on electroweak breaking. There must be more still to be discovered.

The way to get to the bottom of the question is a programme of precision measurements on the Higgs boson properties. This is already underway, but future high-energy colliders will be able to bring these precision studies to unprecedented levels of precision. Much can be learned from these studies about the Higgs boson, the phenomenon of electroweak breaking, the fundamental laws that govern nature, the early stages of the universe and its ultimate fate.

The recent history of experimental particle physics has been characterised by large projects that had the exploration of the unknown as the primary goal but, at the same time, had also a clear objective within reach. It was the discovery of the W and Z for the SPS, the top for Tevatron, precision studies of the gauge sector for LEP, the Higgs for LHC. However, the story may change in the future and experimental physics during the post-naturalness era must reckon with situations in which clear objectives within reach are difficult to be determined. Any result from the LHC or other experiments can suddenly drive a change in directions and in experimental priorities. Krisis is a phase dominated by uncertainty. But this makes research only more exciting: a daring programme of exploration with ambitious goals is what the post-naturalness era needs.

Today we are facing many big open problems in the understanding of the particle world, but we are uncertain about where their solutions lie. In the present situation, exploration is the only possible answer to our thirst for knowledge. There is nothing new in this. The very soul of particle physics has always been exploration. Our history is a history of pushing frontiers, crossing boundaries between the known and unknown, exploring virgin territories. But it has never been a blind exploration. We don't pursue a direction just because we have the technology to do so. We do it because we believe that there is something fundamental to learn by pursuing that direction. The measurements that matter are those from which we learn something that we didn't know before. The fantastic success of particle physics has been the result of wise choices about the best frontiers to push, about the right barriers to break. This success was possible only because of the right mixture of theoretical and experimental physics, of daring ideas and fabulous technologies.

Our scientific priorities are likely to shift in the post-naturalness era, but the high-energy frontier seems, more than ever, the most promising direction for us to gain new knowledge in fundamental physics. In the post-naturalness era we will have to tackle new questions, both theoretically and experimentally. All indications point towards higher energy as the right frontier to push in order to answer many fundamental questions about our universe. This is what we can do with future high-energy collider projects.

The role of theory

"The truth is, the Science of Nature has been already too long made only a work of the Brain and the Fancy: It is now high time that it should return to the plainness and soundness of Observations on material and obvious things." These words are not the latest attack against the abstractness of modern theoretical physics or the inability of string theory to find a criterion for empirical falsifiability. No, these words are by Robert Hooke and date back to 1665 [Hooke (1665)], when times were changing in England and a new approach to science was emerging: a radically different use of abstract mathematics in formulating physical theories. Only one year later, in the *annus mirabilis* 1666, Isaac Newton, who had no sympathy for Hooke — sincerely reciprocated by Hooke — started to revolutionise physics by making use of calculus in his formulation of mechanics and gravitation. A masterpiece of a work of the Brain.

The criticism for excessive mathematical abstractness has accompanied theoretical physics throughout all its major breakthroughs. It was the case for Maxwell's highly mathematical treatise *A Dynamical Theory of the Electromagnetic Field*, which gives the foundation of modern electromagnetism and predicts electromagnetic waves travelling at the speed of light. "For more than twenty years, his theory of electromagnetism was largely ignored," recounts Dyson. "It was regarded as an obscure speculation without much experimental evidence to support it." [Dyson (2007)] Then came Einstein's relativity theory, which was rejected by many, including the Nobel Laureate Philipp Lenard, for "lacking the illustrativeness of classical physics" [Lenard (1920)] and, in later darker times, for contradicting the principles of "Aryan physics by man of Nordic kind [...] which only accepts scientific knowledge based on experiments, and only if accessible to the senses." [Lenard (1936)] At the beginning of last century, the abstract developments of quantum mechanics — from Schrödinger's complex analysis describing wavefunctions to Heisenberg's non-commutative operator algebra — were targets of innumerable grievances about how theoretical physics had departed from physical reality. Today similar lamentations about the status of theoretical physics are not uncommon.

The role of logical deduction in physics and the need to develop a mathematical language for the description of nature have always been clear to the great minds of our field, starting from Galileo and Newton at the very beginning of modern science and continuing throughout the history of physics. The concept was expressed by Einstein during a lecture at Oxford in 1933: "It is my conviction that pure mathematical construction enables us to discover the concepts and the laws connecting them which give us the key to the understanding of the phenomena of Nature. Experience can of course guide us in our choice of serviceable mathematical concepts; it cannot possibly be the source from which they are derived; experience of course remains the sole criterion of the serviceability of a mathematical construction for physics, but the truly creative principle resides in mathematics. In a certain sense, therefore, I hold it to be true that pure thought is competent to comprehend the real, as the ancients dreamed." [Einstein (1934)]

Dirac pushed the concept further, elaborating on the creative role of theoretical physics in employing mathematics: "The steady progress of physics requires for its theoretical formulation a mathematics that gets continually more advanced. This is only natural and to be expected. What, however, was not expected by the scientific workers of the last century was the particular form that the line of advancement of the mathematics would take, namely, it was expected that the mathematics would get more and more complicated, but would rest on a permanent basis of axioms and definitions, while actually the modern physical developments have required a mathematics that continually shifts its foundations and gets more abstract. Non-Euclidean geometry and non-commutative algebra, which were at one time considered to be purely fictions of the mind and pastimes for logical thinkers, have now been found to be very necessary for the description of general facts of the physical world. It seems likely that this process of increasing abstraction will continue in the future and that advance in physics is to be associated with a continual modification and generalisation of the axioms at the base of mathematics rather than with a logical development of any one mathematical scheme on a fixed foundation." [Dirac (1931)]

Physics doesn't simply borrow the language from mathematics, but manipulates it, adapts it, and reinvents it according to the needs for a description of physical reality. The magic about physics is transforming the pure logic of mathematics into a beautiful narrative about nature. A poet couldn't write poetry without a language, but language is not sufficient to make poetry. The same happens in science: mathematics is the language, but it takes physics to do poetry.

I leave the punchline to Dirac: "I think there is a moral to this story, namely that it is more important to have beauty in one's equations than to have them fit experiment." [Dirac (1963)]

It is Popperianly evident that empirical confirmation is the ultimate judge on any physical theory; only a deranged scientist would deny this truth. It is also clear that experiments are indispensable in guiding theorists to ask the right questions. But denying the speculative nature of physics is betraying the very spirit of our discipline. Logical deduction is the engine that has always driven physicists to discern the inner structure of physical reality and the laws that govern the universe. Most of the sparks that ignited major advances in physics can be traced back to pure thought, and not to the motivation of explaining a certain observation. It is difficult to imagine how Einstein could have discovered general relativity, had he started by trying to fix the observed discrepancy in the precession of Mercury's perihelion.[3] Logical deduction, based on arguments like the equivalence principle, is what brought Einstein to his monumental theoretical construction.

While it is always good to keep in mind the role of experiments as inspirational guideline and ultimate arbiter, the call to abandon "the Brain and the Fancy" and return "to the plainness and soundness of Observations on obvious things" can have destructive consequences. In times of krisis, more than ever, we need pure unbridled speculation, driven by imagination and vision. Paradigm changes require unbiased thinking and the freedom to blaze new trails. Support for an intense and diverse programme in theoretical physics — from the most speculative areas to those closely-connected with experiment — is the central element for success in the post-naturalness era.

---

[3]If one takes the discrepancy in Mercury's data as starting point, the most straightforward solution would be postulating a new planet causing a perturbation in Mercury's orbit. This is the approach followed by Urbain Le Verrier, who had been previously successful in predicting the existence of Neptune, and who tried his luck with a new planet he named Vulcan. The discovery of Vulcan was announced by Le Verrier in 1860, on the basis of observations made by an amateur astronomer, Edmond Modeste Lescarbault. The debate over the observations lasted for decades and, when he died in 1877, Le Verrier was still convinced about Vulcan's existence [Baum and Sheehan (1997)].

Exchange and communication between neighbouring fields is also an essential fertiliser for theoretical physics. When in search of new directions, one must look beyond the boundaries of a single discipline. The transfer of ideas is a key to breakthroughs. A good historical example is the concept of spontaneously broken symmetry, which has so radically influenced particle physics — from pion interactions to the Standard Model — but was pioneered in condensed matter physics. Today, at the dawn of the post-naturalness theory, we start observing a revival in the transfer of ideas, not only with fields traditionally akin to particle physics — such as mathematics, astrophysics and cosmology — but with other areas as well. Ideas originally formulated in condensed matter, quantum information, and atomic physics are finding unexpected applications in particle physics (examples are the use of tensor networks in holography and quantum gravity (see e.g. [Perimeter Institute for Theoretical Physics (2015)]), and the quantum simulation of field theories with ultracold atoms or photonic systems [Cirac and Zoller (2012)]). At the same time, the concept of the AdS/CFT correspondence [Maldacena (1999, 2005)] developed from string theory is finding its way into condensed matter [Sachdev (2010)] and heavy-ion physics [Casalderrey–Solana *et al.* (2011)].

What matters is not the difference between "abstract" and "real" physics, but between "good" and "bad" physics: the difference between physics that is imaginative, original, ambitious in the pursuit of fundamental questions; and physics that is misguided, ill grounded, or simply repetitive and unimaginative. Of course, the distinction between the two is not always easy to discern at the beginning and, since research will never be an efficient process, a large number of unproductive attempts and false starts must be expected and taken into account. But this cannot be used as an argument to accept lower scientific standards. On the contrary, logical rigour and verifiable truth remain the ultimate guiding principles to discriminate between "good" and "bad" physics.

Theoretical physics is the driving force needed in the search for a new paradigm during the post-naturalness era. The aim for our community is to create an environment that supports and encourages free thinking in theoretical physics, preparing a fertile ground able to nurture the single idea that will eventually become the paradigm changer.

13.5 Conclusions

> *It was the best of times, it was the worst of times, it was the age of wisdom, it was the age of foolishness, it was the epoch of belief, it was the epoch of incredulity, it was the season of Light, it was the season of Darkness, it was the spring of hope, it was the winter of despair, we had everything before us, we had nothing before us, we were all going direct to Heaven, we were all going direct the other way.*
>
> — Charles Dickens [Dickens (1859)]

When I started studying physics, I fell in love with the era when quantum mechanics was born. An era of confusion, inexplicable experimental results and crazy theoretical ideas, rebellion against the sacred laws of physics and heated debates at the Solvay conferences. How much I wished I could have been there: I missed my chance, I thought.

As I advanced in my studies, I was captivated by the excitement of the post-war period, when physicists had to deal with untameable infinities and a zoo of particles, produced by cosmic rays or popping out from pioneering accelerators, and whose meaning was obscure. The era of Shelter Island and the birth of QED, the attempts with Regge trajectories, S-matrix, bootstrap to end up with gauge quantum field theory. How much I wished I could have been there: I missed my chance, I thought.

Well, now I realise that my chance is today. As a scientist, I have the privilege to live in a new era of krisis. Ideas thrive in the periods of krisis dominated by uncertainty and confusion, when physicists are in search of a paradigm change able to deal with the puzzles they are confronted with. There is no lack of open fundamental questions we must tackle today: the nature of the Higgs boson, the structure of quarks and leptons, inflation, the cosmic baryon asymmetry, dark matter, dark energy, quantum gravity, and more. But there is also a widespread feeling that our theoretical tools — which have been so successful in bringing particle physics to its present stage of maturity — are becoming inadequate to address the next layer of open questions. A new paradigm change seems to be necessary.

Experimental physics is reacting to the present status of krisis with a broad and ambitious programme that will enable humanity to cross the borders of knowledge on many fronts. Theoretical physics is exploring new directions and looking beyond the boundaries of traditional particle physics, across different disciplines. Revolutions in science don't happen overnight: it took thirty years for quantum mechanics to develop from Planck's black-body radiation to Dirac's equation; twenty-five years for the Standard Model to go from QED to the asymptotic freedom of QCD. We can't expect to find all answers today. But we are experiencing all the right symptoms — unresolved fundamental questions, an old paradigm that seems to run out of mileage, bold experimental projects, revived theoretical curiosity — that indicate we are living in the dawn of a new era. This is the post-naturalness era, which may soon become a new chapter in the continuing story of human exploration of knowledge.

Acknowledgements

I thank Nima Arkani–Hamed, Eckhard Elsen, Debra Giudice, Michelangelo Mangano, Matthew McCullough, and Riccardo Rattazzi for comments on the manuscript. I thank all my friends and colleagues for the many discussions that continuously reshape my views on particle physics. And I thank Stefano Forte, Aharon Levy, and Giovanni Ridolfi for taking the initiative of creating this tribute in the memory of Guido Altarelli.

Chapter 14

The paradox of the Standard Model

Riccardo Barbieri
Scuola Normale Superiore
Piazza dei Cavalieri 7 56126 Pisa, Italy and INFN, Sezione di Pisa, Italy

As a way to recall and honour Guido Altarelli I contribute to the discussion on the prospects for the future of particle physics. I do this because I am sure that Guido would have liked the discussion, even though I am not equally sure that he would have agreed on all the opinions I express here.

The triumph of reductionism represented by the Standard Model (SM) is evident, as made particularly manifest by its Lagrangian in Fig. 14.1. In three lines one describes in details the behaviour of the constituents of matter with the greatest empirical adequacy at all distances between about 10^{-18} and 10^{-7} metres and with the possibility to be extended much further. The synthetic character of particle physics (PP) emerges at full strength. Yet there are strong reasons to think that the SM is not a complete theory. This is the paradox alluded to in the title. From a practical point of view and in a simplified way the contrast can be summarised by considering two apparently alternative directions of research:

- Take the SM for granted and "look elsewhere".
- Keep testing the SM to learn how to complete it.

I actually think that this is a false alternative. The "look elsewhere" attitude is at the heart of fundamental physics and does not need any defence. On the other hand I see reasons of poor understanding and reasons of incompleteness to keep pursuing the second direction as well. In the first part of this discussion I illustrate these reasons in succession.

14.1 The case for precision

Although the three different lines in Fig. 14.1 are no doubt interdependent, I think that the distinction in the dates — so to say — of their respective experimental shining, as indicated in the figure, is totally meaningful. The tests of the gauge sector of the SM are overwhelming in extension and precision, both in the strong and in the electroweak interactions. The constrained structure of a general gauge Lagrangian plays an important role in this, in contrast to the case of the second and third lines in Fig. 14.1.

From the experimental side the case is also different. On the second line the tests started indirectly with the electroweak precision measurements and, in a sense, culminated in 2012 with the discovery of the Higgs boson [Aad *et al.* (2012c); Chatrchyan *et al.* (2012)]. Yet the precision at which the Higgs couplings are known at present does not exceed the 20% level, limited to a few cases [Aad *et al.* (2016d)]. Precision in PP is a key to understanding. As an example the current measurements of the Higgs couplings allow a compositeness scale for the Higgs boson as low as about 600 GeV, compared to the tens of TeVs in the case, e.g., of the electron, whose couplings are much better known. There are reasons to think

$$\mathcal{L}_{\sim SM} = -\frac{1}{4}F^a_{\mu\nu}F^{a\mu\nu} + i\bar{\psi}\not{D}\psi \qquad (\sim 1975\text{-}2000)$$

$$+|D_\mu h|^2 - V(h) \qquad (\sim 1990\ \text{-2012-}\ \text{now})$$

$$+\psi_i\lambda_{ij}\psi_j h + h.c. \qquad (\sim 2000\ \text{-}\ \text{now})$$

Fig. 14.1 The Lagrangian of the SM with the indication, in correspondence to each line, of the approximate dates of their experimental progressive shining.

that the Higgs boson is the least understood particle of the SM. It cannot be the one that is less precisely measured.

The case of the flavour line, the third one in Fig. 14.1, is again different. The test of unitarity of the Cabibbo–Kobayashi–Maskawa (CKM) matrix has strongly progressed in the last twenty years or so, starting from the measurement of the ϵ'/ϵ parameter in K-physics [Fanti *et al.* (1999)]. Yet the pure parametric description of the spectrum and mixings of quarks and leptons is a strong drawback of the SM model. Admittedly one does not know if further measurements will allow to improve on this situation, but the case for better precision in flavour physics is also very strong. About this, Fig. 14.2 shows an interesting comparison. The figure on the left side shows that the genuine electroweak loops in the SM, of about $5 \cdot 10^{-3}$ in both ϵ_1 and ϵ_3 [Altarelli and Barbieri (1991)], are currently measured at about 20% level, as shown in the figure by their deviations from the SM values [Ciuchini *et al.* (2016)]. Interestingly enough the deviations of the Wilson coefficients of the leading SM operators, arising from the dominant Flavour Changing Neutral Currents loops, are bound to be at most of similar 10–20% relative size. This is shown in the figure on the right, where it is assumed that possible new physics contributions in ΔB_s and in ΔB_d scale, relative to each other, like V_{ts}/V_{td} as in the SM. It is well appreciated how important it would be to improve by one order of magnitude the precision on the electroweak loops, as it might be possible in a future e^+e^- facility. The goal of reaching similar relative precision in flavour tests might be attainable by an aggressive flavour programme that aimed at collecting $\mathcal{O}(10)$ times the LHCb upgrade luminosity and, in parallel, at significantly improving the efficiency in suitable low p_T events in ATLAS and CMS.

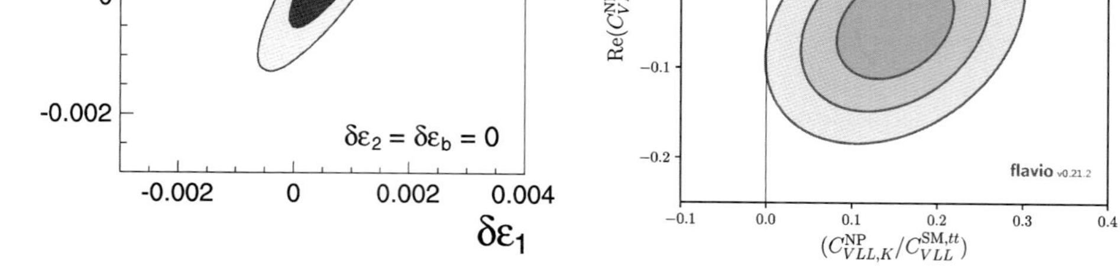

Fig. 14.2 Left: Deviations of the electroweak precision parameters from the SM values, $\epsilon_1^{SM} = 5.21 \cdot 10^{-3}$ and $\epsilon_3^{SM} = 5.28 \cdot 10^{-3}$ (courtesy of the HEPfit group); Right: Relative deviations of the Wilson coefficients of the leading SM operators from the SM values in $\Delta B_{d,s} = 2$, vertical axis, and in $Im(\Delta S = 2)$, horizontal axis (courtesy of D. Straub).

14.2 The incompleteness of the Standard Model

It is often recalled, rightly so, that the SM does not account for some major phenomena, like neutrino masses, the existence of dark matter or the matter-antimatter asymmetry. I find it useful to insert these issues in the broader perspective of seeing the various incompletenesses of the SM all at once, as attempted in Fig. 14.3. Most of the entries are well known, to the point that we think that we know the solutions for many of them. The trouble is that we do not know if any or almost any of these solutions is right. We lack the corresponding experimental evidence, whose search is at the core of the PP programme. For reasons of space here I limit myself to some comments on the hierarchy problem, once again, as due after the (temporary) blank of the LHC on BSM searches.

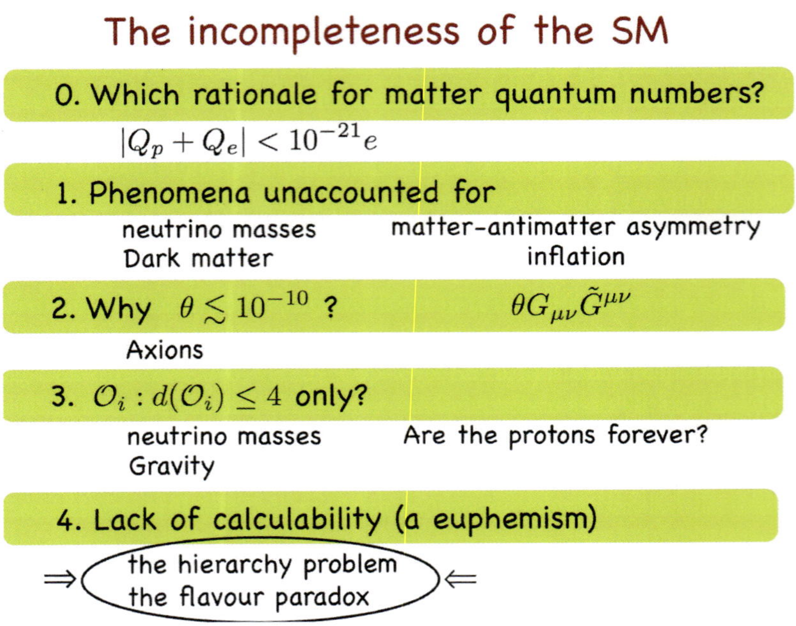

Fig. 14.3 A list of the incompletenesses of the SM.

14.2.1 *The hierarchy problem, once again*

Not the least drawback of the SM is that we cannot compute or not even estimate the Higgs mass, which is quadratically divergent. The classic way to address this issue has consisted in relating the Higgs mass to some other physical scale, not far from the Fermi scale itself and not subject on its own to power-law divergences. From an aesthetic and general theoretical point of view supersymmetry remains the relatively best solution of the hierarchy problem. All the history of PP seems to cry out for it, as is synthetically but also effectively illustrated in Fig. 14.4.

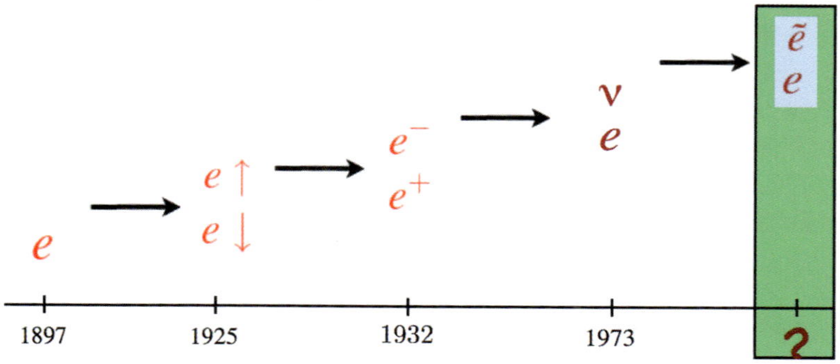

Fig. 14.4 A synthetic illustration of the role of symmetries in the history (and perhaps the future history) of the electron.

The problem, not for supersymmetry in itself but for knowing if it is true in nature, is that the relation between the Higgs mass, or the Higgs vev, and the mass of the s-partners is a quantitative one only if one bars accidental cancellations among otherwise unrelated parameters. Although depending on peculiar configurations of the s-particle spectrum or on particular extensions of the Minimal Supersymmetric Standard Model, some level of accidental cancellations is now needed to be consistent with the LHC's negative searches so far. This motivates the question: Is "low energy" supersymmetry still alive? The answer cannot be a sharp one. The judgement is suspended, although with pretty clear reasons of concern, in my view.

The absence, so far, of new physics signals at LHC raises in fact more general questions:

- Is the quest for "naturalness" still relevant?
- How about: Naturalness = "low energy" New Physics?

The answer to the last question, I think, is clear. It is not a "theorem" anymore, as we used to say before or even at the time of LEP, because the experiments so far do not support it. Are we lacking a clever "IR-UV connection", perhaps capable to solve as well the cosmological constant problem, without the need for "nearby" physics? Maybe, but the difficulties of decoupling the "low energy" scales from the Planck scale or from the Landau poles of non-asymptotically free couplings in the SM or from whatever "high energy" physical scale could exist, remain there as a stumbling block. Lastly, does the multiverse emerge, supplemented by anthropic considerations, as the "only solution" of the naturalness problem? A "solution" is such, in my view, only if it is supported by independent evidence, like the s-partners or the Higgs form factors in the case of the standard putative solutions. I do not yet see independent evidences in the case of the multiverse. In a way or another, therefore, I think that the quest for naturalness remains relevant more than ever.

14.3 Examples of interesting directions to be followed

This is a time of healthy uncertainty in PP, which causes the emergence of new directions of research. Among the many possibilities in the second part of this discussion I want to describe briefly two such directions that look relevant and, most importantly, have clear experimental implications.

14.3.1 *Minimal Mirror Twin Higgs*

The hypothesis of the existence of a Mirror Sector, an identical copy of the SM, has a long history [Lee and Yang (1956); Kobzarev *et al.* (1966)]. Two key results follow from introducing an approximate spacetime parity symmetry, P, that exchanges the two sectors. First, the SM Higgs boson can be understood as a pseudo-Goldstone boson via the Twin Higgs mechanism [Chacko *et al.* (2006)], thus explaining the absence of BSM signals at LHC with a modest amount of fine-tuning. Second, and in my view even more importantly, dark matter (DM) may be mirror baryons with a density expected to be the same order as the baryon density, since both are due to similar matter-antimatter asymmetries [Goldberg and Hall (1986)].

For all this to work, P must be broken, in particular to avoid inconsistencies with observations both in DM and in dark radiation (DR). Interestingly this can be done all at once by breaking P in the Yukawa couplings [Barbieri *et al.* (2005, 2016a)]. Suppose in particular that the observed fermion mass hierarchy is described by suppression factors ϵ^{n_i} for charged fermion i, as can arise in Froggatt–Nielsen [Froggatt and Nielsen (1979)] and extra-dimensional theories of flavour [Kaplan and Tait (2001)]. The corresponding flavour factors in the mirror sector are $(\epsilon')^{n_i}$, so that spontaneous breaking of the parity P arises from a single parameter ϵ'/ϵ. An overall consistent picture emerges [Barbieri *et al.* (2017a)] for simple values of the n_i that properly describe the observed fermion mass hierarchy.

The key experimental implications are:

- The existence of DR, in the form of mirror photons and mirror neutrinos, with a density between the 1σ and the 2σ upper bounds set by the current Planck data [Ade *et al.* (2016)].
- A universal reduction of the Higgs signal strengths by 5 to 20 %, due to a mixing between the original Higgs and mirror Higgs fields.
- The existence of hydrogen-like, helium-like or neutron-like DM, depending on which one is the lightest, with properties broadly predicted in terms of ϵ'/ϵ and of an $\mathcal{O}(1)$ parameter δ, as summarised in Fig. 14.5 together with various astrophysical and cosmological constraints, taken from ref. [Barbieri *et al.* (2017a)].

14.3.2 *Tantalising deviations from the SM in flavour physics*

Starting from the 2012 ICHEP conference in Melbourne up to very recent results, numerous hints have emerged of Lepton Flavour Universality (LFU) violations

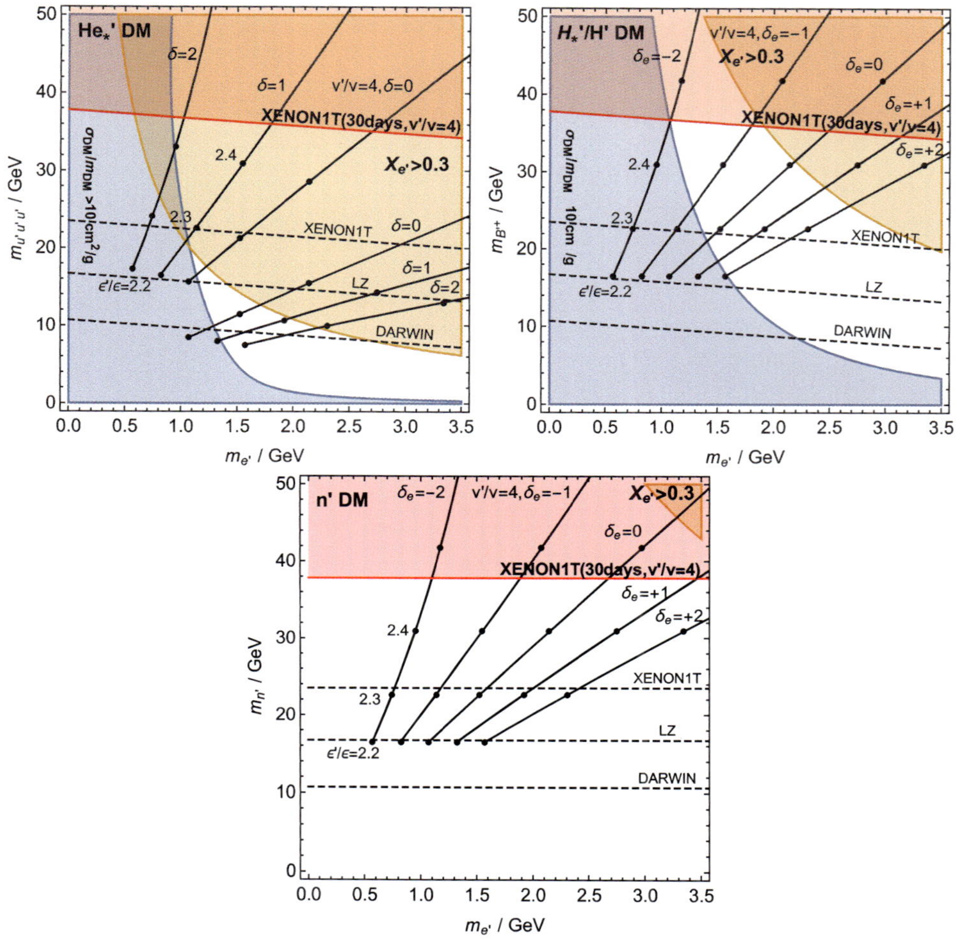

Fig. 14.5 The masses of: (i) charge-2 mirror baryon $B_{u'u'u'}$ (top left panel); (ii) charge-1 mirror baryon $B^+ = B_{d'd'd'}/B_{u'u'd'}$ (top right panel); (iii) charge-0 mirror baryon $B_{u'd'd'} = n'$ (lower panel) shown as full lines against the mass of the mirror electron e'. Different coloured areas give the excluded portions of the planes by self-interactions of mirror atoms, by the mirror ionisation fraction $X_{e'}$ and by direct detection [Aprile *et al.* (2017)]. Dashed curves give expected reaches of future direct detection experiments.

observed in semi-leptonic B-decays, as recently summarised in ref. [Buttazzo *et al.* (2017)] and references therein. While not a single LFU ratio measurement exhibits a deviation with respect to the SM above the 3σ level, the overall set of observables is pretty much consistent and, once combined, the probability of a mere statistical fluctuation is very low. At the underlying quark level, the evidences collected so far can be grouped into:

- Deviations from τ/μ (and τ/e) universality in $b \to cl\nu$ charged currents;
- Deviations from μ/e universality in $b \to sl\bar{l}$ neutral currents,

both at $\mathcal{O}(10\%)$ level relative to the corresponding SM amplitudes. Caution in accepting evidence for LFU violation, not seen before in whatever process, is never enough. Yet the overall set of observables justifies the strong reaction of the theoretical community.

The explanation of any putative deviation from the SM in flavour physics must be well designed enough to respect the full body of experimental constraints in flavour physics, by now significantly extended. To this purpose I advocate a weak form of Minimal Flavour Violation, as arising from a minimally broken $U(2)^n$ flavour symmetry, with the different $U(2)$ factors referring to the different irreducible representations of the gauge group. Under $U(2)^n$ the fermions of the third generation transform as singlets and the first two generations as doublets [Barbieri *et al.* (2011, 2012)].

$U(2)^n$ introduces a natural distinction between semi-leptonic K and π decays on one side, where tests of LFU are at the per-mille level, and B semi-leptonic decays on the other side, where LFU violations seem to emerge. Furthermore $U(2)^n$ makes plausible that a massive bosonic mediator around the TeV scale, responsible for the new effect, be predominantly coupled to the third generation of fermions, with the couplings to the lighter generations only coming after small $U(2)^n$ breaking. Suppose that the mediator is a leptoquark, singlet under $U(2)^n$. Since leptons and quarks of the first two generations transform as doublets of different $U(2)$ factors, in the unbroken $U(2)^n$ limit such a leptoquark indeed couples to the third generation only. This hypothesis provides a natural first-order explanation for the different size of the charged current versus the neutral current effects, about equally deviating from the tree-level versus loop-level SM amplitudes respectively: $b \to c\tau\nu$ only involves a single second generation particle, whereas $b \to s\mu\bar{\mu}$ has three light generation fermions [Barbieri *et al.* (2016b)].

Already analysed from an EFT point of view [Bordone *et al.* (2017)], this broad picture leads to several testable predictions, among which:

- $b \to c(u)l\nu$: Universal deviation from unity of the Branching Ratios, normalised to the SM, for $B \to D^*\tau\nu, B \to D\tau\nu, \Lambda_b \to \Lambda_c\tau\nu, B \to \pi\tau\nu, \Lambda_b \to p\tau\nu, B_u \to \tau\nu$;
- $b \to s\tau\tau$: a large enhancement, relative to the SM, made plausible;
- $b \to s\nu\nu$: $\mathcal{O}(1)$ deviations from the SM rate expected;
- $K \to \pi\nu\nu$: $\mathcal{O}(1)$ deviations from the SM rate expected.

As already mentioned, the size of the effect in charged current calls for the existence of a massive mediator around the TeV scale, more likely in a strong coupling regime. From the theory side this will require a non-trivial UV completion of the phenomenological picture [Buttazzo *et al.* (2016); Barbieri *et al.* (2017b); Megias *et al.* (2017)]. Experimentally a sizeable (broad) excess in $pp \to \tau\tau$ and $pp \to bb$, tt is expected [Buttazzo *et al.* (2016); Faroughy *et al.* (2017)].

14.4 Conclusions

Looking in part at the crystal ball, I think that in the future of PP a deeper theory than the SM is likely to emerge, of which theory the SM itself will in any case be a relevant limit. This comes from considering the incompletenesses of the SM, as opposed to its great experimental successes: a kind of paradoxical situation, in my view.

The very nature of fundamental physics and the current uncertain situation of PP require and are already generating highly diverse directions of research. I have argued that precision measurements in Higgs and flavour physics are to be vigorously pursued. Imagine, at least as an example, what it would mean if the hints of LFU violation described in Section 14.3.2 were confirmed. As I have already recalled, in the tradition of PP, precision is a key to understanding. On the other hand many directions of research indicate that the boundaries between PP, cosmology and astrophysics are fading away more and more. The Mirror World briefly discussed in Section 14.3.1 provides a motivated example.

I am grateful to Stefano Forte, Aharon Levy and Giovanni Ridolfi for taking the initiative of this book in memory and honour of Guido Altarelli.

Chapter 15

Future deep inelastic scattering with the LHeC

Max Klein

University of Liverpool

For nearly a decade, Guido Altarelli accompanied the Large Hadron electron Collider project, as invited speaker, referee and member of the International Advisory Committee. This text summarises the status and prospects of the development of the LHeC, with admiration for a one-time scientist and singular leader whom I met first nearly 40 years ago under the sun shining for the "Herceg Novi School" in Kupari, where we both lectured about the beautiful science of deep inelastic scattering and enjoyed life under a yellow moon.

15.1 Introduction

The time is now 50 years after the birth of deep inelastic scattering (DIS) with the discovery of partons [Bloom *et al.* (1969); Breidenbach *et al.* (1969)] and 40 years after the paper of Altarelli and Parisi, cited over six thousand times, on "Asymptotic Freedom in Parton Language" [Altarelli and Parisi (1977)]. We are now located in a time 20 years after the approval of the huge LHC detectors, ATLAS and CMS. These have now taken about 100 fb^{-1} of data each. Adding the publications of all LHC collaborations, nearly two thousand papers have appeared, based on the superb performance of the Large Hadron Collider. The Higgs boson was discovered five years ago. About this most impressive harvest Guido Altarelli had noted, *we expected complexity but we found a maximum of simplicity [Altarelli (2014)]*, commenting not on the hugely complex experiments nor ingenuitive, novel analysis techniques but referring to the surprising absence of new physics, besides the exciting observation of the Higgs boson. No evidence was observed so far in support of extra dimensions, symmetries, particles nor a grand unified theory. Guido thought about this deeply: ... *it is true that the SM theory is renormalisable and completely finite and predictive. If you forget the required miraculous fine-tuning you are not punished, you find no catastrophe! ... The possibility that the Standard Model holds well beyond the electroweak scale must now be seriously considered.. We are experiencing a very puzzling situation but, to some extent, this is good because big steps forward in fundamental physics have often originated from paradoxes. We highly hope that the continuation of the LHC experiments will bring new light on these problems...* [Altarelli (2014)].

The LHC is now projected to operate for two decades hence, with a major interruption for accelerator and detector upgrades in between. A new horizon for these explorations would be opened with an additional *ep*/*eA* experiment, i.e. with electron-proton (and electron-ion) collisions concurrent to the default LHC operation in the thirties. The Large Hadron electron Collider (LHeC) was principally designed in an extended concept report (CDR) [Fernandez and the LHeC Study Group (2012)]. This was published in June 2012, a few weeks before the Higgs boson discovery was announced. It considers the addition to the LHC of a 60 GeV energy electron beam accelerator, arranged tangential to the main ring of 27 km circumference. This configuration is thought to form a novel *ep* and *eA* experiment which would be able to collect O(1) ab^{-1} of integrated luminosity, exceeding the HERA result by a factor of thousand and its kinematic range by nearly 20. Its salient feature will be the synchronous operation of *ep* with *pp* such that no principal loss of LHC luminosity shall occur. The LHeC would enable electron-hadron energy frontier physics explorations at about 1 TeV centre-of-mass (cms) energy, which is four times higher than future Higgs facilities, such as in China (CepC) or Japan (ILC), may achieve in electron-positron collisions. The question such a programme of deep inelastic electron-hadron scattering at the LHC has to answer consists of how that may lead to new insight into the Standard Model (SM) and beyond, on its own and in complementing the *pp* and heavy ion physics pursued at the Large Hadron Collider. This paper has been written in memory of Guido Altarelli who advised the LHeC development in his unforgettable manner as a member of the International Advisory Committee, a referee to the CDR and with many illuminating presentations and discussions. Together with Lev Nikolaevitch Lipatov, whom we lost so early also, the LHeC had two monumental theoretical physicists on its side, together with their inspiration, far-reaching theoretical insight and high demand for quality.

The article presents brief accounts of the achievements of HERA (Section 15.2) as well as lessons from the first years of LHC physics (Section 15.3). There follows a description of the LHeC accelerator design including an introduction to its planned test facility [Angal-Kalinin *et al.* (arXiv:1705.08783, 2017.)] and a sketch of its detector concept (Section 15.4). The LHeC would be the fifth major experiment at the LHC and it has five major themes which are illustrated in five subsequent sections:

- based on the unique hadron beams of the LHC and employing a point-like probe, it would represent the world's cleanest, high resolution microscope for exploring the substructure of and dynamics inside matter (Section 15.5);
- with concurrent *ep* and *pp* operation, the LHeC would transform the LHC into a 3-beam, twin-collider, energy frontier accelerator facility. Through ultra-precise strong and electroweak interaction measurements, the *ep* results would make the HL LHC facility a much more powerful search and measurement laboratory than current expectations based on *pp* only could possibly entail (Section 15.6);
- the clean DIS final state in neutral (NC) and charged currents (CC) enables a high precision Higgs physics programme with the LHeC if it reaches $O(10^{34})$ $cm^{-2}s^{-1}$ luminosity. The joint *pp*/*ep* LHC facility can become a Higgs factory of unprecedented impact. It would combine high, percent precision for abundant Higgs decay channels in *ep* with improved precision on rare channels, high *pp* luminosity and the resolution of QCD uncertainties through *ep*. This transforms the view on the Higgs physics potential at the LHC (Section 15.7);
- as a new, unique, luminous TeV scale collider, the LHeC has an outstanding opportunity to discover new physics, such as in the exotic Higgs, dark matter, heavy neutrino and QCD areas, which is actively being studied and will be illustrated in Section 15.8;
- the LHeC leads into the region of high parton densities at small Bjorken x. It extends the kinematic range in lepton-nucleus scattering by nearly four orders of magnitude. It is expected to transform nuclear particle physics completely, by establishing a QCD base for Quark Gluon Plasma (QGP) phenomena and resolving the hitherto hidden parton dynamics and structure of nuclei (Section 15.9).

Guido Altarelli delivered his first lecture to the LHeC community at the first CERN–ECFA workshop in 2008 [Altarelli (2008)], in which he posed questions to work on. In June 2015 he delivered the last scientific presentation of his lifetime, actually to the annual workshop on the LHeC, when all of us had no clue about his health situation. There followed a session of the International Advisory Committee for the LHeC, chaired by CERN DG emeritus, Herwig Schopper, in which Guido had asked: *The physics case of the LHeC is essentially made, can we proceed and build it?* He knew there would not be a straight answer possible then as we were all aware of things still to be done. This article is an attempt to describe the background of this statement and to sketch how indeed we can build a TeV scale ep/eA experiment, for the High Luminosity (HL) LHC lifetime and beyond into the High Energy (HE) LHC, should our community find the courage and means for supporting it. Some concluding remarks are made in Section 15.10.

Guido Altarelli often raised the questions about the value and determination of the strong coupling constant α_s and also, since his youth [Altarelli and Martinelli (1978)], why the longitudinal structure function F_L was so hard to obtain. Having spent many years of my life to measure both, I have added an Appendix, written from the perspective of HERA and the LHeC, to this otherwise more general overview.

15.2 Lessons from HERA

It has become almost a habit to present pp as the discovery machines and e^+e^- as the clean machines, and to forget to mention ep. Be it intentional or sloppiness, this has been so often repeated that one may get tired in raising the concern that such a view is indeed unjust with respect to the roles of Tevatron, LEP and HERA, or the role of HERA for the LHC physics, and that it will hinder the prosperity of our science if its future is unwarily narrowed. Guido Altarelli knew that a $2 \to 2$ scattering of e and p could be viewed as ee, ep and pp, just as it had three Mandelstam variables (s, t and u), and all three reactions had their merits and complementary roles because the electron is a point-like lepton and the proton a composite hadron. The principal recognition of the LHeC development is that pp and ep are a symbiotic unit of different entities, which together would reach out much further than alone. A combined LH(e)C facility, for example, would challenge a possible future e^+e^- collider as ep/pp combined with e^+e^- would unravel the Higgs properties and the possible Higgs link to new physics particularly thoroughly. The HERA ep collider could have discovered the Higgs boson had it had a thousand times higher luminosity, which is the luminosity goal now. The LHeC will surpass its physics by an enormous extent, but it rests on HERA's foundations.

The "Hochenergie Ringanlage" (HERA) was the first electron-proton collider ever built. Its proposal was endorsed in 1984. HERA was operated between 1992 and 2007 with colliding electron (also positron) and proton beams, of energy, for most of the time, $E_e = 26.7\,\mathrm{GeV}$ and $E_p = 920\,\mathrm{GeV}$. The cms energy was $\sqrt{s} = 2\sqrt{E_e E_p} = 319\,\mathrm{GeV}$ with which HERA was suited to explore the physics at the Fermi scale, set by the vacuum expectation value of the Higgs field, $v = (G_F\sqrt{2})^{-1/2} = 246\,\mathrm{GeV}$. The luminosity eventually reached values of up to

$4 \cdot 10^{31}\,\mathrm{cm}^{-2}\,\mathrm{s}^{-1}$, and a total integrated luminosity was $0.5\,\mathrm{fb}^{-1}$ collected by H1 and as well by ZEUS in 15 years. HERA established the *ep* scattering energy frontier and served to a large extent as a search laboratory for new physics, especially through configurations characteristic for the spacelike *ep* configuration such as leptoquarks or $R-$parity violating supersymmetry. Albeit the appearance of certain fluctuations, especially on leptoquark and exotic multi-lepton signatures, one concluded that in the mass range explored by HERA, new particles, with certain couplings, would not exist. H1 and ZEUS confirmed the unification of the electromagnetic and weak interactions at 4-momentum transfers, $\sqrt{Q^2} \leq v$, through high Q^2 measurements of photon and Z exchange in NC and of W exchange in CC [Abramowicz *et al.* (2015)]. HERA discovered a rise of the sea quark, $x\Sigma$, and gluon, xg, densities in the newly accessed region of small Bjorken x, by precisely measuring the structure function $F_2(x, Q^2)$ and its derivative $\partial F_2/\partial \ln Q^2$, respectively. Despite some peculiarities in the QCD analysis of the NC cross section at low x and Q^2 [Abramowicz *et al.* (2015)], the linear Q^2 evolution DGLAP equations, the monumental predictions of Guido Altarelli and Lev Lipatov with collaborators, were established to hold in the full HERA range which could be seen as indeed a surprise as the $\ln 1/x$ terms, characteristic to the BFKL evolution, were large at small x. Often overlooked, but of crucial relevance for future low x physics, as $x \propto Q^2/s$, we also discovered that the gluon distribution would not rise towards low x but, on the contrary, disappear when Q^2 was approaching the mass of the proton squared region $M_p^2 \sim 1\,\mathrm{GeV}^2$, at the edge of the DIS kinematic region. Following some initial EMC data on charm production in DIS, HERA also measured the charm and bottom quark densities in the proton and provided an experimental base for their theoretical study in a variety of dynamic schemes of changing flavour numbers which now turn out to be quite relevant for the description of *pp* data such as on associated b and vector boson production at the LHC. Another striking observation at HERA was the finding that the proton remained intact in about 10 % of the violent *ep* collisions which gave birth to the field of diffractive DIS. Instrumentation and physics at HERA were largely innovative, and many more important measurements were done [Klein and Yoshida (2008)].

Seen from the LHC, HERA has been the indispensable, only base for adequately describing the parton-parton interactions in *pp*, such as the gluon-gluon fusion to generate the Higgs boson. The field of parton distribution functions, PDFs, was born which prior to HERA was at its infancy [Klein (2010)]. For QCD, HERA demonstrated that the structure of matter is much richer than the collinear approximation leading to the classic proton PDFs suggests. Measurements of deeply virtual Compton scattering, diffraction, jets, photons, neutrons opened new areas, supported by impressive theory developments, for generalised, pomeron, unintegrated, photon, pion and neutron structure functions. All of these ought to be studied much deeper for understanding the universe at truly microscopic scales.

HERA kept electron-proton scattering as an integral part of high energy particle physics. It demonstrated the richness of DIS physics and the feasibility of constructing and operating energy frontier *ep* colliders. It is a testimony of the vision and authority of Bjoern Wiik. What did we learn to take into a next higher energy *ep* collider design? Perhaps there were three lessons about i) the need for higher energy, for three reasons: to make charged currents a real, precision part of *ep* physics, for instance for the complete unfolding of the flavour composition of the sea and valence quarks, to produce heavier mass particles (Higgs, top, exotics) with favourable cross sections and, a third reason, to discover or disprove the existence of gluon saturation for which one needs to measure at lower $x \propto Q^2/s$ than HERA could; ii) the need for much, much higher luminosity: the first almost ten years of HERA provided just a hundred pb^{-1}. As a consequence, HERA could not accurately access the high x region, and it was inefficient and short of statistics in resolving puzzling event fluctuations; iii) the complexity of the interaction region design when a bent electron beam caused synchrotron radiation while the opposite proton beam generated quite some halo background through beam-gas and beam-wall proton-ion interactions. This we had not seen clearly enough prior to and during the initial phase of the HERA luminosity upgrade.

A key question to the future of deep inelastic scattering, to both its particle and nuclear physics components, is the choice of energy. This is an issue also in the debate [Aschenauer *et al.* (2017)] about the current EIC designs which have a maximum of luminosity at a $\sqrt{s}$ of either 35 (for the Jlab EIC [Abeyratne *et al.* (2015)]) or 100 GeV (for eRHIC [Aschenauer *et al.* (2014)]). This discussion, however, too often leaves the comparison with the LHeC out,[1] which has a $\sqrt{s}$ of 1300 GeV. A complete view is desirable, extending first attempts such as [Klein (2016a)], as to what can be learned from going back by one or two orders of magnitude in Q^2 below HERA, as with the US EICs, and what can be achieved by a large increase, as with LHeC or even FCC-eh, and, furthermore, which energy range was crucial to be covered in electron-ion scattering. It would then become clearer as to where real synergy resides. A particular question regards the puzzle on the composition of the proton spin which at HERA was investigated by the HERMES experiment, and has recently been claimed to be resolved in lattice QCD [Alexandrou *et al.* (2017)]. That does *not* require maximum energies, as the asymmetries of polarised scattering cross sections to measure vanish proportional to x. Very high energy in *eh* scattering is crucial for a DIS Higgs, top, BSM, PDF, vector meson and low x programme. One crudely may call the low energy EICs extended continuations of the NMC, COMPASS and HERMES research programmes, on nuclear structure, QCD and spin, while the LHeC is in the line of H1 and ZEUS with the important

[1]One here considers, for example, an extension of saturation scales Q_s^2 from 1 to 2 GeV^2 while it is known that at so low Q^2 the gluon density at lower x is practically zero, and at the edge of having a meaning anyhow since one is near the limit of application of pQCD. A true discovery of the saturation of a rise of xg requires to be at *very* low x where densities are large but also in the perturbative region where α_s is small, i.e. $Q^2 \geq 10\,\mathrm{GeV}^2$. One cannot discover a saturating gluon density in a region where it is non-existing and no nuclear environment amplification, as speculative as it is by itself, can change that conclusion.

extension to electron-deuteron and electron-ion scattering which HERA was not able to study. A question is whether the global nuclear and particle physics communities have the intention and influence to shape a common future of DIS which HERA has opened and which can only be rich and convincing if the programme as a whole is considered.

15.3 Lessons from the LHC

15.3.1 *Proton-proton scattering*

Expectations for the observation at the LHC of new physics beyond the Standard Model had been almost unlimited as is illustrated, for example, by the comprehensive study of the ATLAS Collaboration on its physics potential which appeared prior to data taking [Aad *et al.* (2009)]. And yet, the first sizeable data sets revealed no new particle nor symmetry. At the EPS Conference 2011 at Grenoble, Guido Altarelli, in a memorable plenary talk, stated it was "too early for desperation but enough for despair". In 2012 ATLAS and CMS discovered the Higgs boson, termed the last and a most crucial component of the SM gauge field theory. As more data had been collected and analysed (at the time of this article LHC delivered a record integrated luminosity of $100\,\text{fb}^{-1}$) Guido later noted: *It is now less inconceivable that no new physics will show up at the LHC... We expected complexity and instead we have found a maximum of simplicity... The possibility that the Standard Model holds well beyond the electroweak scale must now be seriously considered [Altarelli (2014)].*

If today one talks about lessons from the LHC one may emphasise the following observations: (i) the LHC and the general purpose detectors ATLAS and CMS, which just celebrate the 25 years since the approval of their letters of intent, and also the other experiments as LHCb, have operated with high efficiency and reliable success owing to the ingenuitive work of thousands of physicists and engineers. The LHC and its experiments comprise the overwhelming majority of all HEP efforts and had they not succeeded, particle physics would be in danger of its existence; (ii) there has been no observation of supersymmetric particles of mass up to O(1) TeV and no observation of any of the considered extensions of the SM, such as Kaluza Klein excitations of gravitons or new vector bosons Z' or W', currently up to masses of O(5) TeV. The absence of physics Beyond the SM (BSM)[2] had been the biggest

[2]This has had major consequences especially for the layout of new electron-positron colliders: at CERN and in China circular machine designs reappeared when it was most likely that an electron-positron collider above the scale set by Higgs and top production was an "empty machine". Interesting enough, both the Chinese CEPC and the most recent version of the Japanese ILC set $\sqrt{s} = 2E_e \simeq v$, in order to measure the Higgs-strahlung from Z, $e^+e^- \to Z^* \to ZH$. The coming energy frontier in e^+e^- is nothing but the Fermi scale with a question mark about CLIC which could principally reach 3 TeV at high cost should a new spectrum of particles still be discovered at subsequent LHC runs or/and new theories inevitably led to the conclusion that new physics was in reach at a few TeV mass.

surprise at the LHC so far; (iii) a next lesson at the LHC was the confirmation of the SM at a striking level of precision, if one considers the measurement of the inclusive Drell–Yan scattering cross sections of the W and Z bosons [Aaboud *et al.* (2017h)] to about half a percent precision, apart from the also striking 2 % luminosity uncertainty, or the recent measurement of the mass of the W boson to 19 MeV precision [Aaboud *et al.* (2018)]; iv) QCD and electroweak theory had been extended to high orders, with the N^{3}LO calculation of the Higgs production at the LHC as an outstanding example. The success of the LHC is its challenge. Globally a huge effort is undertaken to upgrade the LHC accelerator and its experiments to highest reasonable, in terms of event pile-up, luminosities. The HL LHC is projected to operate for two more decades hence, and a fundamental question is that about its science base. The addition of ep adds new possibilities for the discovery of physics beyond the SM *and* it transforms the LHC facility eventually into a precision particle physics laboratory of wider and deeper range than a sole e^+e^- or pp machine can possibly represent.

15.3.2 *Scattering of heavy ions*

The $PbPb$ runs at the LHC have confirmed the picture coming from RHIC that the medium created in such collisions behaves very early like a very low viscosity fluid whose behaviour can be well described by relativistic viscous hydrodynamics. It is very opaque to particles traversing it, a phenomenon known as jet quenching. The first statement is supported by extensive studies of azimuthal asymmetries far beyond, both in kinematic reach and order of correlations, those that could be studied at RHIC. The second one is provided by a wealth of measurements, including jet yields and shapes both isolated and accompanied by an electroweak boson, which have triggered a huge theoretical activity to understand the modifications of the QCD branching inside a coloured medium — for a review see [Armesto and Scomparin (2016)]. Finally, a great surprise came from the measurements of smaller systems, pp and pPb, which revealed a not expected smooth transition when increasing the colliding system size for observables (ridge and azimuthal asymmetries, particle species, interferometry,...) that have been traditionally considered as signatures of the creation of deconfined partonic matter close to thermodynamical equilibrium — Quark Gluon Plasma — in high-energy nucleus-nucleus collisions. In short, all hadron-hadron collisions at the LHC were observed to be of collective nature. It therefore needs the electron-ion configuration to (i) establish a new level of understanding for the parton dynamics inside nuclei in the kinematic range covered by the HI experiments at the LHC and to (ii) provide the theoretically sound initial conditions for the macroscopic evolution of the system leading eventually to an understanding of the emergence of that hydrodynamical macroscopic behaviour from the QCD microscopic dynamics.

15.4 The LHeC project

Owing to the high energy, intense hadron beams of the LHC, CERN has the unique potential for discovery and precision DIS at the energy frontier and *it would be a waste* not to utilise that as Guido had formulated already ten years ago [Altarelli (2008)]. The idea of *ep* collisions with the LHC was considered already at the first LHC workshop in 1984, again at the Aachen workshop in 1990, and *ep* was presented *as a bonus* when Chris Llewellyn Smith, then Director General of CERN, announced the approval of the LHC now 25 years ago. For a brief account of *ep* collider history see the introduction to [Fernandez and the LHeC Study Group (2012)]. The idea was revived when CERN, ECFA and later NuPECC invited a study of how one may realise *ep* and *eA* collisions in view of the LHC physics findings and accelerator and detector technology advances. This was the base for the extended conceptual design report (CDR) of the LHeC [Fernandez and the LHeC Study Group (2012)]. It was published in summer 2012. At just that time the Higgs boson was discovered. The CDR had presented a machine of $10^{33}\,\text{cm}^{-2}\text{s}^{-1}$ luminosity, about a hundred times more intense than HERA I. The CDR had recognised [Fernandez and the LHeC Study Group (2012)] that the H boson, predominantly produced in CC DIS, was generated with a cross section of 200 fb in polarised $ep \to \nu HX$ as large as in e^+e^-. The H production is through radiation from a spacelike W (CC) or Z (NC), two production mechanisms which unlike in *pp* are uniquely distinguishable. The final state is clean as there is no pile-up at the LHeC. This caused a reconsideration of the accelerator design assumptions, especially a modest enlargement of the electron beam current and reduction of the β^* parameter. Since, moreover, the beam brightness, i.e. the ratio of number of protons per bunch and the proton beam emittance, was found to be almost three times larger than the so-called "ultimate beam parameters", predicted at that time to guard the future LHC and adopted in the LHeC CDR, a path had been found for the LHeC to become a genuine and competitive Higgs physics facility at atobarn reach of integrated luminosity [Zimmermann *et al.* (2013); Brüning and Klein (2013)]. It is interesting to recall that Guido Altarelli, in all his LHeC presentations, nevertheless emphasised the high precision, large kinematic range character of the genuine DIS programme, see below, which he knew to value so much that Higgs and BSM, as interesting as they could be, were not required by him to justify an *ep*/*eA* energy frontier experiment at CERN.[3]

[3]This often reminded me of the proposal for what later became the quark discovery *ep* experiment at SLAC [Bloom *et al.* (1969); Breidenbach *et al.* (1969)] which had stated the main aim would be "to collect data which could be of use for future experiments". One is struck by such reasons when nowadays no less than a discovery of something like dark matter is requested for a project to proceed. The two mile SLAC linac was longer than the two times 1 km linacs of the LHeC, but the physics programme requirements are orders of magnitude harder to satisfy. One is tempted to note certain arrogance of ourselves towards nature and its predictability. The present "no BSM physics" situation of the LHC may lead to a reset of our ambitions and expectations towards a more humble understanding of what we may expect and what we should promise.

15.4.1 *Principal design considerations*

The LHeC should provide an intense, high-energy electron beam to collide with the LHC. The intensity is gauged through the luminosity goal of O(1) ab^{-1}, the energy chosen to achieve a TeV energy collision. At the time of this write up, a cost-physics-energy evaluation takes place with the goal to be able to realise the LHeC with a most far-reaching physics scope. The wall plug power of the default design [Fernandez and the LHeC Study Group (2012)] had been constrained to 100 MW. In that configuration two superconducting linacs, opposite to each other, accelerate the passing electrons by 10 GeV each. This leads to a final electron beam of 60 GeV in a 3-turn racetrack configuration as is illustrated in Fig. 15.1. For measuring at very low Q^2 and/or determining F_L, see below, the electron beam energy may be reduced to a minimum of perhaps 20 GeV. For maximising the acceptance at large Bjorken x, the proton beam energy, E_p, may be reduced to about 1 TeV. If the Energy recovery linac (ERL) may be combined in the further future with a double energy (HE) LHC, E_p could reach 14 TeV. One therefore considers a DIS scattering complex with an energy range between $\sqrt{s} \simeq 300$ to 1800 GeV. It thus covers a range from HERA to the TeV region, at hugely increased luminosity and much more sophisticated experimental techniques.

The ERL arrangement, Fig. 15.1, is located inside the LHC ring but outside its tunnel, which minimises any interference with the main hadron beam infrastructure. The electron accelerator may thus be built independently, to a considerable

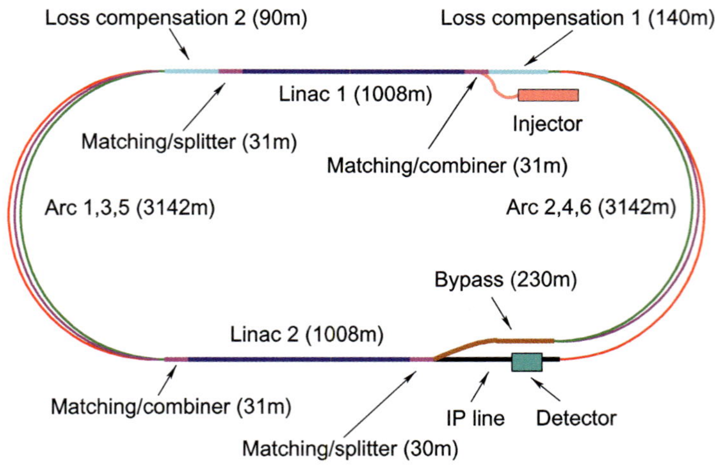

Fig. 15.1 Schematic view of the default LHeC configuration. Each linac accelerates the beam to 10 GeV, which leads to a 60 GeV electron energy at the interaction point after three passes through the opposite lying linac structures made of 60 cavity-cryo modules each. The cryomodules comprise four 5-cell cavities at 802 MHz frequency. The arc radius is about 1 km, and the circumference chosen to be 1/3 of that of the LHC. The beam is decelerated for recovering the beam power after having passed the IP. The dump therefore is very ambient friendly.

extent, of the status of operation of the proton machine. This is the principal advantage of the Linac-Ring over the Ring-Ring configuration which had been studied to considerable detail in the CDR also.

The chosen energy of 60 GeV leads to a circumference U of the electron racetrack of 8.9 km. This length is a fraction $1/n$ of the LHC circumference, for $n = 3$, as is required for the e and p matching of bunch patterns. It is chosen also in order to limit the energy loss in the last return arc and as a result of a cost optimisation between the fractions of the circumference covered by SRF and by return arcs. That configuration is adopted also as the default for the FCC-he. Tentatively, for the LHC the ERL would be tangential to IP2 which, according to the current plans, is taken by the ALICE detector until the first long shutdown following the three year pause of the LHC operation for upgrading the luminosity performance and detectors. This shutdown is termed LS4 and may begin in 2030. For FCC-he the preferred position is IP L for geological reasons mainly.

The LHeC operation is transparent to the LHC collider experiments owing to the low lepton bunch charge and resulting small beam-beam tune shift experienced by the protons. The LHeC is thus designed to run simultaneously with pp (or AA) collisions. This tames the cost and optimises the physics return since a concurrent operation with the LHC will have a direct impact on the HL LHC physics programme as is sketched subsequently.

15.4.2 *Parameters*

Following [Brüning, O and others (2017)] the parameters of the current LHeC default design may be summarised as follows. The luminosity L of the linac-ring ep collider may be approximated as

$$L = \frac{N_p N_e f \gamma_p}{4\pi \epsilon_p \beta_p} \cdot H_{geo} H_{bb} H_{fill}. \tag{15.1}$$

Here, N_p is the number of protons per bunch and ϵ_p and β_p are the proton emittance and beta-functions, respectively. The main proton beam parameters N_p and ϵ_p are defined by the LHC General Purpose Detector (GPD) experiments in concurrent ep and pp operation. The proton beta-function in the electron-proton collision point may be as small as $\beta_p = 7$ cm.

Furthermore, $f = 1/\Delta$ denotes the bunch frequency, which for the default bunch spacing of $\Delta = 25\,\text{ns} = 40\,\text{MHz}$. N_e is the number of electrons per bunch which determines the electron current $I_e = eN_e f$ with a target value of 15 mA, a twofold increase compared to the conservative CDR. For 60 GeV, this will yield a total synchrotron radiation of about 40 MW in the return arcs. To compensate for this power loss through the beam, a grid power of the order of 65 MW may be required. This load may be substantially reduced if the electron energy is diminished.

The factors H_{geo}, H_{bb} and H_{fill} are geometric correction factors with values typically close to unity. H_{geo} is the reduction of the luminosity due to the hourglass

Table 15.1 Baseline parameters and estimated peak luminosities of future electron-proton collider configurations for the electron ERL when used in concurrent *ep* and *pp* operation mode, taken from [Brüning, O and others (2017)].

parameter [unit]	LHeC CDR	ep at HL-LHC	ep at HE-LHC	FCC-he
E_p [TeV]	7	7	12.5	50
E_e [GeV]	60	60	60	60
$\sqrt{s}$ [TeV]	1.3	1.3	1.7	3.5
bunch spacing [ns]	25	25	25	25
protons per bunch [10^{11}]	1.7	2.2	2.5	1
$\gamma\epsilon_p$ [μm]	3.7	2	2.5	2.2
electrons per bunch [10^9]	1	2.3	3.0	3.0
electron current [mA]	6.4	15	20	20
IP beta function β_p^* [cm]	10	7	10	15
hourglass factor H_{geo}	0.9	0.9	0.9	0.9
pinch factor H_{bb}	1.3	1.3	1.3	1.3
proton filling H_{fill}	0.8	0.8	0.8	0.8
luminosity [$10^{33}\mathrm{cm}^{-2}\mathrm{s}^{-1}$]	1	8	12	15

effect, H_{bb} is the increase of the luminosity by the strong attractive beam-beam forces and H_{fill} is a factor that takes the filling patterns of the electron and the proton beam into account. Estimates for these parameters are shown in Table 15.1. Compared to the CDR of the LHeC from 2012, it seems indeed possible to achieve peak luminosities near to or larger than $10^{34}\,\mathrm{cm}^{-2}\mathrm{s}^{-1}$, which makes these future *ep* colliders most exciting and efficient machines for the study of new physics at the accelerator energy frontier.

15.4.3 *The energy recovery development facility PERLE at Orsay*

When Guido Altarelli had asked, at the 2015 LHeC workshop, whether we could build the LHeC, the IAC, of which he was a very prominent member, discussed the importance of developing the basic technology of the LHeC. Since then a conceptual design report was published [Angal-Kalinin *et al.* (arXiv:1705.08783, 2017.)] of a low-energy, energy recovery linac facility. A collaboration has been founded with the intent to realise a "Powerful Energy Recovery Linac for Experiments" (PERLE) at the site of LAL Orsay where a hall and parts of the infrastructure already exist. Initially participating institutes are BINP Novosibirsk, CERN, Daresbury, the University of Liverpool, Jefferson Laboratory and the two now merging Orsay Institutes, LAL and IPN. The main parameters of PERLE are derived from the LHeC, such as the electron current of 20 mA chosen to achieve its $10^{34}\,\mathrm{cm}^{-2}\mathrm{s}^{-1}$ luminosity target.

The PERLE layout considers two accelerating superconducting cavity structures, cryogenic modules each housing four 802 MHz frequency 5-cell cavities, which are embedded at opposite linear parts in a racetrack configuration with two triple arcs. Operation for ERL is in Continuous Wave (CW) mode at typical gradients of 18 MV/m. An electron beam of up to 20 mA current, derived from a DC photocathode, is injected with 7 MeV energy into the PERLE lattice which is an ellipse of main dimensions $5.5 \times 24\,\text{m}^2$. With three turns the beam is accelerated to about 500 MeV and, after a $\pi/2$ phase shift, decelerated and dumped at the injection energy. PERLE therefore represents a unique facility with its multi-turn ERL and a new, high-power level of 10 MW which distinguishes it from various other facilities at lower energy or different current, frequency or return arc magnet technology.

The frequency of 802 MHz is chosen to comply with the LHC constraint following an optimisation involving beam breakup stability for the LHeC, cost of the structure and RF power as well as cryogenic capacity [Jensen (2016); Marhauser (2017)]. A 5-cell Niobium cavity prototype for PERLE, designed by CERN and Jlab [Angal-Kalinin *et al.* (arXiv:1705.08783, 2017.)], has been built at Jefferson Laboratory in 2017 and successfully passed all checks for mechanical stability, tolerance and flatness. Very recently its Q_0 factor, which determines the maximum of stored energy relatively to the input power, has been initially measured and was observed to safely surpass the requirement of a value in excess of 10^{10}, see Fig. 15.2. The development of PERLE is thus passing its first and a crucial milestone, with further investigations as on the field emission to possibly follow and elements for higher order mode extractions under design.

Currently the PERLE Collaboration works on the technical design of the facility with emphasis on building the first cryomodule, using the considerable experience gained with the existing SPL, ESS and SNS modules developed by CERN, IPN Orsay and Jlab, respectively. It is hoped that PERLE can start operation within a few years' time, also building on the large DC photocathode experience at Daresbury Lab and on the art of magnet fabrication at BINP Novosibirsk. Primarily, PERLE

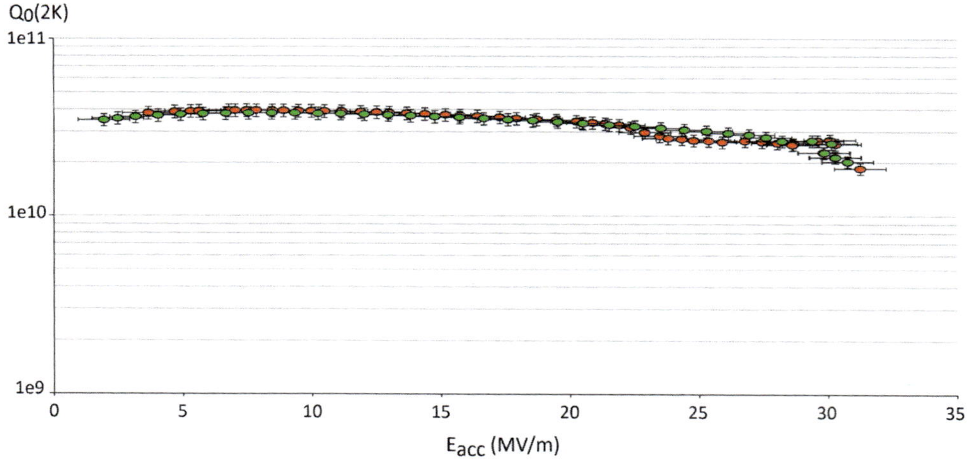

Fig. 15.2 Quality factor Q_0 of the first 5-cell, 801.58 MHz frequency, Niobium, PERLE prototype cavity determined as a function of the acceleration gradient. The observed behaviour surpasses by far the expectation, with a Q_0 well exceeding 10^{10} and an apparent very weak degradation towards gradients extending much beyond the considered CW operation working point of 18 MV/m, from [Marhauser *et al.* (2018)].

is built as an accelerator R&D facility. Questions such as on the time structure, transverse and longitudinal matching, source limitations, space charge, beam break up, halo and other effects ought to be investigated in order to master ERL technology and dynamics at the 10 MW power level.

The high intensity and chosen energy make PERLE in the mid term future a unique facility for the study of photo-nuclear reactions and also particle and nuclear electron beam physics on, for example, the study of the scale dependence of the weak mixing angle, a precise measurement of the proton radius or the search for dark light photons. Other applications, also described in the PERLE CDR [Angal-Kalinin *et al.* (arXiv:1705.08783, 2017.)], comprise the beam based development of SCRF, the test of accelerator components, such as the quench behaviour of superconducting magnets, or detector developments. With a photon intensity exceeding by orders of magnitude that of the ELI facility [ELI (2017)] currently under construction in Southern Europe, PERLE represents a far-reaching new accelerator base for the laboratories in the Orsay and Saclay area, the further collaborating institutes and future possible users for fundamental and applied physics investigations. PERLE may also be configured as a powerful laser physics facility when complemented with an undulator system. At the same time, experience and technology emerging from PERLE are enablers for the LHeC, the highest-energy ERL application so far considered.

15.4.4 *Detector*

The arrangement of the electron ERL tangential to the LHC defines one interaction region (IR) for one detector. By LS4, now scheduled for 2030, the ion-ion programme of the LHC ends in order to maximise the *pp* luminosity collection, and currently the operation of the ALICE detector at IP2 is considered to then be complete. This opens the opportunity for replacing it with a new detector for precision electron-hadron physics. A study has been made [Gaddi (2017)] of the sequence of installing the new detector in IP2. Based on experience with the pre-mounting of the CMS detector and exploiting the modular structure of the LHeC detector it was established that a time of less than 2 years shall be sufficient for such an "exchange of detectors". Typical shutdown periods of the LHC range from one to nearly three years, and one may expect that the first long shutdown after the upgrade of machine and detectors may not be short for reasons independent of *ep*. This opens the real opportunity for DIS physics with the LHC in the thirties.

In considering the further future one may expect the LHC to be replaced by its double energy version, the HE LHC, following the termination of the LHC operation currently envisaged to take place in ~ 2038 after accumulating 3 or $4\,\mathrm{ab}^{-1}$ of integrated *pp* luminosity. It thus is conceivable that the *ep* detector may also be used for operation during the HE LHC phase, from likely the end of the forties to two decades hence. It then seemed not illogical to design an *ep* detector which could cope with first 7 and later $\simeq 13\,\mathrm{TeV}$ proton beams. The current status of its design

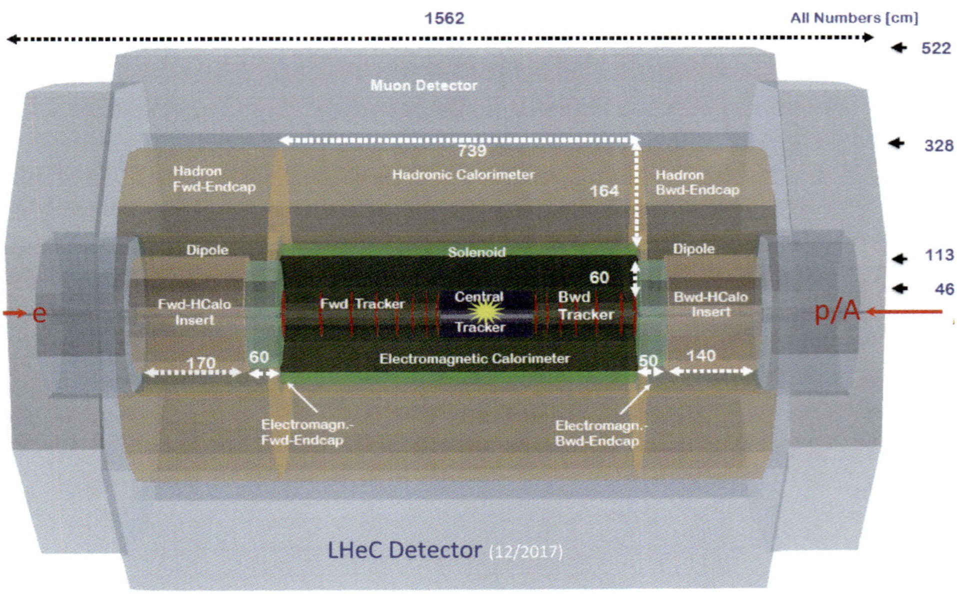

Fig. 15.3 View on the LHeC detector under design to operate with both the high luminosity (HL) and the high energy (HE) LHC. The proton beam comes from the right, pointing to the forward region to which jets and electrons may be scattered of multi-TeV energies. The electron beam points into the backward region which has to measure hadron and electromagnetic energies not exceeding the electron beam energy. The detector is therefore asymmetric. Special attention is required for forward calorimetry and the polar angle acceptance, essential for forward b tagging and for low-Q^2 backward scattered electrons. The beam pipe is eccentric to permit the synchrotron radiation fan of the bent electron beam to pass through the interaction region. The HE (HL) central LHeC detector shown here has a length of 15.6 (13.2) m and a radius of 5.2 (4.4) m. This may be compared with the CMS dimensions of $21 \times 7.5\,\mathrm{m}^2$.

is sketched in Fig. 15.3. The detector has a classic collider detector structure and its basic design considerations as well as simulations are summarised in the LHeC CDR [Fernandez and the LHeC Study Group (2012)]. Its characteristic feature is a combined magnet structure which has to be inserted between the electromagnetic and hadronic calorimeters, i.e. a dipole magnet for bending the electron beam into and out of the proton beam axis and a 3 T solenoid currently combined with that dipole. Two versions exist of warm and LAr electromagnetic calorimetry.

Recent work focusses on the adaption to evolving detector technology, such as low material, high resolution Silicon detectors, mainly in connection with the LHC luminosity upgrade. The LHeC conditions are comparatively easier with an *ep* event pile-up of only 0.1 and a radiation level about hundred times lower than the LHC. Strong efforts are made to update the IR design from the CDR to new types of quadrupoles of which high field Nb_3Sn actively shielded quadrupoles are currently most promising [Parker (2017)]. The design of the IR is most challenging, and at the end of 2017, still work in progress in collaboration among BNL,

CERN and Liverpool. Much care and time are also dedicated to the development of DD4HEP/DDG4 based design, simulation and reconstruction software. The central LHeC detector must be completed by forward proton, neutron and possibly deuteron taggers while the backward region has to be instrumented with near axis photon and electron detectors to measure the *ep* luminosity from Bethe–Heitler scattering and to tag and normalise the photoproduction background to genuine DIS. These taggers have all been considered in the CDR. It is intended to update the CDR detector design within 2018 and to possibly assemble an LHeC Collaboration which would certainly significantly move these design concept studies, currently also involving service routing and material minimisation, forward.

15.5 Resolving the substructure of matter

DIS is the means to explore the substructure of matter using photons and W, Z bosons as probes with the kinematics fixed by the scattered electron (NC) or neutrino (CC). Both the squared (negative) 4-momentum, Q^2, and the parton momentum fraction x are determined externally to the virtual γ (W or Z) interaction with the parton. The kinematics can equally well be determined with the hadronic final state. The resolution is $1/\sqrt{Q^2}$, where 1 GeV corresponds to a distance of 0.2 fm, which is how Hofstatter discovered a finite proton radius of about 0.8 fm with an electron beam of 200 MeV. Following a series of fixed target lepton scattering DIS experiments, with $Q^2 \leq s = 2M_pE_l$, HERA made a big step forward by increasing the squared cms energy $s = 4E_eE_p$ to $10^5\,\mathrm{GeV}^2$ in the first *ep* collider ever built. That range could not be fully explored because of the low luminosity of HERA. The LHeC is designed to achieve a factor of 1000 higher integrated luminosity, and extends the squared energy to $1.7 \cdot 10^6\,\mathrm{GeV}^2$. This will lead to quantitatively more valuable and much deeper reaching insight into the structure and dynamics of parton interactions inside matter.

15.5.1 *The proton's partonic structure and dynamics*

The momentum distributions of partons inside the proton, $xP(x, Q^2)$, must be determined from experiment. The most suited process to obtain an unbiased determination of the quark and gluon distributions is inclusive deep inelastic NC and CC scattering. "PDFs" as they are now termed have two meanings in that i) they represent a probability view on the substructure of the proton, or similarly of other hadrons, at a given distance $1/\sqrt{Q^2}$, and ii) they describe, supposedly universally through cross section factorisation theorems [Collins *et al.* (1989)], the hard scattering processes involving partons.[4] Their determination therefore is crucial both for

[4] In his referee report on the LHeC CDR, in 2012, Guido Altarelli noted on the factorisation theorem in QCD for hadron colliders that: *many people still advance doubts. Actually this question could be studied experimentally, in that the LHeC, with its improved precision, could put bounds*

QCD and for searches for new states which makes them so fashionable nowadays at the LHC. DIS is *the* process to determine PDFs, unlike Drell–Yan scattering, because it is theoretically clean, free of colour interactions in the initial or final state, and it has the prescribable variation of Q^2 and finally the determination of x external to the intrinsic scattering process.

The PDF programme of the LHeC, as is briefly illustrated below, is of unprecedented depth for the following reasons:

- it will resolve the partonic structure of the proton (and nuclei, see Section 15.9.2) for the first time completely, i.e. determine $P = u_v,\ d_v,\ u,\ d,\ s,\ c,\ b$ and the top and gluon momentum distributions through NC and CC cross section and direct heavy quark measurements in a huge kinematic range, from below $x = 10^{-6}$ to $x = 0.9$ and in Q^2 from below the DIS region to almost $Q^2 = 4E_eE_p = 1.7 \cdot 10^6\,\text{GeV}^2$;
- a thousand-fold increase of the HERA luminosity, unprecedented precision from new detector technology and the redundant evaluation of the event kinematics from the lepton and hadron final state components will lead to extremely high precision and, technically important, to the fixation of the various PDF analysis parameters from the LHeC data themselves;
- because the high-energy CC, unlike at HERA or low energy EICs, becomes equivalent to NC and thus no other data will be required: that is, there is no influence from higher twists or nuclear uncertainties, and no other experiment needed, i.e. LHeC will be the unique base for PDFs, for predictions, discovery and novel tests of theory.

Given the impressive theoretical progress on pQCD, see e.g. [Moch *et al.* (2017)], one will have these PDFs available at N^{3}LO and only then have the PDFs available to respond to the N^{3}LO Higgs in pp cross-section calculations demand. For QCD, this will resolve open issues (and probably create new ones) on α_s, discussed in the Appendix, answer the question on the persistence (or not) of the linear parton evolution equations at small x, see Section 15.9.1, and also decisively test whether factorisation holds or not between DIS and Drell–Yan scattering.

The prospect for the determination of the proton-parton distributions with the LHeC has been studied in detail. It relies on the careful simulation of NC and CC measurements, including a full account of their systematic uncertainties and expected correlations, as described in [Fernandez and the LHeC Study Group (2012); Klein and Radescu (2013)]. Two examples are given on the expected precision. Surprisingly after 50 years of DIS, there is no certainty about the behaviour of the valence-quark distributions at large (and small) x. This is illustrated in Fig. 15.4 which displays the uncertainties and central values for the up and down valence quarks as a function of x as obtained in recent PDF analyses. One finds

on the allowed amount of possible factorisation violations (e.g. by measuring in DIS the gluon at large x and then comparing with jet production at large p_T in hadron colliders).

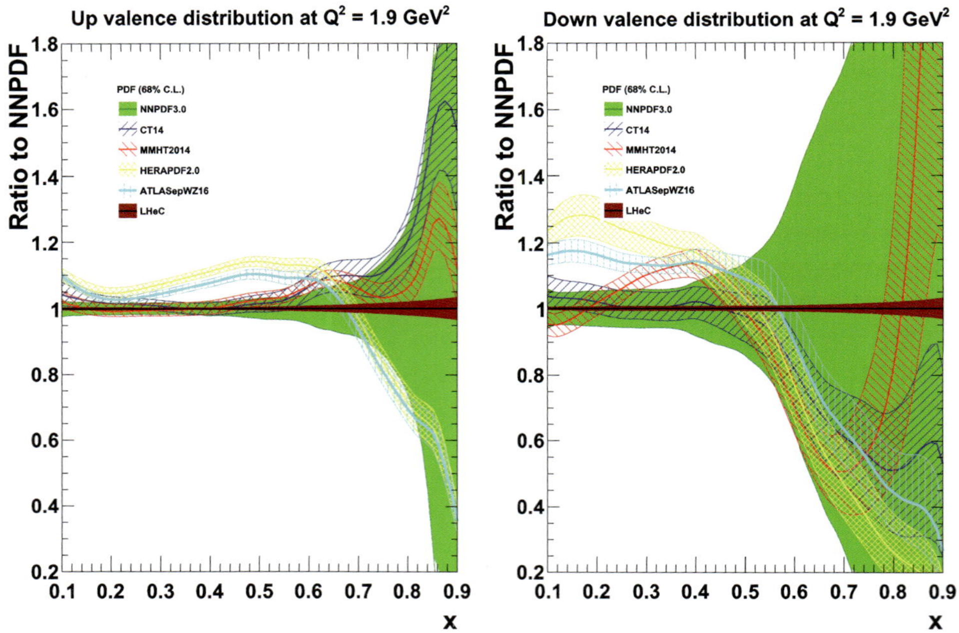

Fig. 15.4 Determination of the valence quark distributions as functions of Bjorken x. Plotted are the ratios to the NNPDF result with uncertainties displayed as are provided by the individual sets, left for the up-valence quark and right the down-valence quark distribution. For the LHeC the total uncertainty is plotted and the central value assumed to agree with NNPDF. As non-singlet quantities, the valence quark distributions are approximately the same with varying Q^2.

that the up-valence quark is better determined than the down-valence quark, which is related to the proton structure function F_2/x having the large x limit of $4u_v + d_v$. However, at large $x \geq 0.5$ where the distributions decrease, nuclear corrections rise and higher twist effects enter most, we are far from understanding the valence quark distributions. Moreover, there is an apparent difference of $\simeq 20\,\%$ between various expectations even at medium x. The LHeC will determine both u_v and d_v with very high precision, see Fig. 15.4, in the full range of x. This is of high interest for QCD, for example concerning the long standing issue of the high x limit of the d/u ratio. For the LHC searches resolving these huge uncertainties is crucial as the luminosity-upgraded LHC will explore the high-mass Drell–Yan regions which currently are masked by the PDF uncertainties, see Section 15.6.

One should note that the distributions of the up and down quarks are related, via sum rules, to those of the heavier quarks. This has been demonstrated recently by the ATLAS Collaboration which discovered [Aad *et al.* (2012b)] that there should be a light flavour democracy including the strange quark: finding that the ratio $(s + \overline{s})/(\overline{u} + \overline{d})$ was about 1 and not suppressed, it was derived [Aad *et al.* (2012a)] that the light sea, $x\Sigma = 2x(\overline{u} + \overline{d} + \overline{s})$ was enhanced by $8\,\%$. While ATLAS has derived this result from essentially a moment measurement of xs, the LHeC will determine the strange quark distribution from charm tagging in CC directly and as a function of x and Q^2 over several orders of magnitude as has been demonstrated in the CDR. This is an example of what the difference is between pp, the LHC which can still lead to valuable constraints and insight for PDFs, and ep, the LHeC which is the ultimate machine to measure them. Too often that is misunderstood and hopes are expressed that one could determine PDFs from pp. This is not the case in terms of principles and precision, and it has the philosophical problem, that new physics, hopefully still appearing, as for example a contact interaction effect in pp, shall not be misinterpreted as a PDF effect. Two unknowns, PDFs and new physics, require two independent sources of input.

Another important example for the singular PDF prospects with the LHeC regards the gluon distribution. That is of fundamental importance because, as we discovered with HERA, the proton's momentum below $x \simeq 0.01$ is almost completely occupied by gluons, xg is much larger than any of the quark distributions, in the DIS region of Q^2 exceeding M_p^2 and small x. Moreover, the dominant Higgs production channel in pp is gluon-gluon to Higgs fusion, via top loops. The intent to measure the Higgs properties as precisely as possible at the LHC requires control of xg in the region of x between 0.001 and 0.1 to percent precision, which is not achieved. The large x behaviour of the gluon distribution is particularly uncertain but especially relevant for testing QCD, as with the factorisation theorem discussed above, and for controlling the production of new states as in gluino pair production, see Section 15.6. For reliably measuring $xg(x, Q^2)$ one needs to accurately measure the Q^2 derivative of the DIS cross section, which, not surprisingly, came out as the only way to accurately access xg at HERA, not diffraction, not J/ψ production, neither open charm which are all theoretically less clean and experimentally not competitive with inclusive DIS.

The current status on the determination of xg is illustrated in Fig. 15.5. A HERA physicist recognises the limits: the determination of xg from $\partial F_2/\partial \ln Q^2$ at HERA reaches not below $x_{min} \simeq 2 \cdot 10/(0.7s) = 3 \cdot 10^{-4}$ where the factor of 2 is due to the splitting function, 10 is the minimum Q^2, 0.7 the maximum y and $s = 10^5\,\mathrm{GeV}^2$. There arises consequently a huge uncertainty of xg below such x_{min} as is obvious from Fig. 15.5 (left). Practically there is a very large uncertainty for x below 0.001. We therefore had not been able to establish the gluon density and neither small x theory subtleties as resummation nor BFKL effects at HERA. There are attempts to constrain xg at lower x through heavy flavour production at the LHC. The real determination of the gluon density at small x in the appropriate theory context has to wait for the first LHeC data. The prospect is seen in Fig. 15.5. There, one also recognises the HERA limits on the xg determination at large x. The gluon density at $x = 0.7$ is uncertain by two orders of magnitude, it could be equal to 0.01 but also equal 0.001 or 0.1. The prospects to derive xg with the LHeC are outstanding, numerically one finds a total (statistical and systematic) uncertainty, at $Q^2 = 1.9$ (M_Z^2) of 5.3 (0.2), 0.6 (0.2) and 21 (2) % at $x = 5 \cdot 10^{-5}, 0.01$ and 0.7, respectively. The Q^2 evolution helps in reducing the uncertainty, it yet needs

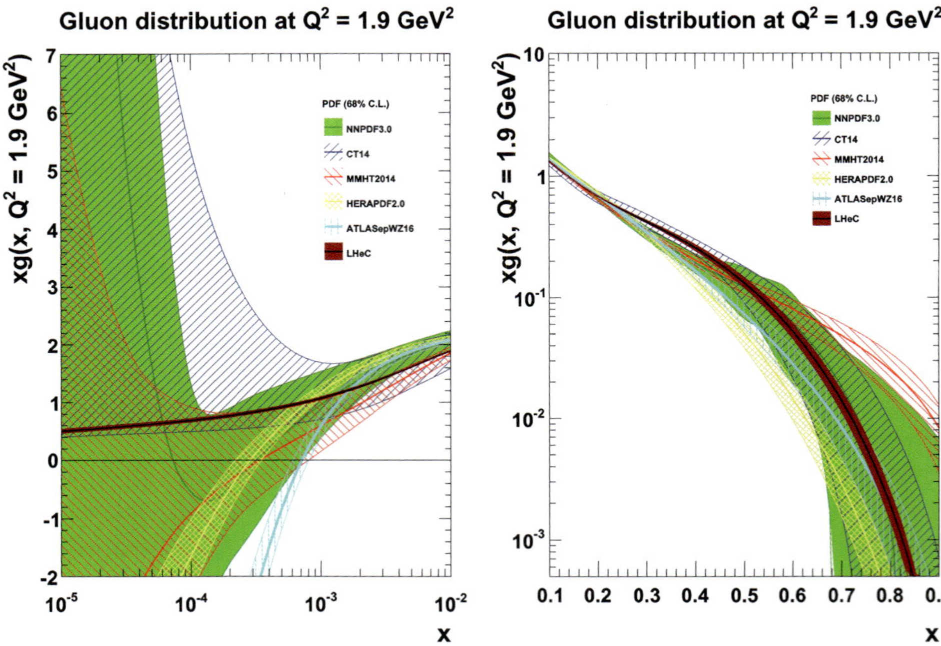

Fig. 15.5 Determination of the gluon momentum distribution in the proton. The expected total experimental uncertainty on xg from the LHeC (dark purple bands) is compared with the most recent global PDF determinations which include the final HERA data, covering for xg a range from $x \simeq 5\ 10^{-4}$ to $x \simeq 0.6$, and much of the LHC data from Run I. Left: xg at small x; Right at large x.

to be verified at the corresponding level for which, at small x one will need the complementary access on xg via F_L, see the Appendix.

15.5.2 *DIS and confinement*

The LHeC nowadays is often reduced to a PDF delivery machine or a Higgs facility. While these may currently indeed be its most striking characteristics, there is a wealth of far-reaching, genuine DIS physics ahead with the LHeC. QCD may break, become embedded in a higher symmetry; odderons and instantons await to be discovered. Many fundamental questions are not answered as with HERA. That regards, for example, the structure of the deuteron, neutron, pion, nuclei, photon and that of the diffractive exchange often called Pomeron. It comprises the many ways to modify and extend the collinear picture of parton distributions such as the transverse momentum dependent unintegrated parton distributions, the amplitude based generalised parton distributions or the non-DGLAP evolving partons. Jets and jet's substructure, characteristics of quark and gluon jets, vector mesons — a huge field of unexplored physics is awaiting an LHeC experiment. The top quark will appear light when Q^2 exceeds M_t^2 which holds for the LHeC. The theory of heavy flavour production in *ep* is not tested well enough, such that fixed flavour or variable flavour number schemes coexist which causes important principal uncertainty in QCD and practical uncertainty for predictions for LHC processes. DIS at high energies is a means to precisely test the electroweak theory in the spacelike region with unique measurements as of the scale dependence of the weak mixing angle from below the Z mass up to 1 TeV, or of the weak neutral current couplings of light quarks. The LHeC therefore will open a new laboratory for deep inelastic physics of hitherto unseen importance, being well comparable to the present GPD LHC experiments which it is designed to complement. Two examples may illustrate this rich potential.

QCD, as much as it seems to work for the description of high-energy interactions as at the LHC, still has a very fundamental problem: the explanation of the confinement of partons inside the proton. This caused Jaffe and Witten, in 2000, to propose the inclusion of Quantum Yang–Mills Theory as one of the seven millenium prize questions, in which they indicated that QCD indeed shall be a consistent non-perturbative Quantum Field Theory which has to have three features: i) a mass gap between confined massless gluons and their massive bound states, ii) confinement of partons compatible with SU(3) invariant free hadrons and iii) chiral symmetry breaking to incorporate current algebra [Jaffe and Witten (2000)]. Thus there is much more behind the strong interaction theory than is naively known. Deep inelastic scattering is ideally suited to approach the confinement puzzle by i) studying the stunning phenomenon of DIS diffraction where the proton remains intact despite a hugely energetic collision, and ii) by exploring the process of hadronisation, i.e. the propagation of partons in spacetime, their interaction and the coalescence into colour-singlet states. The role the LHeC can play in these investigations has been discussed in the CDR, see also [Accardi *et al.* (2010)]. The LHeC allows to follow the particle formation, sketched in Fig. 15.6, over a long distance which can be prescribed with the scattering kinematics. Measurements of Deeply Virtual Compton Scattering (DVCS) processes can establish a dynamic picture of the 3D structure of the proton as is entailed in Generalised Parton Distributions. In this area, the polarised-polarised *eh* scattering configuration at low energies provided by a future EIC and the high-energy LHeC microscopy may indeed well complement each other. Important further insight is gained with the study of the hadronisation process in protons and nuclei, especially with the variation of the energy transfer $\nu = E_e - E'$.

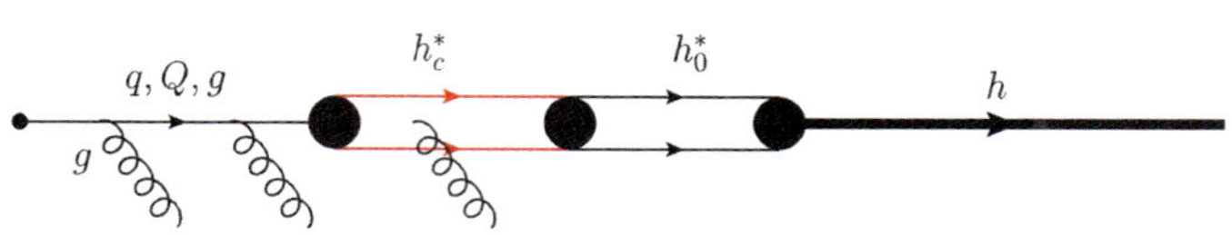

Fig. 15.6 Time evolution of hadronisation from a parton created in an interaction on to the formation of coloured and white states preceding the emission of a hadron, from [Accardi *et al.* (2010)].

15.6 Empowering physics at the LHC

The LHC has currently delivered 100 fb^{-1} of integrated luminosity to ATLAS and CMS. In 2012 both discovered the Higgs boson, with about 10 fb^{-1}, but since then, despite many expectations and predictions no further particle or symmetry beyond the SM. LHC physics therefore focusses on two avenues, both requiring a substantial increase of the luminosity: i) searches for new phenomena at the edges of phase space, i.e. states of high mass $M \propto \sqrt{sx_1x_2}$ initiated by the interaction of two partons carrying momentum fractions $x_{1,2}$ of their parent protons, and ii) measurements of maximum precision to test the SM at a new level. The addition of the LHeC is the ideal complement for both tasks, and it would be crucial also in the interpretation of new physics should that appear. Empowering the physics at the LHC through *ep* is related to the ultimate precision and clarity with which PDFs and the strong coupling constant will be determined at the LHeC as is discussed in this article.

The physics motivation, as Guido taught us, *for the LHeC is tightly related to the future continuation of the hadron collider experiments, HL LHC and FCC* [Altarelli (2015b)]. With increasing luminosity the LHC accesses larger and larger masses M. A severe problem into which it runs is that the discovery reach is thereby moved into the region of large Bjorken x where the gluon as well as the valence and sea quark distributions are basically unknown. This is illustrated in Fig. 15.7. The left side shows the uncertainty of a prediction of gluino pair production in the MSSM $gg \to \tilde{g}\tilde{g}$. At the HL LHC one may reach values up to 4 TeV corresponding to an average $\sqrt{x_1x_2}$ of about 0.5, or even higher depending on couplings to new physics. One observes how the uncertainty rises with M which degrades the limit values one may wish to set and complicates an interpretation should there be new physics appearing. The right side shows the uncertainty of the prediction for heavy W bosons as are expected in grand unified theories. The uncertainty, of the quark initiated process, rises very strongly towards large masses. This effectively acts like a barrier for the high luminosity upgrade: huge efforts are made to access high masses, but the proton structure is too uncertain for the new and most promising area of phase space to be safely explored. The PDF uncertainty is already now the largest one in such searches which currently set limits of 5.1 TeV, obtained for sequential coupling and ATLAS data of 36.1 fb^{-1} [Aaboud *et al.* (2017i)]. The luminosity upgrade necessitates an electron beam upgrade for exploiting its real potential. *We often hear the statement that all the relevant information on PDFs can directly be obtained from the LHC without need of the LHeC, that is not really true, certainly not at the same level of precision* [Altarelli (2015b)].

The second avenue of LHC physics, more and more recognised, is "precision", that is the desire and need to test the Standard Model as severely as possible. This concerns primarily the physics of the Higgs boson because its characteristics are not well determined and it may be the portal to new physics, see below. Moreover, the LHC is surprisingly able to perform certain stringent tests of the electroweak theory,

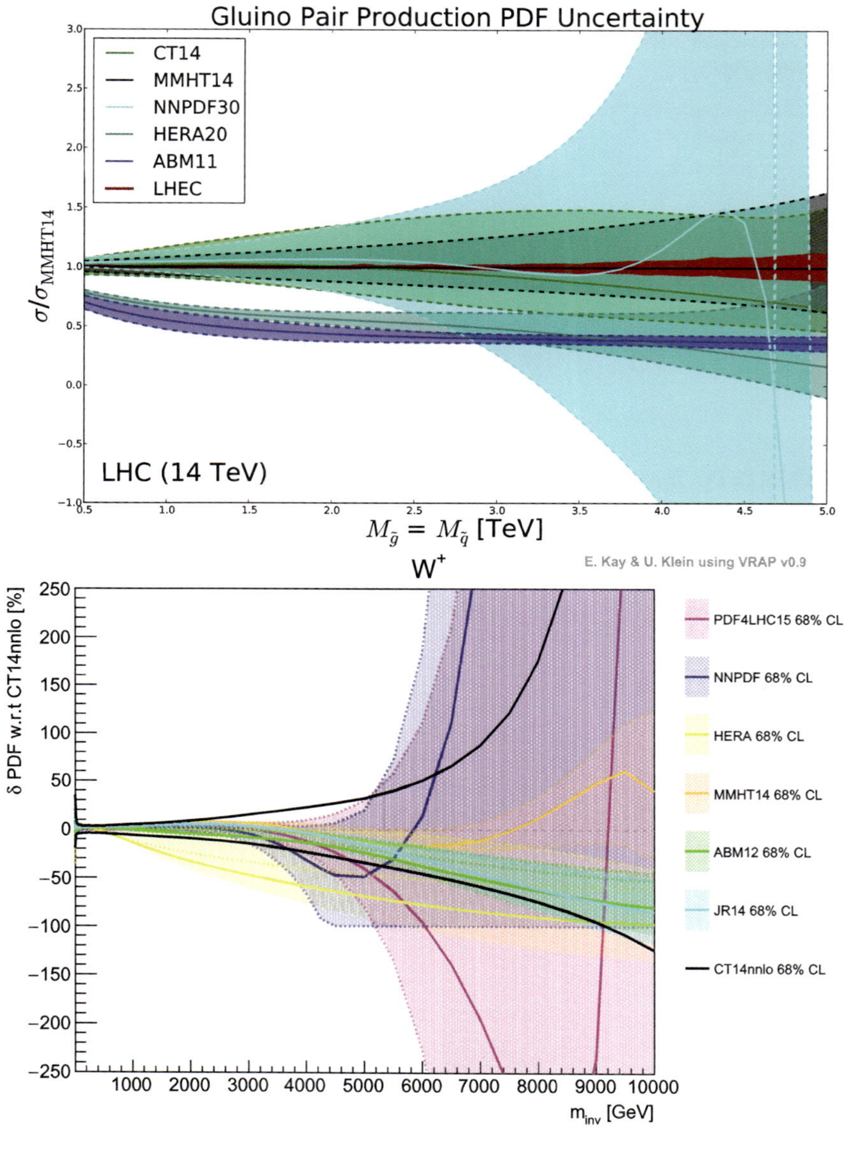

Fig. 15.7 Uncertainties of LHC production cross sections from recent determinations of parton distributions. Top: Gluino pair production as functions of the gluino mass. The inner red band represents the projected LHeC PDF uncertainty [Borschensky and Kramer (2016)]. Bottom: High mass W^+ production [Kay and Klein (2017)].

with the recent ATLAS measurement of the mass of the W boson [Aaboud *et al.* (2018)] to 19 MeV precision as a most striking example. In this result, the PDFs are the single largest systematic uncertainty, of 9.2 MeV, resulting from quite a magic cross correlation analysis, which works through the constraint DIS data impose on the total light quark sea, and by considering only a restricted set of PDFs. The LHeC is estimated to predict this M_W measurement to 2.8 MeV precision as derived from the uncertainties of the LHeC PDF set [Camarda (2017)] which eliminated the PDF problem in the M_W mass measurement at the LHC. It is moreover interesting to note that M_W can also be extracted from the LHeC data as a fundamental parameter of the electroweak theory, with an estimated total uncertainty of (14 (exp) and 10 (pdf)) MeV [Britzger and Klein (2017)]. This is an example of the ultra high precision the LHeC will provide in the development of the SM and its possible eventual failure. Many other examples could be discussed here too, such as the scale dependence of the weak mixing angle, the determination of the V_{cs} and V_{tb} CKM matrix elements or the per mille accurate measurement of α_s (cf Appendix).

15.7 The LHC as a precision Higgs physics facility

It is crucial for future experiments at the LHC and elsewhere to confirm the properties of the Higgs and the absence of new physics [Altarelli (2013a)]. The Higgs mechanism generates the mass of the weak gauge bosons and the elementary fermions. Its simplest realisation contains a scalar field, $J^{CP} = 0^{++}$, which is referred to as the Higgs boson. Following the discovery of this particle, in 2012, much effort has been spent by ATLAS and CMS to pin down its quantum numbers and couplings to its decay products which are expected to depend linearly on their squared mass. The mass of the Higgs boson of 125 GeV implies that a particularly large number of channels is open dominated by the $H \to b\bar{b}$ decay with a relative probability of 58 % in the SM. Due to the fundamental role of the Higgs phenomenon, and in the absence of other new physics, its exploration has become the centre of attention at the LHC and for planning the future of collider particle physics.

In this context the question arises about the potential of the HL LHC, for probing the Higgs properties. The LHC is the only existing Higgs facility and for very long the only one with reliable access to very rare decays such as $H \to \mu\mu$. At the same time it has considerable difficulties to accurately measure certain abundant decays, for which one needs to employ the associate vector boson production, which just recently allowed the $H \to b\bar{b}$ decay to be recognised [Sirunyan *et al.* (2017a); Aaboud *et al.* (2017a)]. There are, moreover, considerable theoretical uncertainties, partially related to the insufficient control of xg, α_s and non-availability of genuine N^3LO PDFs which can be cured with the LHeC. It has been argued that the LHC can be "turned into a precision Higgs facility" [Klein (2017a)] with the addition of the LHeC which has no pile-up and a clean final state signature which distinguishes $WW \to H$ from $ZZ \to H$ production. Considerable work is ongoing to study this at the required level, and initial results and the direction of this exploration are illustrated here. In addition, as sketched in Section 15.8 below, there are unique opportunities with the LHeC to access new Higgs physics, in the production

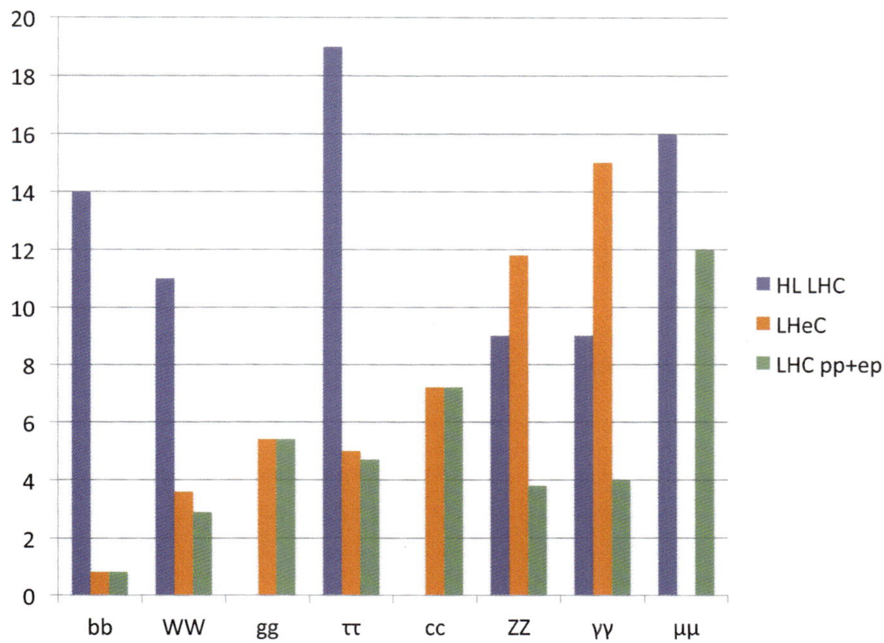

Fig. 15.8 Preliminary estimates of the expected determinations of the relative signal strength $\delta\mu/\mu$, in %, of the Higgs decay to various particles at HL LHC. Blue: the ATLAS expectation for the HL LHC [ATLAS Collaboration (2014b)]; Brown: LHeC prospect [Klein and Klein (2018)]; Green: LHC prospect for an ep/pp combination, see text. The decay channels are arranged in descending branching ratio order, from 57.7 % for $H \to b\bar{b}$ to 0.02 % for $H \to \mu\mu$. The improvement for all channels wrt. the HL LHC is due to new information on various decays from the LHeC and to an essential removal of the QCD uncertainties with the provision of N^{3}LO precision PDF and α_s measurements by the LHeC. An empty column, such as for the LHeC (red) in the $\mu\mu$ channel, indicates that this may not be accessible to notable precision.

(as of charged Higgs bosons) and decay (as into dark matter), which would lead beyond the SM.

A particular question is whether the total decay of the Higgs boson is exhausted by its branchings into the known particles, such as pairs of b, W, g, τ, c, Z, γ and μ and a few mixed combinations such as $Z\gamma$. Figure 15.8 illustrates the expected precision of the measurement of the signal strength μ for the eight most abundant decay modes in the SM. The plot shows the total uncertainty, in blue, as estimated by ATLAS [ATLAS Collaboration (2014b)] for the HL LHC, i.e. $3\,\mathrm{ab}^{-1}$ of integrated luminosity. A second column shows the expected precision on μ for the LHeC for an assumed luminosity of $1\,\mathrm{ab}^{-1}$ collected in electron-proton scattering. The results refer to a charged current interaction where the Higgs boson is emitted from a W exchanged in the t channel. The third, green column illustrates the expected accuracy for the combined pp and ep result at the LHC. It is to be noted for the LHeC that so far only the $b\bar{b}$ and $c\bar{c}$ results for the LHeC are due to complete simulations [Klein (2017b)] while the other numbers are still of tentative nature.[5] For the LHC itself one would expect also improvements related to progress in the analysis techniques and a possible final combination of the two experiment results.

Based on the numbers presented in Fig. 15.8 one may calculate the sum B_H of the Higgs boson branching fractions, br_i, $i = bb, WW, ...$, and their total uncertainty. The LHC result, assuming complete independence of the br_i measurements, comes out to be $B_H(pp) = 0.89 \pm 0.12$. The LHeC, assuming it indeed established the $H \to gg$ decay, which has a branching fraction of 8.6 %, to about 10 % precision, would reconstruct this sum as $B_H(ep) = 1.00 \pm 0.02$. The combination of *ep* and *pp*, with the additional reduction of the QCD uncertainty in *pp*, can be expected to accurately determine the total sum as

$$B_H(pp + ep) = \Sigma_i \; br_i = 1.00 \pm 0.01. \tag{15.2}$$

This prospective result means that the LHC can indeed be turned into a very powerful Higgs facility through the addition of *ep*. This is possible at moderate cost compared to other new facilities currently considered to which *pp*/*ep* would be the ideal complement. The precise measurements on $b\bar{b}$ and $c\bar{c}$, the identification of the *gg* and other decays, still under study, the reduction of QCD uncertainties for the *pp* results, as well as the experimental and the theoretical cleanliness of the *ep* Higgs production process [Blumlein *et al.* (1993)] are crucial improvements the LHeC entails. The result of Eq. (15.2) would narrow any window for exotic Higgs decays, which currently is wide open (to about 30 %), to a per cent and therefore represent one of the most crucial new tests of the SM. Failure to reconstruct B_H to 1 would, on the contrary, be an ultimate hint to the existence of new physics. Further interpretations of that experimental input on $\delta\mu_i/\mu_i$ may be made in joint fits of the couplings as are frequently done in prospective Higgs studies, such as in the recent EFT based analysis for the Linear Collider prospects [Barklow *et al.* (2017)].

It should be seen that a SM independent measurement of the sum of branching ratios B_H would require access to the total width Γ. That is considered to be an important advantage of e^+e^- colliders. It is intended for the LHeC to study this. The neutral current reaction $e^-p \to e^-HX$ can be kinematically fully constrained. The sizeable total NC Higgs cross section corresponds to $25k$ events which ought to be useful for a yet challenging measurement of Γ. One should further note that only the LHC (and its possible extensions HE LHC and FCC), neither the LHeC

[5]The signal strength result for the prominent $H \to b\bar{b}$ decay is 0.8 %, or a 0.4 % for the coupling determination. This is a most stunning precision and used to optimise the LHeC detector design, especially for forward b tagging acceptance and optimum resolution of the hadronic calorimeter and joint track-cluster reconstruction. The *bb* decay was initially studied for *ep* when the physics of LEP x LHC was explored [Grindhammer *et al.* (1990)]. Looking back at that analysis one recognises the impressive progress in the experimental and multi-variant analysis techniques and values the 10^{34} luminosity prospect particularly highly. The $c\bar{c}$ decay is considered to not be accessible at the LHC while the preliminary *ep* result [Klein (2017b)] is 7.2 %.

nor a future e^+e^- collider, has the possibility to access rare Higgs decay channels. The total H production cross section at the LHeC, very similarly for the ILC or CepC, is 200 fb^{-1}. For a luminosity of 1 ab^{-1} this provides 40 $H \to \mu\mu$ events only. It is obvious that the combination of *pp* and *ep* at the LHC form a unique, very powerful facility to lead Higgs physics in the thirties to an unprecedented art even if some of the prospects assumed here may alter in the course of the ongoing further study.

15.8 New physics at the LHeC

Compared to the LHC proton-proton interactions, the LHeC is a cleaner configuration because it not only has no pile-up of events but it has a unique NC or CC final state free of colour connection effects. Compared to the ILC and CepC, with perhaps initially $\sqrt{s} = 0.25$ TeV energy, it has a cms energy about five times higher. The *ep* configuration would fuse electrons and partons to leptoquarks, but it primarily is viewed as a spacelike probe, with photons and Z bosons in neutral currents and $W^\pm$ in charged currents, of the substructure and dynamics of matter. In these interactions, however, new heavy particles, such as the Higgs boson, may be produced and new phenomena beyond the Standard Model revealed. With its luminosity prospect [Brüning and Klein (2013); Brüning, O and others (2017)] of O(10^{34}) cm^{-2}s^{-1} the LHeC brings huge potential for new physics to be discovered in *ep*. One should note that such a luminosity allows the 15 years of HERA data taking to actually be pursued in just one day. None of the two e^+e^- colliders mentioned, which currently have certain probability of being realised, promise to have significantly higher luminosity. The HL LHC, with *pp* and *ep*, and a new e^+e^- machine set to ZH production would be a strong triple for the next phase of energy frontier collider physics, in the third decade of our century. This would substantially strengthen CERN's leading role in particle physics while a healthy competition could rise in Asia, prior to the realisation of a next, more demanding step with the raise of energy for *pp* collisions at CERN.

With increasing intensity and novel approaches, the genuine BSM potential of the LHeC is now being studied, in often comparative analyses, and it is generally observed that the BSM potential in *ep* is much stronger than thought to be hitherto. A new generation of theorists challenges the conventional beliefs. It is regretful we cannot discuss this with Guido Altarelli who had put to question whether *ep* could indeed be a competition with e^+e^- or hadron colliders by calling it, in contrast to those, *not a main discovery machine* [Altarelli (2015b)]. A few examples are here given subsequently which illustrate new avenues and results towards a new exploration of the BSM physics potential in *ep* which leads further than the CDR, because of the challenges set with the absence of obvious new phenomena at the LHC.

A central question in particle physics is that of the origin of mass. The LHeC provides fundamental insight to the origin of baryonic mass, with its far-reaching exploration of gluon dynamics in hadrons, of dark mass, through Higgs-to-invisible measurement prospects, of boson and top mass, through unique contributions to their determination and to the Higgs mass, through the approach to the Higgs self-coupling [Kumar *et al.* (2017)]. A recent study [Antusch *et al.* (2017)] draws the attention to the origin of neutrino masses which may require an extension of the Higgs sector or the existence of neutral fermions. Such heavy neutrino mass eigenstates, often termed sterile neutrinos, through their interaction with weak bosons, give rise to a rich phenomenology for their possible different production and decay modes, the dependence on mixing parameters or the occurrence of lepton number or lepton flavour violation effects to which the three collider configurations exhibit different sensitivity. Figure 15.9 displays the result of a comparative study, here focussed on HL LHC and LHeC, of the sensitivity of future collider configurations to the existence of heavy neutrinos, which in *ep* are primarily produced in the CC reaction $ep \to NX$. The LHeC reaches a sensitivity of neutrino mass up to $\sim 1\,\text{TeV}$ at mixing angles squared down to about 10^{-3} extending the Θ^2 range of the HL LHC by an order of magnitude to smaller values, owing to the cleanliness of the final state. For lighter masses, up to about 30 GeV, very small mixing angles of Θ^2 down to $2 \cdot 10^{-8}$ are reached. In this area, from an initial study, the HL LHC sensitivity looks stronger than that in *ep*, see Fig. 15.9.

Another example for the unique potential to discover new physics at the LHeC has just been presented with a study of the search for long lived Higgsinos [Curtin *et al.* (2017)]. These occur in the MSSM together with Wino and Bino states. Their mixing generates 4 neutralino and 2 chargino mass eigenstates. It has been pointed out that while such particles may be detected at hadron colliders for large mixing, a particularly clean final state is required to discover Higgsinos for small mixing which leads to mass degeneration. The result of quite a thorough analysis of a t-channel

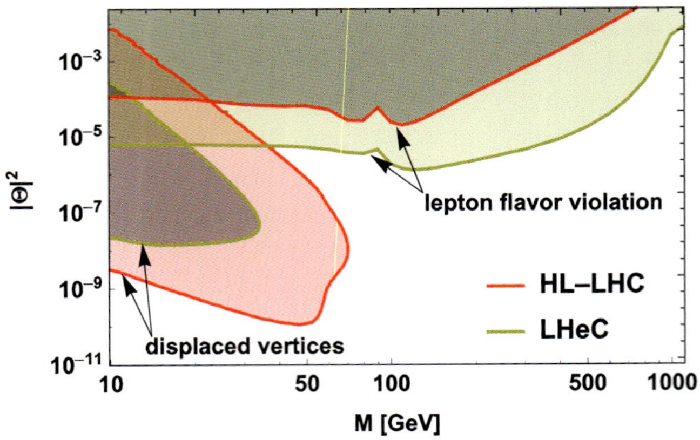

Fig. 15.9 Sensitivity of the HL LHC and the LHeC to the existence of sterile neutrinos in the plane of the active sterile mixing angle Θ and the heavy neutrino mass M. The left, lower mass sensitivity is due to displaced vertex sensitivity while the large mass sensitivity is due to lepton flavour violating multi-jet signatures [Antusch *et al.* (2017)].

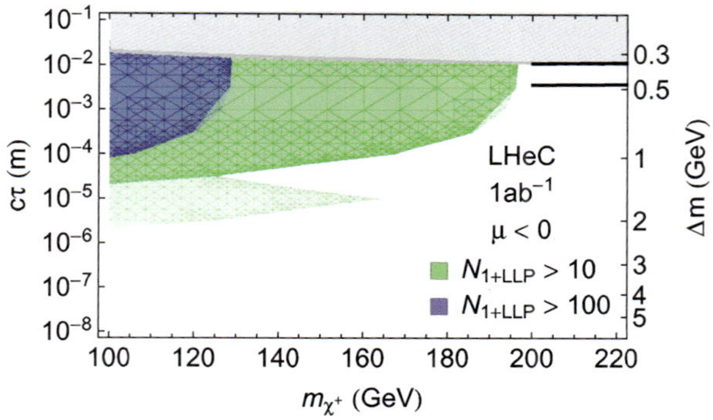

Fig. 15.10 Regions in the mass-lifetime Higgsino parameter plane accessible for LHeC events with at least one LLP. Black curves are projected bounds from searches at the HL LHC, from [Curtin *et al.* (2017)].

$VV \to \chi\chi$ pair production of supersymmetric particles χ from exchanged vector bosons V is reproduced in Fig. 15.10 which displays the number of observed events with at least one long lived particle (LLP) as a function of the chargino mass and lifetime. It can be seen that the LHeC is much more sensitive at small $c\tau$ than the LHC. This advantage holds similarly for exotic Higgs decays which can occur with sizeable branching fractions as is argued in [Curtin *et al.* (2017)] too.

A number of further analyses have been published or are being pursued, as are summarised in [Klein (2017b)], which demonstrate a very high and often unique potential opened by the LHeC for the discovery of exotic Higgs physics. These regard, for example, the exotic Higgs decays, as into dark matter with a branching sensitivity to 5 %, or into two new scalars [Liu *et al.* (2017)] which provides sensitivities which surpass the HL LHC very much. There are also LHeC related papers published on the top-Higgs coupling [Coleppa *et al.* (2017)] on the possible discovery of charged Higgs bosons [Azuelos *et al.* (2017)] and others. These and further ongoing studies lead the examination of BSM *ep* physics much beyond the level reached with the CDR on the LHeC in 2012. They strongly support the conclusion reached in [Curtin *et al.* (2017)] that adding a high luminosity, high-energy DIS programme to a *pp* collider "is necessary to fully exploit its discovery potential for new physics" [Curtin *et al.* (2017)].

The potential to discover new physics at the LHeC reaches far beyond the Higgs sector and its exotic extension. From very high precision measurements of the structure functions at large x new sensitivity is reached to contact interactions into the O(100) TeV region, subject to assumptions on their coupling structure. An analysis of the potential to exploit the general Lagrangian structure as embedded in EFTs has not been done yet for *ep* even though 1053 out of 2499 parameters characterising deviations from the SM are related to lepton-quark interactions [Trott (2014)].

While HERA had too low energies to study top quark physics, the LHeC is actually a top quark factory with single t and $t\bar{t}$ production cross sections of ~ 5 and $\sim 0.02\,\text{pb}$, respectively. This permits high precision measurements and unique searches for new physics in many areas, such as the anomalous $t-q-\gamma$, $t-g-Z$ and triple gauge [Cakir (2017)] couplings, the top contribution to the proton's momentum balance (top PDF), anomalous top-Higgs couplings [Sun and Wang (2016)] and CP nature [Coleppa *et al.* (2017)] and others [Sun (2017); Schwanenberger (2017)]. The prospects for top *ep* physics are a particularly strongly developing field with many interesting results supporting the physics case of the LHeC.

New physics can certainly be expected to be discovered in the strong interaction sector. QCD is a very rich and largely unexplored theory regarding, besides the confinement puzzle cited above, for example i) the AdS/CFT duality, which links supersymmetric modifications of QCD to string theory in anti-de-Sitter space or ii) the instanton fluctuations of the gluon field. The LHeC may also discover odderons, a saturation of the gluon density at small x, or a breaking of factorisation, as has been discussed in detail in [Forte (2012)]. It should clarify the question why all hadron-hadron interactions of protons and ions, AA', pA and pp, exhibit collective phenomena in the expected contrast to *ep*. The LHeC will push the evolution of QCD much forward, such as the higher order (N^3LO and N^4LO) calculations in pQCD, or the physics of non-conventional PDFs. In a list of possible QCD discoveries, presented by C. Quigg with "the unreasonable effectiveness of the SM" at the DIS workshop in 2013 [Quigg (2013)], the existence of free quarks, unconfined colour, quark substructure and the embedding of QCD in a higher symmetric theory are included, the discovery of each/any of which would be of far-reaching consequences for particle physics. There are strong, accumulating reasons to believe that the LHeC, as the second highest energy, high luminosity, asymmetric collider should indeed be among the main configurations for leading beyond the Standard Model.

15.9 LHeC as an EIC

The prime task of an electron-ion collider at high energies, besides a detailed mapping of nuclear dynamics and structure, is to investigate the QCD dynamics at high parton densities. These are observed at small Bjorken x and may be amplified in a nuclear medium.

15.9.1 *Low x physics*

... subleading terms cannot be ignored and must also be resummed. Their effect is to push the onset of the truly asymptotic regime down to smaller values of x, while in an intermediate region, relevant for HERA data, the evolution shows a shallow dip, somewhat lower than the NLO perturbative result. The resummation procedure is determined by solid guiding principles from duality, momentum conservation, symmetry under gluon exchange of the BFKL kernel and accurate implementation

of running coupling effects [Altarelli *et al.* (2008)]. Deep inelastic scattering is the interaction of leptons and hadrons at four-momentum squared values exceeding the squared mass of the proton, $Q^2 > M_p^2$. In this region perturbative QCD may be applied as $\alpha_s(Q^2) \simeq 1/(\beta_0 \ln (Q^2/\Lambda^2)) < 1$, where $\beta_0 = (33 - 2N_f)/12\pi$ is the one-loop renormalisation group equation coefficient, N_f the number of active flavours and $\Lambda \simeq 200\,\text{MeV}$ the famous parameter which quantifies the value of the strong coupling, which is about 0.5 at $Q^2 = 1\,\text{GeV}^2$. Bjorken x is related to Q^2 via $x = Q^2/sy$, where the inelasticity y is related to the energy transfer $(E_e - E')/E_e$ and limited between 0 and 1. The smallest value of x any collider may reach in the DIS region is therefore $x_{min} = 1/s$ with $s = 4E_eE_p$ in GeV^2. There is a sequence of minimum x values of $\log x_{min} \simeq (-2, -3, -4, -5, -6, -7)$ for fixed target DIS, the two US EIC versions, HERA, the LHeC and FCC-eh, respectively. One would call low x the region below $x \simeq 10^{-4}$ down to which the DGLAP equations have been observed at HERA to approximately hold. Low x physics thus is a prime area of research for the LHeC, and for future higher energy DIS colliders under study for CERN and China.

A striking discovery at HERA was the rise of the quark and gluon distributions towards small x. It has been argued that unitarity should tame that rise through non-linear parton-parton interactions [Gribov *et al.* (1983)], predominantly gluons because at small x the gluon density is much larger than that of sea quarks. This observation holds only for large Q^2, in excess of $\simeq 10\,\text{GeV}^2$, because xg turns valence-like at $Q^2 \sim 1\,\text{GeV}^2$.[6] The key theme of the physics at small x is therefore the understanding of parton interaction dynamics at high densities and for small α_s, which is an unexplored dynamic state of matter. The key experimental challenge is the discovery of non-linear effects, often synonymously denoted as BFKL dynamics, which would end the region of validity of the linear DGLAP evolution equations. This is not only a question of theoretical importance but of very practical relevance for the LHC and higher-energy hadron colliders: the gluon cannot be directly observed, its distribution is defined in a theoretical framework. If that changes fundamentally, the PDFs change and the cross-section calculations alter. Note that in Drell–Yan scattering of two partons a state of mass M is produced with a rapidity η related as $\sqrt{x_1x_2s} = Mexp(\pm\eta)$. For Higgs production the central (very forward) rapidity value corresponds to an average $x = \sqrt{x_1x_2}$ of 0.009 (0.0002). These x values are lower by a factor of 7 for the FCC. The hadron colliders already sample very small x values for the Higgs but even more so for lighter states. The clarification of the low x dynamics with the LHeC is therefore crucial for the HL LHC and a necessity for the higher-energy colliders. It is also of much relevance for the understanding of ultra high-energy neutrino physics. These tasks need DIS energies higher than HERA's and not lower.

[6]That observation is key because it enlarges the effective x_{min} values by an order of magnitude such that LHeC will investigate the domain down to $x \simeq 10^{-5}$ and eRHIC to only $\simeq 10^{-3}$ which has already been explored with high precision at HERA.

The decisive test for BFKL dynamics in DIS proceeds through high precision measurements of the salient structure functions F_2 and F_L as has been demonstrated in the LHeC CDR [Fernandez and the LHeC Study Group (2012)]. This holds because these functions are most cleanly theoretically defined. Both, however, are needed because if one had only F_2 one could attribute changes in the evolution law too easily to parametric peculiarities of the collinear PDFs. This causes a huge challenge for a precision measurement of F_L at small x which is discussed in the Appendix.

The establishment of BFKL dynamics at small x would be convincing only if the whole dynamics in this range was consistently described. Consequently there are many other processes under study for the observation of non-linear parton interactions. One example, advocated at a recent LHeC workshop by Mark Strikman, is the production of a vector meson with a rapidity gap in virtual photon-hadron scattering [Strikman (2015)]. This reaction is viewed as small dipole elastic scattering at large t and large energies, small x, and may lead to an observation of BFKL dynamics. Pursued also in electron-ion scattering it will permit the study of the effect of nuclear media with regard to the regulation of shadowing and the eventually possible observation of the black disc regime.

As an experimentalist one admires the considerable theoretical development of low x QCD, indicated also by Guido's remarks cited above, and must admit that the information we have been providing is not enough, HERA being too low in energy and too weak in precision and the LHC being plagued by its nature, the colour connections in the hh interaction. The LHeC, with much less luminosity than $1\,\text{ab}^{-1}$, will change that completely and has, as one of its major and unique tasks, the clarification of the parton dynamics at small x.

15.9.2 *Nuclear parton dynamics and densities*

Electron-ion collisions provide rich perspectives for nuclear chromodynamics. This holds for the low-energy EIC collider [Accardi *et al.* (2016b)] which may reach cms energies of $\leq 100\,\text{GeV}$ as compared to $\simeq 20\,\text{GeV}$ by the CERN muon-hadron scattering experiment NMC. It holds as well for the LHeC reaching the TeV energy scale. A difference, apart from the kinematic reach, is the dedication the EIC may put to electron-ion collisions which at CERN would most likely remain in the shadow of the pp programme. However, recent evaluations of the ePb luminosity at the LHeC, presented in [Brüning, O and others (2017)], result in an estimate of $L = 7\,10^{32}\,\text{cm}^{-2}\text{s}^{-1}$. One can thus collect annually about $1\,\text{fb}^{-1}$ of integrated luminosity which at low Q^2 is ample, also for semi-inclusive analyses, and implies that a complete DIS programme in electron-ion scattering is ahead including, for example, high Q^2 charged current physics and providing $\text{O}(10^3)$ Higgs events emerging from electron-lead scattering.

A recent summary of novel perspectives for nuclear physics includes topics "such as the hidden colour of nuclear form factors, the relation of the nuclear force at short distances to quark interchange interactions, the effects of colour transparency on the baryon-to-meson anomaly in hard heavy-ion collisions, novel exotic multi-quark states, the anomalous nuclear dependence of quarkonium hadroproduction, flavour-dependent antishadowing, and the breakdown of sum rules for nuclear structure functions. I also briefly discuss the insights into hadron physics and colour confinement that one obtains from light-front holography, including supersymmetric features of the hadron spectrum" [Brodsky (2017)], too detailed to be discussed in this contribution. It may suffice to state gratefully that Stan Brodsky has had quite an influence on shaping the LHeC programme and supporting its development as a member of the IAC and in his outstanding capacity as a world leading theorist.

For the search for non-linear parton interactions one often cites the expectation, especially when gluon saturation is discussed in the US EIC context, that the gluon density in a nucleus A is amplified by a factor $A^{1/3}$, which may only work if the underlying proton-parton density is large, i.e. for xg for $Q^2 \geq 10\,\mathrm{GeV}^2$ as mentioned above. Moreover, that amplification may be tamed by flavour dependent shadowing effects. There are, principally, two tasks which require independent input: the exploration of BFKL effects and the clarification of nuclear medium and binding effects. It therefore is strongly believed that one needs higher energies than HERA's in order to probe for non-linear evolution in ep and then to compare the result with eA. The high luminosity in eA at the LHeC and the large kinematic coverage enable the derivation of the flavour decomposed light sea quarks, the gluon and direct measurements of the strange, charm and beauty distribution in nuclei available at the LHC. This programme is being investigated, for a summary see [Armesto (2017a)]. It will completely change the way nuclear PDFs are obtained, end the use of π_0 data, for example, and deliver nPDFs free of any pPDF base. One then measures directly the ratio $R = f_A/A \cdot f_p$, flavour by flavour and for x between a few times 10^{-6} up to 1. This is in contrast to the present method in which R is evaluated using certain set of proton PDFs and a theoretical parameterisation of nuclear effects, with reliable data available only for $x > 0.01$. This is a most exciting prospect as has been outlined in [Klein (2016b); Helenius *et al.* (2016)].

The study of eA collisions in the TeV regime at the LHeC would be crucial for the development of nuclear QCD and also for the interpretation of the data coming from the heavy-ion programme. Its importance would be threefold [Armesto (2017b)]: first, it would greatly reduce our lack of knowledge on nuclear parton densities, that for some observables causes uncertainties of the same size as the eventual signal of high temperature, deconfinement effects. Second, it would establish the QCD dynamics of relevance for the LHC kinematic domain, were it the traditional fixed order perturbative one or a new non-linear perturbative regime. Finally, it would provide theoretically based initial conditions for the macroscopic evolution of the system and contribute to understanding the emergence of the hydrodynamical macroscopic behaviour from the QCD microscopic dynamics.

15.10 Concluding remarks

The future of deep inelastic scattering at high energies can be bright. The opportunity for founding and opening a new laboratory for energy frontier deep inelastic scattering which the LHC provides is indeed unique. It should not be disregarded, nor should the huge LHC investment be thrown away. The LHeC is a one-in-a-few-decades opportunity. Its science programme is unique and rich in its five components sketched here which all are dynamically advancing as a new generation of theorists appears and the LHC physics becomes clearer. The huge step in energy and luminosity, which the LHeC represents, will almost certainly entail surprises, unknown unknowns. It is an important step to the development and exploitation of modern, low power, superconducting accelerator technology as well. The detector represents a challenge many may wish to take up as the design of the HL LHC detector upgrades passes. The LHeC is far more than a mere step for the future of DIS because *ep* and *eA* scattering at TeV energies are recognised to be intimate parts of modern particle and nuclear physics in general. This is underpinned by the observation and design that the *ep* and *pp* collisions at the LHC may be recorded synchronously.

The next generation of machines at the energy frontier should comprise the LHeC for it is the only realistic option for energy frontier *eh* in the coming decades. Not only will the structure of matter be explored deeper, new particles may also be discovered and new dynamics be found. The key carriers of hope and study of modern collider physics, the W, Z bosons, the top quark and the Higgs ought to and can be investigated to sub-percent precision in *ep* too. A future big step in *pp* collider energy requires a big step deeper into the dynamics inside the proton.

In order to proceed, particle physics must utilise its rare opportunities; the LHeC is a particularly attractive one, also for CERN's future. It can be realised prior to the next, higher-energy *pp* collider, most likely the HE LHC, which requires more time, more technology developments and substantially larger funds than the LHeC needs. As Guido had put it there were *the physics case of the LHeC is made, can we proceed to build: it* [Altarelli (2015a)]. So sadly did he say the last words which many of us heard, in his unforgettable style. It is time, 50 years after the discovery of quarks at Stanford, to open a new chapter for high-energy electron-hadron scattering for which the LHC is the single and luminous base. The linac to be built is shorter than the one Panofsky and colleagues mastered to build at SLAC.

Acknowledgement

The development of the LHeC proceeds because it has been supported by so many exceptional young and elder colleagues whom I just cannot mention here. I want to thank Oliver Brüning for a decade of pleasant and fruitful co-coordination, the CERN directorates under Rolf Heuer and Fabiola Gianotti for their supportive attitude, the various active convenors on LHeC subjects, the FCC coordinators Michael Benedikt and Frank Zimmermann, and the members of the International Advisory Committee led by Herwig Schopper for their encouraging and critical evaluation of the progress. The recent move towards the PERLE facility at Orsay would not have been possible without the conviction of Achille Stocchi, shared with many of us and not least the admirable teams of accelerator scientists at CERN, Jlab and other institutes. For this article I enjoyed special help from Claire Gwenlan, Uta Klein, Johannes Blümlein, Peter Kostka, Oliver Fischer and Nestor Armesto.

Appendix 1: The strong coupling constant

There was hardly a talk on deep inelastic scattering where Guido Altarelli would not speak about the importance and difficulty of determining the strong coupling constant. It is indeed amazing to see that we have been working on the determination of α_s, or $\Lambda_{\overline{MS}}$, since the times of the CDHS and BCDMS and other lepton-hadron scattering experiments, while others worked on e^+e^- or pp, that is since 40 years, and yet we are not satisfied with the net result. In Guido's words: *The crucial ingredient of perturbative QCD is the "running" coupling $\alpha_s(Q^2)$... Some particularly simple and clear hard processes must be used to measure the running coupling at a suitable energy scale .. The cleanest processes for measuring α_s are the totally inclusive ones (no hadronic corrections) with light-cone dominance, like Z decay, scaling violations in DIS and perhaps τ decay (but, for τ , the energy scale is dangerously low)... In principle DIS is expected to be an ideal laboratory for the determination of α_s but in practice the outcome is still to some extent open* [Altarelli (2013b)].

Status

The coupling constant of the strong interaction, α_s, is much less well known than the fine-structure constant α and the Fermi constant G_F. Its accurate and more precise determination is a key challenge for experimental and theoretical particle physics, and it is of fundamental importance. This has a number of reasons:

- Approximately, within certain supersymmetric theories, the three couplings approach a point of convergence at the Planck scale. It is the large uncertainty of α_s which prevents strong conclusions drawn about this unification point as is illustrated in Fig. 15.11. This sets a goal to reach an accuracy ten times higher with the LHeC, see below;
- the current average of various, different determinations of α_s is not satisfying, which is discussed below and indicated by Guido's remark above. A single, or two independent, consistent measurements of α_s at per mille precision are the obvious, though difficult task ahead. In particular, future ep and e^+e^- collider experiments must challenge the lattice results, which also will improve;
- deep inelastic scattering has a special role in these determinations. The theory of inclusive DIS is clean, free of hadronic final state phenomenology. In the past

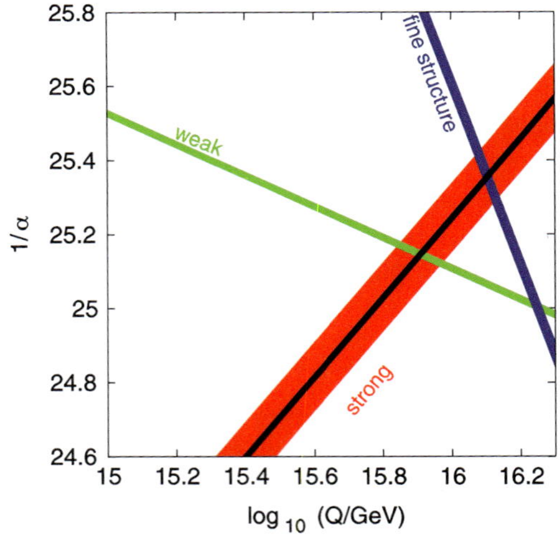

Fig. 15.11 Extrapolation of the inverse coupling constants to the Planck scale, illustrated with a version of MMSM, from [Abelleira Fernandez *et al.* (2012b)]. The coloured bands represent uncertainties as quoted by PDG in 2012, while the black insert to the red band on α_s illustrates the projected LHeC uncertainty, solely from inclusive DIS. The uncertainty of α_s quoted by PDG has recently increased by about 50 %.

it led to peculiarities, due to lack of calibration redundance (BCDMS), smallness of Q^2 which caused larger power (higher twist) corrections and nuclear and deuteron binding corrections. The most precise DIS data for α_s, from BCDMS, point to a too low value. For long a tendency had been observed for jet analyses to provide larger values on α_s than inclusive ones. All this will be overcome with the LHeC. At that time, DIS theory will have been calculated to N^{3}LO, and a new level of testing QCD and its development may be near;

- with the LHC reaching high precision in various areas, the uncertainty of α_s becomes an obstacle. This holds in particular for the derivation of Higgs couplings at the LHC, because the dominant production process is through gluon-gluon fusion, that is the product of $\alpha_s \cdot xg$ enters squared. An uncertainty of $\alpha_s(M_Z^2)$ taken to be ± 0.0015, or ± 1.3 %, translates to a 2.6 % uncertainty on the cross section[7] as obtained in [Anastasiou *et al.* (2016)]. This assumed uncertainty does not account for the different result obtained by the ABM team, see below. One clearly wishes to determine PDFs and the strong coupling in a process suited to perform a genuine N^{3}LO pQCD analysis at a precision which should render effects from α_s and the parton distributions, discussed

[7]This is consistent with the result presented in [Brüning and Klein (2013)]: using the iHix code (v1.3) the Higgs production cross section in 14 TeV pp Drell–Yan scattering was calculated for NNPDF2.1 using two values of α_s which differed by 1.6 %. The cross section for 0.121 is larger than the one for 0.119 by about 3.6 %. One observes, as expected, that $\delta\sigma/\sigma \simeq 2\delta\alpha_s/\alpha_s$.

Table 15.2 Mean values of the strong coupling constant $\alpha_s(M_Z^2)$, to next-to-next-to-leading order and for five active flavours. [Patrignani *et al.* (2016)].

Method	$\alpha_s(M_Z^2)$
Lattice QCD	0.1184 ± 0.0012
τ-decays	0.1192 ± 0.0018
DIS	0.1156 ± 0.0021
Hadron Collider	0.1151 ± 0.0028
Electroweak Fits	0.1196 ± 0.0030
e^+e^-	0.1169 ± 0.0034

in Section 15.5.1, negligible. For Higgs physics at the LHC, this would be achieved with a few per mille accuracy. Similar conclusions one derives when studying the PDF-α_s sensitivity of a high precision measurement of the W boson mass [Aaboud *et al.* (2018)] or of the weak mixing angle at the LHC.

There exists a vast amount of literature on α_s due to its fundamental importance, see e.g. [Deur *et al.* (2016); Dissertori (2016); d'Enterria (2015)]. The current status of the determination of α_s has recently been summarised [Bethke (2017)] and is reproduced in Table 15.2 using the values as quoted in the PDG2016 [Patrignani *et al.* (2016)].

A mean value has been derived of $\alpha_s(M_Z^2) = 0.1181 \pm 0.0011$ that obviously is dominated by the result of lattice calculations. These have recently been discussed in quite some detail [Aoki *et al.* (2017)]. In the lattice calculations the role of a measured cross section is taken by suitably defined Euclidean short distance quantities. Lattice calculations have a number of additional, common peculiarities, they need input of the experimental hadronic spectrum and quark masses, they treat only light quarks with perturbative, matching additions of charm and beauty quark effects and they have uncertainties from discretisation and truncation of perturbative theory. There follows quite a range in the resulting α_s values obtained, beyond the simple value of uncertainty quoted, which is achieved by implementing certain quality criteria of the theoretical treatments as are presented in [Aoki *et al.* (2017)]. From a personal point of view one wished to keep the lattice results separate as PDG also does in providing an average of

$$\alpha_s(M_Z^2) = 0.1174 \pm 0.0016 \tag{15.3}$$

for all non-lattice results contained in Table 15.2. The next accurate result is that from τ-decays, which, however, as remarked by Guido above, suffers from the small scale, i.e. largeness of $\alpha_s(\mu)$ at $\mu = M_\tau$ such that fixed order or contour improved perturbation theory lead to 7 % different results, at M_τ, and non-perturbative contributions to similarly sizeable uncertainties [Patrignani *et al.* (2016)]. The next precise result is from deep inelastic scattering which has a number of partially well known peculiarities the LHeC will eventually resolve.

Deep inelastic scattering on α_s — now and with the LHeC

The status of the determination of α_s in DIS is well summarised in Fig. 15.12 taken from a recent comprehensive paper [Alekhin *et al.* (2017)] by the ABMP team. A big discrepancy is observed, of the large $\alpha_s \simeq 0.121$ value obtained with the SLAC data, at low Q^2 and large x, and the BCDMS result of $\alpha_s \simeq 0.11$. The cure to that observation had often been, as in [Alekhin *et al.* (2017)] too, the introduction of a higher twist correction for F_2 which, as a power term $\propto Q^{-4}$, obviously only affects the SLAC result beyond the uncertainty, see Fig. 15.12. The fit to all DIS data by ABMP gives an NNLO average of

$$\alpha_s = 0.1140 \pm 0.0009 \tag{15.4}$$

employing HT corrections and for $N_f = 5$, which is drawn as the red shaded band for the "HERA+fixed target data"-fit. The ABM result thus is (again) lower than the PDG average but consistent with the DIS value quoted from the PDG in Table 15.2. The overall NNLO value of α_s the ABMP group obtains, including hadron collider data, is 0.1147 ± 0.0008. This may be compared, as was done in [Alekhin *et al.* (2017)], with the MSTW value of 0.1172 ± 0.0013 or NNPDF of 0.1174 ± 0.0007.

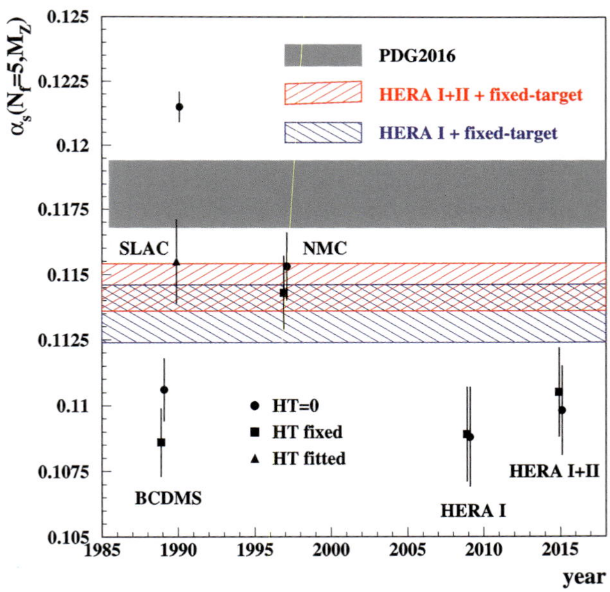

Fig. 15.12 Determinations of α_s by the ABMP group [Alekhin *et al.* (2017)]. One observes quite some spread and a final result (red dashed) below the PDG average value, which includes DIS. Results are shown with the higher twist terms set to zero (circles), HT terms fixed to the values obtained in the ABMP16 fit from considering all data sets (squares) and when fitting the HT terms to the individual data sets (triangles). The bands are obtained by using the combination of the SLAC, BCDMS and NMC samples together with the combined H1 and ZEUS data from HERA I (left-tilted hatches, blue) and from HERA I+II (right-tilted hatches, red).

There have been quite some disputes over time about such a low value, the results of which Guido, above, called *to some extent still open*. It is in my view remarkable that an in depth analysis of the fixed target DIS (SLAC, NMC, BCDMS) and then available data by H1 came to similar conclusions as ABMP recently. The H1 Collaboration developed a dedicated procedure [Adloff *et al.* (2001)] for the determination of α_s which reduced the number of parton distributions to just two for quarks, owing to the observation that F_2 in the proton has a singlet, Σ, and a non-singlet, Δ, component,

$$F_2 = \frac{2}{9} \cdot x\Sigma + \frac{1}{3} \cdot x\Delta = \frac{1}{3} \cdot xV + \frac{11}{9} \cdot xA. \tag{15.5}$$

This introduced two functions V and A, which approximately are valence and sea dominated, respectively. One therefore hoped to be able to reduce the uncertainty on α_s by reducing the PDF correlations to a minimum. Similar observations to Fig. 15.12 were made and in much detail described in [Wallny (2001)]. The cure, however, was chosen to be different: because an unpleasant Q^2_{min} dependence of α_s was observed, and to avoid HT terms, a cut of $Q^2 \geq 7.5\,\text{GeV}^2$ was implemented for the fixed target data. That was built into the bulk BCDMS data anyhow, as the detector had originally left space for a calorimeter to detect CC scattering which was never put in place. It was then seen that the NMC data had no further effect on α_s, due to the more precise BCDMS data. Moreover, inconsistencies were observed and studied at length, see [Wallny (2001)], of the slopes in Q^2 at high x measured by SLAC and BCDMS. This caused the introduction of a cut for the inelasticity y_{BCDMS} to be larger than 0.3. The result was that of full consistency, in the so constrained phase space, such that a $\chi^2 + 1$ criterion was applicable which rarely is the case in global PDF and α_s analyses using unpleasant "tolerance criteria" to screen the incompatibilities of data used. In 2001, H1 obtained an NLO α_s value of

$$\alpha_s(M_Z^2) = 0.1150 \quad \pm \quad 0.0017\,(exp) \ {}^{+\,0.0009}_{-\,0.0005}\,(model), \tag{15.6}$$

which to NNLO pQCD would provide a lower number, probably close to 0.114 as ABMP indeed now obtained. Here, the first error represents the experimental uncertainty on α_s of 1.5 %. The second, lower error includes all uncertainties associated with the construction of the QCD model for the measured cross section. A number to be noted may be that a 100 MeV uncertainty on the value of the charm mass results in an α_s uncertainty of ± 0.0005. A programme to determine α_s to per mille precision thus requires the value of M_c to be fixed to be better than 10 MeV.

When we on H1 designed our QCD fit strategy, was it in 1990, I recall that the admirable Willy van Neerven advised to keep inclusive and jet data QCD analyses separate. It could be, he argued convincingly, that the gluon determined in inclusive DIS told us something when compared to the jet data result, which obviously concerned α_s as well. Therefore, putting it all together was avoided for a long time. In 2017, H1 published a comprehensive NNLO analysis of jet and all the inclusive data [Andreev *et al.* (2017)]. It led to the interesting result, for jets only

Table 15.3 Results of NLO QCD fits to HERA data (top, without and with jets) to the simulated LHeC data alone and to their combination, for details of the fit see [Fernandez and the LHeC Study Group (2012)]. The resulting uncertainty includes all the statistical and experimental systematic error sources taking their correlations into account. The LHeC result does not include jet data.

Case	Cut [Q^2 (GeV^2)]	Uncertainty	Relative precision (%)
HERA only	$Q^2 > 3.5$	0.00224	1.94
HERA+jets	$Q^2 > 3.5$	0.00099	0.82
LHeC only	$Q^2 > 3.5$	0.00020	0.17
LHeC+HERA	$Q^2 > 3.5$	0.00013	0.11
LHeC+HERA	$Q^2 > 7.0$	0.00024	0.20
LHeC+HERA	$Q^2 > 10.$	0.00030	0.26

and to NNLO, of

$$\alpha_s(M_Z^2) = 0.1157 \pm 0.0020\,(exp) \pm 0.0029\,(thy). \tag{15.7}$$

A joint fit to jet and inclusive DIS H1 data led to $\alpha_s = 0.1142 \pm 0.0028$ (tot), in remarkable agreement with ABMP. There again is a tendency to be noted that α_s from jets is somewhat higher than from inclusive DIS. This is about as far as HERA will lead and so we studied comprehensively the potential to obtain α_s from simulated LHeC data.

Two independent simulations and fit approaches have been undertaken in order to verify the potential of the LHeC to determine α_s, see the CDR [Fernandez and the LHeC Study Group (2012)] for details. Table 15.3 summarises the main results. It can be seen that the total experimental uncertainty on α_s is 0.2 % from the LHeC and 0.1 % when combined with HERA. This determination is free of higher twist, hadronic and nuclear corrections relying solely on inclusive DIS *ep* data at high Q^2. There are known further parametric uncertainties in DIS determinations of α_s. These can safely be expected to be much reduced also by the LHeC, which promises to determine the charm mass, for example, by 3 MeV, as compared to 40 MeV at HERA, corresponding to an α_s uncertainty of 0.04 %. Matching the experimental uncertainty requires that when the LHeC operates such analyses were performed in N^{3}LO pQCD in order to reduce the renormalisation and factorisation scale uncertainty. Due to the huge range in Q^2 and the high precision of the data, new decisive tests will also become available for answering the question whether the strong coupling determined with jets and in inclusive DIS are the same. If confirmed, as is demonstrated in Table 15.3 with the HERA data, a joint inclusive and jet analysis has the potential to even further reduce the uncertainty of α_s derived from future LHeC data. The ambition to measure α_s to per mille accuracy therefore represents a vision for a renaissance of the physics of deep inelastic scattering, which is a major goal of the whole LHeC enterprise.

Appendix 2: The longitudinal structure function and R

In 2012, Guido Altarelli published an article [Altarelli (2012)], in honour of Mario Greco, about "The Early Days of QCD (as seen from Rome)" in which he describes one of his main achievements, and long term irritation, regarding the so-called longitudinal structure function, F_L, and its measurement. He writes: *Back to Rome I met Guido Martinelli, then a post-doc with a contract for doing accelerator physics at Frascati, and I rescued him into particle physics, with a work on the transverse momentum distributions for jets in lepto-production final states [Altarelli and Martinelli (1978)]. In the same paper we derived an elegant formula for the longitudinal structure function F_L, also an effect of order $\alpha_s(Q^2)$, as a convolution integral over $F_2(x, Q^2)$ and the gluon density $g(x, Q^2)$. I find it surprising that it took 40 years since the start of deep inelastic scattering experiments to get meaningful data on the longitudinal structure function. The present data, recently obtained by the H1 experiment at DESY, are in agreement with this LO QCD prediction but the accuracy of the test is still far from being satisfactory for such a basic quantity.*

The measurement of F_L concerned me for more than twenty years, from first studies of the sensitivity at HERA in 1988 to the first announcement of results at in Philadelphia ICHEP 2008 [Klein (2008)] and its final publication. Why is F_L a quantity of special interest, and why was its determination so difficult at HERA and how would LHeC surpass it?

The inclusive, deep inelastic electron-proton scattering cross section at low Q^2, written in its reduced form,

$$\sigma_r = \frac{d^2\sigma}{dxdQ^2} \cdot \frac{Q^4 x}{2\pi\alpha^2 Y_+} = F_2(x, Q^2) - \frac{y^2}{Y_+} \cdot F_L(x, Q^2), \tag{15.8}$$

is defined by two proton structure functions, F_2 and F_L, where Q^2 is the negative four-momentum transfer squared, y the inelasticity $y = Q^2/sx \leq 1$, $Y_+ = 1+(1-y)^2$ and α is the fine structure constant. The cross section may also be expressed [Hand *et al.* (1963)] as a sum of two contributions, $\sigma_r \propto (\sigma_T + \epsilon\sigma_L)$, referring to the transverse and longitudinal polarisation state of the exchanged boson, with ϵ characterising the ratio of the longitudinal to the transverse polarisation. The ratio of the longitudinal to transverse cross sections is termed

$$R(x, Q^2) = \frac{\sigma_L}{\sigma_T} = \frac{F_L}{F_2 - F_L}, \tag{15.9}$$

which is related to F_2 and F_L as given above. Due to the positivity of the cross sections $\sigma_{L,T}$ one observes that $F_L \leq F_2$. The reduced cross section σ_r, for Q^2 much below the weak boson masses, is therefore a direct measure of F_2. That holds apart from a limited region of high y where a contribution of F_L may be sizeable. The principal task to measure F_L thus requires precisely measuring the inclusive DIS cross section near to $y = 1$, and then to disentangle the two structure functions by utilising the y^2/Y_+ variation which depends on x, Q^2 and s, that is one may vary s, and obtains F_2 and F_L at a common, fixed point of x, Q^2. That is challenging not only because the F_L part is small but because $y = 1 - E'/E_e$ is large only for scattered electron energies E' much smaller than the electron beam energy E_e, e.g. $E' = 2.7\,\text{GeV}$ for $y = 0.9$ at HERA. In the region where E' is a few GeV only, the electron identification becomes a major problem and the electromagnetic ($\pi^0 \to \gamma\gamma$) and hadronic background, mainly from unrecognised photo-production, rises enormously.

The interest in a measurement of F_L, especially at low x, is related to the uncertainty in the determination of the gluon distribution, $xg(x, Q^2)$, in pQCD fit extractions of the quark and gluon densities, besides α_s. On more theoretical grounds, there is an expectation that at small x the linear DGLAP Q^2 evolution equations, which give xg a certain meaning, may eventually have to be replaced by a BFKL type of law. The gluon density is defined in a theoretical framework which needs to be confirmed or disproven which on the basis of only F_2 is just not possible. With F_2 and F_L, however, one has two principal handles for an extraction of xg and the main motivation for our proposal [H1 Collaboration (2005)] to operate HERA at lower energies was to test higher-order QCD at small x.

The theoretically most sound and experimentally most precise determination of the gluon density xg is through the measurement of the Q^2 dependence of F_2 in DIS[8]. Their relation is prescribed by the DGLAP equations

$$\frac{\partial F_2(x, Q^2)}{\partial \ln Q^2} = \frac{\alpha_s(Q^2)}{2\pi} \int_x^1 dz \left[F_2(\frac{x}{z}) P_{qq}(z) + 2 \sum_{i=1}^{N_f} e_i^2 \cdot G(\frac{x}{z}) P_{qG}(z) \right] \qquad (15.10)$$

with the splitting functions P, $xg = G$, e_i the electric charge of quark i, and N_f the number of active flavours. At small x the first term is almost negligible. One thus finds that the measurement of the $\ln Q^2$ derivative of F_2 determines xg, i.e. one needs very high precision in $F_2(x, Q^2)$ combined with a large lever arm in Q^2 at each value of x. The extraction of the F_2 derivative had been evaluated in quite some detail by the H1 Collaboration when first precise measurements on F_2 became available, see [Adloff *et al.* (2001)]. It was also instrumental to the discovery [Abt *et al.* (1994)] of the rise of xg towards low x.

The second method to cleanly access xg is through the relation mentioned by Guido

$$F_L(x, Q^2) = \frac{\alpha_s}{\pi} x^2 \int_x^1 \frac{dz}{z^3} \left[\frac{4}{3} F_2(z, Q^2) + 2 \sum_i^{N_f} e_i^2 \cdot G(z, Q^2) \left(1 - \frac{x}{z}\right) \right] \qquad (15.11)$$

derived in [Altarelli and Martinelli (1978)], see also [Gluck *et al.* (1982)]. One can calculate that the integrand at small x is dominated by the gluon distribution term which directly relates F_L and xg.

The final measurement of F_L at HERA [Andreev *et al.* (2014)] is illustrated in Fig. 15.13. The data were obtained in an evolved separation procedure which principally determined F_L as the intercept, at $y = 0$, and F_L as the negative slope of

[8]This is transparent from the sudden increase in uncertainty for xg below $x \simeq 0.0005$, see Fig. 15.5, where HERA's access to $\partial F_2 / \partial \ln Q^2$ ends.

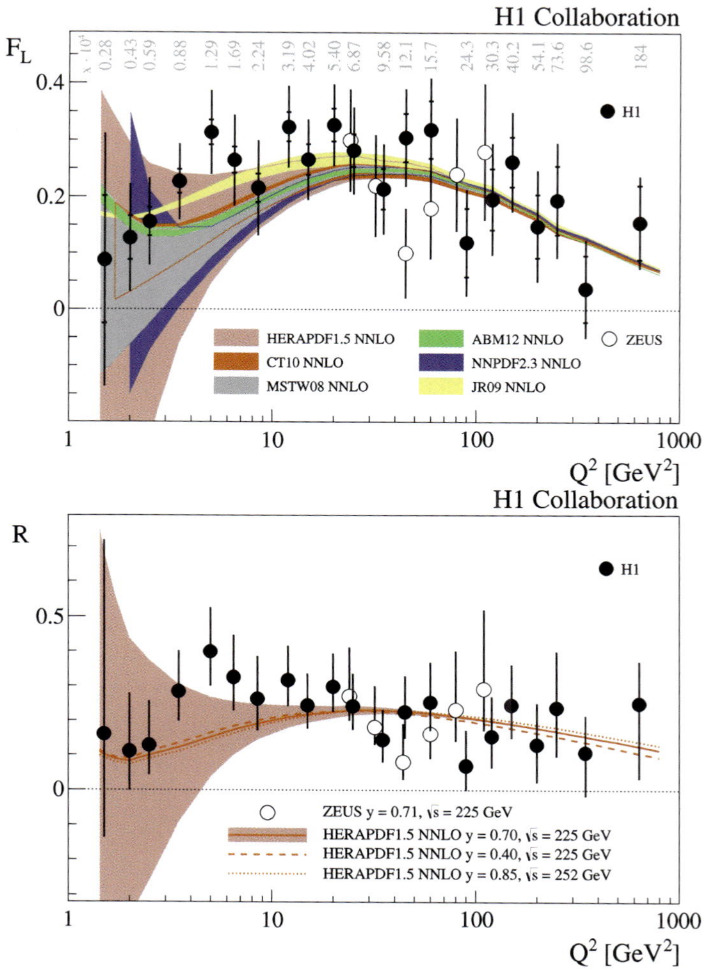

Fig. 15.13 Top: measurement of the structure function $F_L(x, Q^2)$ at H1 (solid points) and ZEUS (open circles), from a variation of proton beam energy in the final half year of HERA operation. The curves and their uncertainty bands illustrate predictions on F_L obtained to NNLO pQCD by various groups engaged in the extraction of PDFs. Bottom: measurement of the structure function ratio $R = F_L/(F_2 - F_L)$ at H1 (solid points) and ZEUS (open circles), from a variation of proton beam energy in the final half year of HERA operation. The curve represents an NNLO QCD fit analysis of the other HERA data. This becomes uncertain for Q^2 below $10\,\text{GeV}^2$ where the Q^2 dependence of F_2 does not permit an accurate determination of the gluon density which dominates the prediction on F_L, see text.

the linear dependence of σ_r on y^2/Y_+. The variation in y was achieved at HERA by comparing high statistics data at highest energy, $E_p = 920\,\text{GeV}$, with about $13\,\text{pb}^{-1}$ of data at $460\,\text{GeV}$ and $7\,\text{pb}^{-1}$ at $575\,\text{GeV}$. The values of E_p had been chosen for about equidistant separation of the measurements in y^2/Y_+. The low energy runs took place from March to June 2007, with only a few days of setup time for the machine. One notices that the absolute, total uncertainty of the H1 and ZEUS data is about ± 0.1, with notable statistical uncertainties for $Q^2 \geq 20\,\text{GeV}^2$. The ZEUS data do not extend to low Q^2 values, which could be accessed by H1 [Aaron *et al.* (2011a)] thanks to the existence of the Backward Silicon Tracker which my group had developed [Eick *et al.* (1997)], built and operated over a time of 15 years. A clear track-cluster link was required to identify low-energy scattered electrons near the beam pipe. The determination of the electron's charge, in a demanding, small radius curvature measurement, permitted in addition the removal of a large part of background with a residual account for charge asymmetry effects as are caused by the difference in energy depositions of anti-protons and protons at low energy. As is described in [Aaron *et al.* (2011a)] quite refined techniques had to be found and to work for accessing F_L.

In Fig. 15.13 the data are compared with predictions, calculated in NNLO pQCD, from various groups who perform comprehensive analyses of parton distribution functions. One observes a generally good agreement and recognises the origin for Guido Altarelli's remark cited above: the data are impressive, however, they are not precise enough for crucial tests of pQCD. In the specially interesting region of low x where the PDF uncertainties, mainly related to xg (see Section 15.5.1), are substantial, the data hint to a positive and regular F_L but lack discriminative power also. This can also be illustrated well with the measurement of σ_L/σ_T.

The first measurements of R at SLAC, in the range of Q^2 between 1 and about $10\,\text{GeV}^2$ and $0.1 \leq x \leq 0.5$, found the ratio to be about 0.18, i.e. small enough to be taken as evidence for quarks to carry spin 1/2. The HERA data determine this ratio as a function of Q^2, and correspondingly varying $x = Q^2/(4E_eE_py) \simeq Q^2/(4 \cdot 10^4 GeV^2)$, for $E_p = 575\,\text{GeV}$ and $y \simeq 0.7$. The result is displayed in Fig. 15.13. To good approximation, $R(x, Q^2)$ is a constant which was determined as $R = 0.23 \pm 0.04$, in astonishing good agreement with the SLAC values. The resulting value also obeys the condition, $R \leq 0.37$, that had been obtained in a rigorous attempt to derive the dipole model for inelastic DIS [Ewerz and Nachtmann (2007)].

The LHeC will lead to the first precision measurement of F_L and R. This is due to three main improvements: (i) the luminosity is expected to be so large that high statistics, fb^{-1} measurements can be performed at even more than two low beam energy points, compared with the two $10\,\text{pb}^{-1}$ sets at HERA; (ii) the apparatus in the backward region is being designed as a high precision device with more Silicon detector planes of higher acceptance and resolution and a hadronic backward calorimeter which was basically absent on H1; (iii) the increased electron beam energy implies that high y may be achieved at larger scattered electron energy E'. Both the improved detector and the enlarged E_e will enable highest y values to be reached at much reduced background.

A simulation had been performed for the LHeC CDR [Fernandez and the LHeC Study Group (2012)] which is illustrated in Fig. 15.14. In order to be conceptually independent of the LHC operation, for the LHeC the electron beam energy is lowered

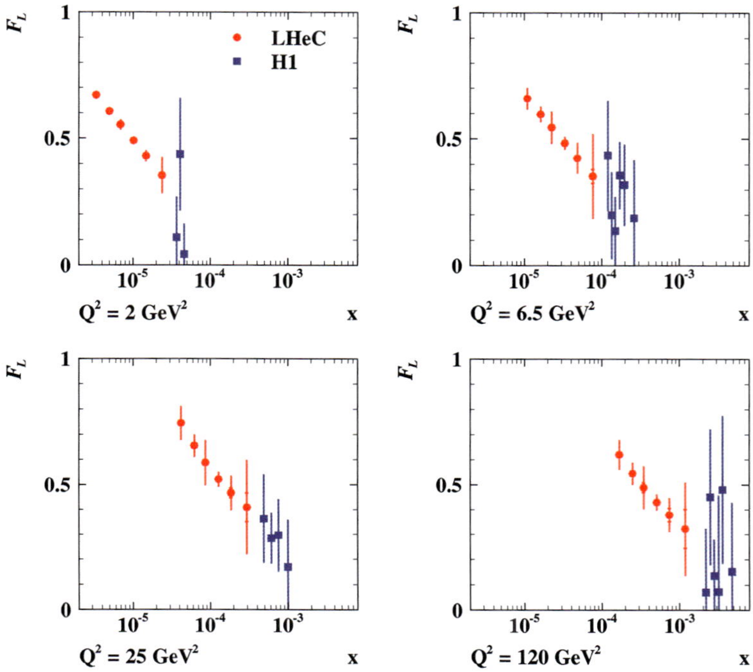

Fig. 15.14 Simulations of F_L measurements with the LHeC (red circles) compared with measurements at H1 (blue squares), see text.

as opposed to HERA. The point-by-point precision is impressively improved, from at best $\delta F_L \simeq \pm 0.1 - 0.2$ with H1 to typically a 0.02 total uncertainty for the LHeC. Based on the invaluable experience gained with H1 at HERA and on the design prospects for the LHeC and its *ep* experiment, one can indeed be optimistic that Guido Altarelli's wish for a precise determination of F_L will eventually be fulfilled. The simulated data, with their exceptional determinations of F_2 and F_L, were used in a study, presented in the CDR, to illustrate the unique potential in discriminating theory at small x.

Chapter 16

A future for very high-energy QCD studies

Allen Caldwell
Max Planck Institute for Physics
Munich, Germany

I had the great fortune to interact with Guido Altarelli on several occasions. As everyone knows, he was a leader in the theoretical physics community and a founder of the modern approach to understanding hadronic structure. His evolution equation for parton densities as a function of momentum scale [Altarelli and Parisi (1977); Gribov and Lipatov (1972a); Dokshitzer (1977)] will be forever enshrined in our textbooks on particle physics. As an experimental physicist active in the HERA electron-proton collider programme at the DESY laboratory, I had many opportunities to discuss with Guido a wide range of topics related to structure functions, diffractive scattering, and other topics in strong interactions. But there was also a second area where we discussed in depth, and that was related to an early fluctuation in HERA data that had us excited about the possibility of having seen leptoquarks, a new form of matter. Guido came to DESY and explained to us why our data was most likely not a sign of new physics, and he was right. His manner of explaining was at once completely understandable and also completely convincing, on this topic and in general. He was a real pleasure to be around and discuss with, and we miss him greatly.

In this article, I have decided to look to the future and to consider an option for a new round of electron-proton and electron-ion colliders based on a completely new technology — plasma wakefield acceleration. Guido was always forward looking and I believe, would have been a great supporter of this project. I dedicate this article to him.

16.1 Where is the new physics?

Now that the Higgs has been discovered, the question 'what are the most promising directions of research in particle physics?' is being posed with ever increasing urgency. We all know the list of 'known questions', such as 'what is the nature of dark matter and dark energy, what is the cause of the matter-antimatter imbalance, $\cdots$?' and there are surely many 'unknown questions' — the questions that we should be posing to discover new physics. J.D. Bjorken wrote in 1990: 'I never would have dreamed in the 1960s that things would have progressed to the highly orderly state we now have, where a theorist working within the Standard Model feels like an engineer, and one working beyond it feels like a crackpot.' [Bjorken (1991)]. If that was the case then, what about today? And what is the role of novel accelerator technology development in this context?

An oft-stated goal of novel accelerator technology is to provide a means to build a high-energy electron-positron linear collider capable of reaching high luminosity at TeV and beyond energy scales at a much lower cost than could be achieved with existing radio frequency based accelerating structures. While achieving this would indeed provide an exciting outlook for particle physics, I do not consider this as the primary or only goal of developing novel accelerator technology. Rather, the technology development should be seen as a generic way of providing research opportunities. As has often been the case in the history of physics, new technologies bring discoveries — often not along the lines for which they were initially developed. Novel accelerators based on plasma wakefields, dielectric structures or direct laser acceleration can bring scientific opportunities that reach far beyond colliders, and should therefore be pursued with the highest priority as enablers of discoveries.

A few examples of research directions that could be created by these technologies are studies of the non-linear regime of electromagnetism and electron-positron pair creation from the vacuum, studying scenarios for dark matter interactions in the very strong electromagnetic fields generated in wakefields or by lasers, or studying particle acceleration mechanisms in plasmas that could be relevant for understanding the generation of very high-energy cosmic ray particles. More conventional collider topics would be studying the nature of the strong force that binds quarks and gluons in nuclei in a new regime in very high-energy electron-proton collisions and searches for new forms of matter in single beam experiments. There are surely many more topics for new research directions for particle physics that could be opened up with the new technologies. And of course there are also the well known applications in photon science and medical applications.

So, in the absence of discoveries that point the way for particle physics, and in particular high-energy particle physics, it is important to push technology developments that can open new research directions, and to think broadly about the research topics that would become available. In particular, it is important to consider what physics can be probed by reaching much higher energies even if the luminosity is modest. Here, I will focus on a kind of 'super-HERA', with a kinematic reach that goes three orders of magnitude beyond that available at HERA. The material in the following borrows heavily from the publications [Caldwell *et al.* (2016)] and [Caldwell and Wing (2016b)].

16.2 Electron-proton and electron-ion physics

The HERA electron-proton accelerator was the first and so far only lepton-hadron collider worldwide. With its centre-of-mass energy of about 300 GeV, HERA dramatically extended the kinematic reach [Abramowicz *et al.* (2015)] for the deep inelastic scattering process compared to fixed-target experiments. A broad range of physics processes were studied and new insights were gleaned from HERA which complemented the $p\bar{p}$ and e^+e^- colliders, the Tevatron and LEP. The LHeC project [Newman and Stasto (2013); Abelleira Fernandez *et al.* (2012a)] is a proposed ep collider with significantly higher energy and luminosity than HERA with a program to investigate Higgs physics and QCD, to search for new physics, etc. This will use significant parts of the LHC infrastructure at CERN with different configurations, such as eA, also possible. In this article, the possibility of having a very high-energy electron-proton collider (VHEeP) is considered with an ep centre-of-mass energy of about 9 TeV, a factor of six higher than proposed for the LHeC and a factor of 30 higher than HERA.

The VHEeP machine would strongly rely on the use of the LHC beams and the technique of plasma wakefield acceleration to accelerate electrons to 3 TeV over relatively short distances. Given such an acceleration scheme, the luminosity will be relatively modest with $10-100\,\text{pb}^{-1}$ expected over the lifetime of the collider. With such an increase in centre-of-mass energy, the VHEeP collider will probe a new regime in deep inelastic scattering and QCD in general. The kinematic regime accessible will be extended by three orders of magnitude compared to that measured at HERA.

I begin with a brief review of the kinematics of deep inelastic scattering.

16.3 Deep inelastic scattering

Deep inelastic scattering (see, for example, [Devenish and Cooper–Sarkar (2004)]) can be classified as either neutral current (NC) or charged current (CC), depending on whether the exchanged boson is a photon or Z^0 boson (NC) or a $W^\pm$ boson (CC). The NC reaction, which dominates the cross section for the processes considered, in electron-proton collisions can be written as $e^-(k)+p(p)\to e^-(k')+X(p')$ with the four vectors given in the brackets and X referring to a hadronic final state. In the case of CC, the final state electron is replaced by a neutrino. The four-momentum of the exchanged boson, q, is given by $q = k - k'$. The events can be described by the following quantities: the squared centre-of-mass energy, s, is given by

$$s = (k + p)^2 \,; \tag{16.1}$$

the virtuality, Q^2, of the exchanged boson is given by

$$Q^2 = -q^2 \,; \tag{16.2}$$

the Bjorken-x variable, interpreted as the fraction of proton's momentum carried by the struck quark in the proton's infinite-momentum frame, is given by

$$x = \frac{Q^2}{2p \cdot q} \,; \tag{16.3}$$

the inelasticity, y, interpreted in the proton rest frame as the fraction of energy transferred from the lepton to the proton, is given by

$$y = \frac{p \cdot q}{p \cdot k} \,. \tag{16.4}$$

The above variables are related, $Q^2 = s\,x\,y$, where particle masses can be ignored at the very high energies considered here. Another quantity of importance is the exchanged-boson-proton centre-of-mass energy, W, which is given by

$$W^2 = (q + p)^2 \,. \tag{16.5}$$

The neutral current cross section for e^-p scattering, $\frac{d^2\sigma_{\rm NC}^{e^-p}}{dx\,dQ^2}$, can be written in terms of these variables and the structure functions of the proton as

$$\frac{d^2\sigma_{\rm NC}^{e^-p}}{dx\,dQ^2} = \frac{2\,\pi\,\alpha^2}{x\,Q^4}\left(Y_+F_2 - Y_-xF_3 - y^2F_{\rm L}\right), \tag{16.6}$$

where α is the fine structure constant and $Y_\pm = 1 \pm (1 - y)^2$. The structure function F_2 is sensitive to the quark and anti-quark distributions in the proton and dominates at low Q^2; the structure function xF_3 is sensitive to the difference in the quark and anti-quark distributions in the proton and arises due to the interference of the photon- and Z^0-exchange contributions at Q^2 values around the mass of the W and Z bosons. The longitudinal structure function $F_{\rm L}$ is sensitive to the gluon distribution in the proton and becomes important at high values of y.

The double-differential cross section can also be written in terms of the photon-proton cross section, $\sigma^{\gamma p}$, and the photon flux, ϕ. The equivalent photon approximation [Williams (1934); von Weizsäcker (1934); Gribov *et al.* (1962)] relates these as

$$\frac{d^2\sigma^{e^-p}}{dy\,dQ^2} = \phi(y, Q^2)\sigma^{\gamma p}(y, Q^2)\,, \tag{16.7}$$

where all quantities depend on both Q^2 and y.

The event kinematics are calculated using the initial beam energies and the kinematics of the scattered electron and hadronic final state. Of particular importance are the scattered electron energy, E'_e, polar angle, θ_e, and the hadronic angle, γ_{had}, where both angles are measured with respect to the proton beam direction.

16.4 Proton-driven plasma wakefield acceleration

Proton-driven plasma wakefield acceleration has been proposed as an approach capable of pushing the energy frontier for electron beams [Caldwell *et al.* (2009b)]. The significant advantage of proton bunches as drivers of plasma wakes is the much greater energy of the driver compared to laser or electron bunch drivers. This idea has generated considerable interest in the community and we have built a collaboration of scientists intent on carrying out a demonstration of the concept. A proposal for an experiment using the SPS proton beam at CERN, the AWAKE experiment [Assmann *et al.* (2014)], has been approved in August 2013, and first data taking took place in 2016. The experiment will run in three phases: the first phase will demonstrate the self-modulation of a long proton bunch in a plasma; the second phase will demonstrate the acceleration of electrons in the wake generated by the modulated proton bunch and a third phase will demonstrate scalability of the process.

Ionised gases, or plasmas, with densities over a thousand times lower than that of the atmosphere, can sustain accelerating gradients (several 10 GV/m) orders of magnitude larger than RF structures [Tajima and Dawson (1979)]. These large fields are due to the collective response of the plasma electrons to the electric field of a laser pulse or charged particle bunch driver. The plasma supports waves or wakes travelling at velocities near the speed of light, ideal to accelerate particles to relativistic energies. These wakefields have a longitudinal component able to extract energy from the driver and transfer it to a trailing witness bunch (of e.g. electrons). The wakes also have transverse wakefield components with focussing strength orders of magnitude larger than that of conventional magnets, allowing for the two beams to remain transversely small over long distances. This combination of large plasma fields and long confined propagation distances can lead to the large energy gain necessary for high-energy physics applications, but over much shorter distances than with today's RF and magnet technology.

The potential of plasma as the medium for high gradient acceleration has been demonstrated with short and intense laser pulse drivers yielding electron bunches of up to 2 GeV energy gain in cm-long channels [Leemans *et al.* (2006); Wang *et al.* (2013)] that corresponds to about 100 GV/m average accelerating fields. High gradients have also been demonstrated with a short, high charge electron bunch driver with an energy gain of 42 GeV in 85 cm, corresponding to 52 GV/m [Blumenfeld *et al.* (2007)].

However, in both of these pioneering experiments the energy gain was limited by the energy carried by the driver and the propagation length of the driver in the plasma (< 1 m). The laser pulse and electron bunch driver schemes therefore require staging [Chattopadhyay and Jones (2011); Leemans (2009)]; i.e., the stacking of many 10–25 GeV acceleration, or plasma, stages to reach the ~ 1 TeV energy per particle or equivalently ~ 2 kJ of energy in $\sim 2 \times 10^{10}$ electrons and positrons. Our approach solves these propagation-length and energy limitations by using a proton bunch to drive the wakefields.

Bunches with 3×10^{11} protons and 19 kJ of energy (the CERN SPS 400 GeV beam), and with 1.7×10^{11} protons and 180 kJ of energy (the CERN LHC 6.5 TeV beam) are produced routinely today. Because of their high energy and mass, proton bunches can drive wakefields over much longer plasma lengths than other drivers. They can take a witness bunch to the energy frontier in a single plasma stage, as was demonstrated in simulations [Caldwell *et al.* (2009b)]. This proton-driven scheme therefore greatly simplifies and shortens the accelerator. In addition, because there is no gap between the accelerator stages, this scheme avoids gradient dilution.

16.4.1 *Self-modulation instability of particle beams in plasmas*

The major challenge in realising proton-driven plasma wakefield acceleration today is the length of the existing proton bunches. A plasma can be understood as an ensemble of oscillators swinging at the plasma frequency. To enforce the resonant swinging of these oscillators, the driver must contain a Fourier component close to the plasma frequency. The maximum field of a plasma wake scales as

$$E_{\max} \approx \sqrt{\frac{n_e[\mathrm{cm}^{-3}]}{10^{14}}}\,\mathrm{GV/m}, \tag{16.8}$$

where n_e is the plasma electron density. Consequently, plasma densities of at least $n_e \approx 10^{14}\,\mathrm{cm}^{-3}$ are required to reach accelerating gradients of GV/m and above. The corresponding plasma wavelength, calculated from the plasma oscillation frequency and assuming a relativistic driver, is

$$\lambda_p \approx \sqrt{\frac{10^{15}}{n_e[\mathrm{cm}^{-3}]}}\,\mathrm{mm}. \tag{16.9}$$

At these densities, the plasma wavelength is of the order of a millimetre. On the other hand, proton bunches available today are much longer, $\sigma_z = 3-12$ cm. Having a nearly Gaussian shape, they are not resonant and thus no strong plasma wakefields can be excited directly.

Fortunately, a mechanism has been discovered that automatically splits the proton bunch propagating in plasma into a number of micro-bunches: the self-modulation instability (SMI) [Lotov (1998); Kumar *et al.* (2010)]. The instability starts from a seeding wave whose transverse field acts on the beam and modulates its radius. Its amplitude grows exponentially from head to tail of the bunch and along the propagation distance. The physics of the instability is now well understood,

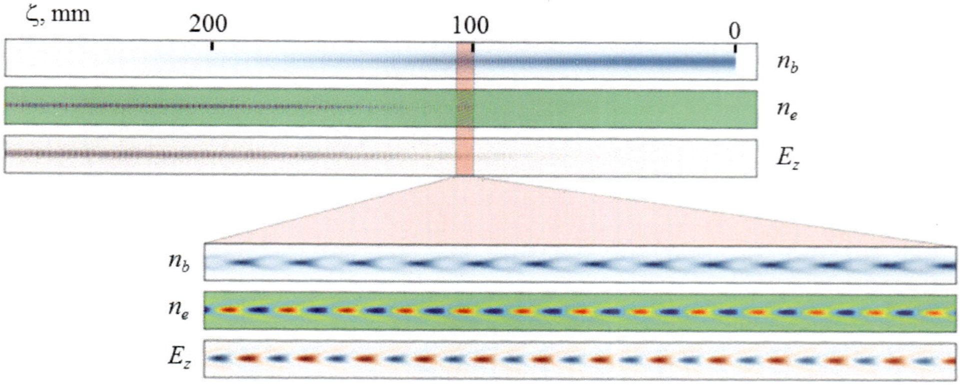

Fig. 16.1 Proton beam density n_b, plasma electron density n_e, and the longitudinal component of the wakefield E_z after 4 m of propagation in plasma. The coordinate $\zeta = ct - z$ is counted from the bunch head. The region 10 cm behind the bunch head is zoomed in.

and theoretical predictions agree well with results of simulations [Pukhov *et al.* (2011); Schroeder *et al.* (2011); Lotov (2011); Caldwell and Lotov (2011); Vieira *et al.* (2012a)]. At saturation, the initially long and smooth beam is split into a train of micro-bunches that resonantly excite a strong plasma wave. This plasma wave is inherently weakly non-linear [Lotov (2013)], so its excitation is almost independent of the charge sign of the drive beam. An example of a self-modulated bunch as observed in 3D simulations using the particle-in-cell code VLPL [Pukhov (1999); Tückmantel *et al.* (2010)] is shown in Fig. 16.1.

The proton bunch with $\sigma_z \approx 10$ cm must be split into some 100 micro-bunches to drive the high amplitude wake. All these micro-bunches must work constructively. This translates into a very strict requirement on the plasma density uniformity [Lotov *et al.* (2013)]. Simulations show that electron trapping and subsequent stable acceleration are mainly affected by density gradients. The relative plasma density variations must be controlled down to

$$\frac{\delta n_e}{n_e} \approx \frac{\lambda_p}{2\pi\sigma_z}. \tag{16.10}$$

This means that for realistic proton bunches we have to control the plasma density to better than 0.5% over distances of many metres. The best option to achieve the high density uniformity is to fill an evacuated vessel with a neutral gas and ionise it instantaneously with a laser [Öz and Muggli (2014)]. The laser pulse must be short, shorter than λ_p. If the laser co-propagates together with the proton bunch, the fast creation of plasma inside the bunch has the same effect as a sharp leading edge of the bunch would have; it reliably seeds the self-modulation.

For long-distance acceleration, other plasma cell designs will be necessary. In this context, a helicon cell prototype has been successfully developed [Buttenschön B. and Kempkes P., and Grülke O. and , Klinger T. (2014)]. A metre-scale prototype has reached the density needed for AWAKE, and further tests on the uniformity of the plasma are now underway.

16.5 AWAKE experiment at CERN

The AWAKE experiment at CERN [Gschwendtner *et al.* (2016)] is the first proton-driven plasma wakefield experiment worldwide. The conceptual design of the proposed AWAKE experiment is shown in Fig. 16.2. The laser ionises the metal vapour in the first plasma section and seeds the self-modulation instability. The self-modulated proton bunch (shown in the left-hand side inset) enters a second plasma section where it drives the plasma wakefield structure (shown in the right-side inset). The electrons are injected in the wakefields and their multi-GeV energy is measured with an electron spectrometer.

The AWAKE experiment is installed in the CERN Neutrinos to Gran Sasso (CNGS) facility [Gschwendtner *et al.* (2010)]. A design of the experimental setup is shown in Fig. 16.3. Nominally a proton bunch with intensity of the order of 3×10^{11} protons, a length of 12 cm and energy of 400 GeV is extracted every 30 s from the SPS and transported along more than 800 m of beam line towards the AWAKE experimental area that is installed in the upstream part of the (previous) CNGS facility. The focussed transverse beam-size at the plasma cell is $\sigma_r = 0.2$ mm and the transverse normalised emittance is $\epsilon_N = 3.5$ mm mrad. The laser beam is merged with the proton beam ~ 20 m upstream of the entrance of the plasma cell in a junction system. The area downstream of the plasma cell houses the beam diagnostics systems and the electron spectrometer. The proton beam is dumped in the existing CNGS hadron stop, ~1,000 m downstream of the experimental area, thus avoiding any backscattering of particles and radiation into the AWAKE experimental area.

The plasma cell technology which best fulfills the requirements for the proton bunch modulation is an alkali metal vapour source, as alkali metals have been used

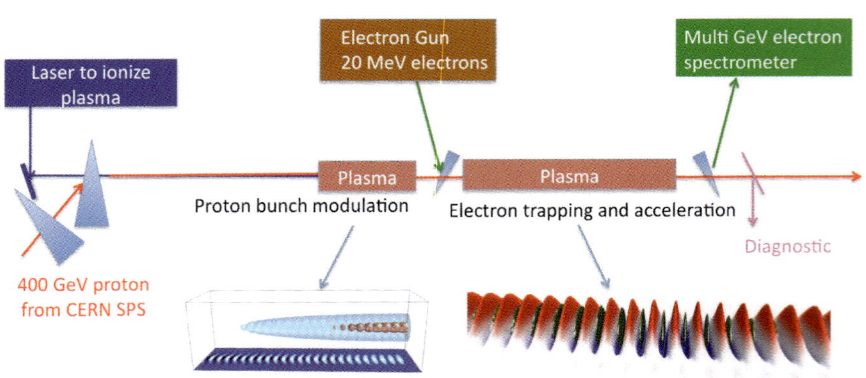

Fig. 16.2 Conceptual design of the AWAKE experiment, showing the major sections and description of expected effects.

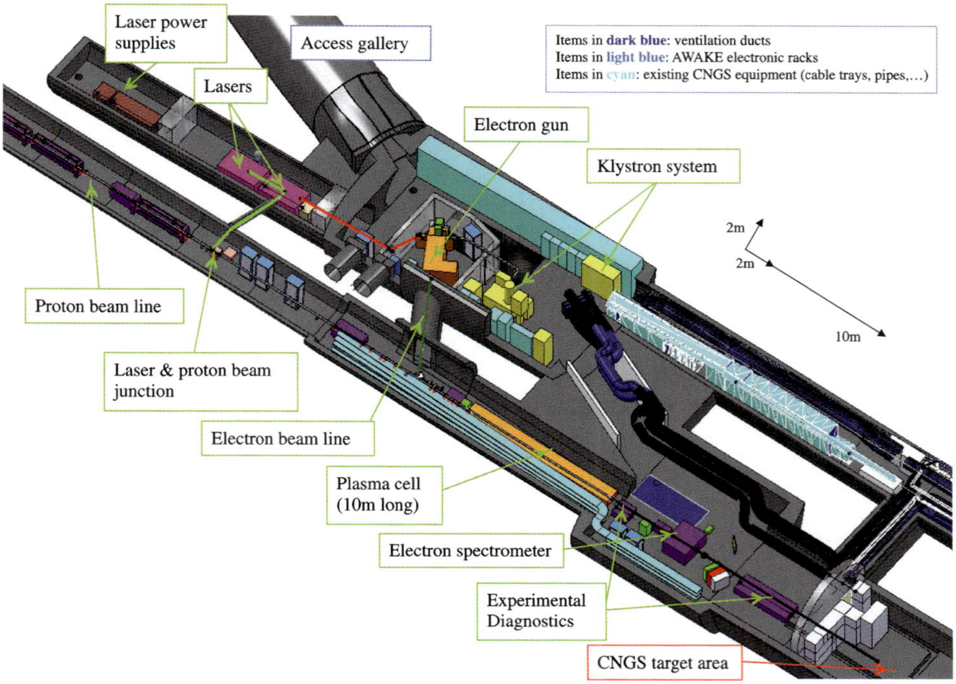

Fig. 16.3 Design of the layout of the AWAKE experiment.

in previous experiments [Muggli *et al.* (1999)] and have low ionisation potentials. The vapour is therefore relatively easy to ionise with a laser pulse. Rubidium has a large ion mass, which makes the plasma less sensitive to ion motion [Vieira *et al.* (2012b)].

In addition to an electron spectrometer, several other diagnostic systems are used to characterise the beams and plasma so as to better understand the physics of self-modulation and acceleration of electrons in the wake of the proton beam. Examples are shown in Fig. 16.4(a) in which a proton beam passes through a metal foil, thereby producing a cone of optical transition radiation that is measured using a streak camera.

Figure 16.4(b) shows an example of an electron bunch energy spectrum as obtained from numerical simulations using the simulation tools LCODE [Lotov (2003)] and OSIRIS [Fonseca R. A. *et al.* (2002)] for the nominal proton bunch parameters. The peak gradient seen in this simulation is above 1 GeV/m, and the average gradient witnessed by the electrons is 350 MeV/m. The acceleration of electrons begins after the full modulation of the proton bunch, and is effective over the last 4–5m of the plasma cell.

First protons to the experiment were delivered at the end of 2016 and AWAKE is now in the midst of a two-year experimental programme. The AWAKE experiment

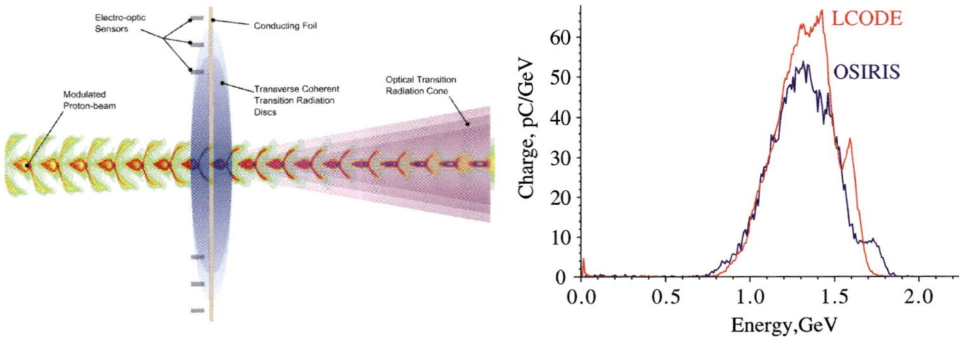

(a) (b)

Fig. 16.4 a) A modulated proton beam passing through a conducting foil, leading to transition radiation, detected using a streak camera and electro-optic sensors; b) The electron energy spectrum expected from the exit of the 10m AWAKE plasma cell from two different simulation codes.

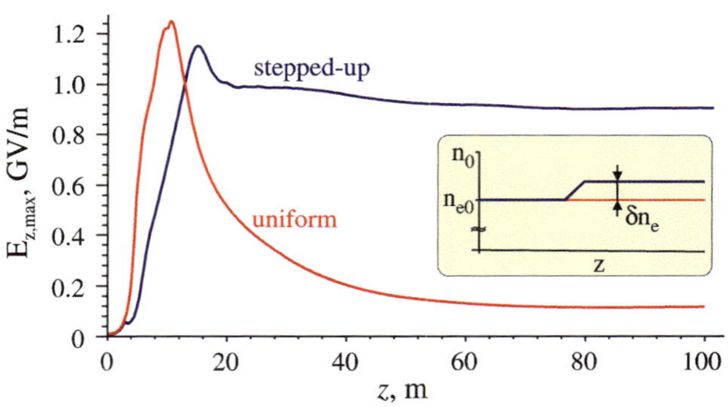

Fig. 16.5 The maximum wakefield amplitude versus the propagation distance for the stepped-up (blue line) and uniform (red line) plasmas for a simulation with an LHC bunch. The step δn_e is 1.6% [Caldwell and Lotov (2011)]. The inset illustrates the change in the plasma density profile at $z = 3\,\text{m}$.

at CERN is the first proton-beam driven plasma wakefield acceleration experiment worldwide. Its success will open a pathway towards a revolutionary plasma-based TeV lepton accelerator.

As a long-term prospect, acceleration of electrons in the wake of a self-modulating 7 TeV LHC beam was also studied [Caldwell and Lotov (2011)]. A test electron bunch was accelerated to 6 TeV, thus proving the capability of the self-modulation scheme to reach a multi-TeV energy scale with state-of-the-art proton beams. The high energy gain is only possible in a longitudinally non-uniform plasma with a small density step in the region of instability growth [Lotov (2011)]. The field evolution for the stepped plasma profile is shown in Fig. 16.5 in comparison with the uniform plasma case for the LHC beam. With the density step, the wakefield is preserved for a long distance at a large fraction of the maximum amplitude. It is particularly remarkable that long acceleration distances are possible without additional focussing of the proton beam by external quadrupoles; these were an essential part of the initial concept [Caldwell *et al.* (2009b); Lotov (2010)]. The realisation of the plasma density step is thus considered as a likely upgrade of the AWAKE experiment.

16.6 VHEeP accelerator complex

The very high-energy electron proton collider (VHEeP) is based on current LHC infrastructure and a new tunnel to house the plasma accelerator. The facility uses one of the LHC proton beams to generate wakefields and accelerate a trailing electron bunch which then collides with the other proton beam. This is shown in a simple schematic in Fig. 16.6 in which the electron beam is chosen to have an energy of 3 TeV, achieved in a plasma accelerator of $\lesssim 4$ km, and the proton beams have an energy of 7 TeV. Separation of the drive proton beam and witness electron beam will be needed to avoid pp collisions; as well as the temporal difference, the beams will need to be separated transversely by O(mm). Further study of this important issue is needed.

Although the energy of VHEeP will be very high, obtaining high luminosities will be, as with all plasma wakefield acceleration schemes, a challenge. In these initial studies, an integrated luminosity over the lifetime of VHEeP of 10–100 pb^{-1} is considered [Caldwell and Wing (2015); Wing and Caldwell (2016)], based on the expected capabilities of the LHC and pre-accelerators. The lower limit will be sufficient for measurements at low x where the cross section is expected to rise with

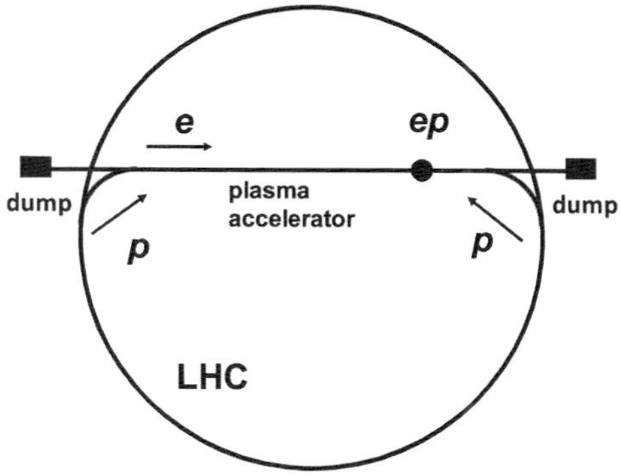

Fig. 16.6 Simple schematic of the VHEeP accelerator complex, showing the LHC ring. Protons from the LHC are extracted into the VHEeP plasma accelerator and used to accelerate bunches of electrons. Proton bunches rotating in the other direction in the LHC are extracted into the VHEeP tunnel and collided with electrons. Both proton and electron dumps could be used for fixed-target (beam-dump) experiments.

decreasing x. A higher integrated luminosity will aid the search for physics beyond the Standard Model, typically at high Q^2.

16.7 QCD physics at VHEeP

Electron-proton collisions at $\sqrt{s} \sim 9\,\text{TeV}$ give access to a completely new kinematic regime for deep inelastic scattering with, in particular, a reach in x a factor of about 1000 lower than at HERA and, depending on the luminosity, a similar increase in the reach at high Q^2. The energy dependence of hadronic cross sections, such as the total photon-proton cross section, is poorly understood. Predictions calculated from first principles are often not available and so phenomenological models are used to describe the dependence. Being able to measure the energy dependence, particularly with the long lever arm presented by VHEeP, will deepen our understanding of QCD and the structure of matter. In this section, some highlight physics measurements which particularly benefit from this extended kinematic regime are discussed.

16.7.1 *Total photon-proton cross section*

Measurements of the total γp cross section are shown in Fig. 16.7 compared to phenomenological models. A projection for a measurement at the expected maximum W at VHEeP is also shown; further measurements of similar precision at lower W are also expected. The many cross sections at low energies can be fitted in Regge phenomenology in which the dominant contributions arise from Reggeon exchange, which falls with increasing centre-of-mass energy, and Pomeron exchange, which rises. As an example, two such fits from Donnachie and Landshoff [Donnachie and Landshoff (1992, 2004)] are shown in the figure. The 1992 fit predates HERA data and is made over the range $6 < W < 20\,\text{GeV}$. It has a single Pomeron term and a single Reggeon term. The 2004 fit includes HERA data (photoproduction, as well as DIS data for $Q^2 < 45\,\text{GeV}^2$) and allows an additional second Pomeron term. Up to the highest HERA energies, both fits give good descriptions of the data. At higher energies, the second Pomeron term starts to become dominant and the cross-section predictions differ significantly. Also shown is an extrapolation based on the Froissart bound [Froissart (1961)], $\sigma_{\gamma p} \propto \ln^2(s)$. Assuming measurements from VHEeP up to $W \sim 6\,\text{TeV}$, even with very low luminosities ($\mathcal{L} = 49\,\text{nb}^{-1}$, as used in the ZEUS measurement [Chekanov *et al.* (2002b)]), the data will be able to strongly constrain the energy dependence of the total cross section and hence provide a clearer picture of QCD. It should be noted that the multi-TeV energies and hence large lever arm attainable at VHEeP are necessary to do this.

A photon-proton collision of $W = 6\,\text{TeV}$, corresponding to photon and proton energies of, respectively, $1.3\,\text{TeV}$ and $7\,\text{TeV}$, is equivalent to a $20\,\text{PeV}$ photon on a fixed target. This extends significantly into the region of ultra high-energy cosmic rays. Therefore VHEeP data could be used to constrain cosmic-ray air-shower

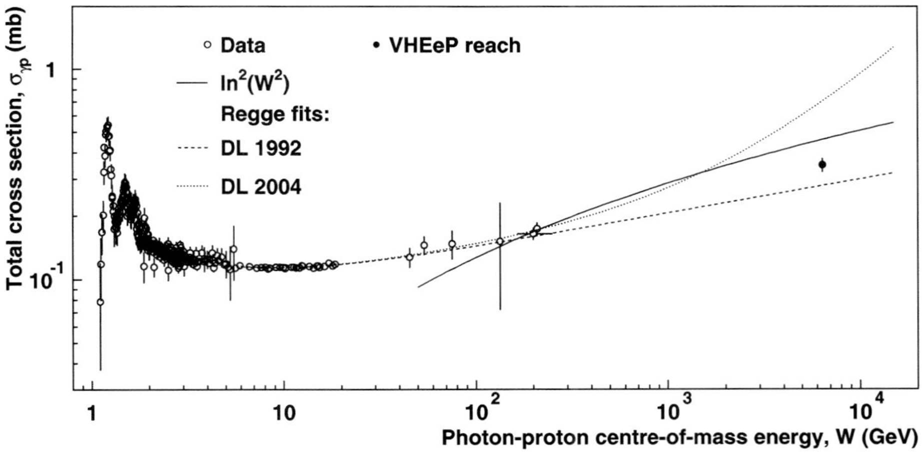

Fig. 16.7 Total γp cross section versus photon-proton centre-of-mass energy, W, shown for data compared to various models. The data is taken from the PDG [Yao *et al.* (2006)], with references to the original papers given therein. The VHEeP data point is shown at the same W value relative to $\sqrt{s}$ as the HERA results. The VHEeP cross section is assumed to be double the ZEUS value and the same uncertainties are assumed. The ZEUS measurement is at $\sqrt{s}$ =209 GeV and used a luminosity of 49 nb^{-1}. The Regge fits shown are those of Donnachie–Landshoff (1992 [Donnachie and Landshoff (1992)] and 2004 [Donnachie and Landshoff (2004)]). The $\ln^2(W^2)$ form is based on [Froissart (1961)].

simulations and so will be of benefit to understanding the nature of cosmic rays at the highest energies (for example, see [Cornet *et al.* (2015)]).

16.7.2 *Vector meson production*

Vector meson production is dependent on the partonic distributions in the proton and given the need for two gluons from the proton to create a vector meson, see Fig. 16.8, is particularly sensitive to saturation of the parton densities or other effects. For high vector meson masses, QCD calculations are expected to be more reliable and so measurements of J/ψ production can constrain the gluon density in a complementary way to fits to inclusive cross section in deep inelastic scattering (see [Chekanov *et al.* (2002a); Alexa *et al.* (2013)] and references therein).

Vector meson production has been measured extensively at HERA and fixed-target experiments (see, for example, ref. [Favart *et al.* (2016); Ivanov *et al.* (2006)] and references therein) and the cross sections are shown in Fig. 16.9 as a function of the photon-proton centre-of-mass energy, W. Photoproduction cross sections extracted from LHC data [Aaij *et al.* (2013, 2014, 2015); Abelev *et al.* (2014)] are also shown, which extend up to the TeV scale. The cross sections all rise with increasing W or equivalently decreasing x. These cross-section dependences, as well as the total cross section from Fig. 16.7 which is also shown, can be parametrised in terms of a power of W, as also shown in Fig. 16.9. The simple power-law behaviour

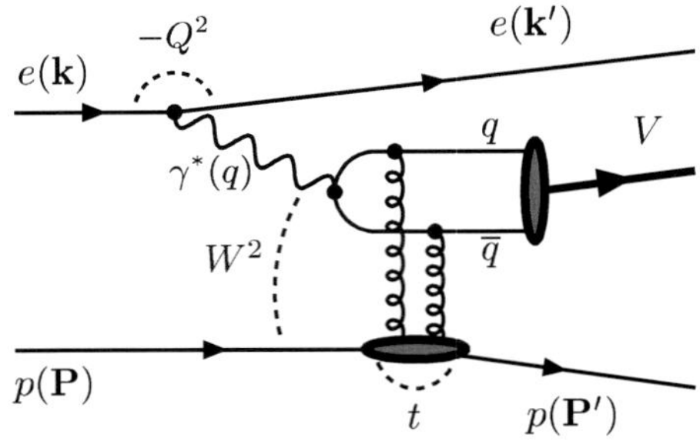

Fig. 16.8 Feynman representation of vector meson production in *ep* collisions.

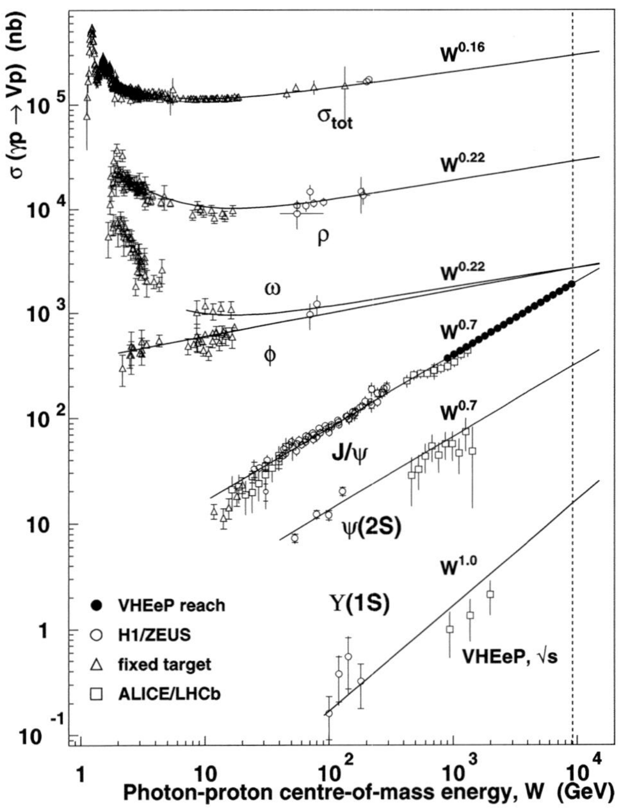

Fig. 16.9 The total γp cross section compared to the cross sections for exclusive vector meson production. The data in each of the reactions is shown along with a function depending on a power of W. The ALICE and LHCb data are extracted from measurements of vector meson production in $p - Pb$ and pp collisions, respectively [Aaij *et al.* (2013, 2014, 2015); Abelev *et al.* (2014)]. The reach of VHEeP is shown with simulated points for J/ψ production up to the centre-of-mass energy, indicated as the vertical dashed line.

of W with particle mass describe well the data from fixed-target, HERA and LHC experiments. However, such functions lead to nonsensical results at VHEeP energies, also indicated, in which e.g. $\sigma_{J/\psi}$ approaches $\sigma_\omega, \sigma_\phi$ and will at some point be larger. Clearly, the cross sections must take on another form and start to level out, be it through saturation or some other mechanism. As shown, data from VHEeP will be able to determine this behaviour.

16.7.3 *Physics at low Bjorken x*

As discussed above, the energy dependence of the total photoproduction cross section at high energies is of great interest, both on fundamental grounds and for understanding cosmic-ray events in the atmosphere. The energy dependence of scattering cross sections for virtual photons on protons is also of fundamental interest, and its study at different virtuality is expected to bring insight into the processes leading to the observed universal behaviour of cross sections at high energies.

In deep inelastic scattering of electrons on protons at HERA, the strong increase in the proton structure function F_2 with decreasing x for fixed, large, Q^2 is usually interpreted as an increasing density of partons in the proton, providing more scattering targets for the electron. This interpretation relies on choosing a particular reference frame to view the scattering — the Bjorken frame. In the frame where the proton is at rest, it is the state of the photon or weak boson that differs with varying kinematic parameters. For the bulk of the electron-proton interactions, the scattering process involves a photon, and we can speak of different states of the photon scattering on a fixed proton target. What is seen is that the photon-proton cross section rises quickly with W for fixed Q^2 [Breitweg *et al.* (1997)]. In the proton rest frame, we interpret this as follows: as the energy of the photon increases, time dilation allows shorter lived fluctuations of the photon to become active in the scattering process, thereby increasing the scattering cross section.

Figure 16.10 shows the results of extrapolation of fits to the energy dependence of the photon-proton cross section for different virtualities as given in the caption [Caldwell (2016)], for two different assumptions on the energy behaviour. In one instance (blue curves), the energy dependence is assumed to follow a simple behaviour at small values of x:

$$\sigma^{\gamma p} \propto x^{-\lambda(Q^2)}\,, \tag{16.11}$$

while in the second instance a form inspired by double asymptotic scaling [Ball and Forte (1994)] was used

$$\sigma^{\gamma p} \propto e^{B(Q^2)\cdot\sqrt{\log 1/x}}\,. \tag{16.12}$$

The simple behaviour is what has been used in most fits to HERA data [Abramowicz *et al.* (2015)] to date, and it deviates strongly from the expectations of double asymptotic scaling in the VHEeP kinematic range. It is found that this simple behaviour cannot continue to ever smaller values of x as this would result in large-Q^2

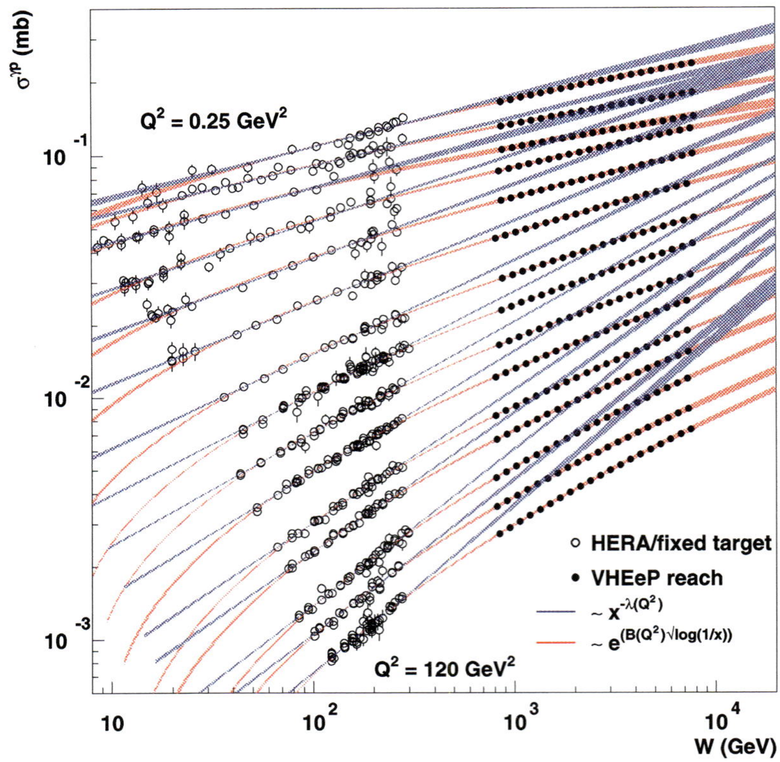

Fig. 16.10 Measurements (open points) of $\sigma^{\gamma p}$ versus W for $0.25 < Q^2 < 120\,\mathrm{GeV}^2$ from HERA and fixed-target experiments. The blue lines show fits to the data, performed separately for each Q^2 value, of the form given in Eq. (16.11). The red lines show fits of the form given in Eq. (16.12). The reach of VHEeP is shown as projected data points (closed points). The points are placed on the red curve. The uncertainties are assumed to be of order 1%, given the increased cross section expected and similar systematics to those at HERA and are not visible as error bars on this plot.

cross sections becoming larger than small-Q^2 cross sections. A change of the energy dependence is therefore expected to become visible in the VHEeP kinematic range. This should yield exciting and unique information on the fundamental underlying physics at the heart of the high-energy dependence of hadronic cross sections.

16.8 Physics beyond the Standard Model

Let me also briefly discuss the prospects for the discovery of new physics beyond the Standard Model (BSM). The key features of VHEeP are the very large centre-of-mass energy and the unique combination of electron and proton scattering. I therefore focus on new physics where these features are essential. Two topics were chosen for this initial study [Caldwell and Wing (2016b)]: the search for quark substructure, and the search for production of leptoquarks. This is by no means an exhaustive list of topics, but is intended to show the capabilities of VHEeP for BSM physics.

16.8.1 *Quark substructure*

In order to look for quark substructure, an effective finite radius of the quark can be assigned. The Standard Model prediction, $\frac{d\sigma^{\rm SM}}{dQ^2}$, is modified using a semi-classical form-factor approach [Kopp *et al.* (1995)]:

$$\frac{d\sigma}{dQ^2} = \frac{d\sigma^{\rm SM}}{dQ^2}\left(1 - \frac{R_e^2}{6}Q^2\right)^2\left(1 - \frac{R_q^2}{6}Q^2\right)^2, \tag{16.13}$$

where R_e^2 and R_q^2 are the mean-square radii of the electron and quark radius, respectively. As has been done in a recent analysis of HERA data [Abramowicz *et al.* (2016)], the radius of the electron is assumed to be zero, i.e. the electron is point-like.

The existence of a finite quark radius primarily changes the Q^2 distribution; in a more complete analysis, the full $x - Q^2$ dependence of the cross section must be analysed to separate the effects of a finite quark radius from effects due to uncertain parton density distributions. For this first study, it was assumed that the parton densities were known with negligible uncertainty, and only the modification of the Q^2 dependence was considered. The effect of the finite quark radius grows with Q^2, as seen in Eq. (16.13) but this is counterbalanced by the rapidly falling cross section with Q^2. Deep inelastic scattering events were generated with the ARIADNE Monte Carlo program, corresponding to a luminosity of $100\,\text{pb}^{-1}$. The event distribution as a function of Q^2 for the range $10^5 < Q^2 < 10^7\,\text{GeV}^2$ was analysed and an upper limit on R_q was determined by reweighting the known cross section using Eq. (16.13) and performing a fit to the simulated data using the BAT package [Caldwell *et al.* (2009a)]. The limits correspond to 68 % credibility upper limits where a flat prior was taken for $0 \leq R_q < 5 \cdot 10^{-19}\,\text{m}$.

The extracted limit on R_q is $R_q \leq 1 \times 10^{-19}\,\text{m}$, which can be compared with the 95 % Confidence Level limit extracted from HERA data [Abramowicz *et al.* (2016)], $R_q < 4 \times 10^{-19}\,\text{m}$. The limit extracted from the HERA data was the result of a much fuller analysis; it is expected that the limits from VHEeP would become considerably stronger if lower Q^2 data were included in the analysis, but this would require a much more complete analysis using detailed information on systematic uncertainties. The limit would also improve by about a factor 3 for a factor 10 increase in luminosity.

16.8.2 *Leptoquark production*

Electron-proton collisions are particularly sensitive to leptoquark production as the leptoquark is produced resonantly in the s-channel. This is shown pictorially in Fig. 16.11, where an electron and quark fuse, with a coupling λ. The leptoquark subsequently decays to a lepton-quark system, again with a coupling λ, and this effect can be searched for by reconstructing the invariant mass of the final states or looking for a resonant deviation from the Standard Model in the x distribution which is related to the mass of the leptoquark.

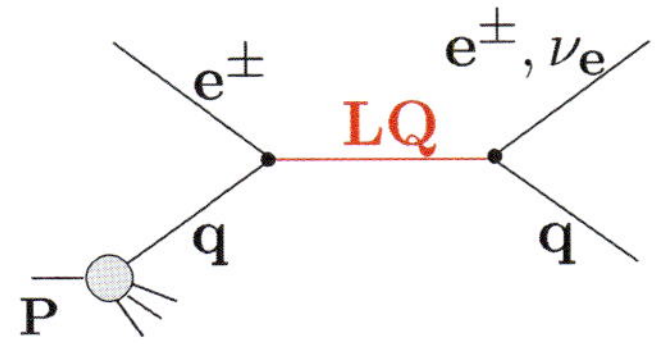

Fig. 16.11 Feynman representation of s-channel production of a leptoquark in ep collisions.

In this analysis, deep inelastic scattering events were generated with the ARIADNE Monte Carlo program and the x distribution plotted. This is the same sample as used to extract the limit on R_q, corresponding to a luminosity of 100 pb^{-1}, with events up to about $x \sim 0.5$, see Fig. 16.12(a). Cuts $Q^2 > 10\,000\,\text{GeV}^2$ and $y > 0.1$ were applied to enhance the possible signal over background. A much larger independent sample was generated, again using ARIADNE, and used as the Standard Model prediction, also shown in Fig. 16.12(a). The 90% probability upper limit on the number of signal events, ν, above the Standard Model prediction was then extracted based on this pseudo-data sample and is shown as a function of the leptoquark mass in Fig. 16.12(b).

In order to extract a signal or limit on leptoquark production, the Standard Model prediction is convoluted with the prediction for leptoquark production according to the Buchmüller–Rückl–Wyler (BRW) model [Buchmüller *et al.* (1987)]. The Born-level cross section for resonant s-channel leptoquark production in the narrow-width approximation (NWA), σ^{NWA}, is

$$\sigma^{\text{NWA}} = (J+1)\frac{\pi}{4\,s}\lambda^2\, q(x_0, M_{\text{LQ}}^2) \tag{16.14}$$

where $q(x_0, M_{\text{LQ}}^2)$ is the initial-state quark (or antiquark) parton-density function in the proton for a Bjorken-x value of $x_0 = M_{\text{LQ}}^2/s$, where M_{LQ} is the mass of the leptoquark, and J is the spin of the leptoquark. Given the limit on ν, a limit on the coupling λ as a function of the mass of scalar leptoquarks was extracted and is shown in Fig. 16.12.

These results show that VHEeP has sensitivity up to the kinematic limit of 9 TeV. The HERA limits are $\lambda = 0.01$ just below the kinematic limit of 0.3 TeV, rising rapidly to $\lambda = 1$ at about 1 TeV [Aaron *et al.* (2011c); Abramowicz *et al.* (2012)]. Limits from the LHC experiments are also $\lambda = 1$ at about 1–2 TeV for pp collisions at $\sqrt{s} = 8$ TeV for $\mathcal{L} \sim 20\,\text{fb}^{-1}$ [Aad *et al.* (2016b); Khachatryan *et al.* (2016f)]. Given the increased centre-of-mass energy and higher luminosities the sensitivity to leptoquark production at the LHC will extend to $2-3$ TeV. Hence

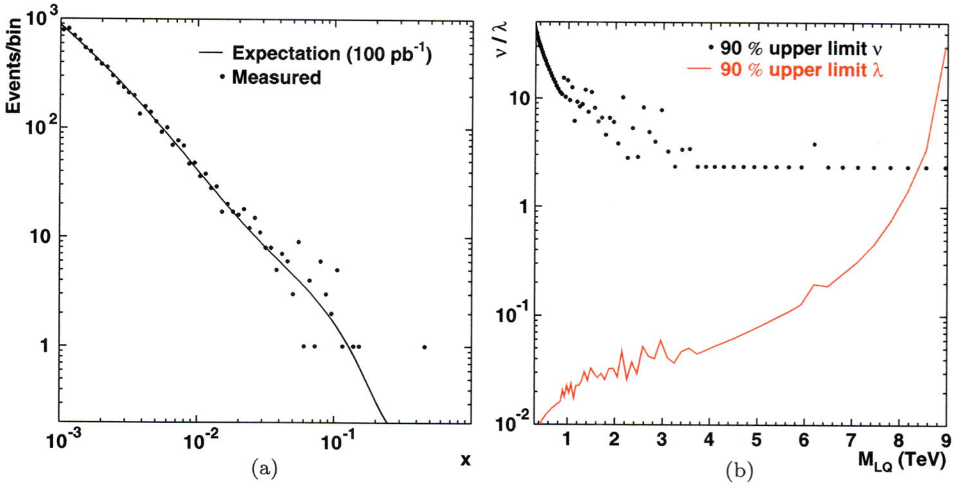

Fig. 16.12 (a) Simulated deep inelastic scattering data with a luminosity of 100 pb^{-1} (points) and the expectation, both generated with the ARIADNE Monte Carlo program with $Q^2 > 10\,000\,\text{GeV}^2$ and $y > 0.1$. (b) Upper limits on the number of events, ν, and leptoquark coupling parameter, λ, versus mass of the leptoquark, M_{LQ}.

VHEeP has a sensitivity to leptoquark production significantly beyond the HERA limits and LHC limits, both measured and expected.

16.9 Further physics areas

As well as the areas discussed in the previous sections, many other areas, in particular in QCD, will be open to investigation at VHEeP. Standard tests of QCD, performed at HERA and other colliders, will be possible in this new kinematic regime, such as measurements of the strong coupling constant, jet and heavy flavour production and properties of the hadronic final state. The structure of the proton and photon can be further investigated and fits of the parton densities to the data performed. The results will be related to the measurements above, sensitive to effects such as saturation, but more conventional determinations of the inclusive deep inelastic scattering cross section can be made in a region less sensitive to the more exotic QCD effects. In particular, given the possibility to change the beam energy and to also vary this as widely as possible, the longitudinal structure function can be well measured, which was done with only limited precision at HERA given the relatively small lever arm in centre-of-mass energy [Aad *et al.* (2016f); Abramowicz *et al.* (2014b)]. Given the possibility to run with other heavy ions as well as protons in the LHC, VHEeP will also be able to investigate the properties of electron-ion, eA, scattering. Finally, diffraction, an area of QCD re-invigorated at HERA, and particularly sensitive to low-x dynamics can be studied.

16.10 Summary and outlook

The concept of proton-driven plasma wakefield acceleration will be tested in the next few years in the AWAKE experiment at CERN. Recent simulations of the scheme indicate that acceleration of electrons with 7 TeV LHC proton bunches could be used to bring electrons to 3 TeV with an average gradient of about 1 GeV/m. This opens up tremendous opportunities for considering very high energy colliders; in this article, I discussed the possibility of colliding 3 TeV electrons accelerated in this way with 7 TeV protons from the LHC. This will allow an extension of the kinematic reach of the HERA electron-proton collider by three orders of magnitude, albeit with moderate luminosities. This is surely a prospect that Guido Altarelli would have been very excited about.

Chapter 17

Physics at future circular colliders

Michelangelo L. Mangano
CERN, Theoretical Physics Department,
1211 Geneva 23, Switzerland

17.1 Foreword

Guido Altarelli has been a key catalyst of the physics programmes of LEP, HERA and the LHC. He contributed to the definition of the physics goals, through his own research, coordinating several workshop activities, and inspiring with his vision and wisdom the young and senior physicists who made these grandiose scientific enterprises successful. As the scientific community continues brainstorming about possible future initiatives, to push the field beyond the reach of the LHC, his absence leaves a vacuum that cannot be easily filled. With the discovery of the Higgs boson [Aad *et al.* (2012c); Chatrchyan *et al.* (2012)] and the completion of the Standard Model (SM), now more than ever in the past 40 years the field would need his sharp common sense, to identify priorities and define the path that will help us moving beyond. Let the legacy of his work and his vision guide us in our next steps.

17.2 Introduction

A future generation of circular colliders, modelled on the successful sequence of LEP and LHC, has recently joined the established and well studied linear e^+e^- colliders as a strong candidate to address the current challenges of high-energy physics. These challenges emerge from the leading priorities of our field: the exploration of the physics of electroweak symmetry breaking (EWSB), and the identification of the origin of the known departures from the SM (dark matter, neutrino masses, the baryon asymmetry of the universe). The key justification for future circular colliders (FCCs) builds on the belief that the origin of EWSB and its puzzles, and the origin of (at least some of) the departures from the SM, are deeply intertwined, and worth investigating together.

One of the main questions we face in planning our future is "why don't we see as yet any sign of the new physics that we confidently expected to be present around the TeV scale?". The question admits two possible answers: the mass scale of the new physics lies beyond the LHC reach, or, while being within LHC's reach, its manifestations are elusive and have escaped the direct search so far. These two scenarios are a priori equally likely, but they clearly impact in different ways the future of our field, and thus the assessment of the physics potential of possible future facilities. Our safest hedge is therefore the readiness to cover both scenarios, via an experimental programme relying on higher precision and sensitivity (to address possible elusive signatures), and an extended energy and mass reach relative to the LHC.

A future circular collider provides this hedge. Current projects, at CERN (FCC [Benedikt (2017)]) and in China (CepC/SppC [CEPC (2017)]), rely on a 100km tunnel, hosting, in sequence, an e^+e^- and a pp collider. The FCC e^+e^- accelerator (FCC-ee) would operate at energies ranging from the Z pole to above the $t\bar{t}$ threshold, including operations at the peak of the Higgs production rate and at the W^+W^- threshold. The pp collider (FCC-hh), relying on a new generation of dipoles with field in the range of 16 T (based on Nb_3Sn superconducting technology), would reach $\sqrt{S} \sim 100\,\text{TeV}$. The addition of a 60 GeV electron energy recovery linac (ERL) could furthermore enable ep collisions at $\sqrt{S} \sim 3.5\,\text{TeV}$, operating in parallel with the pp collider [Zimmermann (2017); Klein (2018)]. Preliminary studies of the physics potential of these facilities can be found in refs. [Bicer *et al.* (2014); Arkani-Hamed *et al.* (2016b); Mangano (2017); Abelleira Fernandez *et al.* (2012a)].

The physics potential of an FCC builds on three pillars:

- **The guaranteed deliverables:** these include the study of the properties of the Higgs, gauge bosons and of the top quark, and the exploration of EWSB phenomena, with unmatchable precision and sensitivity.
- **The discovery potential:** on one side the direct discovery potential is extended with respect to the LHC by a factor as large as 7 (i.e. 100 TeV/ 14 TeV); on the other, discovery prospects are enhanced by the increased

precision brought by the e^+e^- runs, and by the immense statistics and opportunities to extend the indirect sensitivity at large Q^2 in pp collisions. Should the LHC give evidence for new phenomena, the FCC would also greatly extend the ability to pin down the origin of this new physics.

- **Definitive answers to key questions:** with the Higgs discovery and the completion of the SM spectrum, there is no remaining discovery that can unequivocally be promised by any future facility. Most new particles or phenomena predicted by theories beyond the SM (BSM) can have mass scales just above whatever energy threshold the new accelerators can deliver, their discovery cannot be guaranteed, and new limits would often be inconclusive on the question of whether these BSM theories are right or not. On the other hand, there are key questions that FCC could give definitive yes/no answers to, thus proving, or ruling out, vast classes of BSM scenarios. The first example, is the issue of whether the electroweak phase transition that took place in the early universe has been the source of the baryon asymmetry (a scenario called EW baryogenesis, EWBG [Assamagan *et al.* (2016)]). In the SM, we know that this would not have been possible: a strong enough first order phase transition would have required a Higgs mass $\lesssim$80 GeV, and a source of CP violation larger than that given by the CKM mixing. To modify the SM predictions, and enable the occurrence of a phase transition strong enough and with a source of CP violation large enough, new particles and interactions must have been at work. In particular, a modification of the Higgs potential would be necessary. Preliminary studies [Contino *et al.* (2017)] support the conclusion that any such scenario would give rise to signatures visible at the FCC. Lack of such signals would rule out EWBG, information that, however negative, would have a great impact on our exploration of the origin of matter. A second example is dark matter (DM). There are many models for DM, including some — such as axions — that could not be tested by the FCC. But a large class of models, with weakly interacting massive particles (WIMPS), can be tested directly at accelerators, starting from the LHC. The LHC has not enough energy, however, to probe the upper mass range for thermal WIMPS predicted by the observed DM density, this range extending to the few TeV region. Even in this case, current studies [Golling *et al.* (2017)] suggest that the mass reach of a 100 TeV collider is sufficient to detect, or exclude, the existence of such WIMPS. Once again a possible firm yes/no answer, which would redefine our ideas about DM.

While the studies of both EWBG and DM WIMP scenarios are still preliminary, and more exercises in model building could identify possible loopholes, the emergence of such possible "no-lose" theorems in the context of some of the most pressing open questions of our field provides a strong justification for the realisation of an FCC facility. One should also add that the extensive exploration of the energy scale up to several TeV, via precision measurements and direct searches, would help clarify whether the hierarchy problem admits a *natural* solution at the EW scale, or whether our ideas about naturalness require a radical revision.

17.3 The FCC operation scenarios

The e^+e^- collider would operate at different energies, with an evolution suited to match the physics needs for statistics at the different energies and the gradual availability and installation of accelerating cavities. The preliminary plan [Janot (2017)], which reflects the current assumptions on the overall machine performance (optics, running efficiency, ...) and the existence of two detectors, is shown in Table 17.1. The runs at the Z include scans at 88 and 94 GeV (for a precise determination of $\alpha_{QED}(M_Z)$ [Janot (2016)]), and those at the top include a small fraction at the production threshold, and the rest at ~370 GeV. Additional runs at the Higgs pole (to probe directly the Higgs coupling to the electron [d'Enterria (2017)]) are being considered. The gradual decrease in luminosity is due to the constraint of keeping constant, in the range of few hundred MW, the overall radiofrequency power needed to compensate for the synchrotron radiation losses, which grow as the fourth power of the beam energy.

The running conditions of the *pp* collider are outlined in Table 17.2. The overall current is limited by the cooling power needed to evacuate the energy deposited by synchrotron radiation, while the peak luminosity is limited by the beam lifetime, which requires frequent refills, and by the time required to reinject the beams. A total integrated luminosity in the range of 20–30ab^{-1} is expected, over a lifetime of 20–30 years. During phase II, a possible reduction of the bunch crossing time to 5ns can be envisaged, to reduce the amount of pile-up to the levels expected at the HL-LHC.

Table 17.1 Operation model for the FCC e^+e^- collider (2 interaction regions and baseline optics).

FCC-ee run	Z	WW	HZ	$t\bar{t}$
$\sqrt{s}$ (GeV)	88/91/94	161	240	350/370
$\mathcal{L}$(ab^{-1}/year)	15-30	4	1	0.3
Events/yr	1.5×10^{12} (at 91 GeV)	1.5×10^7	2×10^5	2×10^5
Runtime (yrs)	6	2	5	5

Table 17.2 Operation scenarios for the FCC *pp* collider. A total integrated luminosity of 20-30ab^{-1} is envisaged.

	β^*(m)	$t_{turnaround}$ (hr)	Bunch crossing (ns)	Pile-up	$\mathcal{L}$/year (fb^{-1})
Phase I	1	5	25	170	250
Phase II	0.3	4	25(5)	1000 (200)	1000

17.4 Studies of Higgs couplings

The Higgs boson is a totally new and peculiar object in the spectrum of fundamental particles. At tree level, it couples to everything that has a mass, including itself, with strengths that are precisely determined, in the SM, by the EW gauge couplings and by the particles' masses. At the loop level, virtual transitions of the Higgs to any particle with charge or colour induce also couplings to photons or gluons. The Higgs is therefore a powerful probe of the possible existence of new particles or interactions: its total width could be altered by decays to otherwise invisible particles, its decays to photons or gluons could receive contributions from yet undetected heavy particles, its own self-coupling could be modified by corrections to the SM quartic potential, and the Yukawa pattern of interactions to fermions could fail in the case of the lightest generations or in view of a more complex flavour structure. Existing data, direct [Aad *et al.* (2016d)] or indirect, support the SM-like properties of the Higgs, but only a minimal fraction of all Higgs properties is actually probed, and incisive tests will require a precision at the level of percent, or better. Achieving this goal of precision is one of the primary deliverables of the FCC, and a thorough measurement programme is being defined by the ongoing studies.

17.4.1 *Higgs studies at FCC-ee*

Over a million H bosons will be produced by FCC-ee, via associated ZH production, and vector boson fusion (VBF) $e^+e^- \to \nu\bar{\nu}(e^+e^-)H$. Even excluding the neutrino decays of the Z, for most of the ZH events it is possible to measure the Z momentum, and therefore reconstruct the missing mass ($\sqrt{(p_{e+e-} - p_Z)^2}$). This exposes a clear H mass peak, allowing a measurement of the $e^+e^- \to ZH$ cross section with a statistical precision in the determination of the HZZ coupling at the permille level. A detailed study is reported in ref. [Bicer *et al.* (2014)], with the results shown in Table 17.3.[1] The absolute measurement of the HWW coupling will benefit from the VBF production channel. For the other Higgs couplings, one must mostly rely on the decays to final states with a branching ratio large enough to leave sufficient statistics. This will cover the couplings to the bottom and charm quarks, to the tau lepton, and to gluons. More limited precision, due to smaller statistics, is expected for the coupling to photon and muon pairs. Exclusive Higgs decays to

Table 17.3 Expected precision on Higgs couplings with FCC-ee. Last two columns: limit on the Higgs decay to invisible particles, and precision in the measurement of its total decay width.

XY	ZZ	WW	bb	cc	gg	$\tau\tau$	$\gamma\gamma$	$\mu\mu$	tt	HH	BR_{inv}	Γ_{tot}
$\delta g_{HXY}(\%)$	0.15	0.19	0.42	0.71	0.8	0.54	1.5	6.2	13	30	<0.45%	1%

[1] The assumptions about luminosity and number of detectors have evolved since the original study, but the goal of the operation plan is to preserve the precision targets.

hadronic resonances with lighter quarks (u, d, s) are being explored, with the goal of at least establishing strong evidence for the Higgs interaction with these quarks. The precision measurement of the HZZ coupling, and of the $e^+e^- \to t\bar{t}$ cross section, can furthermore constrain the Higgs self-coupling [McCullough (2014)], and the coupling to the top, in an indirect way. FCC-ee, furthermore, can directly measure with percent precision the total decay width, and probe to a fraction of percent the invisible decay rate.

17.4.2 *Higgs studies at FCC-hh*

The immense statistics of Higgs production at FCC-hh (see Table 17.4) provides the ideal complement to the precision of FCC-ee. While the hadron collider has an inferior precision in the determination of absolute rates (whether because of backgrounds, or of theoretical systematics in the calculation of absolute cross section and distributions, or because of experimental limitations in the measurement of the absolute luminosity or detection efficiencies), the large statistics gives access to final states with very small branching ratios, and potentially allows for very precise determinations of ratios of branching ratios. These ratios have a value of their own, since different decay channels can be affected in different ways by possible new physics. For example, the existence of new weakly interacting particles could alter to leading order the loop-induced $H \to \gamma\gamma$ decay, while having only a subleading effect on the HZZ coupling, leading to measurable effects in a precise determination of $BR(H \to \gamma\gamma)/BR(H \to ZZ)$. A further measurement of $BR(H \to Z\gamma)/BR(H \to \gamma\gamma)$ would probe the existence and couplings of these new particles in a complementary way. But the precise measurement of such ratios, in conjunction with the absolute measurements from FCC-ee, would also enable the knowledge of the ratios to be promoted to an absolute measurement of couplings for the rarer channels that are statistically limited at FCC-ee.

Higgs production rates for the dominant channels at 100 TeV are shown in Table 17.4. The comparison with the rates at LHC energies shows, by naive scaling, a potential for improvements in the statistical precision in the range of few hundreds wrt to run 1, and of order 10-20 wrt HL-LHC. As is well known, the HL-LHC itself will already be systematics-dominated for several measurements. But with such

Table 17.4 First row: Cross sections at a 100 TeV collider for the production of a SM Higgs boson in gg fusion, vector boson fusion, associated production with W and Z bosons, and associated production with a $t\bar{t}$ pair. Second row: Statistics for 20ab^{-1}. Third row: Statistics increase for 20ab^{-1} at 100 TeV, relative to HL-LHC (3ab^{-1} at 14 TeV).

	$gg \to H$	VBF	$HW^\pm$	HZ	$t\bar{t}H$
σ(pb)	802	69	15.7	11.2	32.1
N_{events}/20ab^{-1}	1.6×10^{10}	1.4×10^9	3.1×10^8	2.2×10^8	6.4×10^8
N(100 TeV)/N(14 TeV)	110	120	65	85	420

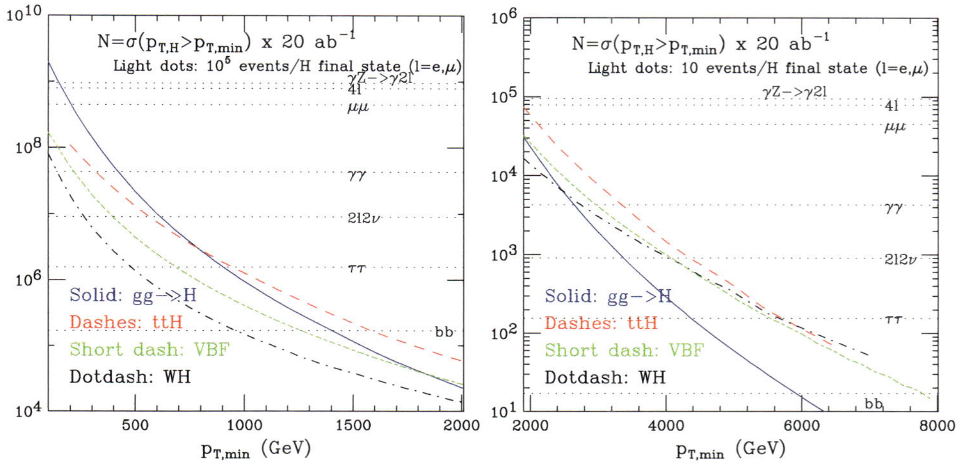

Fig. 17.1 Integrated Higgs transverse momentum rates, for various production channels, with 20 ab^{-1}. The light-dotted horizontal lines in the left (right) panel correspond to the production of 10^5 (10) events with a Higgs decay to the indicated final states.

a huge increase in rate and, as we shall see, in kinematic reach, one can envisage new approaches to both precision measurements and to the exploration of new phenomena in the production dynamics. Furthermore, these rates will push the search for rare or forbidden Higgs decays well beyond the LHC reach.

The most remarkable feature of Higgs production at 100 TeV is not just the rate increase wrt the LHC, but the extreme kinematical range over which the Higgs bosons are distributed. Figure 17.1 shows the integrated p_T spectra for the dominant production processes, and prompts several important remarks. To start with, we highlight the remarkable statistics. Horizontal light-dotted lines in the figures show the $p_T(H)$ values corresponding to samples of 10^5 (left) and 10 (right) events, for various final states. The former statistics are possibly suitable for percent-level measurements, the latter indicate the most extreme p_T values at which measurements of Higgs production dynamics are in principle possible. This could have relevance, for example, in the context of searches for new physics, where Higgses could either be part of a signal, or a background.

Secondly, we note that the hierarchy of rates among the different processes changes significantly as a function of the Higgs p_T. As $p_T(H)$ grows above $\sim$700 GeV, $t\bar{t}H$ emerges as the most abundant source of large-p_T Higgses. Moving to yet larger p_T, even VBF and eventually associated VH production come to be more important than $gg \to H$. The key reason for this is the form-factor-like suppression of the ggH vertex at large virtuality, when the finite-m_{top} effects are properly accounted for.

This observation has important implications for the measurements. For example, while dedicated cuts are needed to extract the VBF Higgs-production signal from the inclusive $gg \to H + X$ Higgs sample, at large p_T the dominant source of irreducible background is top production. The separation of $t\bar{t}H$ from VBF when $p_T(H) > 1$ TeV can rely on kinematic and event-shape discriminators, which are likely more powerful and efficient than the usual VBF cuts. This may also have important implications on the detector, since optimal acceptance to VBF cuts requires instrumentation in the very difficult forward η region.

Large Higgs p_T values, furthermore, make it possible to consider using the otherwise disfavoured $H \to b\bar{b}$ decay mode, thanks to the ever-improving discrimination power of jet-substructure techniques. The ability to use this high-BR decay extends considerably the accessible $p_T(H)$ range. Lower-BR final states, such as $H \to \gamma\gamma$, ZZ^*, $Z\gamma$ or $\mu^+\mu^-$, remain nevertheless usable for precision measurements (i.e. event rates in excess of 10^4), over a broad range of p_T.

Some explicit examples [Contino *et al.* (2017)] of the statistical reach are shown in Figs. 17.2 and 17.3. Limiting the analyses to events with Higgs produced with p_T above a few hundred GeV, and assuming the knowledge of g_{HZZ} from FCC-ee, leads to the estimates of precision given in Table 17.5. In the table we also report the projected precision for the top Yukawa and for the Higgs trilinear self-coupling, obtained in the analyses of refs. [Mangano *et al.* (2016)] and [Contino *et al.* (2017)] respectively. Preliminary studies [Harris (2017)] of the SM missing transverse energy ($\not{E}_T$) backgrounds, in the region $\not{E}_T \gtrsim 1$ TeV, suggest a sensitivity to invisible Higgs decays with BR as small as 10^{-4}, sufficient to observe the SM $H \to 4\nu$ rate and to probe a vast range of parameter space for DM scenarios.

The prospects for precision Higgs physics at FCC-eh are described in ref. [Klein (2018)].

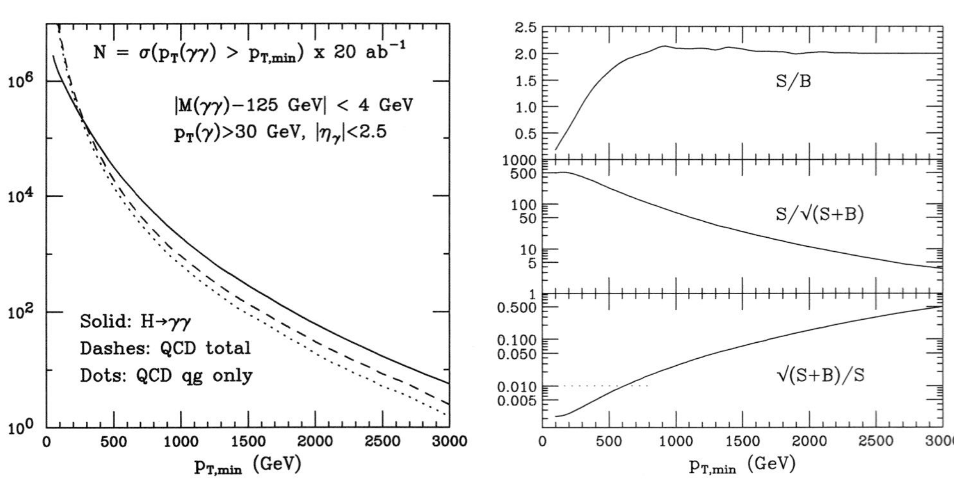

Fig. 17.2 Left: Integrated transverse momentum rates (20 ab^{-1}) for a photon pair with mass close to the Higgs mass: signal and QCD background. Right: S/B, significance of the signal, and potential statistical accuracy of the sample.

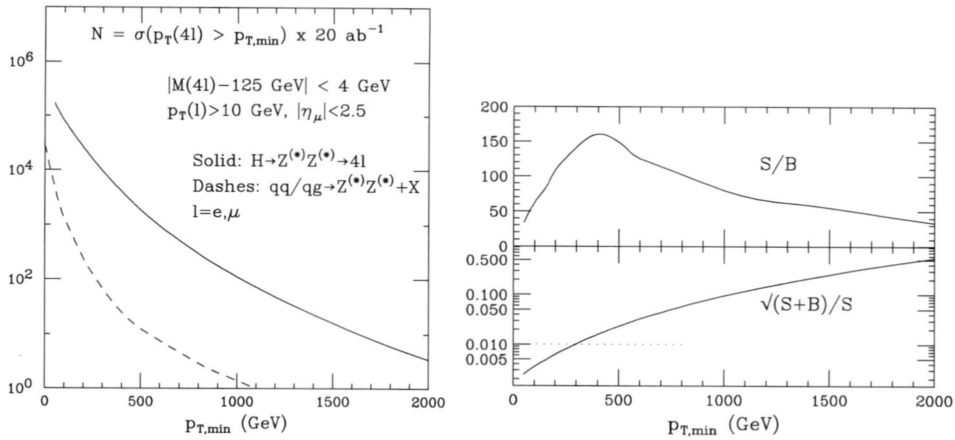

Fig. 17.3 Integrated transverse momentum rates (20 ab^{-1}) for a four-lepton final state ($\ell = e, \mu$), with mass close to the Higgs mass: signal and QCD background.

Table 17.5 Expected statistical precision on Higgs couplings with FCC-pp, assuming the absolute knowledge of g_{HZZ} from FCC-ee.

XY	$\gamma\gamma$	tt	HH	$\mu\mu$	$Z\gamma$
$\delta g_{HXY}(\%)$	0.5	1	3.5	1	1

17.5 The precision frontier at the FCC

The in-depth study of EW processes and the detection of possible small SM deviations demands a new step in the precise determination of both EW observables and EW fundamental parameters ($M_{Z,W,top}$, $\sin^2\theta_W$, $g_{V,A}$ couplings of fermions, $\alpha_{QED,s}(M_Z)$, etc). FCC-ee runs at the Z peak and at the W^+W^- and $t\bar{t}$ thresholds will lead to potential improvements in precision by factors ranging from 10 to 100 for both observables and parameters [Bicer *et al.* (2014); Azzi *et al.* (2017)]. The detector design [Rolandi (2017)] aims at addressing the challenge of controlling the experimental systematics required to match this potential precision, and the needs are being outlined for a new generation of higher order theoretical calculations [Heinemeyer (2017)]. When combined with the projected Higgs precision, a global analysis would lead to a sensitivity, or constraints, on the scale of possible new physics reaching up to several tens of TeV. This is shown for example in Fig. 17.4 from ref. [De Blas (2017)], which displays the scale sensitivity for a large set of dimension-6 effective operators, described in detail in refs. [de Blas *et al.* (2016, 2017)]. The various bands represent the contribution of successive additions of data from several FCC-ee runs, and the colour shadings describe the impact of parametric and theoretical uncertainties. The energy reach arising from this type of studies is well consistent with the mass reach for direct observation of new phenomena provided by 100 TeV pp collisions, underlying the strong synergy between the two components of the FCC programme.

FCC-hh itself can directly contribute to the EW precision programme, exploiting its energy lever arm. The contribution of higher-dimension operators grows with energy, and provided one retains sensitivity to deviations at the level of a few 10%, it is possible to indirectly explore large mass scales while remaining in the domain where an effective field theory description is justified [Contino *et al.* (2016)]. If the correction to some observable $\mathcal{O}$ induced by a certain dim-6 operator evolves with the measurement scale Q like $\delta_{\mathcal{O}} \propto Q^2/\Lambda^2$, then a 10% measurement at $Q = 1\,\text{TeV}$ has the same sensitivity of a per mille measurement at $Q = 100\,\text{GeV}$. As an example, this observation was explored in detail in ref. [Farina *et al.* (2017)], considering the constraining power of high-mass neutral- (Z/γ^*) and charged-current (W) Drell–Yan (DY) measurements on the presence of the following dimension-6 operators:

$$\mathbf{W} = -\frac{W}{4m_W^2}\left(D_\rho W^a_{\mu\nu}\right)^2, \quad \mathbf{Y} = -\frac{Y}{4m_W^2}\left(\partial_\rho B_{\mu\nu}\right)^2. \tag{17.1}$$

The results are summarised in Table 17.6, showing, for these particular operators, the strength of pp colliders in relation to e^+e^-. Another possible use of high-mass

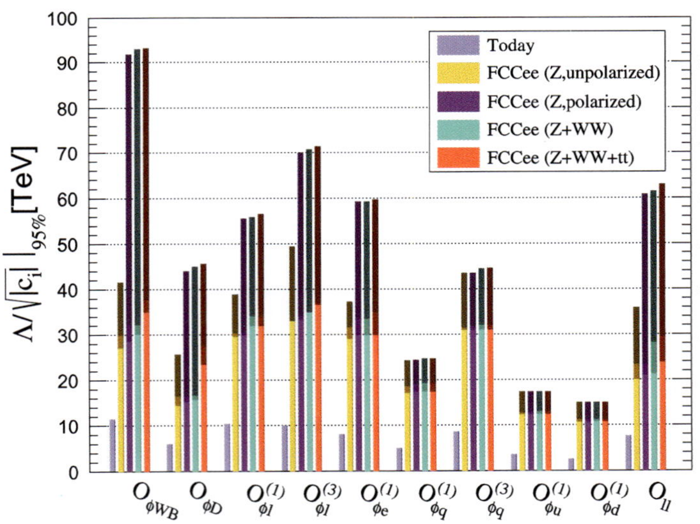

Fig. 17.4 High mass scale sensitivity, for various dim-6 operators, with the reach of FCC-ee through precision EW and Higgs observables. For each operator, the bars from left to right represent: the current status; the result of unpolarised runs at the Z peak; the addition of polarised Z runs; the addition of runs at the WW thereshold; the addition of runs at the $t\bar{t}$ threshold. The light lower part of each bar is the reach including the systematics estimated at the time of operations; the darker (darkest) shaded additions correspond to the neglect of the theoretical (theoretical and parametric) systematics. See refs. [De Blas (2017); de Blas *et al.* (2016, 2017)].

Table 17.6 FCC sensitivity to the dim-6 **W** and **Y** operators (see Eq. (17.1)), compared to LEP and LHC [Farina *et al.* (2017)]. NC: neutral current DY processes; CC: charged current DY processes.

	LEP	LHC (300fb^{-1})	LHC (3ab^{-1})	FCC-pp (10ab^{-1})	FCC-ee (10^{12} Z)
NC: $W \times 10^4$	$[-19, 3]$	± 1.5	± 0.8	± 0.04	± 0.7
NC: $Y \times 10^4$	$[-17, 4]$	± 2.3	± 1.2	± 0.06	± 1
CC: $W \times 10^4$	—	± 0.7	± 0.45	± 0.02	—

DY production at FCC-pp, in the context of precision EW studies, was proposed in ref. [Alves *et al.* (2015)]: monitoring the Q^2 dependence of the DY cross section allows one to measure the running of the weak coupling constants α_W, with sufficient sensitivity to detect the onset of new thresholds in the EW β-functions, a signal of the existence of new massive particles with EW couplings. This measurement could therefore be sensitive to massive weakly-interacting particles otherwise elusive to direct searches.

In addition to the precision programme of EW and Higgs physics, FCC will greatly enhance the knowledge of top quark properties. The experimental precision of top mass measurement, carried out at the production threshold in FCC-ee, could reach ± 15 MeV, and is therefore limited only by the theoretical systematics, estimated to be in the range of ± 50 MeV. Measurements of the lepton energy spectrum and angular distribution, in semileptonic $t\bar{t}$ decays at $\sqrt{S} \sim 370$ GeV, will determine [Janot (2015)] the SM vector (axial) couplings to the photon and the Z with a precision of $O(10^{-3})$ ($O(10^{-2})$). Production of $t\bar{t}$ pairs with invariant mass in the multi-TeV region at FCC-hh, furthermore, can greatly extend the future LHC constraints on the anomalous chromomagnetic moments [Aguilar-Saavedra *et al.* (2015)]. The statistics of over a billion of $t\bar{t}$ pairs produced at FCC-hh [Mangano *et al.* (2017)] provide new opportunities to search for rare or forbidden top decays.

The huge statistics of Z production, with clean samples of tagged fermions such as tau leptons or bottom quarks, allow furthermore for a rich programme of flavour physics [FCC (2017)], including searches for rare or forbidden decays, massive or sterile neutrinos, and lepton-flavour violating processes. As for many other topics in FCC physics, studies have only started now uncovering the full spectrum of opportunities for a broad range of incisive tests of the flavour and EW sectors.

17.6 The high-mass reach of the FCC

For massive particles at the edge of the available phase space, the energy growth at FCC-hh relative to LHC obviously leads to an increase in mass reach by a factor of 5–7, depending on the total integrated luminosity [Hinchliffe *et al.* (2015)]. This is the case, for example, for new gauge bosons such as W' or Z', or for supersymmetric particles such as squarks and gluinos. Some examples are given in Figs. 17.5 and 17.6. Many more examples and studies are given in ref. [Golling

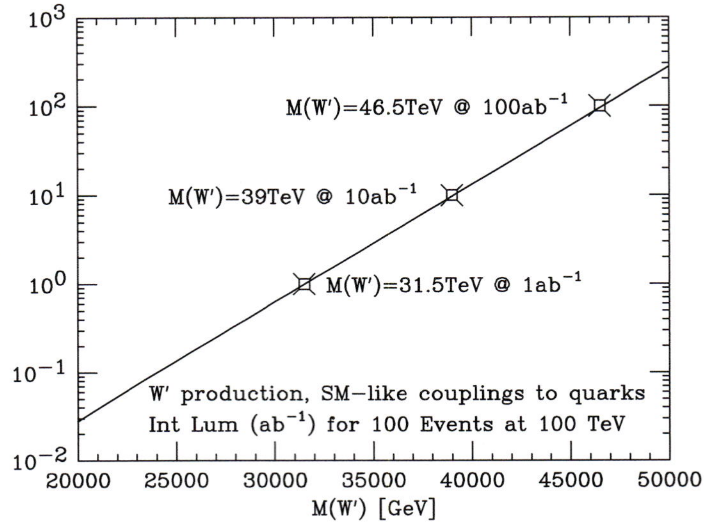

Fig. 17.5 Mass reach sensitivity, as a function of integrated luminosity, for production of 100 events of a new W' gauge boson with SM-like couplings.

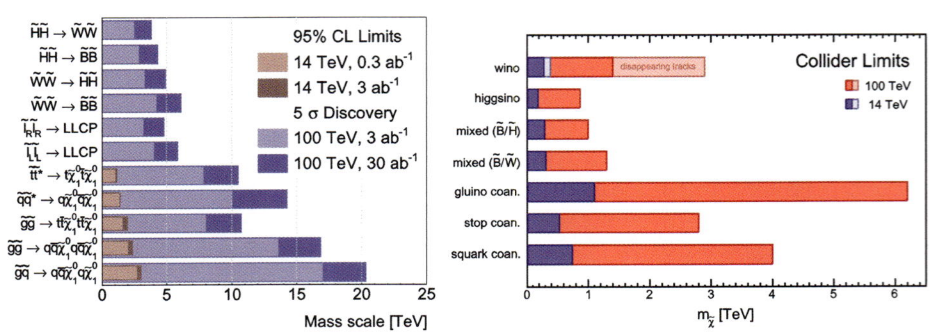

Fig. 17.6 Mass reach of FCC-hh for supersymmetric (left) and WIMP DM particles (right), studied in the context of simplified models. Details for both sets of studies are given in ref. [Golling *et al.* (2017)]

et al. (2017)]. This huge explorative power could allow discovery of, or exclude, many scenarios of new physics. As mentioned in the Introduction, for example, the reach for WIMPS could extend to the upper limit of the expected mass window, which lies within a range of a few TeV, depending on the EW charge. This is shown explicitly in Fig. 17.6, where higgsino-like states (EW doublets) could be probed up to $m \sim 1$ TeV, and wino-like states (EW triplets) up to $m \sim 3$ TeV. In both cases, these values indeed represent the upper end of the allowed mass range. These conclusions are the start of the effort to prove a no-lose scenario for the discovery or exclusion of WIMPs as a DM candidate.

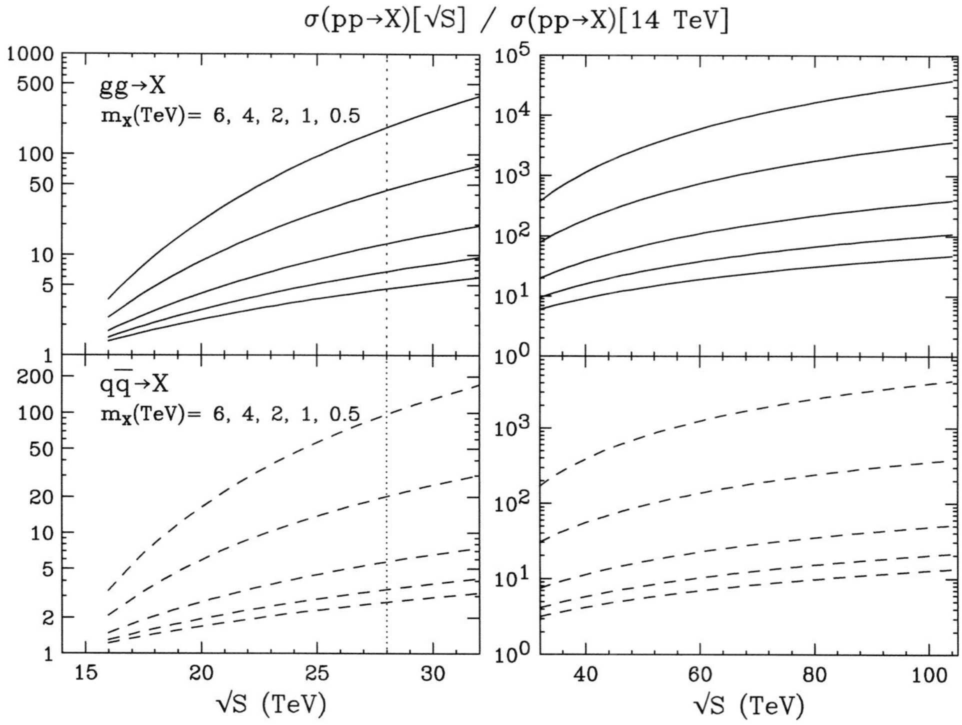

Fig. 17.7 Increase with energy of the partonic luminosities, with respect to $\sqrt{S} = 14\,\text{TeV}$, for the production of massive final states via the gg and $q\bar{q}$ production channels.

Other new discovery scenarios might emerge from the results of the LHC data. More in general, FCC-hh would greatly increase the statistics of possible discoveries of new phenomena by the LHC, and allow a thorough study of their origin. To illustrate this, Fig. 17.7 shows the increase with energy of the partonic luminosities, with respect to $\sqrt{S} = 14\,\text{TeV}$, for the production of massive final states X via the gg and $q\bar{q}$ production channels. $m_X = 6\,\text{TeV}$, for example, would correspond to the production of a single 6 TeV particle in the s-channel, or of a pair of $\lesssim 3\,\text{TeV}$ particles. The increase in rate at FCC-hh, relative to the LHC, could reach 3 or 4 orders of magnitude!

17.7 The possible role of the HE-LHC

The development of 16 T dipoles, required to meet the goal of 100 TeV pp collisions in a 100 km ring, could lead to collisions in the range of 28 TeV in the LHC tunnel, a possible future evolution of the LHC known as HE-LHC [Zimmermann (2017)]. The timescale required for mass-scale production of 16 T magnets could allow the construction of HE-LHC to start right after the completion of the HL-LHC programme. The scale of the project, while smaller than the FCC in terms of time, cost and difficulties, is likely to exceed that of the LHC, and this therefore cannot be considered as a mere upgrade like the HL-LHC. HE-LHC will require a complete removal of the current hardware (costly in terms of time and money), the replacement of the magnets (likely to be more expensive than the LHC ones) and of most of the LHC elements (radiofrequency system, cryogenics, collimation, etc.). The full exploitation of the possible high luminosity (which could almost quadruple with respect to the HL-LHC [Zimmermann (2017)]) will also require an important upgrade of the SPS injector and transfer lines, to increase the injection energy to the TeV level. And, unavoidably, the detectors will need major maintenance and upgrades after the HL-LHC phase. Once undertaken, this project would likely engage the community for a period of no less than 30 years, including construction and full data taking, with a consequent postponement of the start of FCC.

While some studies of the physics potential of the HE-LHC have appeared since the early 2000's, a systematic analysis, also in view of the Higgs discovery and of the status of BSM searches today, is still lacking. For what concerns the general discovery reach, one expects the obvious factor of $\sim$2 extension relative to the LHC. In complete absence of LHC discoveries, or of hints of possible phenomena just behind the corner from the LHC, it is debatable whether such a step forward in energy justifies the effort. There are nevertheless scenarios, emerging from LHC discoveries, which could make the HE-LHC particularly interesting. Figure 17.7 can help orient this discussion. Let us consider, for example, a possible LHC discovery of a final state X with $m_X \sim 6$ TeV in the gg channel. This is close to the borderline of what the LHC can reach. The rate for such a final state at 28 TeV would grow by a factor of $\sim$200.[2] This is smaller than the over 10^4 gain obtained by FCC-hh at 100 TeV, but is sufficient, united with the higher luminosity of HE-LHC, to improve by a factor of $\gtrsim$20 the statistical precision of the measurement of its properties, and to gain access to possible additional nearby new states. Learning more about these objects, on a timescale shorter than the FCC-hh would allow, could be a good way to prepare on more solid ground the big jump to 100 TeV, and to maintain continuity in the activity of the HEP community at large. Needless to say the choice could depend on the first interpretations of an LHC discovery, and on the anticipation of what else might accompany it, whether it should be nearby, or for example equally spaced like a Kaluza–Klein resonance. Such a scenario of sequential resonances could be hard to test at 28 TeV.

As a second scenario, one could consider the discovery at the LHC of a state X with m_X in the 0.5–1 TeV range, in the $q\bar{q}$ channel. This is the typical HL-LHC reach for non-resonant production of a pair of weakly coupled particles, such as gauginos or sleptons. The cross-section increase at 28 TeV is a mere factor of $\sim$3–4 relative to the LHC, growing to only 10–20 at 100 TeV. The impact of the higher

[2]Notice that, at this high mass, even a mere upgrade from 14 to 16 TeV would increase the rate by a factor of 4.

energy here is therefore limited, and, all things considered, other inputs would be needed for a clear choice between HE-LHC and FCC-hh (costs, timescales, potential presence of other phenomena at the relevant energies, etc.).

A more systematic set of studies, addressing in particular the potential of HE-LHC to deepen our understanding of the Higgs boson, starting from the determination of its self-coupling, is an important priority for the planning of future facilities.

17.8 Conclusions

We still lack today direct and concrete experimental clues on the nature of the theory that will eventually extend the SM and accommodate its shortcomings. This makes it hard to outline a development path for our field that will lead, with certainty, to discovery. Chances of success can, however, be increased by investing on several complementary and synergetic fronts, to extend the upper end of the achievable mass scale via both direct discovery and via precision measurements, and to sharpen the sensitivity at lower masses to possible elusive signals. A key goal of the next large-scale facility must be to leave no stone unturned in the scrutiny of phenomena at the EW scale, up to several TeVs.

The studies of the FCC physics potential focus today on exploring possible scenarios, emerging from data and from new theoretical ideas, defining benchmarks for the accelerator and detector design and performance, in order to better inform the discussions that will take place when the time to take a decision will come. The existing studies already show that the interplay of the three colliders (*ee*, *hh* and *eh*) is key to the full exploitation of the FCC physics potential. The most ambitious component of the programme, the 100 TeV *pp* collider, is a clear long-term goal for the field, simply because no other proposed or foreseeable project can have direct sensitivity to such large mass scales: for all we know today, high-energy physics will continue seeking higher and higher energies as its most effective way to learn about fundamental phenomena at the shortest distances. Nevertheless, the precise route followed to get to 100 TeV must take account of the fuller picture, to emerge from the LHC as well as from other current and future experiments in areas ranging from flavour physics and neutrinos, to searches for dark matter and axions. For example, an intermediate step at the HE-LHC might well be the best choice if LHC discoveries pointed to that energy range, and if the financial envelope of the FCC required to stretch the project over a very long timescale. The proper assessment of the FCC technological feasibility and cost will require several years. For example, the first full prototypes of the 16 T magnets may not be ready before 2026, just before the start of HL-LHC. We therefore have time to carry on our physics evaluation, with a constant reference to the status of the ongoing progress. The next 10 years will be crucial for the definition of the far future of our research.

Chapter 18

Flashes from Guido's life

Monica Pepe Altarelli
CERN, EP Department
CH 1211 Geneva 23, Switzerland

13 July 2011: the day after Guido's 70th birthday. This is when our life was turned upside down. I was by chance in Guido's office at CERN, at the moment in which he casually opened a pdf file with the results of a routine blood test that he had taken the day before. Everything was perfect, as usual, with many parameters literally as good as those of a young man, which always triggered the compliments of our general practitioner and the comment that he was destined to a very long life, just like his parents. Except for ... in the middle of the last page, in bold characters, a PSA level of 14.72 ng/mL, while normal values should be between 0 and 4. We were flabbergasted .... How could it be? Guido had checked his PSA only one and a half years before and it was completely normal. I remember he said to me, "Mop (a funny contraction between my first and last names, which he often used), I'm doomed" ("Mop, sono fottuto"). Guido used to keep a journal, in which he jotted down a few lines every day. On that day he wrote "Terrore". A doctor told us "But cancer does not manifest itself like this!", and another "The PSA is too low for the tumour to have metastasised". In the following days, we went through the stress of a second PSA test, and then biopsy, skeletal scintigraphy, MRI, PET ... . Before every test, doctors described possible options, some of which were much more favourable than others, giving us hope. At some point we made plans for an operation. But Guido was not lucky and at every step the worst-case scenario always prevailed. The tumour was particularly aggressive, had already metastasised to lymph nodes and bones, and could not be operated on nor irradiated. It could only be treated pharmacologically with a limited number of weapons: a couple of hormonal treatments and two types of chemo (therapy).

Guido became a member of the tribe of the sick.

He decided to only tell the kids, his brother and a few close friends. Despite this dramatic reminder of human mortality, we tried to live as normal a life as we could. However, it was a life marked by the emotional and psychological distress of the regular PSA tests ("I numerini", Guido used to call them, in a gentle, almost familiar way, "The small numbers"), which inexorably gave us a measure of the progression of the disease.

Guido had a controlled and rational attitude towards his cancer, sometimes almost as if it were someone else's disease. He tried to take his destiny in his own hands, started to study all the most recent research papers in specialised medical journals, including presentations at the most respected scientific urological congresses, and very quickly became very familiar with the subject. He was more up-to-date on the latest research than his oncologist and often steered the therapy in the direction which he thought was best. His oncologist was impressed and explicitly told him that he in his long career had never encountered a patient like him. He even managed to meet one of the top researchers on urologic oncology at the Sloan–Kettering Institute in New York, who was also surprised by his knowledge, and the lucid way in which he presented his personal file, documented by graphs, fits and a quantitative analysis of the cancer evolution. In the end, it all made little difference. Doctors had told him from the very beginning that he would live four to five years. This time they were right. He lived for four. Always with great dignity and self-restraint. In the early hours of September 30th 2015, Guido had a stroke, related to his very degraded health and to the nasty effects of the chemo and all other medications that he took to control pain. He died a few hours later. By that time, the cancer had already spread to the liver (when he found out, in his journal he wrote "End of the story") and a scan of his skeleton looked like a terrifying Christmas tree. He had just started the second and last type of chemo, purely at my insistence, as he was tired.

Still, death managed to take us by surprise and Guido did depart without a goodbye. The evening before, he was relatively well, had driven his daughter Claudia to the airport, and had watched with growing irritation a Champions League match lost by his beloved football team, A.S. Roma. The images of that fatal night will never abandon me.

The news of Guido's death produced an instant outpouring of emotion, with hundreds and hundreds of e-mails, letters, phone-calls. It was clear that he had touched the hearts of many people. Also, many had not realised that he was gravely ill and were in shock. Most of those messages were not just a formal expression of condolences; they were clearly strongly felt, sincere, sometimes recalling anecdotes, or things that he had done, said or written as a mentor, teacher, friend, and that had touched the lives of the people who wrote them. I knew that Guido was held in high regard as a scientist and as a person, but I did not expect such a strong, almost tangible, collective emotion. He was modest and would have been surprised and very moved. In my terrible loss, this was a consolation, even if essentially for one year I could not walk around CERN or even sit at my desk in the office, without being stopped by people expressing their sympathy and sorrow. Then, there were several, very moving memorial celebrations, attended by friends and colleagues from all over the world, articles, a yearly award created to honour Guido's memory in the context of the Deep Inelastic Scattering conference series, an effort to collect and compile his written notes, the publication of his Primer on Collider Physics within the Standard Model, and this beautiful book, for which I am very grateful to Aharon, Giovanni, Stefano and all the authors.

I only wish he could see all this!

I have chosen to write about our last four years, as they were so hard and intense and their memory still so vivid in my mind. Guido had a full, beautiful life. From my vast repertoire of photographs, I have selected a few to share with you. Guido was a cultivated man and an avid reader, especially of history books (he liked Roman history, the French revolution, more recently the great Asian civilisations: India, Japan, China) and biographies (e.g. Napoleon, the Roman emperors, the French generals of the Revolutionary and Napoleonic Wars, great scientists). He had a wry sense of humor and a very creative and witty use of the Italian language. His play on words could be hilarious. He loved classical music and the arts. Together we started a collection of Japanese woodblock prints (ukiyo-e), which now decorate the walls of our house and of the apartment in Rome. Before studying physics, he had a classical education and could still recite a lot of poetry by heart. I enclose a beautiful poem that he liked very much, written by Vincenzo Cardarelli (1887–1959).

Dying, yes,
not to be assaulted by death!
Dying, persuaded
that such a journey is the best.
And in that ultimate instant to be cheerful
like when one counts the minutes
of the wall clock at the station
and each of them is worth a century.
For death is the faithful spouse
that takes over the deceitful mistress,
we do not want to receive her as an intruder,
nor run away with her.
Too many a time we departed
without a goodbye!
On the verge of stepping, in a moment,
into the time
when even our memory
will desert us,
let us, oh Death, bid farewell to the world,

allow us to linger a moment.
The incommensurable step ought not be
precipitous.
The sheer thought of a sudden death
freezes my very blood.
Oh Death, do not clutch me away
rather announce yourself from afar
and, as a friend, take me
like the ultimate of my habitudes.

Vincenzo Cardarelli

Morire sì,
non essere aggrediti dalla morte.
Morire persuasi
che un siffatto viaggio sia il migliore.
E in quell'ultimo istante essere allegri
come quando si contano i minuti
dell'orologio della stazione
e ognuno vale un secolo.
Poi che la morte è la sposa fedele
che subentra all'amante traditrice,
non vogliamo riceverla da intrusa,
né fuggire con lei.
Troppo volte partimmo
senza commiato!
Sul punto di varcare
in un attimo il tempo,
quando pur la memoria
di noi s'involerà,
lasciaci, o Morte, dire al mondo addio,
concedici ancora un indugio.
L'immane passo non sia
precipitoso.
Al pensier della morte repentina
il sangue mi si gela.
Morte non mi ghermire
ma da lontano annunciati
e da amica mi prendi
come l'estrema delle mie abitudini.

Vincenzo Cardarelli

Fig. 18.1 Guido, 20 years old.

Fig. 18.2 Elephanta caves, Mumbai, January 1989. This is not a high-quality photograph, but it was our first trip to India together, a real adventure and a source of many stories and conversations. We were heading to New Delhi to tour of the north of India, while the second part of the trip was physics related — a conference at the Tata institute in Mumbai. However, due to thick fog, instead of landing in New Delhi, the plane was forced to land in Mumbai, i.e. over 1000 km south! We spent the following day stranded at Mumbai airport, waiting in frustration. Our flight to New Delhi was postponed until the late afternoon, when we finally departed. Unfortunately, by the time we reached New Delhi, the fog had come back and the aircraft could not land. So, we flew back to Mumbai (yes, a four-hour Mumbai–Mumbai flight), where we arrived late at night, totally exhausted, and still had to queue for hours with many other exhausted passengers to receive our hotel voucher. We lost all connections and were unable to reach any other tourist destination because all flights were full due to the Christmas holidays. The only thing we managed to do was to take a night bus to Ajanta and Ellora, two beautiful archaeological sites, from which we returned by taxi, another very interesting experience.

Fig. 18.3 Patagonia, January 1992: a fabulous trip in which we toured Argentina from North to South, besides spending a week at a physics school in Bariloche.

Fig. 18.4 August 2006, Guido's retirement party at CERN, with Jacques Prentki.

Fig. 18.5 Japan, August 2007, climbing Mount Fuji.

Fig. 18.6 Thoiry, October 2007. This photo was taken "chez" Mariella and Emilio Picasso on the occasion of a reunion party for the g-2 friends.

Fig. 18.7 Marettimo, Egadi islands, August 2011. Guido happy at having caught a big octopus that we cooked for dinner. He enjoyed cooking and we often invited friends over for dinner. He loved going out to restaurants and had a preference for simple, unpretentious, high-quality food. He never skipped a meal and never worked after dinner.

Fig. 18.8 April 2012, at the ceremony for the award of the Sakurai prize. Guido's luggage was lost, so he wore the same shoes and jeans that he had for the trip, a shirt that we bought and Rocky Kolbe's tie and jacket, which miraculously fit to perfection. His outfit was not too bad, given the circumstances!

Fig. 18.9 Crete, September 2013, Guido, happy in Kolymbari.

Fig. 18.10 Genova, May 2013 (photograph by Nanni Darbo).

Fig. 18.11 Krakow, September 2013. Guido being thoughtful, at the ceremony in which he was named Honorary Professor of the Institute of Nuclear Physics.

Fig. 18.12 Rome, December 2014. His last Christmas with children and grand-children.

Fig. 18.13 La Thuile, March 2015. At the Rencontres de Moriond, cutting the 50th anniversary cake with Daniel Treille and Jacques Dumarchez. At that meeting Guido and Daniel both gave a personal recollection on 50 years of particle physics, in theory and experiment, as seen through Les Rencontres. I remember that Guido began by saying “All speakers start their talk by thanking the organisers for their kind invitation (indeed they all did, without exception so that this had become an unintentionally comic refrain) but I believe that the organisers should thank us, Daniel and me, for having accepted to prepare this talk”. Everybody burst into laughter. The proceedings of Moriond 2015 were dedicated by the Organisers to “the memory of our dear friend Guido Altarelli”.

Fig. 18.14 Vienna, August 2015, at the award of the EPS HEPP prize, for an outstanding contribution to high energy physics, with Thomas Lohse, Lev Lipatov, Yuri Dokshitzer and Giorgio Parisi. It was a blessing that Guido's work was recognised with this prestigious prize only a few months before his death and that he was still well enough to attend and enjoy the ceremony.

References

Aaboud, M. *et al.* (2016). Measurement of the top quark mass in the $t\bar{t} \to$ dilepton channel from $\sqrt{s} = 8$ TeV ATLAS data, *Phys. Lett.* **B761**, 350–371, arXiv:1606.02179 [hep-ex].

Aaboud, M. *et al.* (2017a). Evidence for the $H \to b\bar{b}$ decay with the ATLAS detector, *JHEP* **12**, 024, arXiv:1708.03299 [hep-ex].

Aaboud, M. *et al.* (2017b). Measurement of inclusive jet and dijet cross-sections in proton-proton collisions at $\sqrt{s} = 13$ TeV with the ATLAS detector, arXiv:1711.02692 [hep-ex].

Aaboud, M. *et al.* (2017c). Measurement of lepton differential distributions and the top quark mass in $t\bar{t}$ production in pp collisions at $\sqrt{s} = 8$ TeV with the ATLAS detector, *Eur. Phys. J.* **C77**, 11, 804, arXiv:1709.09407 [hep-ex].

Aaboud, M. *et al.* (2017d). Measurement of the cross-section for electroweak production of dijets in association with a Z boson in pp collisions at $\sqrt{s} = 13$ TeV with the ATLAS detector, *Phys. Lett.* **B775**, 206–228, arXiv:1709.10264 [hep-ex].

Aaboud, M. *et al.* (2017e). Measurement of the inclusive jet cross-sections in proton-proton collisions at $\sqrt{s} = 8$ TeV with the ATLAS detector, *JHEP* **09**, 020, arXiv:1706.03192 [hep-ex].

Aaboud, M. *et al.* (2017f). Measurement of the prompt J/ ψ pair production cross-section in pp collisions at $\sqrt{s} = 8$ TeV with the ATLAS detector, *Eur. Phys. J.* **C77**, 2, 76, arXiv:1612.02950 [hep-ex].

Aaboud, M. *et al.* (2017g). Measurements of electroweak Wjj production and constraints on anomalous gauge couplings with the ATLAS detector, *Eur. Phys. J.* **C77**, 7, 474, arXiv:1703.04362 [hep-ex].

Aaboud, M. *et al.* (2017h). Precision measurement and interpretation of inclusive W^+, W^- and Z/γ^* production cross sections with the ATLAS detector, *Eur. Phys. J.* **C77**, 6, 367, arXiv:1612.03016 [hep-ex].

Aaboud, M. *et al.* (2017i). Search for a new heavy gauge boson resonance decaying into a lepton and missing transverse momentum in 36 fb^{-1} of pp collisions at $\sqrt{s} = 13$ TeV with the ATLAS experiment, arXiv:1706.04786 [hep-ex].

Aaboud, M. *et al.* (2018). Measurement of the W-boson mass in pp collisions at $\sqrt{s} = 7$ TeV with the ATLAS detector, *Eur. Phys. J.* **C78**, 2, 110, arXiv:1701.07240 [hep-ex].

Aad, G. *et al.* (2009). Expected performance of the ATLAS experiment — Detector, trigger and physics, Tech. Rep. SLAC-R-980, CERN-OPEN-2008-020, CERN, arXiv:0901.0512 [hep-ex].

Aad, G. *et al.* (2012a). Determination of the strange quark density of the proton from ATLAS measurements of the $W \to \ell\nu$ and $Z \to \ell\ell$ cross sections, *Phys. Rev. Lett.* **109**, 012001, arXiv:1203.4051 [hep-ex].

Aad, G. *et al.* (2012b). Measurement of the inclusive $W^{\pm}$ and Z/γ cross sections in the electron and muon decay channels in pp collisions at $\sqrt{s} = 7$ TeV with the ATLAS detector, *Phys. Rev.* **D85**, 072004, `arXiv:1109.5141 [hep-ex]`.

Aad, G. *et al.* (2012c). Observation of a new particle in the search for the Standard Model Higgs boson with the ATLAS detector at the LHC, *Phys. Lett.* **B716**, 1–29, `arXiv:1207.7214 [hep-ex]`.

Aad, G. *et al.* (2013). Measurements of $W\gamma$ and $Z\gamma$ production in pp collisions at $\sqrt{s} = 7$ TeV with the ATLAS detector at the LHC, *Phys. Rev.* **D87**, 11, 112003, `arXiv:1302.1283 [hep-ex]`, [Erratum: *Phys. Rev.* **D91**, 11, 119901 (2015)].

Aad, G. *et al.* (2014a). Measurement of the electroweak production of dijets in association with a Z-boson and distributions sensitive to vector boson fusion in proton-proton collisions at $\sqrt{s} = 8$ TeV using the ATLAS detector, *JHEP* **04**, 031, `arXiv:1401.7610 [hep-ex]`.

Aad, G. *et al.* (2014b). Measurement of the $t\bar{t}$ production cross-section using $e\mu$ events with b-tagged jets in pp collisions at $\sqrt{s} = 7$ and 8 TeV with the ATLAS detector, *Eur. Phys. J.* **C74**, 10, 3109, `arXiv:1406.5375 [hep-ex]`, [Addendum: *Eur. Phys. J.* **C76**, 11, 642 (2016)].

Aad, G. *et al.* (2015a). Combined measurement of the Higgs boson mass in pp collisions at $\sqrt{s} = 7$ and 8 TeV with the ATLAS and CMS experiments, *Phys. Rev. Lett.* **114**, 191803, `arXiv:1503.07589 [hep-ex]`.

Aad, G. *et al.* (2015b). Evidence of $W\gamma\gamma$ production in pp collisions at $\sqrt{s} = 8$ TeV and limits on anomalous quartic gauge couplings with the ATLAS detector, *Phys. Rev. Lett.* **115**, 3, 031802, `arXiv:1503.03243 [hep-ex]`.

Aad, G. *et al.* (2015c). Study of the spin and parity of the Higgs boson in diboson decays with the ATLAS detector, *Eur. Phys. J.* **C75**, 10, 476, `arXiv:1506.05669 [hep-ex]`, [Erratum: *Eur. Phys. J.* **C76**, 3, 152 (2016)].

Aad, G. *et al.* (2016a). Measurement of the double-differential high-mass Drell–Yan cross section in PP collisions at $\sqrt{s} = 8$ TeV with the ATLAS detector, *JHEP* **08**, 009, `arXiv:1606.01736 [hep-ex]`.

Aad, G. *et al.* (2016b). Measurement of the transverse momentum and ϕ^*_η distributions of Drell–Yan lepton pairs in proton-proton collisions at $\sqrt{s} = 8$ TeV with the ATLAS detector, *Eur. Phys. J.* **C76**, 5, 291, `arXiv:1512.02192 [hep-ex]`.

Aad, G. *et al.* (2016c). Measurements of four-lepton production in pp collisions at $\sqrt{s} = 8$ TeV with the ATLAS detector, *Phys. Lett.* **B753**, 552–572, `arXiv:1509.07844 [hep-ex]`.

Aad, G. *et al.* (2016d). Measurements of the Higgs boson production and decay rates and constraints on its couplings from a combined ATLAS and CMS analysis of the LHC PP collision data at $\sqrt{s} = 7$ and 8 TeV, *JHEP* **08**, 045, `arXiv:1606.02266 [hep-ex]`.

Aad, G. *et al.* (2016e). Measurements of $Z\gamma$ and $Z\gamma\gamma$ production in pp collisions at $\sqrt{s} = 8$ TeV with the ATLAS detector, *Phys. Rev.* **D93**, 11, 112002, `arXiv:1604.05232 [hep-ex]`.

Aad, G. *et al.* (2016f). Searches for scalar leptoquarks in PP collisions at $\sqrt{s} = 8$ TeV with the ATLAS detector, *Eur. Phys. J.* **C76**, 1, 5, `arXiv:1508.04735 [hep-ex]`.

Aaij, R. *et al.* (2013). Exclusive J/ψ and ψ(2S) production in pp collisions at $\sqrt{s} = 7$ TeV, *J. Phys.* **G40**, 045001, `arXiv:1301.7084 [hep-ex]`.

Aaij, R. *et al.* (2014). Updated measurements of exclusive J/ψ and ψ(2S) production cross-sections in PP collisions at $\sqrt{s} = 7$ TeV, *J. Phys.* **G41**, 055002, `arXiv:1401.3288 [hep-ex]`.

Aaij, R. *et al.* (2015). Measurement of the exclusive production cross-section in PP collisions at $\sqrt{s} = 7$ TeV and 8 TeV, *JHEP* **09**, 084, arXiv:1505.08139 [hep-ex].

Aaij, R. *et al.* (2016a). Measurement of forward W and Z boson production in pp collisions at $\sqrt{s} = 8$ TeV, *JHEP* **01**, 155, arXiv:1511.08039 [hep-ex].

Aaij, R. *et al.* (2016b). Measurement of the forward Z boson production cross-section in pp collisions at $\sqrt{s} = 13$ TeV, *JHEP* **09**, 136, arXiv:1607.06495 [hep-ex].

Aaij, R. *et al.* (2016c). Measurements of prompt charm production cross-sections in pp collisions at $\sqrt{s} = 13$ TeV, *JHEP* **03**, 159, arXiv:1510.01707 [hep-ex], [Erratum: *JHEP* **05**, 074 (2017)].

Aamodt, K. *et al.* (2008). The ALICE experiment at the CERN LHC, *JINST* **3**, S08002.

Aarnio, P. A. *et al.* (1991). The DELPHI detector at LEP, *Nucl. Instrum. Meth.* **A303**, 233–276.

Aaron, F. D. *et al.* (2010a). Combined measurement and QCD analysis of the inclusive $e^{\pm}p$ scattering cross sections at HERA, *JHEP* **01**, 109, arXiv:0911.0884 [hep-ex].

Aaron, F. D. *et al.* (2010b). Measurement of the charm and beauty structure functions using the H1 vertex detector at HERA, *Eur. Phys. J.* **C65**, 89–109, arXiv:0907.2643 [hep-ex].

Aaron, F. D. *et al.* (2011a). Measurement of the inclusive $e^+ - p$ scattering cross section at high inelasticity y and of the structure function F_L, *Eur. Phys. J.* **C71**, 1579, arXiv:1012.4355 [hep-ex].

Aaron, F. D. *et al.* (2011b). Search for contact interactions in $e^{\pm}p$ collisions at HERA, *Phys. Lett.* **B705**, 52–58, arXiv:1107.2478 [hep-ex].

Aaron, F. D. *et al.* (2011c). Search for first generation leptoquarks in ep collisions at HERA, *Phys. Lett.* **B704**, 388–396, arXiv:1107.3716 [hep-ex].

Aaron, F. D. *et al.* (2012). Inclusive deep inelastic scattering at high Q^2 with longitudinally polarised lepton beams at HERA, *JHEP* **09**, 061, arXiv:1206.7007 [hep-ex].

Abazov, V. M. *et al.* (2011). Measurement of $\sin^2\theta^{\ell}_{\text{eff}}$ and Z-light quark couplings using the forward-backward charge asymmetry in $p\bar{p} \to Z/\gamma^* \to e^+e^-$ events with $\mathcal{L} = 5.0$ fb^{-1} at $\sqrt{s} = 1.96$ TeV, *Phys. Rev.* **D84**, 012007, arXiv:1104.4590 [hep-ex].

Abbas, G., Ananthanarayan, B., Caprini, I., and Fischer, J. (2013). Perturbative expansion of the QCD Adler function improved by renormalization-group summation and analytic continuation in the Borel plane, *Phys. Rev.* **D87**, 1, 014008, arXiv:1211.4316 [hep-ph].

Abbate, R., Fickinger, M., Hoang, A. H., Mateu, V., and Stewart, I. W. (2011). Thrust at N^{3}LL with power corrections and a precision global fit for $\alpha_s(m_Z)$, *Phys. Rev.* **D83**, 074021, arXiv:1006.3080 [hep-ph].

Abbiendi, G. *et al.* (2000). Precision luminosity for Z^0 line shape measurements with a silicon tungsten calorimeter, *Eur. Phys. J.* **C14**, 373–425, arXiv:hep-ex/9910066 [hep-ex].

Abbiendi, G. *et al.* (2006). Determination of α_s using jet rates at LEP with the OPAL detector, *Eur. Phys. J.* **C45**, 547–568, arXiv:hep-ex/0507047 [hep-ex].

Abbiendi, G. *et al.* (2011). Determination of $alpha_s$ using OPAL hadronic event shapes at $\sqrt{s} = 91 - 209$ GeV and resummed NNLO calculations, *Eur. Phys. J.* **C71**, 1733, arXiv:1101.1470 [hep-ex].

Abbott, B. P. *et al.* (2016). Observation of gravitational waves from a binary black hole merger, *Phys. Rev. Lett.* **116**, 6, 061102, arXiv:1602.03837 [gr-qc].

Abe, H., Choi, K.-S., Kobayashi, T., and Ohki, H. (2009). Non-Abelian discrete flavor symmetries from magnetized/intersecting brane models, *Nucl. Phys.* **B820**, 317–333, arXiv:0904.2631 [hep-ph].

Abe, K. *et al.* (1997). Precision determination of the neutron spin structure function g_1^n, *Phys. Rev. Lett.* **79**, 26–30, `arXiv:hep-ex/9705012 [hep-ex]`.

Abe, K. *et al.* (1998). Measurements of the proton and deuteron spin structure functions g_1 and g_2, *Phys. Rev.* **D58**, 112003, `arXiv:hep-ph/9802357 [hep-ph]`.

Abel, S. A. and Goodsell, M. D. (2007). Realistic Yukawa couplings through instantons in intersecting brane worlds, *JHEP* **10**, 034, `arXiv:hep-th/0612110 [hep-th]`.

Abelev, B. B. *et al.* (2014). Exclusive J/ψ photoproduction off protons in ultraperipheral p-Pb collisions at $\sqrt{s_{\rm NN}} = 5.02$ TeV, *Phys. Rev. Lett.* **113**, 23, 232504, `arXiv:1406.7819 [nucl-ex]`.

Abelev, B. I. *et al.* (2006). Longitudinal double-spin asymmetry and cross section for inclusive jet production in polarized proton collisions at $\sqrt{s} = 200$ GeV, *Phys. Rev. Lett.* **97**, 252001, `arXiv:hep-ex/0608030 [hep-ex]`.

Abelev, B. I. *et al.* (2008). Longitudinal double-spin asymmetry for inclusive jet production in $p+p$ collisions at $\sqrt{s} = 200$ GeV, *Phys. Rev. Lett.* **100**, 232003, `arXiv:0710.2048 [hep-ex]`.

Abelev, B. I. *et al.* (2009). Longitudinal double-spin asymmetry and cross section for inclusive neutral pion production at midrapidity in polarized proton collisions at $\sqrt{s} = 200$ GeV, *Phys. Rev.* **D80**, 111108, `arXiv:0911.2773 [hep-ex]`.

Abelleira Fernandez, J. L. *et al.* (2012a). A large hadron electron collider at CERN: Report on the physics and design concepts for machine and detector, *J. Phys.* **G39**, 075001, `arXiv:1206.2913 [physics.acc-ph]`.

Abelleira Fernandez, J. L. *et al.* (2012b). On the relation of the LHeC and the LHC, Tech. Rep. CERN-LHeC-Note-2012-005-GEN, CERN, `arXiv:1211.5102 [hep-ex]`.

Abeyratne, S. *et al.* (2015). MEIC design summary, Tech. Rep. JLAB-ACP-15-2036, JLAB, `arXiv:1504.07961 [physics.acc-ph]`.

Abramowicz, H. *et al.* (2010). Measurement of beauty production in DIS and $F_2^{b\bar{b}}$ extraction at ZEUS, *Eur. Phys. J.* **C69**, 347–360, `arXiv:1005.3396 [hep-ex]`.

Abramowicz, H. *et al.* (2011). Measurement of beauty production in deep inelastic scattering at HERA using decays into electrons, *Eur. Phys. J.* **C71**, 1573, `arXiv:1101.3692 [hep-ex]`.

Abramowicz, H. *et al.* (2012). Search for first-generation leptoquarks at HERA, *Phys. Rev.* **D86**, 012005, `arXiv:1205.5179 [hep-ex]`.

Abramowicz, H. *et al.* (2013a). Combination and QCD analysis of charm production cross section measurements in deep inelastic *ep* scattering at HERA, *Eur. Phys. J.* **C73**, 2, 2311, `arXiv:1211.1182 [hep-ex]`.

Abramowicz, H. *et al.* (2013b). Measurement of $D^{*\pm}$ production in deep inelastic scattering at HERA, *JHEP* **05**, 097, `arXiv:1303.6578 [hep-ex]`, [Erratum: *JHEP* **02**, 106 (2014)].

Abramowicz, H. *et al.* (2013c). Measurement of $D^{\pm}$ production in deep inelastic *ep* scattering with the ZEUS detector at HERA, *JHEP* **05**, 023, `arXiv:1302.5058 [hep-ex]`.

Abramowicz, H. *et al.* (2013d). Measurement of high-Q^2 neutral current deep inelastic e^+p scattering cross sections with a longitudinally polarized positron beam at HERA, *Phys. Rev.* **D87**, 5, 052014, `arXiv:1208.6138 [hep-ex]`.

Abramowicz, H. *et al.* (2014a). Deep inelastic cross-section measurements at large y with the ZEUS detector at HERA, *Phys. Rev.* **D90**, 7, 072002, `arXiv:1404.6376 [hep-ex]`.

Abramowicz, H. *et al.* (2014b). Deep inelastic cross-section measurements at large y with the ZEUS detector at HERA, *Phys. Rev.* **D90**, 7, 072002, `arXiv:1404.6376 [hep-ex]`.

Abramowicz, H. *et al.* (2014c). Measurement of beauty and charm production in deep inelastic scattering at HERA and measurement of the beauty-quark mass, *JHEP* **09**, 127, `arXiv:1405.6915 [hep-ex]`.

Abramowicz, H. *et al.* (2015). Combination of measurements of inclusive deep inelastic $e^{\pm}p$ scattering cross sections and QCD analysis of HERA data, *Eur. Phys. J.* **C75**, 12, 580, `arXiv:1506.06042 [hep-ex]`.

Abramowicz, H. *et al.* (2016). Limits on the effective quark radius from inclusive ep scattering at HERA, *Phys. Lett.* **B757**, 468–472, `arXiv:1604.01280 [hep-ex]`.

Abt, I. (2016). The proton as seen by the HERA collider, *Ann. Rev. Nucl. Part. Sci.* **66**, 1, 377–399.

Abt, I., Cooper-Sarkar, A. M., Foster, B., Gwenlan, C., Myronenko, V., Turkot, O., and Wichmann, K. (2016). Combined electroweak and QCD fit to HERA data, *Phys. Rev.* **D94**, 5, 052007, `arXiv:1604.05083 [hep-ex]`.

Abt, I. *et al.* (1994). Scaling violations of the proton structure function F_2 at small x, *Phys. Lett.* **B321**, 161–167.

Abt, I. *et al.* (1997a). The H1 detector at HERA, *Nucl. Instrum. Meth.* **A386**, 310–347.

Abt, I. *et al.* (1997b). The tracking, calorimeter and muon detectors of the H1 experiment at HERA, *Nucl. Instrum. Meth.* **A386**, 348–396.

Accardi, A., Arleo, F., Brooks, W. K., D'Enterria, D., and Muccifora, V. (2010). Parton propagation and fragmentation in QCD Matter, *Riv. Nuovo Cim.* **32**, 439–553, `arXiv:0907.3534 [nucl-th]`.

Accardi, A., Brady, L. T., Melnitchouk, W., Owens, J. F., and Sato, N. (2016a). Constraints on large-x parton distributions from new weak boson production and deep-inelastic scattering data, *Phys. Rev.* **D93**, 11, 114017, `arXiv:1602.03154 [hep-ph]`.

Accardi, A. *et al.* (2016b). Electron ion collider: The next QCD frontier, *Eur. Phys. J.* **A52**, 9, 268, `arXiv:1212.1701 [nucl-ex]`.

Acciarri, M. *et al.* (2000). Search for manifestations of new physics in fermion pair production at LEP, *Phys. Lett.* **B489**, 81–92, `arXiv:hep-ex/0005028 [hep-ex]`.

Ackerstaff, K. *et al.* (1997). Measurement of the neutron spin structure function g_1^n with a polarized He^3 internal target, *Phys. Lett.* **B404**, 383–389, `arXiv:hep-ex/9703005 [hep-ex]`.

Ackerstaff, K. *et al.* (1999). Measurement of the strong coupling constant α_s and the vector and axial vector spectral functions in hadronic tau decays, *Eur. Phys. J.* **C7**, 571–593, `arXiv:hep-ex/9808019 [hep-ex]`.

Acosta, D. *et al.* (2005). Measurement of the forward-backward charge asymmetry of electron positron pairs in $p\bar{p}$ collisions at $\sqrt{s} = 1.96$ TeV, *Phys. Rev.* **D71**, 052002, `arXiv:hep-ex/0411059 [hep-ex]`.

Actis, S., Denner, A., Hofer, L., Lang, J.-N., Scharf, A., and Uccirati, S. (2017). RECOLA: Recursive computation of one-loop amplitudes, *Comput. Phys. Commun.* **214**, 140–173, `arXiv:1605.01090 [hep-ph]`.

Adam, J. *et al.* (2017). Enhanced production of multi-strange hadrons in high-multiplicity proton-proton collisions, *Nature Phys.* **13**, 535–539, `arXiv:1606.07424 [nucl-ex]`.

Adamczyk, L. *et al.* (2014). Measurement of longitudinal spin asymmetries for weak boson production in polarized proton-proton collisions at RHIC, *Phys. Rev. Lett.* **113**, 072301, `arXiv:1404.6880 [nucl-ex]`.

Adamczyk, L. *et al.* (2015). Precision measurement of the longitudinal double-spin asymmetry for inclusive jet production in polarized proton collisions at $\sqrt{s} = 200$ GeV, *Phys. Rev. Lett.* **115**, 9, 092002, `arXiv:1405.5134 [hep-ex]`.

Adare, A. *et al.* (2007). Inclusive cross-section and double helicity asymmetry for pi0 production in p + p collisions at $\sqrt{s} = 200$ GeV: Implications for the polarized gluon distribution in the proton, *Phys. Rev.* **D76**, 051106, arXiv:0704.3599 [hep-ex].

Adare, A. *et al.* (2009). The polarized gluon contribution to the proton spin from the double helicity asymmetry in inclusive pi0 production in polarized p + p collisions at $\sqrt{s} = 200$ GeV, *Phys. Rev. Lett.* **103**, 012003, arXiv:0810.0694 [hep-ex].

Adare, A. *et al.* (2011). Cross section and parity violating spin asymmetries of $W^{\pm}$ boson production in polarized $p + p$ collisions at $\sqrt{s} = 500$ GeV, *Phys. Rev. Lett.* **106**, 062001, arXiv:1009.0505 [hep-ex].

Adare, A. *et al.* (2014). Inclusive double-helicity asymmetries in neutral-pion and eta-meson production in $\vec{p} + \vec{p}$ collisions at $\sqrt{s} = 200$ GeV, *Phys. Rev.* **D90**, 1, 012007, arXiv:1402.6296 [hep-ex].

Ade, P. A. R. *et al.* (2016). Planck 2015 results. XIII. Cosmological parameters, *Astron. Astrophys.* **594**, A13, arXiv:1502.01589 [astro-ph.CO].

Ademollo, M., Altarelli, G., and Gatto, R. (1965). $u(6) \otimes u(6)$ and nonleptonic decays, *Phys. Rev. Lett.* **14**, 420.

Adeva, B. *et al.* (1998). Spin asymmetries A_1 and structure functions g_1 of the proton and the deuteron from polarized high-energy muon scattering, *Phys. Rev.* **D58**, 112001.

Adler, S. L. and Bardeen, W. A. (1969). Absence of higher order corrections in the anomalous axial vector divergence equation, *Phys. Rev.* **182**, 1517–1536.

Adler, S. S. *et al.* (2004). Double helicity asymmetry in inclusive mid-rapidity π^0 production for polarized $p + p$ collisions at $\sqrt{s} = 200$ GeV, *Phys. Rev. Lett.* **93**, 202002, arXiv:hep-ex/0404027 [hep-ex].

Adloff, C. *et al.* (2001). Deep inelastic inclusive ep scattering at low x and a determination of α_s, *Eur. Phys. J.* **C21**, 33–61, arXiv:hep-ex/0012053 [hep-ex].

Aggarwal, M. M. *et al.* (2011). Measurement of the parity-violating longitudinal single-spin asymmetry for $W^{\pm}$ boson production in polarized proton-proton collisions at $\sqrt{s} = 500 - GeV$, *Phys. Rev. Lett.* **106**, 062002, arXiv:1009.0326 [hep-ex].

Agrawal, V., Barr, S. M., Donoghue, J. F., and Seckel, D. (1998). The anthropic principle and the mass scale of the Standard Model, *Phys. Rev.* **D57**, 5480–5492, arXiv:hep-ph/9707380 [hep-ph].

Aguilar-Saavedra, J. A., Fuks, B., and Mangano, M. L. (2015). Pinning down top dipole moments with ultra-boosted tops, *Phys. Rev.* **D91**, 094021, arXiv:1412.6654 [hep-ph].

Ahmed, M. A. and Ross, G. G. (1976). Polarized lepton-hadron scattering in asymptotically free gauge theories, *Nucl. Phys.* **B111**, 441–460.

Ahmet, K. *et al.* (1991). The OPAL detector at LEP, *Nucl. Instrum. Meth.* **A305**, 275–319.

Airapetian, A. *et al.* (2005). Quark helicity distributions in the nucleon for up, down, and strange quarks from semi-inclusive deep-inelastic scattering, *Phys. Rev.* **D71**, 012003, arXiv:hep-ex/0407032 [hep-ex].

Airapetian, A. *et al.* (2007). Precise determination of the spin structure function g_1 of the proton, deuteron and neutron, *Phys. Rev.* **D75**, 012007, arXiv:hep-ex/0609039 [hep-ex].

Aktas, A. *et al.* (2006). A determination of electroweak parameters at HERA, *Phys. Lett.* **B632**, 35–42, arXiv:hep-ex/0507080 [hep-ex].

Albino, S. (2010). The hadronization of partons, *Rev. Mod. Phys.* **82**, 2489–2556, arXiv:0810.4255 [hep-ph].

Alekhin, S., Blumlein, J., and Moch, S. (2012). Parton distribution functions and benchmark cross sections at NNLO, *Phys. Rev.* **D86**, 054009, arXiv:1202.2281 [hep-ph].

Alekhin, S., Blumlein, J., and Moch, S. (2014). The ABM parton distributions tuned to LHC data, *Phys. Rev.* **D89**, 5, 054028, arXiv:1310.3059 [hep-ph].

Alekhin, S., Bllein, J., Moch, S., and Placakyte, R. (2017). Parton distribution functions, α_s, and heavy-quark masses for LHC Run II, *Phys. Rev.* **D96**, 1, 014011, arXiv:1701.05838 [hep-ph].

Alekseev, M. G. *et al.* (2010a). Quark helicity distributions from longitudinal spin asymmetries in muon-proton and muon-deuteron scattering, *Phys. Lett.* **B693**, 227–235, arXiv:1007.4061 [hep-ex].

Alekseev, M. G. *et al.* (2010b). The spin-dependent structure function of the proton g_1^p and a test of the Bjorken sum rule, *Phys. Lett.* **B690**, 466–472, arXiv:1001.4654 [hep-ex].

Alexa, C. *et al.* (2013). Elastic and proton-dissociative photoproduction of J/psi mesons at HERA, *Eur. Phys. J.* **C73**, 6, 2466, arXiv:1304.5162 [hep-ex].

Alexakhin, V. Yu. *et al.* (2007). The deuteron spin-dependent structure function g_1^d and its first moment, *Phys. Lett.* **B647**, 8–17, arXiv:hep-ex/0609038 [hep-ex].

Alexandrou, C., Constantinou, M., Hadjiyiannakou, K., Jansen, K., Kallidonis, C., Koutsou, G., Avil-Casco, A. V., and Wiese, C. (2017). The nucleon spin explained using lattice QCD simulations, *Phys. Rev. Lett.* **119**, 14, 142002, arXiv:1706.02973 [hep-lat].

Alighieri, D. (1321). *Inferno*, in *Divina Commedia*, vv 34-36.

Alioli, S., Bauer, C. W., Berggren, C., Tackmann, F. J., and Walsh, J. R. (2015). Drell–Yan production at NNLL'+NNLO matched to parton showers, *Phys. Rev.* **D92**, 9, 094020, arXiv:1508.01475 [hep-ph].

Alioli, S., Fernandez, P., Fuster, J., Irles, A., Moch, S.-O., Uwer, P., and Vos, M. (2013). A new observable to measure the top-quark mass at hadron colliders, *Eur. Phys. J.* **C73**, 2438, arXiv:1303.6415 [hep-ph].

Alonso, R., Gavela, M. B., Isidori, G., and Maiani, L. (2013). Neutrino mixing and masses from a minimum principle, *JHEP* **11**, 187, arXiv:1306.5927 [hep-ph].

Altarelli, G. (1982a). Electroweak processes beyond the tree approximation, *Acta Phys. Austriaca Suppl.* **24**, 229–273.

Altarelli, G. (1982b). Partons in quantum chromodynamics, *Phys. Rept.* **81**, 1.

Altarelli, G. (1985). QCD and experiment, in *International Europhysics Conference on High-Energy Physics Bari, Italy, July 18–24, 1985.*

Altarelli, G. (1987). Beyond the Standard Model, in M. Levy, B. Jean-Louis, M. Jacob, D. Speiser, W. Jacques, and R. Gastmans (eds.), *Proceedings of the Cargèse Summer Institute in Particle Physics: Recent Experimental and Theoretical Results on the Basic Constituents of Matter, Cargèse, France, 15–31 July 1985*, Vol. 150 (Plenum Press (New York)), 387.

Altarelli, G. (1989a). Experimental tests of perturbative QCD, *Ann. Rev. Nucl. Part. Sci.* **39**, 357–406.

Altarelli, G. (1989b). Theory of precision electroweak experiments, in *1989 International Symposium on Lepton and Photon Interactions at High Energies Stanford, California, August 7–12, 1989*, 286–304.

Altarelli, G. (1990). Polarized electroproduction and the spin of the quarks inside the proton, in *International School of Subnuclear Physics: 27th Course: The Challenging Questions Erice, Italy, July 26–August 3, 1989*, 33–74.

Altarelli, G. (1998). LHCC meeting.

Altarelli, G. (2002). A QCD primer, in *New states of matter in hadronic interactions. Proceedings, Pan-American Advanced Study Institute, PASI 2002, Campos do Jordao,*

Sao Paulo, Brazil, January 7–18, 2002, Vol. 631, 70–111, arXiv:hep-ph/0204179 [hep-ph].

Altarelli, G. (2003). http://conferences.fnal.gov/hadroncollider/talks/Altarelli.pdf, Summary talk at the International Workshop on Future Hadron Colliders: Physics, Detectors, Machines, Fermilab, 16–18 October 2003.

Altarelli, G. (2008). "Deep Inelastic Scattering in the LHC Time", Opening Lecture, ECFA LHeC Workshop at Divonne, .

Altarelli, G. (2011a). Concluding talk: A theorist overview on particle physics, in *Flavor physics and CP violation. Proceedings, 9th International Conference, FPCP 2011, Maale HaChamisha, Israel, May 23–27, 2011,* arXiv:1108.3514 [hep-ph], http://inspirehep.net/record/924255/files/arXiv:1108.3514.pdf.

Altarelli, G. (2011b). Concluding talk: A theorist overview on particle physics, in *Flavor physics and CP violation. Proceedings, 9th International Conference, FPCP 2011, Maale HaChamisha, Israel, May 23–27, 2011,* arXiv:1108.3514 [hep-ph], http://inspirehep.net/record/924255/files/arXiv:1108.3514.pdf.

Altarelli, G. (2011c). Particle physics at the LHC start, *Prog. Theor. Phys. Suppl.* **187**, 305–321, arXiv:1010.5637 [hep-ph].

Altarelli, G. (2012). The early days of QCD (as seen from Rome), *Nuovo Cim.* **C035N1**, 1–8, arXiv:1106.3189 [hep-ph].

Altarelli, G. (2013a). The Higgs: So simple yet so unnatural, *Phys. Scripta* **T158**, 014011, arXiv:1308.0545 [hep-ph], also in Proceedings, 2nd International Conference on New Frontiers in Physics (ICNFP 2013): Kolymbari, Crete, Greece, August 28–September 5, 2013, *EPJ Web Conf.*, **71**, 5 (2014).

Altarelli, G. (2013b). The QCD running coupling and its measurement, in *Proceedings, 12th Hellenic School and Workshops on Elementary Particle Physics and Gravity (CORFU2012-ST), (CORFU2012-PU) and (CORFU2012-SM): Corfu, Greece, September 8–27, 2012*, Vol. Corfu2012, 002, arXiv:1303.6065 [hep-ph].

Altarelli, G. (2014). The Higgs and the excessive success of the Standard Model, *Frascati Phys. Ser.* **58**, 102, arXiv:1407.2122 [hep-ph].

Altarelli, G. (2015a). Remark at the Chavannes-de-Bogis session of the LHeC Advisory Committee.

Altarelli, G. (2015b). "Importance of Deep Inelastic Scattering in the 21st Century", Talk presented at the LHeC Workshop, Geneva, CERN", .

Altarelli, G., Ball, R. D., and Forte, S. (2000a). Resummation of singlet parton evolution at small x, *Nucl. Phys.* **B575**, 313–329, arXiv:hep-ph/9911273 [hep-ph].

Altarelli, G., Ball, R. D., and Forte, S. (2000b). Singlet parton evolution at small x: A theoretical update, in *QCD: Perturbative or nonperturbative? Proceedings, 17th Autumn School, Lisbon, Portugal, September 29–October 4, 1999*, 249–278, arXiv:hep-ph/0001157 [hep-ph], http://www.slac.stanford.edu/spires/find/books/www?cl=QCD161:A85:1999.

Altarelli, G., Ball, R. D., and Forte, S. (2001). Small x resummation and HERA structure function data, *Nucl. Phys.* **B599**, 383–423, arXiv:hep-ph/0011270 [hep-ph].

Altarelli, G., Ball, R. D., and Forte, S. (2002). Factorization and resummation of small x scaling violations with running coupling, *Nucl. Phys.* **B621**, 359–387, arXiv:hep-ph/0109178 [hep-ph].

Altarelli, G., Ball, R. D., and Forte, S. (2003). An anomalous dimension for small x evolution, *Nucl. Phys.* **B674**, 459–483, arXiv:hep-ph/0306156 [hep-ph].

Altarelli, G., Ball, R. D., and Forte, S. (2006). Perturbatively stable resummed small x evolution kernels, *Nucl. Phys.* **B742**, 1–40, arXiv:hep-ph/0512237 [hep-ph].

Altarelli, G., Ball, R. D., and Forte, S. (2008). Small x resummation with quarks: Deep-inelastic scattering, *Nucl. Phys.* **B799**, 199–240, arXiv:0802.0032 [hep-ph].

Altarelli, G., Ball, R. D., Forte, S., and Ridolfi, G. (1997a). Determination of the Bjorken sum and strong coupling from polarized structure functions, *Nucl. Phys.* **B496**, 337–357, arXiv:hep-ph/9701289 [hep-ph].

Altarelli, G., Ball, R. D., Forte, S., and Ridolfi, G. (1998a). Theoretical analysis of polarized structure functions, *Acta Phys. Polon.* **B29**, 1145–1173, arXiv:hep-ph/9803237 [hep-ph].

Altarelli, G. and Barbieri, R. (1991). Vacuum polarization effects of new physics on electroweak processes, *Phys. Lett.* **B253**, 161–167.

Altarelli, G., Barbieri, R., and Caravaglios, F. (1993a). Nonstandard analysis of electroweak precision data, *Nucl. Phys.* **B405**, 3–23.

Altarelli, G., Barbieri, R., and Caravaglios, F. (1995a). The Epsilon variables for electroweak precision tests: A reappraisal, *Phys. Lett.* **B349**, 145–154.

Altarelli, G., Barbieri, R., and Caravaglios, F. (1998b). Electroweak precision tests: A concise review, *Int. J. Mod. Phys.* **A13**, 1031–1058, arXiv:hep-ph/9712368 [hep-ph].

Altarelli, G., Barbieri, R., and Jadach, S. (1992). Toward a model independent analysis of electroweak data, *Nucl. Phys.* **B369**, 3–32, [Erratum: *Nucl. Phys.* **B376**, 444 (1992)].

Altarelli, G., Brandt, R. A., and Preparata, G. (1971). Light cone analysis of massive mu-pair production, *Phys. Rev. Lett.* **26**, 42–46.

Altarelli, G. and Buccella, F. (1964). Single photon emission in high-energy $e^+ - e^-$ collisions, *Nuovo Cim.* **34**, 1337–1346.

Altarelli, G., Buccella, F., and Gatto, R. (1965). $SU(6)$ and non leptonic hyperon decays, *Phys. Lett.* **14**, 70–72.

Altarelli, G., Cabibbo, N., Corbò, G., Maiani, L., and Martinelli, G. (1982). Leptonic decay of heavy flavors: a theoretical update, *Nucl. Phys.* **B208**, 365–380.

Altarelli, G., Cabibbo, N., and Maiani, L. (1975). Enhancement of nonleptonic decays of charmed particles, *Nucl. Phys.* **B88**, 285–288.

Altarelli, G., Casalbuoni, R., De Curtis, S., Di Bartolomeo, N., Feruglio, F., and Gatto, R. (1991a). Atomic parity violation in extended gauge models and latest CDF and LEP data, *Phys. Lett.* **B261**, 146–152.

Altarelli, G., Casalbuoni, R., De Curtis, S., Di Bartolomeo, N., Feruglio, F., and Gatto, R. (1991b). Improved bounds on extended gauge models from new LEP data, *Phys. Lett.* **B263**, 459–465.

Altarelli, G., Casalbuoni, R., De Curtis, S., Di Bartolomeo, N., Gatto, R., and Feruglio, F. (1993b). Extended gauge models and precision electroweak data, *Phys. Lett.* **B318**, 139–147.

Altarelli, G., Casalbuoni, R., Dominici, D., Feruglio, F., and Gatto, R. (1990a). Z width and branching ratios in extended gauge models, *Mod. Phys. Lett.* **A5**, 495.

Altarelli, G., Casalbuoni, R., Feruglio, F., and Gatto, R. (1990b). Bounds on extended gauge models from LEP data, *Phys. Lett.* **B245**, 669–680.

Altarelli, G., Curci, G., Martinelli, G., and Petrarca, S. (1981). QCD nonleading corrections to weak decays as an application of regularization by dimensional reduction, *Nucl. Phys.* **B187**, 461–513.

Altarelli, G., Diemoz, M., Martinelli, G., and Nason, P. (1988). Total cross-sections for heavy flavor production in hadronic collisions and QCD, *Nucl. Phys.* **B308**, 724–752.

Altarelli, G., Ellis, J. R., Giudice, G. F., Lola, S., and Mangano, M. L. (1997b). Pursuing interpretations of the HERA large Q^2 data, *Nucl. Phys.* **B506**, 3–28, arXiv:hep-ph/9703276 [hep-ph].

Altarelli, G., Ellis, R. K., Greco, M., and Martinelli, G. (1984). Vector boson production at colliders: A theoretical reappraisal, *Nucl. Phys.* **B246**, 12–44.

Altarelli, G., Ellis, R. K., and Martinelli, G. (1978a). Leptoproduction and Drell–Yan processes beyond the leading approximation in chromodynamics, *Nucl. Phys.* **B143**, 521, [Erratum: *Nucl. Phys.* **B146**, 544 (1978)].

Altarelli, G., Ellis, R. K., and Martinelli, G. (1979a). Large perturbative corrections to the Drell–Yan process in QCD, *Nucl. Phys.* **B157**, 461–497.

Altarelli, G., Ellis, R. K., and Martinelli, G. (1985a). Vector boson production at present and future colliders, *Z. Phys.* **C27**, 617.

Altarelli, G., Ellis, R. K., Martinelli, G., and Pi, S.-Y. (1979b). Processes involving fragmentation functions beyond the leading order in QCD, *Nucl. Phys.* **B160**, 301–329.

Altarelli, G. and Feruglio, F. (2005). Tri-bimaximal neutrino mixing from discrete symmetry in extra dimensions, *Nucl. Phys.* **B720**, 64–88, `arXiv:hep-ph/0504165 [hep-ph]`.

Altarelli, G. and Feruglio, F. (2006). Tri-bimaximal neutrino mixing, A_4 and the modular symmetry, *Nucl. Phys.* **B741**, 215–235, `arXiv:hep-ph/0512103 [hep-ph]`.

Altarelli, G. and Feruglio, F. (2010). Discrete flavor symmetries and models of neutrino mixing, *Rev. Mod. Phys.* **82**, 2701–2729, `arXiv:1002.0211 [hep-ph]`.

Altarelli, G., Forte, S., and Ridolfi, G. (1998c). On positivity of parton distributions, *Nucl. Phys.* **B534**, 277–296, `arXiv:hep-ph/9806345 [hep-ph]`.

Altarelli, G. and Franco, E. (1986). Searching for an intermediate mass Higgs boson at large energy e^+e^- linear colliders, *Mod. Phys. Lett.* **A1**, 9, 517–523.

Altarelli, G. and Franzini, P. J. (1988). $B_0\bar{B}_0$ mixing within and beyond the Standard Model, *Z. Phys.* **C37**, 271.

Altarelli, G., Gatto, R., and Maiani, L. (1968). A Regge model for production of axial vector mesons in pion-nucleon collisions, *Nuovo Cim.* **A58**, 259.

Altarelli, G., Gatto, R., Maiani, L., and Preparata, G. (1966). Mixing effects in baryon spectroscopy, *Phys. Rev. Lett.* **16**, 918.

Altarelli, G. and Grunewald, M. W. (2004). Precision electroweak tests of the Standard Model, *Phys. Rept.* **403–404**, 189–201, `arXiv:hep-ph/0404165 [hep-ph]`.

Altarelli, G. and Isidori, G. (1994). Lower limit on the Higgs mass in the Standard Model: An update, *Phys. Lett.* **B337**, 141–144.

Altarelli, G. and Lampe, B. (1990). The light cone expansion and the parton model in polarized muon production, *Z. Phys.* **C47**, 315–324.

Altarelli, G. and Lampe, B. (1993). Second order QCD corrections to heavy quark forward-backward asymmetries, *Nucl. Phys.* **B391**, 3–22.

Altarelli, G., Lampe, B., Nason, P., and Ridolfi, G. (1994a). The Burkhardt–Cottingham sum rule in perturbative QCD, *Phys. Lett.* **B334**, 187–191.

Altarelli, G. and Maiani, L. (1974). Octet enhancement of nonleptonic weak interactions in asymptotically free gauge theories, *Phys. Lett.* **52B**, 351–354.

Altarelli, G. and Maiani, L. (1982). Identical particle interference in D^+ meson lifetime, *Phys. Lett.* **118B**, 414.

Altarelli, G. and Martinelli, G. (1978). Transverse momentum of jets in electroproduction from Quantum Chromodynamics, *Phys. Lett.* **76B**, 89–94.

Altarelli, G., Martinelli, G., Mele, B., and Rückl, R. (1985b). Electroproduction of supersymmetric particles and gauge bosons, *Nucl. Phys.* **B262**, 204–234.

Altarelli, G., Martinelli, G., Petrarca, S., and Rapuano, F. (1996). Failure of local duality in inclusive nonleptonic heavy flavor decays, *Phys. Lett.* **B382**, 409–414, `arXiv:hep-ph/9604202 [hep-ph]`.

Altarelli, G., Mele, B., and Petrarca, S. (1985c). Light gluinos at the CERN $p\bar{p}$ collider, *Phys. Lett.* **160B**, 317–321.

Altarelli, G., Mele, B., and Pitolli, F. (1987). Heavy Higgs production at future colliders, *Nucl. Phys.* **B287**, 205–224.

Altarelli, G., Mele, B., and Ruiz-Altaba, M. (1989). Searching for new heavy vector bosons in $p\bar{p}$ colliders, *Z. Phys.* **C45**, 109, [Erratum: *Z. Phys.* **C47**, 676 (1990)].

Altarelli, G., Nason, P., and Ridolfi, G. (1994b). On the Q^2 dependence of the measured polarized structure functions, *Phys. Lett.* **B320**, 152–158, `arXiv:hep-ph/9311255 [hep-ph]`, [Erratum: *Phys. Lett.* **B325**, 538 (1994)].

Altarelli, G., Nason, P., and Ridolfi, G. (1995b). A study of ultraviolet renormalon ambiguities in the determination of α_s from τ decay, *Z. Phys.* **C68**, 257–268, `arXiv:hep-ph/9501240 [hep-ph]`.

Altarelli, G. and Parisi, G. (1977). Asymptotic freedom in parton language, *Nucl. Phys.* **B126**, 298–318.

Altarelli, G., Parisi, G., and Petronzio, R. (1978b). Transverse momentum in Drell–Yan processes, *Phys. Lett.* **76B**, 351–355.

Altarelli, G., Parisi, G., and Petronzio, R. (1978c). Transverse momentum of muon pairs produced in hadronic collisions, *Phys. Lett.* **76B**, 356–360.

Altarelli, G. and Ridolfi, G. (1995). Understanding the proton spin structure: A status report, *Nucl. Phys. Proc. Suppl.* **39BC**, 106–117.

Altarelli, G. and Ross, G. G. (1988). The anomalous gluon contribution to polarized leptoproduction, *Phys. Lett.* **B212**, 391–396.

Altarelli, G. and Rubinstein, H. R. (1969). Asymptotic properties of virtual Compton amplitudes in the ladder model, *Phys. Rev.* **187**, 2111–2119.

Altarelli, G. and Stirling, W. J. (1989). Phenomenology of the anomalous gluon contribution to polarized leptoproduction, *Part. World* **1**, 2, 40–52.

Altheimer, A. *et al.* (2014). Boosted objects and jet substructure at the LHC. Report of BOOST2012, held at IFIC Valencia, 23rd–27th of July 2012, *Eur. Phys. J.* **C74**, 3, 2792, `arXiv:1311.2708 [hep-ex]`.

Alvarez-Gaumé, L. (2017). Private communication.

Alves, A. A., Jr. *et al.* (2008). The LHCb detector at the LHC, *JINST* **3**, S08005.

Alves, D. S. M., Galloway, J., Ruderman, J. T., and Walsh, J. R. (2015). Running electroweak couplings as a probe of new physics, *JHEP* **02**, 007, `arXiv:1410.6810 [hep-ph]`.

Alwall, J., Frederix, R., Frixione, S., Hirschi, V., Maltoni, F., Mattelaer, O., Shao, H. S., Stelzer, T., Torrielli, P., and Zaro, M. (2014). The automated computation of tree-level and next-to-leading order differential cross sections, and their matching to parton shower simulations, *JHEP* **07**, 079, `arXiv:1405.0301 [hep-ph]`.

Anastasiou, C., Dixon, L. J., Melnikov, K., and Petriello, F. (2004). High precision QCD at hadron colliders: Electroweak gauge boson rapidity distributions at NNLO, *Phys. Rev.* **D69**, 094008, `arXiv:hep-ph/0312266 [hep-ph]`.

Anastasiou, C., Duhr, C., Dulat, F., Furlan, E., Gehrmann, T., Herzog, F., Lazopoulos, A., and Mistlberger, B. (2016). High precision determination of the gluon fusion Higgs boson cross-section at the LHC, *JHEP* **05**, 058, `arXiv:1602.00695 [hep-ph]`.

Anastasiou, C., Duhr, C., Dulat, F., Herzog, F., and Mistlberger, B. (2015). Higgs boson gluon-fusion production in QCD at three loops, *Phys. Rev. Lett.* **114**, 212001, `arXiv:1503.06056 [hep-ph]`.

Anastasiou, C., Herzog, F., and Lazopoulos, A. (2012). The fully differential decay rate of a Higgs boson to bottom-quarks at NNLO in QCD, *JHEP* **03**, 035, `arXiv:1110.2368 [hep-ph]`.

Anastasiou, C. and Melnikov, K. (2002). Higgs boson production at hadron colliders in NNLO QCD, *Nucl. Phys.* **B646**, 220–256, `arXiv:hep-ph/0207004 [hep-ph]`.

Andersen, J. R., Bartle, S., Bern, Z., Febres Cordero, F., Höche, S., Kosower, D. A., Ita, H., Lo Presti, N. A., Maître, D., and Ozeren, K. (2014). High multiplicity processes with BlackHat and Sherpa, in *Proceedings, 12th DESY Workshop on Elementary Particle Physics: Loops and Legs in Quantum Field Theory (LL2014): Weimar, Germany, April 27–May 2, 2014*, Vol. LL2014, 079, `arXiv:1407.1621 [hep-ph]`.

Andreassen, A., Frost, W., and Schwartz, M. D. (2017). Scale invariant instantons and the complete lifetime of the Standard Model, `arXiv:1707.08124 [hep-ph]`.

Andreassen, A. and Schwartz, M. D. (2017). Reducing the top quark mass uncertainty with jet grooming, *JHEP* **10**, 151, `arXiv:1705.07135 [hep-ph]`.

Andreev, V. *et al.* (2014). Measurement of inclusive ep cross sections at high Q^2 at $\sqrt{s} = 225$ and 252 GeV and of the longitudinal proton structure function F_L at HERA, *Eur. Phys. J.* **C74**, 4, 2814, `arXiv:1312.4821 [hep-ex]`.

Andreev, V. *et al.* (2017). Determination of the strong coupling constant $\alpha_s(m_Z)$ in next-to-next-to-leading order QCD using H1 jet cross section measurements, *Eur. Phys. J.* **C77**, 11, 791, `arXiv:1709.07251 [hep-ex]`.

Angal-Kalinin, D. *et al.* (arXiv:1705.08783, 2017.). PERLE: Powerful Energy Recovery Linac for Experiments — conceptual design report, `arXiv:1705.08783 [physics.acc-ph]`.

Anthony, P. L. *et al.* (1996). Deep inelastic scattering of polarized electrons by polarized He^3 and the study of the neutron spin structure, *Phys. Rev.* **D54**, 6620–6650, `arXiv:hep-ex/9610007 [hep-ex]`.

Anthony, P. L. *et al.* (1999). Measurement of the deuteron spin structure function $g_1^d(x)$ for $1\ (GeV/c)^2 < Q^2 < 40\ (GeV/c)^2$, *Phys. Lett.* **B463**, 339–345, `arXiv:hep-ex/9904002 [hep-ex]`.

Anthony, P. L. *et al.* (2000). Measurements of the Q^2 dependence of the proton and neutron spin structure functions g_1^p and g_1^n, *Phys. Lett.* **B493**, 19–28, `arXiv:hep-ph/0007248 [hep-ph]`.

Antoniadis, I., Kumar, A., and Panda, B. (2009). Fermion wavefunctions in magnetized branes: Theta identities and Yukawa couplings, *Nucl. Phys.* **B823**, 116–173, `arXiv:0904.0910 [hep-th]`.

Antusch, S., Cazzato, E., and Fischer, O. (2017). Sterile neutrino searches at future e^-e^+, pp, and e^-p colliders, *Int. J. Mod. Phys.* **A32**, 14, 1750078, `arXiv:1612.02728 [hep-ph]`.

Aoki, S. *et al.* (2009). Precise determination of the strong coupling constant in $N_f = 2 + 1$ lattice QCD with the Schrodinger functional scheme, *JHEP* **10**, 053, `arXiv:0906.3906 [hep-lat]`.

Aoki, S. *et al.* (2014). Review of lattice results concerning low-energy particle physics, *Eur. Phys. J.* **C74**, 2890, `arXiv:1310.8555 [hep-lat]`.

Aoki, S. *et al.* (2017). Review of lattice results concerning low-energy particle physics, *Eur. Phys. J.* **C77**, 2, 112, `arXiv:1607.00299 [hep-lat]`.

Appuhn, R. D. *et al.* (1997). The H1 lead / scintillating fiber calorimeter, *Nucl. Instrum. Meth.* **A386**, 397–408.

Aprile, E. *et al.* (2017). First dark matter search results from the XENON1T experiment, *Phys. Rev. Lett.* **119**, 18, 181301, `arXiv:1705.06655 [astro-ph.CO]`.

Argyropoulos, S. and Sjöstrand, T. (2014). Effects of color reconnection on $t\bar{t}$ final states at the LHC, *JHEP* **11**, 043, `arXiv:1407.6653 [hep-ph]`.

Arkani-Hamed, N., Cohen, T., D'Agnolo, R. T., Hook, A., Kim, H. D., and Pinner, D. (2016a). Solving the hierarchy problem at reheating with a large number of degrees of freedom, *Phys. Rev. Lett.* **117**, 25, 251801, `arXiv:1607.06821 [hep-ph]`.

Arkani-Hamed, N. and Dimopoulos, S. (2005). Supersymmetric unification without low energy supersymmetry and signatures for fine-tuning at the LHC, *JHEP* **06**, 073, `arXiv:hep-th/0405159 [hep-th]`.

Arkani-Hamed, N., Dimopoulos, S., Giudice, G. F., and Romanino, A. (2005). Aspects of split supersymmetry, *Nucl. Phys.* **B709**, 3–46, `arXiv:hep-ph/0409232 [hep-ph]`.

Arkani-Hamed, N., Han, T., Mangano, M., and Wang, L.-T. (2016b). Physics opportunities of a 100 TeV proton-proton collider, *Phys. Rept.* **652**, 1–49, `arXiv:1511.06495 [hep-ph]`.

Arkani-Hamed, N., Motl, L., Nicolis, A., and Vafa, C. (2007). The string landscape, black holes and gravity as the weakest force, *JHEP* **06**, 060, `arXiv:hep-th/0601001 [hep-th]`.

Armesto, N. (2017a). "Electron-Ion Physics: Status Report", Talk at the LHeC Workshop at CERN.

Armesto, N. (2017b). Private communication.

Armesto, N. and Scomparin, E. (2016). Heavy-ion collisions at the Large Hadron Collider: A review of the results from Run 1, *Eur. Phys. J. Plus* **131**, 3, 52, `arXiv:1511.02151 [nucl-ex]`.

Arnison, G. *et al.* (1983a). Experimental observation of isolated large transverse energy electrons with associated missing energy at $\sqrt{s}$ = 540 GeV, *Phys. Lett.* **122B**, 103–116.

Arnison, G. *et al.* (1983b). Experimental observation of lepton pairs of invariant mass around $95(GeV/c)^2$ at the CERN SPS Collider, *Phys. Lett.* **126B**, 398–410.

Arvanitaki, A., Dimopoulos, S., Gorbenko, V., Huang, J., and Tilburg, K. (2017). A small weak scale from a small cosmological constant, *JHEP* **05**, 071, `arXiv:1609.06320 [hep-ph]`.

Aschenauer, E. C., Burton, T., Martini, T., Spiesberger, H., and Stratmann, M. (2013). Prospects for charged current deep-inelastic scattering off polarized nucleons at a future Electron-Ion Collider, *Phys. Rev.* **D88**, 114025, `arXiv:1309.5327 [hep-ph]`.

Aschenauer, E. C., Fazio, S., Lee, J. H., Mantysaari, H., Page, B. S., Schenke, B., Ullrich, T., Venugopalan, R., and Zurita, P. (2017). The EIC: Assessing the energy dependence of key measurements, Tech. rep., BNL, `arXiv:1708.01527 [nucl-ex]`.

Aschenauer, E. C., Sassot, R., and Stratmann, M. (2012). Helicity parton distributions at a future Electron-Ion Collider: A quantitative appraisal, *Phys. Rev.* **D86**, 054020, `arXiv:1206.6014 [hep-ph]`.

Aschenauer, E. C. *et al.* (2014). eRHIC Design Study: An Electron-Ion Collider at BNL. Tech. rep., BNL, `arXiv:1409.1633 [physics.acc-ph]`.

Ashman, J. *et al.* (1988). A measurement of the spin asymmetry and determination of the structure function g_1 in deep inelastic muon-proton scattering, *Phys. Lett.* **B206**, 364.

Assamagan, K. *et al.* (2016). The Higgs portal and cosmology, `arXiv:1604.05324 [hep-ph]`, `http://inspirehep.net/record/1449094/files/arXiv:1604.05324.pdf`.

Assmann, R., Lamont, M., and Myers, S. (2002). A brief history of the LEP collider, *Nucl. Phys. Proc. Suppl.* **109B**, 17–31.

Assmann, R. *et al.* (1999). Calibration of center-of-mass energies at LEP-1 for precise measurements of Z properties, *Eur. Phys. J.* **C6**, 187–223.

Assmann, R. *et al.* (2005). Calibration of centre-of-mass energies at LEP 2 for a precise measurement of the W boson mass, *Eur. Phys. J.* **C39**, 253–292, `arXiv:hep-ex/0410026 [hep-ex]`.

Assmann, R. *et al.* (2014). Proton-driven plasma wakefield acceleration: A path to the future of high-energy particle physics, *Plasma Phys. Control. Fusion* **56**, 8, 084013, arXiv:1401.4823 [physics.acc-ph].

Astill, W., Bizon, W., Re, E., and Zanderighi, G. (2016). NNLOPS accurate associated HW production, *JHEP* **06**, 154, arXiv:1603.01620 [hep-ph].

ATLAS Collaboration (2008). The ATLAS experiment at the CERN Large Hadron Collider, *JINST* **3**, S08003.

ATLAS Collaboration (2014a). First combination of Tevatron and LHC measurements of the top-quark mass, Tech. Rep. ATLAS-CONF-2014-008, CDF-NOTE-11071, CMS-PAS-TOP-13-014, D0-NOTE-6416, FERMILAB-TM-2582-E, CERN, arXiv:1403.4427 [hep-ex].

ATLAS Collaboration (2014b). Projections for measurements of Higgs boson signal strengths and coupling parameters with the ATLAS detector at the HL-LHC, Tech. Rep. ATL-PHYS-PUB-2014-016, CERN, Geneva.

ATLAS Collaboration (2017a). https://atlas.web.cern.ch/Atlas/GROUPS/PHYSICS/CombinedSummaryPlots/SM/index.html, Summary plots.

ATLAS Collaboration (2017b). Combined measurements of Higgs boson production and decay in the $H \to ZZ^* \to 4\ell$ and $H \to \gamma\gamma$ channels using $\sqrt{s} = 13$ TeV proton–proton collision data collected with the ATLAS experiment, Tech. Rep. ATLAS-CONF-2017-047, CERN, https://cds.cern.ch/record/2273854.

ATLAS Collaboration (2017c). Measurement of the top quark mass in the $t\bar{t} \to$ lepton+jets channel from $\sqrt{s} = 8$ TeV ATLAS data, Tech. Rep. ATLAS-CONF-2017-071, CERN, Geneva, http://cds.cern.ch/record/2285809.

Ayala, C., Cvetič, G., and Pineda, A. (2014). The bottom quark mass from the $\Upsilon(1S)$ system at NNNLO, *JHEP* **09**, 045, arXiv:1407.2128 [hep-ph].

Azuelos, G., Sun, H., and Wang, K. (2017). Search for singly charged Higgs in vector boson scattering at the ep colliders, arXiv:1712.07505 [hep-ph].

Azzi, P., Azzurri, P., Biswas, S., Blekman, F., Corcella, G., De Curtis, S., Erler, J., Foppiani, N., Helenius, I., Jadach, S., Janot, P., Jegerlehner, F., Langacker, P., Locci, E., Margaroli, F., Mele, B., Piccinini, F., Reuter, J., Steinhauser, M., Tenchini, R., Vos, M., and Zhang, C. (2017). Physics behind precision, Tech. Rep. arXiv:1703.01626, CERN, http://cds.cern.ch/record/2255661.

Baak, M., Cth, J., Haller, J., Hoecker, A., Kogler, R., Mnig, K., Schott, M., and Stelzer, J. (2014a). The global electroweak fit at NNLO and prospects for the LHC and ILC, *Eur. Phys. J.* **C74**, 3046, arXiv:1407.3792 [hep-ph].

Baak, M., Cth, J., Haller, J., Hoecker, A., Kogler, R., Mnig, K., Schott, M., and Stelzer, J. (2014b). The global electroweak fit at NNLO and prospects for the LHC and ILC, *Eur. Phys. J.* **C74**, 3046, arXiv:1407.3792 [hep-ph].

Babu, K. S., Ma, E., and Valle, J. W. F. (2003). Underlying A_4 symmetry for the neutrino mass matrix and the quark mixing matrix, *Phys. Lett.* **B552**, 207–213, arXiv:hep-ph/0206292 [hep-ph].

Badger, S., Biedermann, B., Uwer, P., and Yundin, V. (2014a). Next-to-leading order QCD corrections to five jet production at the LHC, *Phys. Rev.* **D89**, 3, 034019, arXiv:1309.6585 [hep-ph].

Badger, S., Guffanti, A., and Yundin, V. (2014b). Next-to-leading order QCD corrections to di-photon production in association with up to three jets at the Large Hadron Collider, *JHEP* **03**, 122, arXiv:1312.5927 [hep-ph].

Badger, S., Mogull, G., Ochirov, A., and O'Connell, D. (2015). A complete two-loop, five-gluon helicity amplitude in Yang-Mills theory, *JHEP* **10**, 064, arXiv:1507.08797 [hep-ph].

Baglio, J. and Djouadi, A. (2010). Predictions for Higgs production at the Tevatron and the associated uncertainties, *JHEP* **10**, 064, `arXiv:1003.4266 [hep-ph]`.

Bagnaia, P. *et al.* (1983). Evidence for $Z^0 \to e^+e^-$ at the CERN $\bar{p}p$ collider, *Phys. Lett.* **129B**, 130–140.

Bagnaschi, E., Harlander, R. V., Mantler, H., Vicini, A., and Wiesemann, M. (2016). Resummation ambiguities in the Higgs transverse-momentum spectrum in the Standard Model and beyond, *JHEP* **01**, 090, `arXiv:1510.08850 [hep-ph]`.

Bai, Z. *et al.* (2015). Standard Model prediction for direct cp violation in $K\pi\pi$ decay, *Phys. Rev. Lett.* **115**, 21, 212001, `arXiv:1505.07863 [hep-lat]`.

Baikov, P. A., Chetyrkin, K. G., and Kuhn, J. H. (2008). Order $alpha_s^4$ QCD corrections to Z and τ decays, *Phys. Rev. Lett.* **101**, 012002, `arXiv:0801.1821 [hep-ph]`.

Baikov, P. A., Chetyrkin, K. G., Kuhn, J. H., and Rittinger, J. (2012). Adler function, sum rules and Crewther relation of order $O(\alpha_s^4)$: The singlet case, *Phys. Lett.* **B714**, 62–65, `arXiv:1206.1288 [hep-ph]`.

Baikov, P. A., Chetyrkin, K. G., and Khn, J. H. (2017). Five-loop running of the QCD coupling constant, *Phys. Rev. Lett.* **118**, 8, 082002, `arXiv:1606.08659 [hep-ph]`.

Bak, P., Tang, C., and Wiesenfeld, K. (1987). Self-organized criticality: An explanation of $1/f$ noise, *Phys. Rev. Lett.* **59**, 381–384.

Bali, G. S., Bauer, C., Pineda, A., and Torrero, C. (2013). Perturbative expansion of the energy of static sources at large orders in four-dimensional SU(3) gauge theory, *Phys. Rev.* **D87**, 094517, `arXiv:1303.3279 [hep-lat]`.

Balitsky, I. I. and Lipatov, L. N. (1978). The Pomeranchuk singularity in quantum chromodynamics, *Sov. J. Nucl. Phys.* **28**, 822–829, [*Yad. Fiz.* **28**, 1597 (1978)].

Ball, R. D. (1989). Chiral gauge theory, *Phys. Rept.* **182**, 1.

Ball, R. D. (1995). Perturbative evolution of polarized structure functions, in *The spin structure of the nucleon. Proceedings, International School of Nucleon Structure, 1st Course, Erice, Italy, August 3–10, 1995, 371–383,* `arXiv:hep-ph/9511330 [hep-ph]`, `http://inspirehep.net/record/402278/files/arXiv:hep-ph_9511330.pdf`.

Ball, R. D. (2016). Global parton distributions for the LHC run II, *Nuovo Cim.* **C38**, 4, 127, `arXiv:1507.07891 [hep-ph]`.

Ball, R. D., Bertone, V., Bonvini, M., Marzani, S., Rojo, J., and Rottoli, L. (2017a). Parton distributions with small-x resummation: Evidence for BFKL dynamics in HERA data, `arXiv:1710.05935 [hep-ph]`.

Ball, R. D., Bertone, V., Del Debbio, L., Forte, S., Guffanti, A., Latorre, J. I., Lionetti, S., Rojo, J., and Ubiali, M. (2012). Precision NNLO determination of $\alpha_s(M_Z)$ using an unbiased global parton set, *Phys. Lett.* **B707**, 66–71, `arXiv:1110.2483 [hep-ph]`.

Ball, R. D., Bonvini, M., Forte, S., Marzani, S., and Ridolfi, G. (2013a). Higgs production in gluon fusion beyond NNLO, *Nucl. Phys.* **B874**, 746–772, `arXiv:1303.3590 [hep-ph]`.

Ball, R. D., Del Debbio, L., Forte, S., Guffanti, A., Latorre, J. I., Rojo, J., and Ubiali, M. (2010). Fitting parton distribution data with multiplicative normalization uncertainties, *JHEP* **05**, 075, `arXiv:0912.2276 [hep-ph]`.

Ball, R. D., Deshpande, A., Forte, S., Hughes, V. W., Lichtenstadt, J., and Ridolfi, G. (1996a). Measurement of the polarized structure function $g_1^p(x, Q^2)$ at HERA, in *Future physics at HERA. Proceedings, Workshop, Hamburg, Germany, September 25, 1995-May 31, 1996. Vol. 1, 2,* `arXiv:hep-ph/9609515 [hep-ph]`, `https://inspirehep.net/record/423916/files/arXiv:hep-ph_9609515.pdf`.

Ball, R. D. and Forte, S. (1994). Double asymptotic scaling at HERA, *Phys. Lett.* **B335**, 77–86, `arXiv:hep-ph/9405320 [hep-ph]`.

Ball, R. D. and Forte, S. (1997). Asymptotic freedom at small x, in *Proceedings, 5th International Workshop on Deep Inelastic Scattering and QCD (DIS 97): Chicago, Illinois, April 14–18, 1997*, Vol. 407, 1, 937, arXiv:hep-ph/9706459 [hep-ph].

Ball, R. D. and Forte, S. (1999). The small x behavior of Altarelli–Parisi splitting functions, *Phys. Lett.* **B465**, 271–281, arXiv:hep-ph/9906222 [hep-ph].

Ball, R. D., Forte, S., Guffanti, A., Nocera, E. R., Ridolfi, G., and Rojo, J. (2013b). Unbiased determination of polarized parton distributions and their uncertainties, *Nucl. Phys.* **B874**, 36–84, arXiv:1303.7236 [hep-ph].

Ball, R. D., Forte, S., Guffanti, A., Nocera, E. R., Ridolfi, G., and Rojo, J. (2014). Polarized parton distributions at an Electron-Ion Collider, *Phys. Lett.* **B728**, 524–531, arXiv:1310.0461 [hep-ph].

Ball, R. D., Forte, S., and Ridolfi, G. (1995). Scale dependence and small-x behavior of polarized parton distributions, *Nucl. Phys.* **B444**, 287–309, arXiv:hep-ph/9502340 [hep-ph], [Erratum: *Nucl. Phys.* **B449**, 680 (1995)].

Ball, R. D., Forte, S., and Ridolfi, G. (1996b). Next-to-leading order determination of the singlet axial charge and the polarized gluon content of the nucleon, *Phys. Lett.* **B378**, 255–266, arXiv:hep-ph/9510449 [hep-ph].

Ball, R. D. *et al.* (2013c). Parton distributions with LHC data, *Nucl. Phys.* **B867**, 244–289, arXiv:1207.1303 [hep-ph].

Ball, R. D. *et al.* (2015). Parton distributions for the LHC Run II, *JHEP* **04**, 040, arXiv:1410.8849 [hep-ph].

Ball, R. D. *et al.* (2017b). Parton distributions from high-precision collider data, *Eur. Phys. J.* **C77**, 10, 663, arXiv:1706.00428 [hep-ph].

Banfi, A., McAslan, H., Monni, P. F., and Zanderighi, G. (2015). A general method for the resummation of event-shape distributions in e^+e^- annihilation, *JHEP* **05**, 102, arXiv:hep-ph/0011336 [hep-ph].

Banfi, A., Monni, P. F., and Zanderighi, G. (2014). Quark masses in Higgs production with a jet veto, *JHEP* **01**, 097, arXiv:1308.4634 [hep-ph].

Banfi, A., Salam, G. P., and Zanderighi, G. (2005). Principles of general final-state resummation and automated implementation, *JHEP* **03**, 073, arXiv:hep-ph/0407286 [hep-ph].

Banks, T. and Seiberg, N. (2011). Symmetries and strings in field theory and gravity, *Phys. Rev.* **D83**, 084019, arXiv:1011.5120 [hep-th].

Banner, M. *et al.* (1983). Observation of single isolated electrons of high transverse momentum in events with missing transverse energy at the CERN $p\bar{p}$ collider, *Phys. Lett.* **122B**, 476–485.

Barate, R. *et al.* (2003). Search for the Standard Model Higgs boson at LEP, *Phys. Lett.* **B565**, 61–75, arXiv:hep-ex/0306033 [hep-ex].

Barbieri, R., Buttazzo, D., Sala, F., and Straub, D. M. (2012). Flavour physics from an approximate $U(2)^3$ symmetry, *JHEP* **07**, 181, arXiv:1203.4218 [hep-ph].

Barbieri, R. and Giudice, G. F. (1988). Upper bounds on supersymmetric particle masses, *Nucl. Phys.* **B306**, 63–76.

Barbieri, R., Gregoire, T., and Hall, L. J. (2005). Mirror world at the large hadron collider, arXiv:hep-ph/0509242 [hep-ph].

Barbieri, R., Hall, L. J., and Harigaya, K. (2016a). Minimal mirror twin Higgs, *JHEP* **11**, 172, arXiv:1609.05589 [hep-ph].

Barbieri, R., Hall, L. J., and Harigaya, K. (2017a). Effective theory of flavor for minimal mirror twin Higgs, *JHEP* **10**, 015, arXiv:1706.05548 [hep-ph].

Barbieri, R., Isidori, G., Jones-Perez, J., Lodone, P., and Straub, D. M. (2011). $U(2)$ and minimal flavour violation in supersymmetry, *Eur. Phys. J.* **C71**, 1725, arXiv:1105.2296 [hep-ph].

Barbieri, R., Isidori, G., Pattori, A., and Senia, F. (2016b). Anomalies in B-decays and $U(2)$ flavour symmetry, *Eur. Phys. J.* **C76**, 2, 67, arXiv:1512.01560 [hep-ph].

Barbieri, R., Murphy, C. W., and Senia, F. (2017b). B-decay anomalies in a composite leptoquark model, *Eur. Phys. J.* **C77**, 1, 8, arXiv:1611.04930 [hep-ph].

Barklow, T., Fujii, K., Jung, S., Karl, R., List, J., Ogawa, T., Peskin, M. E., and Tian, J. (2017). Improved formalism for precision Higgs coupling fits, arXiv:1708.08912 [hep-ph].

Bartels, J. (1979). High-energy behavior in a nonabelian gauge theory (I), *Nucl. Phys.* **B151**, 293–341.

Batell, B., Giudice, G. F., and McCullough, M. (2015). Natural heavy supersymmetry, *JHEP* **12**, 162, arXiv:1509.00834 [hep-ph].

Battaglieri, M. *et al.* (2017). US cosmic visions: new ideas in Dark Matter 2017: Community report, arXiv:1707.04591 [hep-ph].

Bauer, C. W., Fleming, S., Pirjol, D., and Stewart, I. W. (2001). An effective field theory for collinear and soft gluons: Heavy to light decays, *Phys. Rev.* **D63**, 114020, arXiv:hep-ph/0011336 [hep-ph].

Baum, R. and Sheehan, W. (1997). *In search of planet Vulcan: The ghost in Newton's clockwork machine* (Plenum Press (New York)).

Bazavov, A., Brambilla, N., Garcia i Tormo, X., Petreczky, P., Soto, J., and Vairo, A. (2014). Determination of α_s from the QCD static energy: An update, *Phys. Rev.* **D90**, 7, 074038, arXiv:1407.8437 [hep-ph].

Becher, T., Broggio, A., and Ferroglia, A. (2015). Introduction to soft-collinear effective theory, *Lect. Notes Phys.* **896**, pp. 1–206, arXiv:1410.1892 [hep-ph].

Becher, T., Neubert, M., Rothen, L., and Shao, D. Y. (2016). Factorization and resummation for jet processes, *JHEP* **11**, 019, arXiv:1605.02737 [hep-ph], [Erratum: *JHEP* **05**, 154 (2017)].

Becher, T. and Schwartz, M. D. (2008). A precise determination of α_s from LEP thrust data using effective field theory, *JHEP* **07**, 034, arXiv:0803.0342 [hep-ph].

Becker, S. and Weinzierl, S. (2012). Direct contour deformation with arbitrary masses in the loop, *Phys. Rev.* **D86**, 074009, arXiv:1208.4088 [hep-ph].

Behnke, O., Geiser, A., and Lisovyi, M. (2015). Charm, beauty and top at HERA, *Prog. Part. Nucl. Phys.* **84**, 1–72, arXiv:1506.07519 [hep-ex].

Bellm, J. *et al.* (2016). Herwig 7.0/Herwig++ 3.0 release note, *Eur. Phys. J.* **C76**, 4, 196, arXiv:1512.01178 [hep-ph].

Benedikt, M. (2017). https://indico.cern.ch/event/556692/contributions/2483390/, "FCC study overview and status", talk at the FCC week 2017, Berlin, 29 May–June 2 2017.

Beneke, M. (1995). More on ambiguities in the pole mass, *Phys. Lett.* **B344**, 341–347, arXiv:hep-ph/9408380 [hep-ph].

Beneke, M. (1999). Renormalons, *Phys. Rept.* **317**, 1–142, arXiv:hep-ph/9807443 [hep-ph].

Beneke, M. and Braun, V. M. (1994). Heavy quark effective theory beyond perturbation theory: Renormalons, the pole mass and the residual mass term, *Nucl. Phys.* **B426**, 301–343, arXiv:hep-ph/9402364 [hep-ph].

Beneke, M., Maier, A., Rauh, T., and Ruiz-Femenia, P. (2017a). Non-resonant and electroweak NNLO correction to the e^+e^- top anti-top threshold, arXiv:1711.10429 [hep-ph].

Beneke, M., Marquard, P., Nason, P., and Steinhauser, M. (2017b). On the ultimate uncertainty of the top quark pole mass, *Phys. Lett.* **B775**, 63–70, arXiv:1605.03609 [hep-ph].

Berera, A. (1992). The proton spin content at small x, *Phys. Lett.* **B293**, 445–450.

Berger, E. L., Gao, J., Li, C. S., Liu, Z. L., and Zhu, H. X. (2016a). Charm-quark production in deep-inelastic neutrino scattering at next-to-next-to-leading order in QCD, *Phys. Rev. Lett.* **116**, 21, 212002, arXiv:1601.05430 [hep-ph].

Berger, E. L., Gao, J., Yuan, C. P., and Zhu, H. X. (2016b). NNLO QCD Corrections to t-channel single top-quark production and decay, *Phys. Rev.* **D94**, 7, 071501, arXiv:1606.08463 [hep-ph].

Berger, E. L., Gao, J., and Zhu, H. X. (2017). Differential distributions for t-channel single top-quark production and decay at next-to-next-to-leading order in QCD, arXiv:1708.09405 [hep-ph].

Bertone, V., Carrazza, S., Hartland, N. P., Nocera, E. R., and Rojo, J. (2017). A determination of the fragmentation functions of pions, kaons, and protons with faithful uncertainties, *Eur. Phys. J.* **C77**, 8, 516, arXiv:1706.07049 [hep-ph].

Bethke, S. (2004). QCD studies at LEP, *Phys. Rept.* **403–404**, 203–220, arXiv:hep-ex/0406058 [hep-ex].

Bethke, S. (2017). α_s 2016, *Nucl. Part. Phys. Proc.* **282-284**, 149–152.

Bethke, S., Kluth, S., Pahl, C., and Schieck, J. (2009). Determination of the strong coupling α_s from hadronic event shapes with $O(\alpha_s^3)$ and resummed QCD predictions using JADE data, *Eur. Phys. J.* **C64**, 351–360, arXiv:0810.1389 [hep-ex].

Bethke, S. *et al.* (eds.) (2011). *Workshop on Precision Measurements of α_s*, arXiv:1110.0016 [hep-ph], http://lss.fnal.gov/cgi-bin/find_paper.pl?conf-11-611.

Bicer, M. *et al.* (2014). First look at the physics case of TLEP, *JHEP* **01**, 164, arXiv:1308.6176 [hep-ex].

Binetruy, P. and Dudas, E. (1995). Dynamical mass matrices from effective superstring theories, *Nucl. Phys.* **B451**, 31–52, arXiv:hep-ph/9505295 [hep-ph].

Bird, S., Cholis, I., Muoz, J. B., Ali-Hamoud, Y., Kamionkowski, M., Kovetz, E. D., Raccanelli, A., and Riess, A. G. (2016). Did LIGO detect dark matter? *Phys. Rev. Lett.* **116**, 20, 201301, arXiv:1603.00464 [astro-ph.CO].

Bizon, W., Monni, P. F., Re, E., Rottoli, L., and Torrielli, P. (2017). Momentum-space resummation for transverse observables and the Higgs $p_\perp$ at N^{3}LL+NNLO, arXiv:1705.09127 [hep-ph].

Bjorken, J. D. (1969). Asymptotic sum rules at infinite momentum, *Phys. Rev.* **179**, 1547–1553.

Bjorken, J. D. (1991). The future of experimental high-energy physics, in *High-energy physics. Proceedings, 25th International Conference, Singapore, August 2–8, 1990. Vol. I+II*, I.329–342, http://www-public.slac.stanford.edu/sciDoc/docMeta.aspx?slacPubNumber=SLAC-PUB-5361.

Bjorken, J. D. (1993). Rapidity gaps and jets as a new physics signature in very high-energy hadron hadron collisions, *Phys. Rev.* **D47**, 101–113.

Blanke, M. (2017). Quo vadis flavour physics? — FPCP2017 theory summary and outlook, in *Proceedings, 15th Conference on Flavor Physics and CP Violation (FPCP 2017): Prague, Czech Republic, June 5–9, 2017*, Vol. FPCP2017, 042, arXiv:1708.06326 [hep-ph].

Bloom, E. D. *et al.* (1969). High-energy inelastic *ep* scattering at 6-degrees and 10-degrees, *Phys. Rev. Lett.* **23**, 930–934.

Blossier, B., Boucaud, P., Brinet, M., De Soto, F., Morenas, V., Pene, O., Petrov, K., and Rodriguez-Quintero, J. (2014). High statistics determination of the strong coupling constant in Taylor scheme and its OPE Wilson coefficient from lattice QCD with a dynamical charm, *Phys. Rev.* **D89**, 1, 014507, arXiv:1310.3763 [hep-ph].

Blumenfeld, I. *et al.* (2007). Energy doubling of 42 GeV electrons in a metre-scale plasma wakefield accelerator, *Nature* **445**, 741–744.

Blumenhagen, R., Cvetic, M., Langacker, P., and Shiu, G. (2005). Toward realistic intersecting D-brane models, *Ann. Rev. Nucl. Part. Sci.* **55**, 71–139, arXiv:hep-th/0502005 [hep-th].

Blumenhagen, R., Kors, B., Lust, D., and Stieberger, S. (2007). Four-dimensional string compactifications with D-branes, orientifolds and fluxes, *Phys. Rept.* **445**, 1–193, arXiv:hep-th/0610327 [hep-th].

Blumlein, J., Bottcher, H., and Guffanti, A. (2007). Non-singlet QCD analysis of deep inelastic world data at $O(\alpha_s^3)$, *Nucl. Phys.* **B774**, 182–207, arXiv:hep-ph/0607200 [hep-ph].

Blumlein, J., van Oldenborgh, G. J., and Rückl, R. (1993). QCD and QED corrections to Higgs boson production in charged current *ep* scattering, *Nucl. Phys.* **B395**, 35–59, arXiv:hep-ph/9209219 [hep-ph].

Bodwin, G. T. (1985). Factorization of the Drell–Yan cross-section in perturbation theory, *Phys. Rev.* **D31**, 2616, [Erratum: *Phys. Rev.* **D34**, 3932 (1986)].

Boito, D., Golterman, M., Maltman, K., Osborne, J., and Peris, S. (2015). Strong coupling from the revised ALEPH data for hadronic τ decays, *Phys. Rev.* **D91**, 3, 034003, arXiv:1410.3528 [hep-ph].

Boito, D., Golterman, M., Maltman, K., and Peris, S. (2017). Strong coupling from hadronic τ decays: A critical appraisal, *Phys. Rev.* **D95**, 3, 034024, arXiv:1611.03457 [hep-ph].

Bolzoni, P., Maltoni, F., Moch, S.-O., and Zaro, M. (2010). Higgs production via vector-boson fusion at NNLO in QCD, *Phys. Rev. Lett.* **105**, 011801, arXiv:1003.4451 [hep-ph].

Bonciani, R., Del Duca, V., Frellesvig, H., Henn, J. M., Moriello, F., and Smirnov, V. A. (2016). Two-loop planar master integrals for Higgs$\to$ 3 partons with full heavy-quark mass dependence, *JHEP* **12**, 096, arXiv:1609.06685 [hep-ph].

Bonvini, M., Forte, S., and Ridolfi, G. (2012). The threshold region for Higgs production in gluon fusion, *Phys. Rev. Lett.* **109**, 102002, arXiv:1204.5473 [hep-ph].

Bonvini, M., Marzani, S., Muselli, C., and Rottoli, L. (2016). On the Higgs cross section at N^{3}LO+N^{3}LL and its uncertainty, *JHEP* **08**, 105, arXiv:1603.08000 [hep-ph].

Bordone, M., Isidori, G., and Trifinopoulos, S. (2017). Semileptonic B-physics anomalies: A general EFT analysis within $U(2)^n$ flavor symmetry, *Phys. Rev.* **D96**, 1, 015038, arXiv:1702.07238 [hep-ph].

Borowka, S., Greiner, N., Heinrich, G., Jones, S. P., Kerner, M., Schlenk, J., and Zirke, T. (2016). Full top quark mass dependence in Higgs boson pair production at NLO, *JHEP* **10**, 107, arXiv:1608.04798 [hep-ph].

Borschensky, C. and Kramer, M. (2016). Private communication.

Botts, J., Morfin, J. G., Owens, J. F., Qiu, J.-W., Tung, W.-K., and Weerts, H. (1993). CTEQ parton distributions and flavor dependence of sea quarks, *Phys. Lett.* **B304**, 159–166, arXiv:hep-ph/9303255 [hep-ph].

Boughezal, R., Campbell, J. M., Ellis, R. K., Focke, C., Giele, W. T., Liu, X., and Petriello, F. (2016). Z-boson production in association with a jet at next-to-next-to-leading order in perturbative QCD, *Phys. Rev. Lett.* **116**, 15, 152001, arXiv:1512.01291 [hep-ph].

Boughezal, R., Caola, F., Melnikov, K., Petriello, F., and Schulze, M. (2015a). Higgs boson production in association with a jet at next-to-next-to-leading order, *Phys. Rev. Lett.* **115**, 8, 082003, arXiv:1504.07922 [hep-ph].

Boughezal, R., Focke, C., Giele, W., Liu, X., and Petriello, F. (2015b). Higgs boson production in association with a jet at NNLO using jettiness subtraction, *Phys. Lett.* **B748**, 5–8, `arXiv:1505.03893 [hep-ph]`.

Boughezal, R., Focke, C., Liu, X., and Petriello, F. (2015c). W-boson production in association with a jet at next-to-next-to-leading order in perturbative QCD, *Phys. Rev. Lett.* **115**, 6, 062002, `arXiv:1504.02131 [hep-ph]`.

Boughezal, R., Guffanti, A., Petriello, F., and Ubiali, M. (2017). The impact of the LHC Z-boson transverse momentum data on PDF determinations, *JHEP* **07**, 130, `arXiv:1705.00343 [hep-ph]`.

Bousso, R. and Polchinski, J. (2000). Quantization of four form fluxes and dynamical neutralization of the cosmological constant, *JHEP* **06**, 006, `arXiv:hep-th/0004134 [hep-th]`.

Bozzi, G., Catani, S., de Florian, D., and Grazzini, M. (2003). The q_T spectrum of the Higgs boson at the LHC in QCD perturbation theory, *Phys. Lett.* **B564**, 65–72, `arXiv:hep-ph/0302104 [hep-ph]`.

Brandt, D., Burkhardt, H., Lamont, M., Myers, S., and Wenninger, J. (2000). Accelerator physics at LEP, *Rept. Prog. Phys.* **63**, 939–1000.

Brandt, S., Peyrou, C., Sosnowski, R., and Wroblewski, A. (1964). The principal axis of jets. An attempt to analyze high-energy collisions as two-body processes, *Phys. Lett.* **12**, 57–61.

Brax, P. and Chemtob, M. (1995). Flavor changing neutral current constraints on standard-like orbifold models, *Phys. Rev.* **D51**, 6550–6571, `arXiv:hep-th/9411022 [hep-th]`.

Breidenbach, M., Friedman, J. I., Kendall, H. W., Bloom, E. D., Coward, D. H., DeStaebler, H. C., Drees, J., Mo, L. W., and Taylor, R. E. (1969). Observed behavior of highly inelastic electron-proton scattering, *Phys. Rev. Lett.* **23**, 935–939.

Breitweg, J. *et al.* (1997). Measurement of the proton structure function F_2 and $\sigma^{\text{tot}}(\gamma * p)$ at low Q^2 and very low x at HERA, *Phys. Lett.* **B407**, 432–448, `arXiv:hep-ex/9707025 [hep-ex]`.

Britzger, D. and Klein, M. (2017). "Electroweak physics with inclusive DIS data at the LHeC and FCC-eh", Proceedings DIS2017, to appear.

Brodsky, S. J. (2017). Novel features of nuclear chromodynamics, *Eur. Phys. J.* **A53**, 3, 48.

Brucherseifer, M., Caola, F., and Melnikov, K. (2013). $\mathcal{O}(\alpha_s^2)$ corrections to fully-differential top quark decays, *JHEP* **04**, 059, `arXiv:1301.7133 [hep-ph]`.

Brucherseifer, M., Caola, F., and Melnikov, K. (2014). On the NNLO QCD corrections to single-top production at the LHC, *Phys. Lett.* **B736**, 58–63, `arXiv:1404.7116 [hep-ph]`.

Brüning, O., and Klein, M. (2013). The Large Hadron Electron Collider, *Mod. Phys. Lett.* **A28**, 16, 1330011, `arXiv:1305.2090 [physics.acc-ph]`.

Brüning, O., and others (2017). FCC-eh baseline parameters, Tech. Rep. FCC-ACC-RPT-0012, CERN.

Bruno, M., Dalla Brida, M., Fritzsch, P., Korzec, T., Ramos, A., Schaefer, S., Simma, H., Sint, S., and Sommer, R. (2017). QCD coupling from a nonperturbative determination of the three-flavor Λ parameter, *Phys. Rev. Lett.* **119**, 10, 102001, `arXiv:1706.03821 [hep-lat]`.

Buchmüller, W., Rückl, R., and Wyler, D. (1987). Leptoquarks in lepton-quark collisions, *Phys. Lett.* **B191**, 442–448, [Erratum: *Phys. Lett.* **B448**, 320 (1999)].

Buras, A. J., Jamin, M., and Lautenbacher, M. E. (1993a). Two loop anomalous dimension matrix for $\Delta S = 1$ weak nonleptonic decays. 2. $O(\alpha\alpha_s)$, *Nucl. Phys.* **B400**, 75–102, `arXiv:hep-ph/9211321 [hep-ph]`.

Buras, A. J., Jamin, M., Lautenbacher, M. E., and Weisz, P. H. (1993b). Two loop anomalous dimension matrix for $\Delta S = 1$ weak nonleptonic decays I: $\mathcal{O}(\alpha_s^2)$, *Nucl. Phys.* **B400**, 37–74, `arXiv:hep-ph/9211304 [hep-ph]`.

Burgess, C. P. and Dolan, B. P. (2001). Particle vortex duality and the modular group: Applications to the quantum Hall effect and other 2-D systems, *Phys. Rev.* **B63**, 155309, `arXiv:hep-th/0010246 [hep-th]`.

Burgess, C. P. and Dolan, B. P. (2002). Duality and nonlinear response for quantum Hall systems, *Phys. Rev.* **B65**, 155323, `arXiv:cond-mat/0105621 [cond-mat]`.

Burgess, C. P. and Dolan, B. P. (2007). The quantum hall effect in graphene: Emergent modular symmetry and the semi-circle law, *Phys. Rev.* **B76**, 113406, `arXiv:cond-mat/0612269 [cond-mat.mes-hall]`.

Buschmann, M., Goncalves, D., Kuttimalai, S., Schonherr, M., Krauss, F., and Plehn, T. (2015). Mass effects in the Higgs-gluon coupling: boosted vs off-shell production, *JHEP* **02**, 038, `arXiv:1410.5806 [hep-ph]`.

Butenschoen, M., Dehnadi, B., Hoang, A. H., Mateu, V., Preisser, M., and Stewart, I. W. (2016). Top quark mass calibration for Monte Carlo event generators, *Phys. Rev. Lett.* **117**, 23, 232001, `arXiv:1608.01318 [hep-ph]`.

Buttazzo, D., Degrassi, G., Giardino, P. P., Giudice, G. F., Sala, F., Salvio, A., and Strumia, A. (2013). Investigating the near-criticality of the Higgs boson, *JHEP* **12**, 089, `arXiv:1307.3536 [hep-ph]`.

Buttazzo, D., Greljo, A., Isidori, G., and Marzocca, D. (2016). Toward a coherent solution of diphoton and flavor anomalies, *JHEP* **08**, 035, `arXiv:1604.03940 [hep-ph]`.

Buttazzo, D., Greljo, A., Isidori, G., and Marzocca, D. (2017). B-physics anomalies: A guide to combined explanations, *JHEP* **11**, 044, `arXiv:1706.07808 [hep-ph]`.

Buttenschön, B., Kempkes, P., Grülke, O., and Klinger T. (2014). A high power helicon discharge as a plasma cell for future plasma wakefield accelerators, in *Proceedings of the 41 EPS Conference on Plasma Physics*, P2.102, `http://ocs.ciemat.es/EPS2014PAP/html/contrib.html`.

Butterworth, J. *et al.* (2016). PDF4LHC recommendations for LHC run II, *J. Phys.* **G43**, 023001, `arXiv:1510.03865 [hep-ph]`.

Cabibbo, N. and Gatto, R. (1961). Electron positron colliding beam experiments, *Phys. Rev.* **124**, 1577–1595.

Cabibbo, N. and Maiani, L. (1970). Weak interactions and the breaking of hadron symmetries, in M. Conversi (ed.), *Evolution of particle physics: A volume dedicated to Edoardo Amaldi in his sixtieth birthday* (Academic Press (New York)), 50–80.

Cabibbo, N., Swallow, E. C., and Winston, R. (2003). Semileptonic hyperon decays, *Ann. Rev. Nucl. Part. Sci.* **53**, 39–75, `arXiv:hep-ph/0307298 [hep-ph]`.

Cacciari, M., Dreyer, F. A., Karlberg, A., Salam, G. P., and Zanderighi, G. (2015). Fully differential vector-boson-fusion higgs production at next-to-next-to-leading order, *Phys. Rev. Lett.* **115**, 8, 082002, `arXiv:1506.02660 [hep-ph]`.

Cacciari, M. and Houdeau, N. (2011). Meaningful characterisation of perturbative theoretical uncertainties, *JHEP* **09**, 039, `arXiv:1105.5152 [hep-ph]`.

Cakir, O. (2017). "Anomalous Triple Gauge Couplings", Talk at the FCC Physics Workshop at CERN.

Caldwell, A. (2016). The evolution of $\sigma^{\gamma P}$ with coherence length, *New J. Phys.* **18**, 7, 073019, `arXiv:1601.04472 [hep-ph]`.

Caldwell, A., Kollar, D., and Kröninger, K. (2009a). BAT: The Bayesian Analysis Toolkit, *Comput. Phys. Commun.* **180**, 2197–2209, `arXiv:0808.2552 [physics.data-an]`.

Caldwell, A. and Lotov, K. (2011). Plasma wakefield acceleration with a modulated proton bunch, *Phys. Plasmas* **18**, 103101, `arXiv:1105.1292 [physics.acc-ph]`.

Caldwell, A., Lotov, K., Pukhov, A., and Simon, F. (2009b). Proton driven plasma wakefield acceleration, *Nature Phys.* **5**, 363–367, arXiv:0807.4599 [physics.acc-ph].

Caldwell, A. and Wing, M. (2015). VHEeP: A very high energy electron–proton collider based on proton-driven plasma wakefield acceleration, in *Proceedings, 23rd International Workshop on Deep-Inelastic Scattering and Related Subjects (DIS 2015): Dallas, Texas, USA, April 27–May 01, 2015*, Vol. DIS2015, 240, arXiv:1509.00235 [physics.acc-ph].

Caldwell, A. and Wing, M. (2016a). VHEeP: A very high energy electron–proton collider, *Eur. Phys. J.* **C76**, 8, 463, arXiv:1606.00783 [hep-ex].

Caldwell, A. and Wing, M. (2016b). VHEeP: A very high energy electron–proton collider, *Eur. Phys. J.* **C76**, 8, 463, arXiv:1606.00783 [hep-ex].

Caldwell, A. *et al.* (2016). Path to AWAKE: Evolution of the concept, *Nucl. Instrum. Meth.* **A829**, 3–16, arXiv:1511.09032 [physics.plasm-ph].

Callan, C. G., Jr., Coleman, S. R., Wess, J., and Zumino, B. (1969). Structure of phenomenological Lagrangians. 2. *Phys. Rev.* **177**, 2247–2250.

Callan, C. G., Jr. and Gross, D. J. (1969). High-energy electroproduction and the constitution of the electric current, *Phys. Rev. Lett.* **22**, 156–159.

Camarda, S. (2017). M_W from the LHeC, Private communication.

Camici, G. and Ciafaloni, M. (1997). Irreducible part of the next-to-leading BFKL kernel, *Phys. Lett.* **B412**, 396–406, arXiv:hep-ph/9707390 [hep-ph], [Erratum: *Phys. Lett.* **B417**, 390 (1998)].

Campbell, J. M., Ellis, R. K., Czakon, M., and Kirchner, S. (2016a). Two loop correction to interference in $gg \to ZZ$, *JHEP* **08**, 011, arXiv:1605.01380 [hep-ph].

Campbell, J. M., Ellis, R. K., Li, Y., and Williams, C. (2016b). Predictions for diphoton production at the LHC through NNLO in QCD, *JHEP* **07**, 148, arXiv:1603.02663 [hep-ph].

Campbell, J. M., Ellis, R. K., Nason, P., and Re, E. (2015). Top-pair production and decay at NLO matched with parton showers, *JHEP* **04**, 114, arXiv:1412.1828 [hep-ph].

Campbell, J. M., Ellis, R. K., and Williams, C. (2016c). Associated production of a higgs boson at NNLO, *JHEP* **06**, 179, arXiv:1601.00658 [hep-ph].

Campbell, J. M., Ellis, R. K., and Williams, C. (2017a). Direct photon production at next-to-next-to-leading order, *Phys. Rev. Lett.* **118**, 22, 222001, arXiv:1612.04333 [hep-ph].

Campbell, J. M., Ellis, R. K., and Williams, C. (2017b). Driving missing data at the LHC: NNLO predictions for the ratio of $\gamma + j$ and $Z + j$, *Phys. Rev.* **D96**, 1, 014037, arXiv:1703.10109 [hep-ph].

Caola, F., Dowling, M., Melnikov, K., Röntsch, R., and Tancredi, L. (2016a). QCD corrections to vector boson pair production in gluon fusion including interference effects with off-shell Higgs at the LHC, *JHEP* **07**, 087, arXiv:1605.04610 [hep-ph].

Caola, F. (2017). Talk at the Higgs Cross Section Working group http://indico.com.ch/event/675782/contributions/2766454/attachments/1551025/2436721/boosted_H_theory.pdf

Caola, F., Forte, S., and Marzani, S. (2011). Small x resummation of rapidity distributions: the case of Higgs production, *Nucl. Phys.* **B846**, 167–211, arXiv:1010.2743 [hep-ph].

Caola, F., Forte, S., Marzani, S., Muselli, C., and Vita, G. (2016b). The Higgs transverse momentum spectrum with finite quark masses beyond leading order, *JHEP* **08**, 150, arXiv:1606.04100 [hep-ph].

Caola, F. and Melnikov, K. (2013). Constraining the Higgs boson width with ZZ production at the LHC, *Phys. Rev.* **D88**, 054024, arXiv:1307.4935 [hep-ph].

Caola, F., Melnikov, K., and Röntsch, R. (2017). Nested soft-collinear subtractions in NNLO QCD computations, *Eur. Phys. J.* **C77**, 4, 248, `arXiv:1702.01352 [hep-ph]`.

Caola, F., Melnikov, K., Röntsch, R., and Tancredi, L. (2015a). QCD corrections to ZZ production in gluon fusion at the LHC, *Phys. Rev.* **D92**, 9, 094028, `arXiv:1509.06734 [hep-ph]`.

Caola, F., Melnikov, K., and Schulze, M. (2015b). Fiducial cross sections for Higgs boson production in association with a jet at next-to-next-to-leading order in QCD, *Phys. Rev.* **D92**, 7, 074032, `arXiv:1508.02684 [hep-ph]`.

Capozzi, F., Di Valentino, E., Lisi, E., Marrone, A., Melchiorri, A., and Palazzo, A. (2017). Global constraints on absolute neutrino masses and their ordering, *Phys. Rev.* **D95**, 9, 096014, `arXiv:1703.04471 [hep-ph]`.

Cappelli, A. and Zemba, G. R. (1997). Modular invariant partition functions in the quantum Hall effect, *Nucl. Phys.* **B490**, 595–632, `arXiv:hep-th/9605127 [hep-th]`.

Cardoso, V., Hopper, S., Macedo, C. F. B., Palenzuela, C., and Pani, P. (2016). Gravitational-wave signatures of exotic compact objects and of quantum corrections at the horizon scale, *Phys. Rev.* **D94**, 8, 084031, `arXiv:1608.08637 [gr-qc]`.

Caron-Huot, S. (2015). Resummation of non-global logarithms and the BFKL equation, `arXiv:1501.03754 [hep-ph]`.

Carrazza, S., Ellis, R. K., and Zanderighi, G. (2016). QCDLoop: A comprehensive framework for one-loop scalar integrals, *Comput. Phys. Commun.* **209**, 134–143, `arXiv:1605.03181 [hep-ph]`.

Casalderrey-Solana, J., Liu, H., Mateos, D., Rajagopal, K., and Wiedemann, U. A. (2011). *Gauge/String duality, hot QCD and heavy ion collisions* (Cambridge University Press (Cambridge)), `arXiv:1101.0618 [hep-th]`.

Casas, J. A., Gomez, F., and Munoz, C. (1993). Complete structure of Z_n Yukawa couplings, *Int. J. Mod. Phys.* **A8**, 455–506, `arXiv:hep-th/9110060 [hep-th]`.

Cascioli, F., Gehrmann, T., Grazzini, M., Kallweit, S., Maierhöfer, P., von Manteuffel, A., Pozzorini, S., Rathlev, D., Tancredi, L., and Weihs, E. (2014). ZZ production at hadron colliders in NNLO QCD, *Phys. Lett.* **B735**, 311–313, `arXiv:1405.2219 [hep-ph]`.

Catani, S., Ciafaloni, M., and Hautmann, F. (1990a). Gluon contributions to small x heavy flavor production, *Phys. Lett.* **B242**, 97–102.

Catani, S., Ciafaloni, M., and Hautmann, F. (1991a). High-energy factorization and small x heavy flavor production, *Nucl. Phys.* **B366**, 135–188.

Catani, S., Ciafaloni, M., and Hautmann, F. (1991b). Small x structure functions and heavy flavor production, *Nucl. Phys. Proc. Suppl.* **18C**, 220–225.

Catani, S., Ciafaloni, M., and Hautmann, F. (1993). High-energy factorization in QCD and minimal subtraction scheme, *Phys. Lett.* **B307**, 147–153.

Catani, S., Cieri, L., de Florian, D., Ferrera, G., and Grazzini, M. (2012). Diphoton production at hadron colliders: A fully-differential QCD calculation at NNLO, *Phys. Rev. Lett.* **108**, 072001, `arXiv:1110.2375 [hep-ph]`, [Erratum: *Phys. Rev. Lett.* **117**, 8, 089901 (2016)].

Catani, S., Cieri, L., Ferrera, G., de Florian, D., and Grazzini, M. (2009). Vector boson production at hadron colliders: A fully exclusive QCD calculation at NNLO, *Phys. Rev. Lett.* **103**, 082001, `arXiv:0903.2120 [hep-ph]`.

Catani, S., D'Emilio, E., and Trentadue, L. (1988). The gluon form-factor to higher orders: Gluon gluon annihilation at small Q_t, *Phys. Lett.* **B211**, 335–342.

Catani, S., Dokshitzer, Y. L., Olsson, M., Turnock, G., and Webber, B. R. (1991c). New clustering algorithm for multi-jet cross-sections in e+ e- annihilation, *Phys. Lett.* **B269**, 432–438.

Catani, S., Fiorani, F., and Marchesini, G. (1990b). QCD coherence in initial state radiation, *Phys. Lett.* **B234**, 339–345.

Catani, S., Fiorani, F., and Marchesini, G. (1990c). Small x behavior of initial state radiation in perturbative QCD, *Nucl. Phys.* **B336**, 18–85.

Catani, S., Fiorani, F., Marchesini, G., and Oriani, G. (1991d). All loops calculation of the structure function for $x \to 0$ in perturbative QCD, *Nucl. Phys.* **B361**, 645–672.

Catani, S. and Grazzini, M. (2007). An NNLO subtraction formalism in hadron collisions and its application to Higgs boson production at the LHC, *Phys. Rev. Lett.* **98**, 222002, `arXiv:hep-ph/0703012 [hep-ph]`.

Catani, S. and Seymour, M. H. (1997). A general algorithm for calculating jet cross-sections in NLO QCD, *Nucl. Phys.* **B485**, 291–419, `arXiv:hep-ph/9605323 [hep-ph]`, [Erratum: *Nucl. Phys.* **B510**, 503 (1998)].

Catani, S. and Trentadue, L. (1989). Resummation of the QCD perturbative series for hard processes, *Nucl. Phys.* **B327**, 323–352.

Catani, S., Turnock, G., and Webber, B. R. (1992). Jet broadening measures in e^+e^- annihilation, *Phys. Lett.* **B295**, 269–276.

CEPC (2017). `http://cepc.ihep.ac.cn`.

Chacko, Z., Goh, H.-S., and Harnik, R. (2006). The twin Higgs: Natural electroweak breaking from mirror symmetry, *Phys. Rev. Lett.* **96**, 231802, `arXiv:hep-ph/0506256 [hep-ph]`.

Chakraborty, B., Davies, C. T. H., Galloway, B., Knecht, P., Koponen, J., Donald, G. C., Dowdall, R. J., Lepage, G. P., and McNeile, C. (2015). High-precision quark masses and QCD coupling from $n_f = 4$ lattice QCD, *Phys. Rev.* **D91**, 5, 054508, `arXiv:1408.4169 [hep-lat]`.

Chatrchyan, S. *et al.* (2012). Observation of a new boson at a mass of 125 GeV with the CMS experiment at the LHC, *Phys. Lett.* **B716**, 30–61, `arXiv:1207.7235 [hep-ex]`.

Chatrchyan, S. *et al.* (2014). Determination of the top-quark pole mass and strong coupling constant from the $t\bar{t}$ production cross section in pp collisions at $\sqrt{s} = 7$ TeV, *Phys. Lett.* **B728**, 496–517, `arXiv:1307.1907 [hep-ex]`, [Erratum: *Phys. Lett.* **B738**, 526 (2014)].

Chattopadhyay, S. and Jones, R. M. (2011). Radiative regime of linear colliders, high repetition rate free electron lasers and associated accelerating structures, *Nucl. Instrum. Meth.* **A657**, 168–176.

Chekanov, S. *et al.* (2002a). Exclusive photoproduction of J / psi mesons at HERA, *Eur. Phys. J.* **C24**, 345–360, `arXiv:hep-ex/0201043 [hep-ex]`.

Chekanov, S. *et al.* (2002b). Measurement of the photon proton total cross-section at a center-of-mass energy of 209-GeV at HERA, *Nucl. Phys.* **B627**, 3–28, `arXiv: hep-ex/0202034 [hep-ex]`.

Chekanov, S. *et al.* (2009). Measurement of high-Q^2 neutral current deep inelastic e^-p scattering cross sections with a longitudinally polarised electron beam at HERA, *Eur. Phys. J.* **C62**, 625–658, `arXiv:0901.2385 [hep-ex]`.

Chekanov, S. *et al.* (2010). Measurement of charm and beauty production in deep inelastic *ep* scattering from decays into muons at HERA, *Eur. Phys. J.* **C65**, 65–79, `arXiv:0904.3487 [hep-ex]`.

Chen, X., Cruz-Martinez, J., Gehrmann, T., Glover, E. W. N., and Jaquier, M. (2016). NNLO QCD corrections to Higgs boson production at large transverse momentum, *JHEP* **10**, 066, `arXiv:1607.08817 [hep-ph]`.

Chetyrkin, K. G., Kuhn, J. H., and Sturm, C. (2006). QCD decoupling at four loops, *Nucl. Phys.* **B744**, 121–135, `arXiv:hep-ph/0512060 [hep-ph]`.

Christiansen, J. R. and Skands, P. Z. (2015). String formation beyond leading colour, *JHEP* **08**, 003, `arXiv:1505.01681 [hep-ph]`.

Ciafaloni, M. (1988). Coherence effects in initial jets at small Q^2/s, *Nucl. Phys.* **B296**, 49–74.

Ciafaloni, M. and Camici, G. (1998). Energy scale(s) and next-to-leading BFKL equation, *Phys. Lett.* **B430**, 349–354, `arXiv:hep-ph/9803389 [hep-ph]`.

Ciafaloni, M. and Colferai, D. (1999). The BFKL equation at next-to-leading level and beyond, *Phys. Lett.* **B452**, 372–378, `arXiv:hep-ph/9812366 [hep-ph]`.

Ciafaloni, M., Colferai, D., Colferai, D., Salam, G. P., and Stasto, A. M. (2003a). Extending QCD perturbation theory to higher energies, *Phys. Lett.* **B576**, 143–151, `arXiv:hep-ph/0305254 [hep-ph]`.

Ciafaloni, M., Colferai, D., and Salam, G. P. (1999). Renormalization group improved small x equation, *Phys. Rev.* **D60**, 114036, `arXiv:hep-ph/9905566 [hep-ph]`.

Ciafaloni, M., Colferai, D., and Salam, G. P. (2000). On factorization at small x, *JHEP* **07**, 054, `arXiv:hep-ph/0007240 [hep-ph]`.

Ciafaloni, M., Colferai, D., Salam, G. P., and Stasto, A. M. (2002). Expanding running coupling effects in the hard pomeron, *Phys. Rev.* **D66**, 054014, `arXiv:hep-ph/0204282 [hep-ph]`.

Ciafaloni, M., Colferai, D., Salam, G. P., and Stasto, A. M. (2003b). Renormalization group improved small x Green's function, *Phys. Rev.* **D68**, 114003, `arXiv:hep-ph/0307188 [hep-ph]`.

Ciafaloni, M., Colferai, D., Salam, G. P., and Stasto, A. M. (2004). The gluon splitting function at moderately small x, *Phys. Lett.* **B587**, 87–94, `arXiv:hep-ph/0311325 [hep-ph]`.

Cirac, J. and Zoller, P. (2012). Goals and opportunities in quantum simulation, *Nature Physics*, 264.

Ciuchini, M., Franco, E., Martinelli, G., and Reina, L. (1994). The $\Delta S = 1$ effective Hamiltonian including next-to-leading order QCD and QED corrections, *Nucl. Phys.* **B415**, 403–462, `arXiv:hep-ph/9304257 [hep-ph]`.

Ciuchini, M., Franco, E., Mishima, S., Pierini, M., Reina, L., and Silvestrini, L. (2016). Update of the electroweak precision fit, interplay with Higgs-boson signal strengths and model-independent constraints on new physics, in *Proceedings, 37th International Conference on High Energy Physics (ICHEP 2014): Valencia, Spain, July 2–9, 2014*, Vol. 273-275, 2219–2225, `arXiv:1410.6940 [hep-ph]`.

Clesse, S. and Garca-Bellido, J. (2017). The clustering of massive primordial black holes as dark matter: Measuring their mass distribution with advanced LIGO, *Phys. Dark Univ.* **15**, 142–147, `arXiv:1603.05234 [astro-ph.CO]`.

CMS Collaboration (2008). The CMS experiment at the CERN LHC, *JINST* **3**, S08004.

CMS Collaboration (2015). Measurement of the $t\bar{t}$ production cross section in the $e\mu$ channel in pp collisions at 7 and 8 TeV, Tech. Rep. CMS-PAS-TOP-13-004, CERN, Geneva.

CMS Collaboration (2017a). `https://twiki.cern.ch/twiki/bin/view/CMSPublic/PhysicsResultsCombined`, Summary plots.

CMS Collaboration (2017b). CMS luminosity measurements for the 2016 data taking period, Tech. Rep. CMS-PAS-LUM-17-001, CERN, Geneva.

CMS Collaboration (2017c). Measurement of the top quark mass with lepton+jets final states in pp collisions at $\sqrt{s} = 13$ TeV, Tech. Rep. CMS-PAS-TOP-17-007, CERN, Geneva, `http://cds.cern.ch/record/2284594`.

CMS Collaboration (2017d). Measurement of the weak mixing angle with the forward-backward asymmetry of Drell–Yan events at 8 TeV, Tech. Rep. CMS-PAS-SMP-16-007, CERN, Geneva, `https://cds.cern.ch/record/2273392`.

Coleman, S. R., Wess, J., and Zumino, B. (1969). Structure of phenomenological Lagrangians. 1. *Phys. Rev.* **177**, 2239–2247.

Coleppa, B., Kumar, M., Kumar, S., and Mellado, B. (2017). Measuring CP nature of top-Higgs couplings at the future large hadron electron collider, *Phys. Lett.* **B770**, 335–341, `arXiv:1702.03426 [hep-ph]`.

Collins, J. C., Soper, D. E., and Sterman, G. F. (1984). All order factorization for Drell–Yan cross-sections, *Phys. Lett.* **134B**, 263.

Collins, J. C., Soper, D. E., and Sterman, G. F. (1985). Factorization for short distance hadron-hadron scattering, *Nucl. Phys.* **B261**, 104–142.

Collins, J. C., Soper, D. E., and Sterman, G. F. (1988). Soft gluons and factorization, *Nucl. Phys.* **B308**, 833–856.

Collins, J. C., Soper, D. E., and Sterman, G. F. (1989). Factorization of hard processes in QCD, *Adv. Ser. Direct. High Energy Phys.* **5**, 1–91, `arXiv:hep-ph/0409313 [hep-ph]`.

Collins, J. C., Wilczek, F., and Zee, A. (1978). Low-energy manifestations of heavy particles: Application to the neutral current, *Phys. Rev.* **D18**, 242.

Contino, R., Falkowski, A., Goertz, F., Grojean, C., and Riva, F. (2016). On the validity of the effective field theory approach to SM precision tests, *JHEP* **07**, 144, `arXiv:1604.06444 [hep-ph]`.

Contino, R. *et al.* (2017). Physics at a 100 TeV pp collider: Higgs and EW symmetry breaking studies, Tech. Rep. Yellow Report 3, CERN, `arXiv:1606.09408 [hep-ph]`.

Cornet, F., Garcia Canal, C. A., Grau, A., Pancheri, G., and Sciutto, S. J. (2015). Photoproduction total cross section and shower development, *Phys. Rev.* **D92**, 114011, `arXiv:1510.07279 [hep-ph]`.

Cremades, D., Ibanez, L. E., and Marchesano, F. (2003). Yukawa couplings in intersecting D-brane models, *JHEP* **07**, 038, `arXiv:hep-th/0302105 [hep-th]`.

Cremades, D., Ibanez, L. E., and Marchesano, F. (2004). Computing Yukawa couplings from magnetized extra dimensions, *JHEP* **05**, 079, `arXiv:hep-th/0404229 [hep-th]`.

Cullen, G. *et al.* (2014). *GOSAM*-2.0: A tool for automated one-loop calculations within the Standard Model and beyond, *Eur. Phys. J.* **C74**, 8, 3001, `arXiv:1404.7096 [hep-ph]`.

Cüppers, C. and Sørensen, P. (eds.) (1998). *A collection of Tibetan proverbs and sayings: Gems of Tibetan wisdom and wit* (Franz Steiner Verlag (Stuttgart)).

Curci, G., Furmanski, W., and Petronzio, R. (1980). Evolution of parton densities beyond leading order: The nonsinglet case, *Nucl. Phys.* **B175**, 27–92.

Curci, G. and Greco, M. (1978). Mass singularities and coherent states in gauge theories, *Phys. Lett.* **79B**, 406–410.

Curci, G. and Greco, M. (1981). Soft corrections to the Drell–Yan process in QCD, *Phys. Lett.* **102B**, 280–284.

Curci, G., Greco, M., and Srivastava, Y. (1979a). Coherent quark-gluon jets, *Phys. Rev. Lett.* **43**, 834–837.

Curci, G., Greco, M., and Srivastava, Y. (1979b). QCD jets from coherent states, *Nucl. Phys.* **B159**, 451–468.

Currie, J., Gehrmann, T., Huss, A., and Niehues, J. (2017a). NNLO QCD corrections to jet production in deep inelastic scattering, `arXiv:1703.05977 [hep-ph]`.

Currie, J., Gehrmann, T., and Niehues, J. (2016). Precise QCD predictions for the production of dijet final states in deep inelastic scattering, *Phys. Rev. Lett.* **117**, 4, 042001, arXiv:1606.03991 [hep-ph].

Currie, J., Gehrmann-De Ridder, A., Gehrmann, T., Glover, E. W. N., Huss, A., and Pires, J. (2017b). Precise predictions for dijet production at the LHC, *Phys. Rev. Lett.* **119**, 15, 152001, arXiv:1705.10271 [hep-ph].

Currie, J., Gehrmann-De Ridder, A., Gehrmann, T., Glover, N., Pires, J., and Wells, S. (2014). Second order QCD corrections to gluonic jet production at hadron colliders, in *Proceedings, 12th DESY Workshop on Elementary Particle Physics: Loops and Legs in Quantum Field Theory (LL2014): Weimar, Germany, April 27–May 2, 2014*, Vol. LL2014, 001, arXiv:1407.5558 [hep-ph].

Currie, J., Glover, E. W. N., Gehrmann, T., Gehrmann-De Ridder, A., Huss, A., and Pires, J. (2017c). Single jet inclusive production for the individual jet p_T scale choice at the LHC, *Acta Phys. Polon.* **B48**, 955–967, arXiv:1704.00923 [hep-ph].

Currie, J., Glover, E. W. N., and Pires, J. (2017d). Next-to-next-to leading order QCD predictions for single jet inclusive production at the LHC, *Phys. Rev. Lett.* **118**, 7, 072002, arXiv:1611.01460 [hep-ph].

Curtin, D., Deshpande, K., Fischer, O., and Zurita, J. (2017). New physics opportunities for long-lived particles at electron-proton colliders, arXiv:1712.07135 [hep-ph].

Czakon, M. (2010). A novel subtraction scheme for double-real radiation at NNLO, *Phys. Lett.* **B693**, 259–268, arXiv:1005.0274 [hep-ph].

Czakon, M., Fiedler, P., Heymes, D., and Mitov, A. (2016a). NNLO QCD predictions for fully-differential top-quark pair production at the Tevatron, *JHEP* **05**, 034, arXiv:1601.05375 [hep-ph].

Czakon, M., Fiedler, P., and Mitov, A. (2013). Total top-quark pair-production cross section at hadron colliders through $O(\alpha_S^4)$, *Phys. Rev. Lett.* **110**, 252004, arXiv:1303.6254 [hep-ph].

Czakon, M., Heymes, D., and Mitov, A. (2016b). High-precision differential predictions for top-quark pairs at the LHC, *Phys. Rev. Lett.* **116**, 8, 082003, arXiv:1511.00549 [hep-ph].

Czakon, M., Heymes, D., Mitov, A., Pagani, D., Tsinikos, I., and Zaro, M. (2017). Top-pair production at the LHC through NNLO QCD and NLO EW, *JHEP* **10**, 186, arXiv:1705.04105 [hep-ph].

Czakon, M. and Mitov, A. (2014). Top++: A program for the calculation of the top-pair cross-section at hadron colliders, *Comput. Phys. Commun.* **185**, 2930, arXiv:1112.5675 [hep-ph].

D'Agostini, G. (1994). On the use of the covariance matrix to fit correlated data, *Nucl. Instrum. Meth.* **A346**, 306–311.

Dalla Brida, M., Fritzsch, P., Korzec, T., Ramos, A., Sint, S., and Sommer, R. (2016). Determination of the QCD Λ-parameter and the accuracy of perturbation theory at high energies, *Phys. Rev. Lett.* **117**, 18, 182001, arXiv:1604.06193 [hep-ph].

Dalla Brida, M., Fritzsch, P., Korzec, T., Ramos, A., Sint, S., and Sommer, R. (2017). Slow running of the gradient flow coupling from 200 MeV to 4 GeV in $N_f = 3$ QCD, *Phys. Rev.* **D95**, 1, 014507, arXiv:1607.06423 [hep-lat].

Dasgupta, M., Dreyer, F., Salam, G. P., and Soyez, G. (2015). Small-radius jets to all orders in QCD, *JHEP* **04**, 039, arXiv:1411.5182 [hep-ph].

Dasgupta, M., Fregoso, A., Marzani, S., and Salam, G. P. (2013). Towards an understanding of jet substructure, *JHEP* **09**, 029, arXiv:1307.0007 [hep-ph].

Dasgupta, M. and Salam, G. P. (2001). Resummation of nonglobal QCD observables, *Phys. Lett.* **B512**, 323–330, arXiv:hep-ph/0104277 [hep-ph].

Dasgupta, M. and Salam, G. P. (2004). Event shapes in e^+e^- annihilation and deep inelastic scattering, *J. Phys.* **G30**, R143, arXiv:hep-ph/0312283 [hep-ph].

Davier, M., Hcker, A., Malaescu, B., Yuan, C.-Z., and Zhang, Z. (2014). Update of the ALEPH non-strange spectral functions from hadronic τ decays, *Eur. Phys. J.* **C74**, 3, 2803, arXiv:1312.1501 [hep-ex].

Davies, C. T. H., Hornbostel, K., Kendall, I. D., Lepage, G. P., McNeile, C., Shigemitsu, J., and Trottier, H. (2008). Update: Accurate determinations of $alpha_s$ from realistic lattice QCD, *Phys. Rev.* **D78**, 114507, arXiv:0807.1687 [hep-lat].

Davison, R. A. and Webber, B. R. (2009). Non-perturbative contribution to the thrust distribution in e+ e- annihilation, *Eur. Phys. J.* **C59**, 13–25, arXiv:0809.3326 [hep-ph].

de Adelhart Toorop, R., Feruglio, F., and Hagedorn, C. (2011). Discrete flavour symmetries in light of T2K, *Phys. Lett.* **B703**, 447–451, arXiv:1107.3486 [hep-ph].

de Adelhart Toorop, R., Feruglio, F., and Hagedorn, C. (2012). Finite modular groups and lepton mixing, *Nucl. Phys.* **B858**, 437–467, arXiv:1112.1340 [hep-ph].

De Blas, J. (2017). https://indico.cern.ch/event/550509/contributions/2413788/, "The global EW fit and constraints on new physics at FCC-ee", talk at the First FCC Physics Workshop, CERN, 16-20 January 2017.

de Blas, J., Ciuchini, M., Franco, E., Mishima, S., Pierini, M., Reina, L., and Silvestrini, L. (2016). Electroweak precision observables and Higgs-boson signal strengths in the Standard Model and beyond: Present and future, *JHEP* **12**, 135, arXiv:1608.01509 [hep-ph].

de Blas, J., Ciuchini, M., Franco, E., Mishima, S., Pierini, M., Reina, L., and Silvestrini, L. (2017). Electroweak precision constraints at present and future colliders, in *Proceedings, 38th International Conference on High Energy Physics (ICHEP 2016): Chicago, IL, USA, August 3–10, 2016*, Vol. ICHEP2016, 690, arXiv:1611.05354 [hep-ph].

de Florian, D., Epele, M., Hernandez-Pinto, R. J., Sassot, R., and Stratmann, M. (2017). Parton-to-Kaon fragmentation revisited, *Phys. Rev.* **D95**, 9, 094019, arXiv:1702.06353 [hep-ph].

de Florian, D., Grazzini, M., Hanga, C., Kallweit, S., Lindert, J. M., Maierhöfer, P., Mazzitelli, J., and Rathlev, D. (2016a). Differential Higgs boson pair production at next-to-next-to-leading order in QCD, *JHEP* **09**, 151, arXiv:1606.09519 [hep-ph].

de Florian, D., Sassot, R., Epele, M., Hernndez-Pinto, R. J., and Stratmann, M. (2015). Parton-to-pion fragmentation reloaded, *Phys. Rev.* **D91**, 1, 014035, arXiv:1410.6027 [hep-ph].

de Florian, D., Sassot, R., and Stratmann, M. (2007a). Global analysis of fragmentation functions for pions and kaons and their uncertainties, *Phys. Rev.* **D75**, 114010, arXiv:hep-ph/0703242 [hep-ph].

de Florian, D., Sassot, R., and Stratmann, M. (2007b). Global analysis of fragmentation functions for protons and charged hadrons, *Phys. Rev.* **D76**, 074033, arXiv:0707.1506 [hep-ph].

de Florian, D., Sassot, R., Stratmann, M., and Vogelsang, W. (2008). Global analysis of helicity parton densities and their uncertainties, *Phys. Rev. Lett.* **101**, 072001, arXiv:0804.0422 [hep-ph].

de Florian, D., Sassot, R., Stratmann, M., and Vogelsang, W. (2009). Extraction of spin-dependent parton densities and their uncertainties, *Phys. Rev.* **D80**, 034030, arXiv:0904.3821 [hep-ph].

de Florian, D., Sassot, R., Stratmann, M., and Vogelsang, W. (2014). Evidence for polarization of gluons in the proton, *Phys. Rev. Lett.* **113**, 1, 012001, arXiv:1404.4293 [hep-ph].

de Florian, D. and Vogelsang, W. (2010). Helicity parton distributions from spin asymmetries in W-boson production at RHIC, *Phys. Rev.* **D81**, 094020, arXiv:1003.4533 [hep-ph].

de Florian, D. *et al.* (2016b). Handbook of LHC Higgs cross sections: 4. Deciphering the nature of the Higgs Sector, Tech. Rep. CERN-2017-002-M, CERN, arXiv: 1610.07922 [hep-ph].

de Gouvea, A. and Murayama, H. (2003). Statistical test of anarchy, *Phys. Lett.* **B573**, 94–100, arXiv:hep-ph/0301050 [hep-ph].

de Gouvea, A. and Murayama, H. (2015). Neutrino mixing anarchy: alive and kicking, *Phys. Lett.* **B747**, 479–483, arXiv:1204.1249 [hep-ph].

de Pierre Cardinal de Bernis, F. (1741). *Réflexions sur les passions et sur les goûts* (Didot (Paris)).

de Rham, C. (2014). Massive gravity, *Living Rev. Rel.* **17**, 7, arXiv:1401.4173 [hep-th].

Decamp, D. *et al.* (1990). ALEPH: A detector for electron-positron annnihilations at LEP, *Nucl. Instrum. Meth.* **A294**, 121–178, [Erratum: *Nucl. Instrum. Meth.* **A303**, 393 (1991)].

Degee, A., Ivanov, I. P., and Keus, V. (2013). Geometric minimization of highly symmetric potentials, *JHEP* **02**, 125, arXiv:1211.4989 [hep-ph].

Degrassi, G., Di Vita, S., Elias-Miro, J., Espinosa, J. R., Giudice, G. F., Isidori, G., and Strumia, A. (2012). Higgs mass and vacuum stability in the standard model at NNLO, *JHEP* **08**, 098, arXiv:1205.6497 [hep-ph].

Del Duca, V., Duhr, C., Kardos, A., Somogyi, G., Szr, Z., Trcsnyi, Z., and Tulipnt, Z. (2016a). Jet production in the CoLoRFulNNLO method: Event shapes in electron-positron collisions, *Phys. Rev.* **D94**, 7, 074019, arXiv:1606.03453 [hep-ph].

Del Duca, V., Duhr, C., Kardos, A., Somogyi, G., and Trcsnyi, Z. (2016b). Three-jet production in electron-positron collisions at next-to-next-to-leading order accuracy, *Phys. Rev. Lett.* **117**, 15, 152004, arXiv:1603.08927 [hep-ph].

Del Duca, V., Duhr, C., Somogyi, G., Tramontano, F., and Trcsnyi, Z. (2015). Higgs boson decay into b-quarks at NNLO accuracy, *JHEP* **04**, 036, arXiv:1501.07226 [hep-ph].

Denner, A., Dittmaier, S., and Hofer, L. (2017). Collier: A fortran-based complex one-loop library in extended regularizations, *Comput. Phys. Commun.* **212**, 220–238, arXiv:1604.06792 [hep-ph].

Dent, T. (2001a). CP violation and modular symmetries, *Phys. Rev.* **D64**, 056005, arXiv:hep-ph/0105285 [hep-ph].

Dent, T. (2001b). On the modular invariance of mass eigenstates and CP violation, *JHEP* **12**, 028, arXiv:hep-th/0111024 [hep-th].

d'Enterria, D. (ed.) (2015). *Proceedings, High-Precision α_s Measurements from LHC to FCC-ee*, CERN (CERN, Geneva), arXiv:1512.05194 [hep-ph], https://inspirehep.net/record/1409920/files/arXiv:1512.05194.pdf.

d'Enterria, D. (2017). Higgs physics at the future circular collider, in *Proceedings, 38th International Conference on High Energy Physics (ICHEP 2016): Chicago, IL, USA, August 3–10, 2016*, Vol. ICHEP2016, 434, arXiv:1701.02663 [hep-ex].

Derrick, M., Gacek, D., Hill, N., Musgrave, B., Noland, R., Petereit, E., Repond, J., Stanek, R., and Sugano, K. (1991). Design and construction of the ZEUS barrel calorimeter, *Nucl. Instrum. Meth.* **A309**, 77–100.

Derrick, M. *et al.* (1993). Measurement of the proton structure function F_2 in *ep* scattering at HERA, *Phys. Lett.* **B316**, 412–426.

Deshpande, A., Milner, R., Venugopalan, R., and Vogelsang, W. (2005). Study of the fundamental structure of matter with an electron-ion collider, *Ann. Rev. Nucl. Part. Sci.* **55**, 165–228, arXiv:hep-ph/0506148 [hep-ph].

Deur, A., Brodsky, S. J., and de Teramond, G. F. (2016). The QCD running coupling, *Prog. Part. Nucl. Phys.* **90**, 1–74, `arXiv:1604.08082 [hep-ph]`.

Devenish, R. and Cooper-Sarkar, A. (2004). *Deep inelastic scattering* (Oxford University Press (Oxford)).

Dharmawardane, K. V. *et al.* (2006). Measurement of the x- and Q^2-dependence of the asymmetry A_1 on the nucleon, *Phys. Lett.* **B641**, 11–17, `arXiv:nucl-ex/0605028 [nucl-ex]`.

Dickens, C. (1859). *A tale of two cities* (Chapman & Hall (London)).

Dirac, P. A. M. (1931). Quantized singularities in the electromagnetic field, *Proc. Roy. Soc. Lond.* **A133**, 60–72.

Dirac, P. A. M. (1963). The evolution of the physicist's picture of nature, *Scientific American* **208**, 5, 45–53.

Dissertori, G. (2016). The determination of the strong coupling constant, *Adv. Ser. Direct. High Energy Phys.* **26**, 113–128, `arXiv:1506.05407 [hep-ex]`.

Dissertori, G., Gehrmann-De Ridder, A., Gehrmann, T., Glover, E. W. N., Heinrich, G., Luisoni, G., and Stenzel, H. (2009). Determination of the strong coupling constant using matched NNLO+NLLA predictions for hadronic event shapes in e+e- annihilations, *JHEP* **08**, 036, `arXiv:0906.3436 [hep-ph]`.

Dissertori, G., Gehrmann-De Ridder, A., Gehrmann, T., Glover, E. W. N., Heinrich, G., and Stenzel, H. (2010). Precise determination of the strong coupling constant at NNLO in QCD from the three-jet rate in electron–positron annihilation at LEP, *Phys. Rev. Lett.* **104**, 072002, `arXiv:0910.4283 [hep-ph]`.

Dittmaier, S. (2017). Standard Model theory, in *2017 European Physical Society Conference on High Energy Physics (EPS-HEP 2017) Venice, Italy, July 5–12, 2017*, `arXiv:1709.08564 [hep-ph]`, `https://inspirehep.net/record/1625326/files/arXiv:1709.08564.pdf`.

Dixon, L. J. (1996). Calculating scattering amplitudes efficiently, in *QCD and beyond. Proceedings, Theoretical Advanced Study Institute in Elementary Particle Physics, TASI-95, Boulder, USA, June 4–30, 1995*, 539–584, `arXiv:hep-ph/9601359 [hep-ph]`.

Dixon, L. J., Friedan, D., Martinec, E. J., and Shenker, S. H. (1987). The conformal field theory of orbifolds, *Nucl. Phys.* **B282**, 13–73.

Dokshitzer, Y. L. (1977). Calculation of the structure functions for deep inelastic scattering and e^+e^- annihilation by perturbation theory in quantum chromodynamics, *Sov. Phys. JETP* **46**, 641–653, [*Zh. Eksp. Teor. Fiz.* **73**, 1216 (1977)].

Dokshitzer, Y. L., Diakonov, D., and Troian, S. I. (1978a). Hard semiinclusive processes in QCD, *Phys. Lett.* **78B**, 290–294.

Dokshitzer, Y. L., Diakonov, D., and Troian, S. I. (1978b). On the transverse momentum distribution of massive lepton pairs, *Phys. Lett.* **79B**, 269–272.

Dokshitzer, Y. L., Leder, G. D., Moretti, S., and Webber, B. R. (1997). Better jet clustering algorithms, *JHEP* **08**, 001, `arXiv:hep-ph/9707323 [hep-ph]`.

Dolan, B. P. (1999a). Duality and the modular group in the quantum Hall effect, *J. Phys.* **A32**, L243, `arXiv:cond-mat/9805171 [cond-mat]`.

Dolan, B. P. (1999b). Modular invariance, universality and crossover in the quantum Hall effect, *Nucl. Phys.* **B554**, 487, `arXiv:cond-mat/9809294 [cond-mat.mes-hall]`.

Donnachie, A. and Landshoff, P. V. (1992). Total cross-sections, *Phys. Lett.* **B296**, 227–232, `arXiv:hep-ph/9209205 [hep-ph]`.

Donnachie, A. and Landshoff, P. V. (2004). Does the hard pomeron obey Regge factorization? *Phys. Lett.* **B595**, 393–399, `arXiv:hep-ph/0402081 [hep-ph]`.

D'Onofrio, M. (2017). "Supersymmetry and Exotics Searches", talk at the EPS Conference on High Energy Physics, Venice 5-12 July 2017.

Douglas, M. R. (2003). The statistics of string/M theory vacua, *JHEP* **05**, 046, `arXiv:hep-th/0303194 [hep-th]`.

Drell, S. D. and Yan, T.-M. (1970). Massive lepton pair production in hadron-hadron collisions at high-energies, *Phys. Rev. Lett.* **25**, 316–320, [Erratum: *Phys. Rev. Lett.* **25**, 902 (1970)].

Dreyer, F. A. and Karlberg, A. (2016). Vector-boson fusion Higgs production at three loops in QCD, *Phys. Rev. Lett.* **117**, 7, 072001, `arXiv:1606.00840 [hep-ph]`.

Dudas, E. (1996). Dynamical mass matrices from moduli fields, in *5th Hellenic School and Workshops on Elementary Particle Physics (CORFU 1995) Corfu, Greece, September 3–24, 1995*, 504–508, [(1996)].

Dudas, E., Pokorski, S., and Savoy, C. A. (1996). Soft scalar masses in supergravity with horizontal U(1)-x gauge symmetry, *Phys. Lett.* **B369**, 255–261, `arXiv:hep-ph/9509410 [hep-ph]`.

Dulat, S., Hou, T.-J., Gao, J., Guzzi, M., Huston, J., Nadolsky, P., Pumplin, J., Schmidt, C., Stump, D., and Yuan, C. P. (2016). New parton distribution functions from a global analysis of quantum chromodynamics, *Phys. Rev.* **D93**, 3, 033006, `arXiv:1506.07443 [hep-ph]`.

Dyson, F. J. (2007). Why is Maxwell's theory so hard to understand? in *Proceedings of 2nd European Conference on Antennas and Propagation (EuCAP 2007), Edinburgh 2007*, 619, `http://ieeexplore.ieee.org/stamp/stamp.jsp?arnumber=4458240`.

Eichten, E., Lane, K. D., and Peskin, M. E. (1983). New tests for quark and lepton substructure, *Phys. Rev. Lett.* **50**, 811–814.

Eick, W. *et al.* (1997). Development of the H1 backward silicon strip detector, *Nucl. Instrum. Meth.* **A386**, 81–86.

Einstein, A. (1934). *"On the method of theoretical physics", Herbert Spencer lecture delivered at Oxford on 10 June 1933.*, in *Mein Weltbild* (Querido Verlag (Amsterdam)), *Phil. Science*, **1**, 2, 163 (1934).

ELI (2017). Extreme light infrastructure, `www.eli-np.ro`.

Ellis, R. K. (2017). Guido Altarelli and the evolution of QCD, *Nuovo Cim.* **C39**, 4, 355, `arXiv:1608.05574 [physics.hist-ph]`.

Ellis, R. K., Georgi, H., Machacek, M., Politzer, H. D., and Ross, G. G. (1979). Perturbation theory and the parton model in QCD, *Nucl. Phys.* **B152**, 285–329.

Ellis, R. K., Martinelli, G., and Petronzio, R. (1981a). Second order corrections to the Drell-Yan process at large transverse momentum, *Phys. Lett.* **104B**, 45–48.

Ellis, R. K., Ross, D. A., and Terrano, A. E. (1981b). The perturbative calculation of jet structure in e^+e^- annihilation, *Nucl. Phys.* **B178**, 421–456.

Ellis, R. K. and Zanderighi, G. (2008). Scalar one-loop integrals for QCD, *JHEP* **02**, 002, `arXiv:0712.1851 [hep-ph]`.

Erler, J., Jungnickel, D., and Lauer, J. (1992). Dependence of Yukawa couplings on the axionic background moduli of Z(N) orbifolds, *Phys. Rev.* **D45**, 3651–3668.

Ermolaev, B. I., Greco, M., and Troyan, S. I. (2010). Overview of the spin structure function g_1 at arbitrary x and Q^2, *Riv. Nuovo Cim.* **33**, 57–122, `arXiv:0905.2841 [hep-ph]`.

Espinosa, J. R. (2003). Anarchy in the neutrino sector? `arXiv:hep-ph/0306019 [hep-ph]`.

Espinosa, J. R., Fong, C. S., and Nardi, E. (2013). Yukawa hierarchies from spontaneous breaking of the $SU(3)_L \times SU(3)_R$ flavour symmetry? *JHEP* **02**, 137, `arXiv:1211.6428 [hep-ph]`.

Espinosa, J. R., Giudice, G. F., Morgante, E., Riotto, A., Senatore, L., Strumia, A., and Tetradis, N. (2015). The cosmological Higgstory of the vacuum instability, *JHEP* **09**, 174, arXiv:1505.04825 [hep-ph].

Evans, L. and Bryant, P. (2008). LHC machine, *JINST* **3**, S08001.

Ewerz, C. and Nachtmann, O. (2007). Towards a nonperturbative foundation of the dipole picture. II. High energy limit, *Annals Phys.* **322**, 1670–1726, arXiv:hep-ph/0604087 [hep-ph].

Fabricius, K., Schmitt, I., Schierholz, G., and Kramer, G. (1980). Order α_S^2 correction to jet cross-sections in e^+e^- annihilation, *Phys. Lett.* **97B**, 431–436.

Fadin, V. S., Kuraev, E. A., and Lipatov, L. N. (1975). On the Pomeranchuk singularity in asymptotically free theories, *Phys. Lett.* **60B**, 50–52.

Fadin, V. S. and Lipatov, L. N. (1998). BFKL pomeron in the next-to-leading approximation, *Phys. Lett.* **B429**, 127–134, arXiv:hep-ph/9802290 [hep-ph].

Fanti, V. *et al.* (1999). A new measurement of direct CP violation in two pion decays of the neutral kaon, *Phys. Lett.* **B465**, 335–348, arXiv:hep-ex/9909022 [hep-ex].

Farhi, E. (1977). A QCD test for jets, *Phys. Rev. Lett.* **39**, 1587–1588.

Farina, M., Panico, G., Pappadopulo, D., Ruderman, J. T., Torre, R., and Wulzer, A. (2017). Energy helps accuracy: Electroweak precision tests at hadron colliders, *Phys. Lett.* **B772**, 210–215, arXiv:1609.08157 [hep-ph].

Faroughy, D. A., Greljo, A., and Kamenik, J. F. (2017). Confronting lepton flavor universality violation in B decays with high-p_T tau lepton searches at LHC, *Phys. Lett.* **B764**, 126–134, arXiv:1609.07138 [hep-ph].

Favart, L., Guidal, M., Horn, T., and Kroll, P. (2016). Deeply virtual meson production on the nucleon, *Eur. Phys. J.* **A52**, 6, 158, arXiv:1511.04535 [hep-ph].

FCC (2017). https://indico.cern.ch/event/550509/timetable/#b-235717-top-and-flavour/, Top and Flavour Session, First FCC Physics Workshop, CERN, 16–20 January 2017.

Fernandez, J. A. and the LHeC Study Group (2012). A large hadron electron collider at CERN, *J. Phys.* **39**, G7, 075001, http://stacks.iop.org/0954-3899/39/i=7/a=075001.

Ferrara, S., Lust, D., Shapere, A. D., and Theisen, S. (1989a). Modular invariance in supersymmetric field theories, *Phys. Lett.* **B225**, 363.

Ferrara, S., Lust, D., and Theisen, S. (1989b). Target space modular invariance and low-energy couplings in orbifold compactifications, *Phys. Lett.* **B233**, 147–152.

Ferrera, G., Grazzini, M., and Tramontano, F. (2014). Higher-order QCD effects for associated WH production and decay at the LHC, *JHEP* **04**, 039, arXiv:1312.1669 [hep-ph].

Ferrera, G., Grazzini, M., and Tramontano, F. (2015). Associated ZH production at hadron colliders: The fully differential NNLO QCD calculation, *Phys. Lett.* **B740**, 51–55, arXiv:1407.4747 [hep-ph].

Ferrera, G., Somogyi, G., and Tramontano, F. (2017). Associated production of a Higgs boson decaying into bottom quarks at the LHC in full NNLO QCD, arXiv:1705.10304 [hep-ph].

Feruglio, F. (2015). Pieces of the flavour puzzle, *Eur. Phys. J.* **C75**, 8, 373, arXiv:1503.04071 [hep-ph].

Feruglio, F., Hagedorn, C., and Ziegler, R. (2013). Lepton mixing parameters from discrete and CP symmetries, *JHEP* **07**, 027, arXiv:1211.5560 [hep-ph].

Fong, C. S. and Nardi, E. (2014). Quark masses, mixings, and CP violation from spontaneous breaking of flavor $SU(3)^3$, *Phys. Rev.* **D89**, 3, 036008, arXiv:1307.4412 [hep-ph].

Fonseca, R. M. and Grimus, W. (2014). Classification of lepton mixing matrices from finite residual symmetries, *JHEP* **09**, 033, `arXiv:1405.3678 [hep-ph]`.

Fonseca R. A. *et al.* (2002). *OSIRIS: A three-dimensional, fully relativistic particle in cell code for modeling plasma based accelerators*, in P. M. A. Sloot, A. Hoekstra, G. C. J. K. Tan, and J. J. Dongarra (eds.), *Computational Science ICCS 2002, Lect. Notes Comp. Sci.*, Vol. 2331, *Lect. Notes Comp. Sci.*, Vol. 2331 (Springer (Berlin/Heidelberg)), 342–351.

Forte, S. (1995). Polarized structure functions: A theoretical update, in *Frontiers in strong interactions. Proceedings, 7th Rencontres de Blois, 7th Blois Workshop on Elastic and Diffractive Scattering, Blois, France, June 20–24, 1995*, 359–379, `arXiv:hep-ph/9511345 [hep-ph]`, `http://inspirehep.net/record/402350/files/arXiv:hep-ph_9511345.pdf`.

Forte, S. (2012). "LHeC: The ultimate factorization machine", Talk at the LHeC Workshop at Chavannes-de-Bogis.

Forte, S., Altarelli, G., and Ball, R. D. (2009). Can we trust small x resummation? *Nucl. Phys. Proc. Suppl.* **191**, 64–75, `arXiv:0901.1294 [hep-ph]`.

Forte, S., Garrido, L., Latorre, J. I., and Piccione, A. (2002). Neural network parametrization of deep inelastic structure functions, *JHEP* **05**, 062, `arXiv:hep-ph/0204232 [hep-ph]`.

Forte, S. and Muselli, C. (2016). High energy resummation of transverse momentum distributions: Higgs in gluon fusion, *JHEP* **03**, 122, `arXiv:1511.05561 [hep-ph]`.

Foster, B. *et al.* (1994). The design and construction of the ZEUS central tracking detector, *Nucl. Instrum. Meth.* **A338**, 254–283.

Franc, C. and Mason, G. (2016). Hypergeometric series, modular linear differential equations and vector-valued modular forms, *The Ramanujan Journal* **41**, 233, `arXiv:1503.05519`.

Frederix, R., Frixione, S., Maltoni, F., and Stelzer, T. (2009). Automation of next-to-leading order computations in QCD: The FKS subtraction, *JHEP* **10**, 003, `arXiv:0908.4272 [hep-ph]`.

Frederix, R., Frixione, S., Papanastasiou, A. S., Prestel, S., and Torrielli, P. (2016a). Off-shell single-top production at NLO matched to parton showers, *JHEP* **06**, 027, `arXiv:1603.01178 [hep-ph]`.

Frederix, R., Frixione, S., Vryonidou, E., and Wiesemann, M. (2016b). Heavy-quark mass effects in Higgs plus jets production, *JHEP* **08**, 006, `arXiv:1604.03017 [hep-ph]`.

Fritzsch, H., Gell-Mann, M., and Leutwyler, H. (1973). Advantages of the color octet gluon picture, *Phys. Lett.* **47B**, 365–368.

Frixione, S., Kunszt, Z., and Signer, A. (1996). Three jet cross-sections to next-to-leading order, *Nucl. Phys.* **B467**, 399–442, `arXiv:hep-ph/9512328 [hep-ph]`.

Frixione, S. and Mitov, A. (2014). Determination of the top quark mass from leptonic observables, *JHEP* **09**, 012, `arXiv:1407.2763 [hep-ph]`.

Frixione, S., Nason, P., and Oleari, C. (2007). Matching NLO QCD computations with parton shower simulations: The POWHEG method, *JHEP* **11**, 070, `arXiv:0709.2092 [hep-ph]`.

Frixione, S. and Webber, B. R. (2002). Matching NLO QCD computations and parton shower simulations, *JHEP* **06**, 029, `arXiv:hep-ph/0204244 [hep-ph]`.

Froggatt, C. D. and Nielsen, H. B. (1979). Hierarchy of quark masses, Cabibbo angles and CP violation, *Nucl. Phys.* **B147**, 277–298.

Froissart, M. (1961). Asymptotic behavior and subtractions in the Mandelstam representation, *Phys. Rev.* **123**, 1053–1057.

Furmanski, W. and Petronzio, R. (1980). Singlet parton densities beyond leading order, *Phys. Lett.* **97B**, 437–442.

Gaddi, A. (2017). "Installation of the LHeC detector", Talk at LHeC Workshop, CERN.

Gal, C. (2014). Measurements of the mid-rapidity parity violating spin asymmetries for $W^{\pm}$ bosons at PHENIX, *Phys. Part. Nucl.* **45**, 76–78.

Gamow, G. (1966). *Thirty years that shook physics: The story of quantum theory* (Anchor Books (Garden City)).

Gao, J., Li, C. S., and Zhu, H. X. (2013). Top quark decay at next-to-next-to leading order in QCD, *Phys. Rev. Lett.* **110**, 4, 042001, `arXiv:1210.2808 [hep-ph]`.

Gauld, R. and Rojo, J. (2017). Precision determination of the small-x gluon from charm production at LHCb, *Phys. Rev. Lett.* **118**, 7, 072001, `arXiv:1610.09373 [hep-ph]`.

Gauld, R., Rojo, J., Rottoli, L., and Talbert, J. (2015). Charm production in the forward region: constraints on the small-x gluon and backgrounds for neutrino astronomy, *JHEP* **11**, 009, `arXiv:1506.08025 [hep-ph]`.

Gaunt, J., Stahlhofen, M., Tackmann, F. J., and Walsh, J. R. (2015). N-jettiness subtractions for NNLO QCD calculations, *JHEP* **09**, 058, `arXiv:1505.04794 [hep-ph]`.

Gavin, R., Li, Y., Petriello, F., and Quackenbush, S. (2011). FEWZ 2.0: A code for hadronic Z production at next-to-next-to-leading order, *Comput. Phys. Commun.* **182**, 2388–2403, `arXiv:1011.3540 [hep-ph]`.

Gehrmann, T. (1997). QCD corrections to the longitudinally polarized Drell–Yan process, *Nucl. Phys.* **B498**, 245–266, `arXiv:hep-ph/9702263 [hep-ph]`.

Gehrmann, T. (1998). QCD corrections to double and single spin asymmetries in vector boson production at polarized hadron colliders, *Nucl. Phys.* **B534**, 21–39, `arXiv:hep-ph/9710508 [hep-ph]`.

Gehrmann, T., Glover, N., Huss, A., Niehues, J., and Zhang, H. (2017). NNLO QCD corrections to event orientation in e^+e^- annihilation, `arXiv:1709.01097 [hep-ph]`.

Gehrmann, T., Grazzini, M., Kallweit, S., Maierhöfer, P., von Manteuffel, A., Pozzorini, S., Rathlev, D., and Tancredi, L. (2014). W^+W^- production at hadron colliders in next to next to leading order QCD, *Phys. Rev. Lett.* **113**, 21, 212001, `arXiv:1408.5243 [hep-ph]`.

Gehrmann, T., Henn, J. M., and Lo Presti, N. A. (2016). Analytic form of the two-loop planar five-gluon all-plus-helicity amplitude in QCD, *Phys. Rev. Lett.* **116**, 6, 062001, `arXiv:1511.05409 [hep-ph]`, [Erratum: *Phys. Rev. Lett.* **116**, 18,189903 (2016)].

Gehrmann, T., Luisoni, G., and Monni, P. F. (2013). Power corrections in the dispersive model for a determination of the strong coupling constant from the thrust distribution, *Eur. Phys. J.* **C73**, 1, 2265, `arXiv:1210.6945 [hep-ph]`.

Gehrmann, T. and Remiddi, E. (2000). Differential equations for two loop four point functions, *Nucl. Phys.* **B580**, 485–518, `arXiv:hep-ph/9912329 [hep-ph]`.

Gehrmann, T. and Stirling, W. J. (1995). Spin dependent parton distributions from polarized structure function data, *Z. Phys.* **C65**, 461–470, `arXiv:hep-ph/9406212 [hep-ph]`.

Gehrmann, T. and Stirling, W. J. (1996). Polarized parton distributions in the nucleon, *Phys. Rev.* **D53**, 6100–6109, `arXiv:hep-ph/9512406 [hep-ph]`.

Gehrmann-De Ridder, A., Gehrmann, T., and Glover, E. W. N. (2005). Antenna subtraction at NNLO, *JHEP* **09**, 056, `arXiv:hep-ph/0505111 [hep-ph]`.

Gehrmann-De Ridder, A., Gehrmann, T., Glover, E. W. N., and Heinrich, G. (2007). Infrared structure of $e^+e^- \to 3$ jets at NNLO, *JHEP* **11**, 058, `arXiv:0710.0346 [hep-ph]`.

Gehrmann-De Ridder, A., Gehrmann, T., Glover, E. W. N., and Heinrich, G. (2008). Jet rates in electron-positron annihilation at $O(\alpha_s^3)$ in QCD, *Phys. Rev. Lett.* **100**, 172001, arXiv:0802.0813 [hep-ph].

Gehrmann-De Ridder, A., Gehrmann, T., Glover, E. W. N., Huss, A., and Morgan, T. A. (2016a). NNLO QCD corrections for Drell–Yan p_T^Z and ϕ^* observables at the LHC, *JHEP* **11**, 094, arXiv:1610.01843 [hep-ph].

Gehrmann-De Ridder, A., Gehrmann, T., Glover, E. W. N., Huss, A., and Morgan, T. A. (2016b). Precise QCD predictions for the production of a Z boson in association with a hadronic jet, *Phys. Rev. Lett.* **117**, 2, 022001, arXiv:1507.02850 [hep-ph].

Gehrmann-De Ridder, A., Gehrmann, T., Glover, E. W. N., Huss, A., and Morgan, T. A. (2016c). The NNLO QCD corrections to Z boson production at large transverse momentum, *JHEP* **07**, 133, arXiv:1605.04295 [hep-ph].

Giele, W. T., Glover, E. W. N., and Kosower, D. A. (1993). Higher order corrections to jet cross-sections in hadron colliders, *Nucl. Phys.* **B403**, 633–670, arXiv:hep-ph/9302225 [hep-ph].

Giele, W. T. and Keller, S. (1998). Implications of hadron collider observables on parton distribution function uncertainties, *Phys. Rev.* **D58**, 094023, arXiv:hep-ph/9803393 [hep-ph].

Gimenez, C. and Gale, M. (2004). *Constantin Brancusi: The essence of things* (Tate Gallery Publishing (London)).

Giudice, G. F. (2008). *Naturally speaking: The naturalness criterion and physics at the LHC*, in G. Kane and A. Pierce (eds.), *Perspectives on LHC physics* (World Scientific (Singapore)), arXiv:0801.2562 [hep-ph].

Giudice, G. F. (2013). Naturalness after LHC8, in *Proceedings, 2013 European Physical Society Conference on High Energy Physics (EPS-HEP 2013): Stockholm, Sweden, July 18–24, 2013*, Vol. EPS-HEP2013, 163, arXiv:1307.7879 [hep-ph].

Giudice, G. F., McCullough, M., and Urbano, A. (2016). Hunting for dark particles with gravitational waves, *JCAP* **1610**, 10, 001, arXiv:1605.01209 [hep-ph].

Giudice, G. F. and Rattazzi, R. (2006). Living dangerously with low-energy supersymmetry, *Nucl. Phys.* **B757**, 19–46, arXiv:hep-ph/0606105 [hep-ph].

Giudice, G. F. and Romanino, A. (2004). Split supersymmetry, *Nucl. Phys.* **B699**, 65–89, arXiv:hep-ph/0406088 [hep-ph], [Erratum: *Nucl. Phys.* **B706**, 487 (2005)].

Gluck, M., Hartl, A., and Reya, E. (2001). On the determination of the polarized sea distributions of the nucleon, *Eur. Phys. J.* **C19**, 77–87, arXiv:hep-ph/0011300 [hep-ph].

Gluck, M., Hoffmann, E., and Reya, E. (1982). Scaling violations and the gluon distribution of the nucleon, *Z. Phys.* **C13**, 119.

Gluck, M., Reya, E., and Vogt, A. (1993). Comparing radiatively generated parton distributions with recent measurements of $F_2(x, Q^2)$ in the small x region, *Phys. Lett.* **B306**, 391–394.

Goldberg, H. and Hall, L. J. (1986). A new candidate for dark matter, *Phys. Lett.* **B174**, 151.

Golling, T. *et al.* (2017). Physics at a 100 TeV pp collider: Beyond the standard model phenomena, *CERN Yellow Report*, **3**, 441–634, arXiv:1606.00947 [hep-ph].

Graham, P. W., Kaplan, D. E., and Rajendran, S. (2015). Cosmological relaxation of the electroweak scale, *Phys. Rev. Lett.* **115**, 22, 221801, arXiv:1504.07551 [hep-ph].

Grazzini, M., Kallweit, S., Pozzorini, S., Rathlev, D., and Wiesemann, M. (2016a). W^+W production at the LHC: Fiducial cross sections and distributions in NNLO QCD, *JHEP* **08**, 140, arXiv:1605.02716 [hep-ph].

Grazzini, M., Kallweit, S., and Rathlev, D. (2015a). $W\gamma$ and $Z\gamma$ production at the LHC in NNLO QCD, *JHEP* **07**, 085, arXiv:1504.01330 [hep-ph].

Grazzini, M., Kallweit, S., and Rathlev, D. (2015b). ZZ production at the LHC: Fiducial cross sections and distributions in NNLO QCD, *Phys. Lett.* **B750**, 407–410, arXiv:1507.06257 [hep-ph].

Grazzini, M., Kallweit, S., Rathlev, D., and Wiesemann, M. (2016b). $W^{\pm}Z$ production at hadron colliders in NNLO QCD, *Phys. Lett.* **B761**, 179–183, arXiv:1604.08576 [hep-ph].

Grazzini, M., Kallweit, S., Rathlev, D., and Wiesemann, M. (2017). $W^{\pm}Z$ production at the LHC: Fiducial cross sections and distributions in NNLO QCD, *JHEP* **05**, 139, arXiv:1703.09065 [hep-ph].

Grazzini, M. and Sargsyan, H. (2013). Heavy-quark mass effects in Higgs boson production at the LHC, *JHEP* **09**, 129, arXiv:1306.4581 [hep-ph].

Greco, M. (2017). Remembering Guido Altarelli, in *Proceedings, Vulcano Workshop 2016: Frontier objects in astrophysics and particle physics: Sicily, Italy, May 22–28, 2016*, 164–173, arXiv:1610.08692 [physics.hist-ph], http://inspirehep.net/record/1494827/files/arXiv:1610.08692.pdf.

Greiner, N., Hoeche, S., Luisoni, G., Schoenherr, M., and Winter, J.-C. (2017). Full mass dependence in Higgs boson production in association with jets at the LHC and FCC, *JHEP* **01**, 091, arXiv:1608.01195 [hep-ph].

Gribov, L. V., Levin, E. M., and Ryskin, M. G. (1983). Semihard processes in QCD, *Phys. Rept.* **100**, 1–150.

Gribov, V. N. and Lipatov, L. N. (1971). Deep inelastic electron scattering in perturbation theory, *Phys. Lett.* **37B**, 78–80.

Gribov, V. N. and Lipatov, L. N. (1972a). Deep inelastic *ep* scattering in perturbation theory, *Sov. J. Nucl. Phys.* **15**, 438–450, [*Yad. Fiz.* **15**, 781 (1972)].

Gribov, V. N. and Lipatov, L. N. (1972b). e^+e^- pair annihilation and deep inelastic *ep* scattering in perturbation theory, *Sov. J. Nucl. Phys.* **15**, 675–684, [*Yad. Fiz.* **15**, 1218 (1972)].

Gribov, V. N., Shekhter, V. M., Kolkunov, V. A., and Okun, L. B. (1962). Covariant derivation of the Weizsäcker–Williams formula, *Sov. Phys. JETP* **14**, 1308, [*Zh. Eksp. Teor. Fiz.* **41**, 1839 (1962)].

Grindhammer, G., Haidt, D., Ohnemus, J., Vermaseren, J., and Zeppenfeld, D. (1990). Searching for the Higgs in *ep* collisions at LEP/LHC, in *ECFA Large Hadron Collider Workshop, Aachen, Germany, 4-9 Oct 1990: Proceedings.2.*, 967–985.

Gross, D. J. and Wilczek, F. (1973a). Asymptotically free gauge theories. 1, *Phys. Rev.* **D8**, 3633–3652.

Gross, D. J. and Wilczek, F. (1973b). Ultraviolet behavior of nonabelian gauge theories, *Phys. Rev. Lett.* **30**, 1343–1346.

Gschwendtner, E. *et al.* (2010). Performance and operational experience of the CNGS facility, in *Proceedings, 1st International Particle Accelerator Conference (IPAC'10): Kyoto, Japan, May 23–28, 2010*, Vol. C100523, 4164.

Gschwendtner, E. *et al.* (2016). AWAKE, The advanced proton driven plasma wakefield acceleration experiment at CERN, *Nucl. Instrum. Meth.* **A829**, 76–82, arXiv:1512.05498 [physics.acc-ph].

Gunning, R. (1962). *Lectures on modular forms* (Princeton University Press (Princeton)).

H1 Collaboration (2005). Running at low proton beam energies, Tech. Rep. H1-10/05-622, DESY.

Haba, N. and Murayama, H. (2001). Anarchy and hierarchy, *Phys. Rev.* **D63**, 053010, arXiv:hep-ph/0009174 [hep-ph].

Haberl, P., Schrempp, F., and Martyn, H. U. (1991). Contact interactions and new heavy bosons at HERA: A model independent analysis, in *Workshop on Physics at HERA Hamburg, Germany, October 29–30, 1991*, 1133–1148.

Hagedorn, C. (2017). Theories of leptonic flavor, in *Proceedings, Prospects in Neutrino Physics (NuPhys2016): London, UK, December 12–14, 2016*, arXiv:1705.00684 [hep-ph], http://inspirehep.net/record/1597572/files/arXiv:1705.00684.pdf.

Hagedorn, C., Meroni, A., and Vitale, L. (2014). Mixing patterns from the groups $\Sigma(n\phi)$, *J. Phys.* **A47**, 055201, arXiv:1307.5308 [hep-ph].

Hall, L. J., Murayama, H., and Weiner, N. (2000). Neutrino mass anarchy, *Phys. Rev. Lett.* **84**, 2572–2575, arXiv:hep-ph/9911341 [hep-ph].

Hamberg, R., van Neerven, W. L., and Matsuura, T. (1991). A complete calculation of the order α_s^2 correction to the Drell–Yan K factor, *Nucl. Phys.* **B359**, 343–405, [Erratum: *Nucl. Phys.* **B644**,403 (2002)].

Hamidi, S. and Vafa, C. (1987). Interactions on orbifolds, *Nucl. Phys.* **B279**, 465–513.

Hamilton, K., Nason, P., Re, E., and Zanderighi, G. (2013). NNLOPS simulation of Higgs boson production, *JHEP* **10**, 222, arXiv:1309.0017 [hep-ph].

Hand, L. N., Miller, D. G., and Wilson, R. (1963). Electric and magnetic formfactor of the nucleon, *Rev. Mod. Phys.* **35**, 335.

Harbottle, T. B. and Dalbiac, P. H. (eds.) (1904). *Dictionary of Quotations* (Schwan Sonnenschein (London)).

Harland–Lang, L. A., Martin, A. D., Motylinski, P., and Thorne, R. S. (2015a). Parton distributions in the LHC era: MMHT 2014 PDFs, *Eur. Phys. J.* **C75**, 5, 204, arXiv:1412.3989 [hep-ph].

Harland–Lang, L. A., Martin, A. D., Motylinski, P., and Thorne, R. S. (2015b). Uncertainties on α_s in the MMHT2014 global PDF analysis and implications for SM predictions, *Eur. Phys. J.* **C75**, 9, 435, arXiv:1506.05682 [hep-ph].

Harland–Lang, L. A., Martin, A. D., and Thorne, R. S. (2017). The impact of LHC jet data on the MMHT PDF fit at NNLO, arXiv:1711.05757 [hep-ph].

Harlander, R. V. and Kilgore, W. B. (2002). Next-to-next-to-leading order Higgs production at hadron colliders, *Phys. Rev. Lett.* **88**, 201801, arXiv:hep-ph/0201206 [hep-ph].

Harlander, R. V., Mantler, H., Marzani, S., and Ozeren, K. J. (2010). Higgs production in gluon fusion at next-to-next-to-leading order QCD for finite top mass, *Eur. Phys. J.* **C66**, 359–372, arXiv:0912.2104 [hep-ph].

Harlander, R. V., Neumann, T., Ozeren, K. J., and Wiesemann, M. (2012). Top-mass effects in differential Higgs production through gluon fusion at order α_s^4, *JHEP* **08**, 139, arXiv:1206.0157 [hep-ph].

Harris, P. (2017). https://indico.cern.ch/event/556692/contributions/2413853/, "Experimental reach of dark matter searches at 100 TeV", talk at the FCC week 2017, Berlin, 29 May–June 2 2017.

Harrison, P. F., Perkins, D. H., and Scott, W. G. (2002). Tri-bimaximal mixing and the neutrino oscillation data, *Phys. Lett.* **B530**, 167, arXiv:hep-ph/0202074 [hep-ph].

Hautmann, F. (2002). Heavy top limit and double logarithmic contributions to Higgs production at $m^2(H)/s$ much less than 1, *Phys. Lett.* **B535**, 159–162, arXiv:hep-ph/0203140 [hep-ph].

Hawking, S. (1971). Gravitationally collapsed objects of very low mass, *Mon. Not. Roy. Astron. Soc.* **152**, 75.

Heinemeyer, S. (2017). https://indico.cern.ch/event/550509/contributions/2413292/, "Theory calculations needed to fully exploit the FCC experimental sensitivity", talk at the First FCC Physics Workshop, CERN, 16-20 January 2017.

Heinemeyer, S., Hollik, W., Weiglein, G., and Zeune, L. (2013). Implications of LHC search results on the W boson mass prediction in the MSSM, *JHEP* **12**, 084, arXiv:1311.1663 [hep-ph].

Heinrich, G., Jahn, S., Jones, S. P., Kerner, M., and Pires, J. (2017). NNLO predictions for Z-boson pair production at the LHC, `arXiv:1710.06294 [hep-ph]`.

Heister, A. *et al.* (2004). Studies of QCD at e+ e- centre-of-mass energies between 91-GeV and 209-GeV, *Eur. Phys. J.* **C35**, 457–486.

Helenius, I., Paukkunen, H., and Armesto, N. (2016). nPDF constraints from the Large Hadron Electron Collider, in *Proceedings, 24th International Workshop on Deep-Inelastic Scattering and Related Subjects (DIS 2016): Hamburg, Germany, April 11–15, 2016*, Vol. DIS2016, 276, `arXiv:1606.09003 [hep-ph]`.

Hellstrom, T. (ed.) (2016). *The Dalai Lama book of quotes: A collection of speeches, quotations, essays and advice from His Holiness* (Hatherleigh Press (Hobart)).

Henn, J. M. (2013). Multiloop integrals in dimensional regularization made simple, *Phys. Rev. Lett.* **110**, 251601, `arXiv:1304.1806 [hep-th]`.

Hernandez, D. and Smirnov, A. Yu. (2012). Lepton mixing and discrete symmetries, *Phys. Rev.* **D86**, 053014, `arXiv:1204.0445 [hep-ph]`.

Herzog, F., Ruijl, B., Ueda, T., Vermaseren, J. A. M., and Vogt, A. (2017). The five-loop beta function of Yang–Mills theory with fermions, *JHEP* **02**, 090, `arXiv:1701.01404 [hep-ph]`.

Hinchliffe, I., Kotwal, A., Mangano, M. L., Quigg, C., and Wang, L.-T. (2015). Luminosity goals for a 100-TeV pp collider, *Int. J. Mod. Phys.* **A30**, 23, 1544002, `arXiv:1504.06108 [hep-ph]`.

Hoang, A. H. (2014). The top mass: Interpretation and theoretical uncertainties, in *Proceedings, 7th International Workshop on Top Quark Physics (TOP2014): Cannes, France, September 28–October 3, 2014*, `arXiv:1412.3649 [hep-ph]`, `https://inspirehep.net/record/1333866/files/arXiv:1412.3649.pdf`.

Hoang, A. H., Jain, A., Lepenik, C., Mateu, V., Preisser, M., Scimemi, I., and Stewart, I. W. (2017a). The MSR mass and the $\mathcal{O}(\Lambda_{\rm QCD})$ renormalon sum rule, `arXiv:1704.01580 [hep-ph]`.

Hoang, A. H., Kolodrubetz, D. W., Mateu, V., and Stewart, I. W. (2015). Precise determination of α_s from the C-parameter distribution, *Phys. Rev.* **D91**, 9, 094018, `arXiv:1501.04111 [hep-ph]`.

Hoang, A. H., Lepenik, C., and Preisser, M. (2017b). On the light massive flavor dependence of the large order asymptotic behavior and the ambiguity of the pole mass, *JHEP* **09**, 099, `arXiv:1706.08526 [hep-ph]`.

Hoang, A. H. and Stewart, I. W. (2008). Top mass measurements from jets and the Tevatron top-quark mass, *Nucl. Phys. Proc. Suppl.* **185**, 220–226, `arXiv:0808.0222 [hep-ph]`.

Höche, S., Li, Y., and Prestel, S. (2014). Higgs-boson production through gluon fusion at NNLO QCD with parton showers, *Phys. Rev.* **D90**, 5, 054011, `arXiv:1407.3773 [hep-ph]`.

Höche, S., Li, Y., and Prestel, S. (2015). Drell–Yan lepton pair production at NNLO QCD with parton showers, *Phys. Rev.* **D91**, 7, 074015, `arXiv:1405.3607 [hep-ph]`.

Holm, U. (1993). The ZEUS detector: Status report 1993.

Holthausen, M., Lim, K. S., and Lindner, M. (2013a). Lepton mixing patterns from a scan of finite discrete groups, *Phys. Lett.* **B721**, 61–67, `arXiv:1212.2411 [hep-ph]`.

Holthausen, M., Lindner, M., and Schmidt, M. A. (2013b). CP and discrete flavour symmetries, *JHEP* **04**, 122, `arXiv:1211.6953 [hep-ph]`.

Hooke, R. (1665). *Micrographia: or, some physiological descriptions of minute bodies made by magnifying glasses* (J. Martyn and J. Allestry (London)).

Hubner, K. (2004). Designing and building LEP, *Phys. Rept.* **403-404**, 177–188.

Hui, L., Ostriker, J. P., Tremaine, S., and Witten, E. (2017). Ultralight scalars as cosmological dark matter, *Phys. Rev.* **D95**, 4, 043541, arXiv:1610.08297 [astro-ph.CO].

Ibanez, L. E. (1986). Hierarchy of quark-lepton masses in orbifold superstring compactification, *Phys. Lett.* **B181**, 269–272.

Ibanez, L. E., Martin-Lozano, V., and Valenzuela, I. (2017). Constraining the EW hierarchy from the weak gravity conjecture, arXiv:1707.05811 [hep-th].

Ishimori, H., Kobayashi, T., Ohki, H., Shimizu, Y., Okada, H., and Tanimoto, M. (2010). Non-Abelian discrete symmetries in particle physics, *Prog. Theor. Phys. Suppl.* **183**, 1–163, arXiv:1003.3552 [hep-th].

Ivanov, I. P., Nikolaev, N. N., and Savin, A. A. (2006). Diffractive vector meson production at HERA: From soft to hard QCD, *Phys. Part. Nucl.* **37**, 1–85, arXiv:hep-ph/0501034 [hep-ph].

Jaffe, A. M. and Witten, E. (2000). http://www.claymath.org/sites/default/files/yangmills.pdf, "Quantum Yang–Mills Theory": Clay Mathematics Institute Millenium Prize problem.

Jager, B., Schafer, A., Stratmann, M., and Vogelsang, W. (2003). Next-to-leading order QCD corrections to high p_T pion production in longitudinally polarized pp collisions, *Phys. Rev.* **D67**, 054005, arXiv:hep-ph/0211007 [hep-ph].

Jager, B., Stratmann, M., and Vogelsang, W. (2004). Single inclusive jet production in polarized pp collisions at $O(alpha_s^3)$, *Phys. Rev.* **D70**, 034010, arXiv:hep-ph/0404057 [hep-ph].

Janot, P. (2015). Top-quark electroweak couplings at the FCC-ee, *JHEP* **04**, 182, arXiv:1503.01325 [hep-ph].

Janot, P. (2016). Direct measurement of $\alpha_{QED}(m_Z^2)$ at the FCC-ee, *JHEP* **02**, 053, arXiv:1512.05544 [hep-ph].

Janot, P. (2017). https://indico.cern.ch/event/556692/contributions/2510747/, "FCC-ee review: Physics and Experiments", talk at the FCC week 2017, Berlin, 29 May–June 2 2017.

Jaroszewicz, T. (1982). Gluonic Regge singularities and anomalous dimensions in QCD, *Phys. Lett.* **116B**, 291–294.

Jenkins, E. E. and Manohar, A. V. (2009). Algebraic structure of lepton and quark flavor invariants and CP violation, *JHEP* **10**, 094, arXiv:0907.4763 [hep-ph].

Jensen, E. (2016). "Choice of rf frequency", Talk at LHeC Workshop, Daresbury.

Ježo, T., Lindert, J. M., Nason, P., Oleari, C., and Pozzorini, S. (2016). An NLO+PS generator for $t\bar{t}$ and Wt production and decay including non-resonant and interference effects, *Eur. Phys. J.* **C76**, 12, 691, arXiv:1607.04538 [hep-ph].

Jezo, T. and Nason, P. (2015). On the treatment of resonances in next-to-leading order calculations matched to a parton shower, *JHEP* **12**, 065, arXiv:1509.09071 [hep-ph].

Jimenez-Delgado, P., Accardi, A., and Melnitchouk, W. (2014). Impact of hadronic and nuclear corrections on global analysis of spin-dependent parton distributions, *Phys. Rev.* **D89**, 3, 034025, arXiv:1310.3734 [hep-ph].

Jimenez-Delgado, P. and Reya, E. (2009). Dynamical NNLO parton distributions, *Phys. Rev.* **D79**, 074023, arXiv:0810.4274 [hep-ph].

Jimenez-Delgado, P. and Reya, E. (2014). Delineating parton distributions and the strong coupling, *Phys. Rev.* **D89**, 7, 074049, arXiv:1403.1852 [hep-ph].

Jones, S. P., Kerner, M., and Luisoni, G. (2018). NLO QCD corrections to Higgs boson plus jet production with full top-quark mass dependence, arXiv:1802.00349 [hep-ph].

Kachru, S., Kallosh, R., Linde, A. D., and Trivedi, S. P. (2003). De Sitter vacua in string theory, *Phys. Rev.* **D68**, 046005, arXiv:hep-th/0301240 [hep-th].

Kallosh, R., Linde, A. D., Linde, D. A., and Susskind, L. (1995). Gravity and global symmetries, *Phys. Rev.* **D52**, 912–935, arXiv:hep-th/9502069 [hep-th].

Kaplan, D. E. and Tait, T. M. P. (2001). New tools for fermion masses from extra dimensions, *JHEP* **11**, 051, arXiv:hep-ph/0110126 [hep-ph].

Karlberg, A., Re, E., and Zanderighi, G. (2014). NNLOPS accurate Drell–Yan production, *JHEP* **09**, 134, arXiv:1407.2940 [hep-ph].

Kashlinsky, A. (2016). LIGO gravitational wave detection, primordial black holes and the near-IR cosmic infrared background anisotropies, *Astrophys. J.* **823**, 2, L25, arXiv:1605.04023 [astro-ph.CO].

Kauffman, R. P. (1992). Higher order corrections to Higgs boson p_T, *Phys. Rev.* **D45**, 1512–1517.

Kawabata, S., Shimizu, Y., Sumino, Y., and Yokoya, H. (2015). Weight function method for precise determination of top quark mass at Large Hadron Collider, *Phys. Lett.* **B741**, 232–238, arXiv:1405.2395 [hep-ph].

Kawabata, S. and Yokoya, H. (2017). Top-quark mass from the diphoton mass spectrum, *Eur. Phys. J.* **C77**, 5, 323, arXiv:1607.00990 [hep-ph].

Kay, E. and Klein, U. (2017). private communication.

Kennedy, J. F. (1959). Box 902, Senate Speech Files, Pre-Presidential Papers, John F. Kennedy Presidential Library, "Remarks at the United Negro College Fund", Indianapolis, 12 April 1959.

Kennedy, J. F. (1963). Television report to the American people on civil rights, 11 June 1963.

Khachatryan, V. *et al.* (2015a). Constraints on the spin-parity and anomalous HVV couplings of the Higgs boson in proton collisions at 7 and 8 TeV, *Phys. Rev.* **D92**, 1, 012004, arXiv:1411.3441 [hep-ex].

Khachatryan, V. *et al.* (2015b). Measurement of electroweak production of two jets in association with a Z boson in proton-proton collisions at $\sqrt{s} = 8\,\text{TeV}$, *Eur. Phys. J.* **C75**, 2, 66, arXiv:1410.3153 [hep-ex].

Khachatryan, V. *et al.* (2015c). Measurement of the Z boson differential cross section in transverse momentum and rapidity in proton-proton collisions at 8 TeV, *Phys. Lett.* **B749**, 187–209, arXiv:1504.03511 [hep-ex].

Khachatryan, V. *et al.* (2015d). Measurements of differential and double-differential Drell–Yan cross sections in proton-proton collisions at 8 TeV, *Eur. Phys. J.* **C75**, 4, 147, arXiv:1412.1115 [hep-ex].

Khachatryan, V. *et al.* (2016a). Measurement of electroweak production of a W boson and two forward jets in proton-proton collisions at $\sqrt{s}$ = 8 TeV, *JHEP* **11**, 147, arXiv:1607.06975 [hep-ex].

Khachatryan, V. *et al.* (2016b). Measurement of the differential cross section and charge asymmetry for inclusive pp $\to$ $W^{\pm} + X$ production at $\sqrt{s}$ = 8 TeV, *Eur. Phys. J.* **C76**, 8, 469, arXiv:1603.01803 [hep-ex].

Khachatryan, V. *et al.* (2016c). Measurement of the double-differential inclusive jet cross section in proton-proton collisions at $\sqrt{s} = 13\,\text{TeV}$, *Eur. Phys. J.* **C76**, 8, 451, arXiv:1605.04436 [hep-ex].

Khachatryan, V. *et al.* (2016d). Measurement of the t-tbar production cross section in the $e - \mu$ channel in proton-proton collisions at $\sqrt{s}$ = 7 and 8 TeV, *JHEP* **08**, 029, arXiv:1603.02303 [hep-ex].

Khachatryan, V. *et al.* (2016e). Measurement of the top quark mass using proton-proton data at $\sqrt{(s)}$ = 7 and 8 TeV, *Phys. Rev.* **D93**, 7, 072004, arXiv:1509.04044 [hep-ex].

Khachatryan, V. *et al.* (2016f). Search for single production of scalar leptoquarks in proton-proton collisions at $\sqrt{s} = 8\ TeV$, *Phys. Rev.* **D93**, 3, 032005, `arXiv:1509.03750 [hep-ex]`, [Erratum: Phys. Rev.D95,no.3,039906(2017)].

Khachatryan, V. *et al.* (2017a). Measurement and QCD analysis of double-differential inclusive jet cross sections in pp collisions at $\sqrt{s} = 8$ TeV and cross section ratios to 2.76 and 7 TeV, *JHEP* **03**, 156, `arXiv:1609.05331 [hep-ex]`.

Khachatryan, V. *et al.* (2017b). Measurement of differential cross sections for top quark pair production using the lepton+jets final state in proton-proton collisions at 13 TeV, *Phys. Rev.* **D95**, 9, 092001, `arXiv:1610.04191 [hep-ex]`.

Khachatryan, V. *et al.* (2017c). Measurement of the transverse momentum spectra of weak vector bosons produced in proton-proton collisions at $\sqrt{s} = 8$ TeV, *JHEP* **02**, 096, `arXiv:1606.05864 [hep-ex]`.

Kieseler, J., Lipka, K., and Moch, S.-O. (2016). Calibration of the top-quark Monte Carlo mass, *Phys. Rev. Lett.* **116**, 16, 162001, `arXiv:1511.00841 [hep-ph]`.

King, S. F. (2017). Unified models of neutrinos, flavour and CP violation, *Prog. Part. Nucl. Phys.* **94**, 217–256, `arXiv:1701.04413 [hep-ph]`.

King, S. F. and Luhn, C. (2013). Neutrino mass and mixing with discrete symmetry, *Rept. Prog. Phys.* **76**, 056201, `arXiv:1301.1340 [hep-ph]`.

King, S. F., Merle, A., Morisi, S., Shimizu, Y., and Tanimoto, M. (2014). Neutrino mass and mixing: From theory to experiment, *New J. Phys.* **16**, 045018, `arXiv:1402.4271 [hep-ph]`.

King, S. F., Neder, T., and Stuart, A. J. (2013). Lepton mixing predictions from $\Delta(6n^2)$ family symmetry, *Phys. Lett.* **B726**, 312–315, `arXiv:1305.3200 [hep-ph]`.

Klein, M. (2008). First measurements of F_L at low Bjorken x, in *Proceedings, 34th International Conference on High Energy Physics (ICHEP 2008): Philadelphia, Pennsylvania, July 30–August 5, 2008*, `arXiv:0810.3857 [hep-ex]`, `http://inspirehep.net/record/800228/files/arXiv:0810.3857.pdf`.

Klein, M. (2010). The structure of the proton and HERA, in *Proceedings of PHOTON 2009, May 11–15, 2009 DESY Hamburg, Germany*, 431–444.

Klein, M. (2016a). "EIC and LHeC side by side", Invited talk at POETIC, Philadelphia.

Klein, M. (2016b). Nuclear parton distributions with the LHeC, in *Proceedings, 6th International Conference on Physics Opportunities at an ElecTron-Ion Collider (POETIC 2015): Palaiseau, France, September 7–11, 2015*, Vol. 112, 03002.

Klein, M. (2018). Contribution to this Volume.

Klein, M. and Klein, U. (2018). SM Higgs couplings in *ep*, in preparation.

Klein, M. and Radescu, V. (2013). Partons from the LHeC, Tech. Rep. LHeC-Note-2013-002 PHY, Snowmass Workshop, CERN.

Klein, M. and Yoshida, R. (2008). Collider physics at HERA, *Prog. Part. Nucl. Phys.* **61**, 343–393, `arXiv:0805.3334 [hep-ex]`.

Klein, U. (2017a). `http://lhec.web.cern.ch`, "Turning the LHC into a powerful Higgs facility", Seminar at LAPP Annecy.

Klein, U. (2017b). Talk on Higgs Physics at the LHeC Workshop at CERN.

Klijnsma, T., Bethke, S., Dissertori, G., and Salam, G. P. (2017). Determination of the strong coupling constant $\alpha_s(m_Z)$ from measurements of the total cross section for top-antitop quark production, *Eur. Phys. J.* **C77**, 11, 778, `arXiv:1708.07495 [hep-ph]`.

Kobayashi, T. and Lebedev, O. (2003). Heterotic Yukawa couplings and continuous Wilson lines, *Phys. Lett.* **B566**, 164–170, `arXiv:hep-th/0303009 [hep-th]`.

Kobayashi, T., Nagamoto, S., and Uemura, S. (2017). Modular symmetry in magnetized/intersecting D-brane models, *Proh. Theor. Exp. Phys.* **2017**, 2, 023B02, `arXiv:1608.06129 [hep-th]`.

Kobzarev, I. Yu., Okun, L. B., and Pomeranchuk, I. Ya. (1966). On the possibility of experimental observation of mirror particles, *Sov. J. Nucl. Phys.* **3**, 6, 837–841, [*Yad. Fiz.* **3**, 1154 (1966)].

Kopp, G., Schaile, D., Spira, M., and Zerwas, P. M. (1995). Bounds on radii and magnetic dipole moments of quarks and leptons from LEP, SLC and HERA, *Z. Phys.* **C65**, 545–550, `arXiv:hep-ph/9409457 [hep-ph]`.

Kubar, J., Le Bellac, M., Meunier, J. L., and Plaut, G. (1980). QCD corrections to the Drell–Yan mechanism and the pion structure function, *Nucl. Phys.* **B175**, 251–275.

Kubar-Andre, J. and Paige, F. E. (1979). Gluon corrections to the Drell–Yan model, *Phys. Rev.* **D19**, 221.

Kudashkin, K., Melnikov, K., and Wever, C. (2017). Two-loop amplitudes for processes $gg \to Hg, qg \to Hq$ and $q\bar{q} \to Hg$ at large Higgs transverse momentum, `arXiv:1712.06549 [hep-ph]`.

Kuhn, T. (1962). *The structure of scientific revolutions* (University of Chicago Press (Chicago)).

Kumar, M., Ruan, X., Islam, R., Cornell, A. S., Klein, M., Klein, U., and Mellado, B. (2017). Probing anomalous couplings using di-Higgs production in electronproton collisions, *Phys. Lett.* **B764**, 247–253, `arXiv:1509.04016 [hep-ph]`.

Kumar, N., Pukhov, A., and Lotov, K. (2010). Self-modulation instability of a long proton bunch in plasmas, *Phys. Rev. Lett.* **104**, 255003, `arXiv:1003.5816 [physics.plasm-ph]`.

Kunszt, Z., Nason, P., Marchesini, G., and Webber, B. R. (1989). QCD at LEP, in *LEP Physics Workshop Geneva, Switzerland, February 20, 1989*, 373–453.

Kunszt, Z. and Soper, D. E. (1992). Calculation of jet cross-sections in hadron collisions at order α_s^3, *Phys. Rev.* **D46**, 192–221.

Kuraev, E. A., Lipatov, L. N., and Fadin, V. S. (1977). The Pomeranchuk singularity in nonabelian gauge theories, *Sov. Phys. JETP* **45**, 199–204, [*Zh. Eksp. Teor. Fiz.* **72**, 377 (1977)].

L3 Collaboration (1990). The construction of the L3 experiment, *Nucl. Instrum. Meth.* **A289**, 35–102.

Lai, H.-L., Guzzi, M., Huston, J., Li, Z., Nadolsky, P. M., Pumplin, J., and Yuan, C. P. (2010). New parton distributions for collider physics, *Phys. Rev.* **D82**, 074024, `arXiv:1007.2241 [hep-ph]`.

Laporta, S. (2000). High precision calculation of multiloop Feynman integrals by difference equations, *Int. J. Mod. Phys.* **A15**, 5087–5159, `arXiv:hep-ph/0102033 [hep-ph]`.

Lauer, J., Mas, J., and Nilles, H. P. (1989). Duality and the role of nonperturbative effects on the world sheet, *Phys. Lett.* **B226**, 251–256.

Lauer, J., Mas, J., and Nilles, H. P. (1991). Twisted sector representations of discrete background symmetries for two-dimensional orbifolds, *Nucl. Phys.* **B351**, 353–424.

Leader, E., Sidorov, A. V., and Stamenov, D. B. (2010). Determination of polarized PDFs from a QCD analysis of inclusive and semi-inclusive deep inelastic scattering data, *Phys. Rev.* **D82**, 114018, `arXiv:1010.0574 [hep-ph]`.

Leader, E., Sidorov, A. V., and Stamenov, D. B. (2011). A possible resolution of the strange quark polarization puzzle? *Phys. Rev.* **D84**, 014002, `arXiv:1103.5979 [hep-ph]`.

Leader, E., Sidorov, A. V., and Stamenov, D. B. (2015). New analysis concerning the strange quark polarization puzzle, *Phys. Rev.* **D91**, 5, 054017, `arXiv:1410.1657 [hep-ph]`.

Lebedev, O. (2001). The CKM phase in heterotic orbifold models, *Phys. Lett.* **B521**, 71–78, `arXiv:hep-th/0108218 [hep-th]`.

Lee, B. W., Quigg, C., and Thacker, H. B. (1977a). The strength of weak interactions at very high-energies and the Higgs boson mass, *Phys. Rev. Lett.* **38**, 883–885.

Lee, B. W., Quigg, C., and Thacker, H. B. (1977b). Weak interactions at very high-energies: The role of the Higgs boson mass, *Phys. Rev.* **D16**, 1519.

Lee, J., B. abd Little (1999). *Bruce Lee: Artist of life* (Tuttle Publishing (Boston)).

Lee, T. D. and Yang, C.-N. (1956). Question of parity conservation in weak interactions, *Phys. Rev.* **104**, 254–258.

Leemans, W. (2009). Laser-driven plasma-wave electron accelerators, *Physics Today* **62**, 44.

Leemans, W. P., Nagler, B., Gonsalves, A. J., Toth, C., Nakamura, K., Geddes, C. G. R., Esarey, E., Schroeder, C. B., and Hooker, S. M. (2006). GeV electron beams from a cm-scale accelerator, *Nature Phys.* **2**, 696–699.

Lenard, P. (1920). Vorträge und Diskussionen von der 86. Naturforscherversammlung in Nauheim vom 19.-25. September 1920, 649.

Lenard, P. (1936). *Deutsche Physik in vier Bänden* (J. F. Lehmann (Múnchen)).

Leontaris, G. K. and Tracas, N. D. (1998). Modular weights, U(1)'s and mass matrices, *Phys. Lett.* **B419**, 206–210, `arXiv:hep-ph/9709510 [hep-ph]`.

LEP electroweak working group (2012). `lepewwg.web.cern.ch/LEPEWWG/plots/winter 2012`, Update for the 2012 Winter conferences.

LHCb Collaboration (2012). Inclusive low mass Drell–Yan production in the forward region at $\sqrt{s} = 7$ TeV, Tech. Rep. LHCb-CONF-2012-013, CERN-LHCb-CONF-2012-013, CERN.

Linde, A. D. (1986a). Eternal chaotic inflation, *Mod. Phys. Lett.* **A1**, 81.

Linde, A. D. (1986b). Eternally existing selfreproducing chaotic inflationary universe, *Phys. Lett.* **B175**, 395–400.

Lindert, J. M., Kudashkin, K., Melnikov, K., and Wever, C. (2018). Higgs bosons with large transverse momentum at the LHC, `arXiv:1801.08226 [hep-ph]`.

Lindert, J. M. *et al.* (2017). Precise predictions for $V+$ jets dark matter backgrounds, *Eur. Phys. J.* **C77**, 12, 829, `arXiv:1705.04664 [hep-ph]`.

Lipatov, L. N. (1975). The parton model and perturbation theory, *Sov. J. Nucl. Phys.* **20**, 94–102, [*Yad. Fiz.* **20**, 181 (1974)].

Lipatov, L. N. (1976). Reggeization of the vector meson and the vacuum singularity in nonabelian gauge theories, *Sov. J. Nucl. Phys.* **23**, 338–345, [*Yad. Fiz.* **23**, 642 (1976)].

Lipatov, L. N. (1997). Small x physics in perturbative QCD, *Phys. Rept.* **286**, 131–198, `arXiv:hep-ph/9610276 [hep-ph]`.

Liu, S., Tang, Y.-L., Zhang, C., and Zhu, S.-h. (2017). Exotic Higgs decay $h \to \phi\phi \to 4b$ at the LHeC, *Eur. Phys. J.* **C77**, 7, 457, `arXiv:1608.08458 [hep-ph]`.

Liu, T. and Penin, A. A. (2017). High-energy limit of QCD beyond the Sudakov approximation, *Phys. Rev. Lett.* **119**, 26, 262001, `arXiv:1709.01092 [hep-ph]`.

Lotov, K. (1998). Instability of long driving beams in plasma wakefield accelerators, in *Proceedings, Sixth European Particle Accelerator Conference (EPAC'98)*, 806–808, `https://epac.web.cern.ch/EPAC/stockholm/EPAC98/Welcome/General.html`.

Lotov, K. (2011). Controlled self-modulation of high energy beams in a plasma, *Phys. Plasmas* **18**, 024501.

Lotov, K. V. (2003). Fine wakefield structure in the blowout regime of plasma wakefield accelerators, *Phys. Rev. ST Accel. Beams* **6**, 061301.

Lotov, K. V. (2010). Simulation of proton driven plasma wakefield acceleration, *Phys. Rev. ST Accel. Beams* **13**, 041301.

Lotov, K. V. (2013). Excitation of two-dimensional plasma wakefields by trains of equidistant particle bunches, *Phys. Plasmas* **20**, 083119, `arXiv:1307.3812 [physics.plasm-ph]`.

Lotov, K. V., Pukhov, A., and Caldwell, A. (2013). Effect of plasma inhomogeneity on plasma wakefield acceleration driven by long bunches, *Phys. Plasmas* **20**, 013102, `arXiv:1205.3388 [physics.plasm-ph]`.

Luscher, M., Weisz, P., and Wolff, U. (1991). A numerical method to compute the running coupling in asymptotically free theories, *Nucl. Phys.* **B359**, 221–243.

Lüst, D. and Palti, E. (2017). Scalar fields, hierarchical UV/IR mixing and the weak gravity conjecture, `arXiv:1709.01790 [hep-th]`.

Lutken, C. A. and Ross, G. G. (1992). Duality in the quantum Hall system, *Phys. Rev.* **B45**, 20, 11837.

Lutken, C. A. and Ross, G. G. (2007). Geometric scaling in the quantum Hall system, *Phys. Lett.* **B653**, 363–365.

Lutken, C. A. and Ross, G. G. (2011). Experimental probes of emergent symmetries in the quantum Hall system, *Nucl. Phys.* **B850**, 321–338.

Ma, E. and Rajasekaran, G. (2001). Softly broken A_4 symmetry for nearly degenerate neutrino masses, *Phys. Rev.* **D64**, 113012, `arXiv:hep-ph/0106291 [hep-ph]`.

Maezawa, Y. and Petreczky, P. (2016). Quark masses and strong coupling constant in 2+1 flavor QCD, *Phys. Rev.* **D94**, 3, 034507, `arXiv:1606.08798 [hep-lat]`.

Maitre, D., Bern, Z., Febres Cordero, F., Kosower, D., and Hamilton, K. (2016). High multiplicity processes with BlackHat and Sherpa, in *Proceedings, 12th International Symposium on Radiative Corrections (Radcor 2015) and LoopFest XIV (Radiative Corrections for the LHC and Future Colliders): Los Angeles, CA, USA, June 15–19, 2015*, Vol. RADCOR2015, 099.

Maldacena, J. M. (1999). The large N limit of superconformal field theories and supergravity, *Int. J. Theor. Phys.* **38**, 1113–1133, `arXiv:hep-th/9711200 [hep-th]`, [*Adv. Theor. Math. Phys.* **2**, 231 (1998)].

Maldacena, J. M. (2005). The illusion of gravity, *Scientific American* **293**, 56.

Maltman, K., Leinweber, D., Moran, P., and Sternbeck, A. (2008). The realistic lattice determination of $alpha_s(M_Z)$ revisited, *Phys. Rev.* **D78**, 114504, `arXiv:0807.2020 [hep-lat]`.

Mangano, M. (2017). Physics at the FCC-hh, a 100 TeV *pp* collider, CERN-2017-003-M.

Mangano, M. L., Plehn, T., Reimitz, P., Schell, T., and Shao, H.-S. (2016). Measuring the top Yukawa coupling at 100 TeV, *J. Phys.* **G43**, 3, 035001, `arXiv:1507.08169 [hep-ph]`.

Mangano, M. L. *et al.* (2017). Physics at a 100 TeV pp collider: Standard Model processes, *CERN Yellow Report*, 3, 1–254, `arXiv:1607.01831 [hep-ph]`.

Mantler, H. and Wiesemann, M. (2013). Top- and bottom-mass effects in hadronic Higgs production at small transverse momenta through LO+NLL, *Eur. Phys. J.* **C73**, 6, 2467, `arXiv:1210.8263 [hep-ph]`.

Marchesano, F. (2007). Progress in D-brane model building, *Fortsch. Phys.* **55**, 491–518, `arXiv:hep-th/0702094 [HEP-TH]`.

Marhauser, F. (2017). "Choice of rf Frequency", Talk at PERLE Workshop, Orsay.

Marhauser, F. *et al.* (2018). "802 MHz ERL Cavity Design and Development", Proc. IPAC, to be published.

Marquard, P., Smirnov, A. V., Smirnov, V. A., and Steinhauser, M. (2015). Quark mass relations to four-loop order in perturbative QCD, *Phys. Rev. Lett.* **114**, 14, 142002, `arXiv:1502.01030 [hep-ph]`.

Marquard, P., Smirnov, A. V., Smirnov, V. A., Steinhauser, M., and Wellmann, D. (2016). $\overline{\text{MS}}$-on-shell quark mass relation up to four loops in QCD and a general SU(N) gauge group, *Phys. Rev.* **D94**, 7, 074025, arXiv:1606.06754 [hep-ph].

Martin, A. D., Stirling, W. J., and Roberts, R. G. (1993). Parton distributions updated, *Phys. Lett.* **B306**, 145–150, [Erratum: *Phys. Lett.* **B309**, 492 (1993)].

Martin, A. D., Stirling, W. J., Thorne, R. S., and Watt, G. (2009). Parton distributions for the LHC, *Eur. Phys. J.* **C63**, 189–285, arXiv:0901.0002 [hep-ph].

Marzani, S. and Ball, R. D. (2009). High energy resummation of Drell–Yan processes, *Nucl. Phys.* **B814**, 246–264, arXiv:0812.3602 [hep-ph].

Marzani, S., Ball, R. D., Del Duca, V., Forte, S., and Vicini, A. (2008). Higgs production via gluon-gluon fusion with finite top mass beyond next-to-leading order, *Nucl. Phys.* **B800**, 127–145, arXiv:0801.2544 [hep-ph].

Marzani, S., Ball, R. D., Falgari, P., and Forte, S. (2007). BFKL at next-to-next-to-leading order, *Nucl. Phys.* **B783**, 143–175, arXiv:0704.2404 [hep-ph].

McCullough, M. (2014). An indirect model-dependent probe of the Higgs self-coupling, *Phys. Rev.* **D90**, 1, 015001, arXiv:1312.3322 [hep-ph], [Erratum: *Phys. Rev.* **D92**, 3, 039903 (2015)].

McNeile, C., Davies, C. T. H., Follana, E., Hornbostel, K., and Lepage, G. P. (2010). High-precision c and b masses, and QCD coupling from current-current correlators in lattice and continuum QCD, *Phys. Rev.* **D82**, 034512, arXiv:1004.4285 [hep-lat].

Megias, E., Panico, G., Pujolas, O., and Quiros, M. (2017). A natural extra-dimensional origin for the LHCb anomalies, in *Proceedings, 52nd Rencontres de Moriond on Electroweak Interactions and Unified Theories: La Thuile, Italy, March 18–25, 2017*, arXiv:1705.04822 [hep-ph], https://inspirehep.net/record/1599572/files/arXiv:1705.04822.pdf.

Melnikov, K. and Penin, A. (2016). On the light quark mass effects in Higgs boson production in gluon fusion, *JHEP* **05**, 172, arXiv:1602.09020 [hep-ph].

Melnikov, K. and Petriello, F. (2006). Electroweak gauge boson production at hadron colliders through $O(\alpha_s^2)$, *Phys. Rev.* **D74**, 114017, arXiv:hep-ph/0609070 [hep-ph].

Melnikov, K., Tancredi, L., and Wever, C. (2016). Two-loop $gg \to Hg$ amplitude mediated by a nearly massless quark, *JHEP* **11**, 104, arXiv:1610.03747 [hep-ph].

Melnikov, K., Tancredi, L., and Wever, C. (2017). Two-loop amplitudes for $qg \to Hq$ and $q\bar{q} \to Hg$ mediated by a nearly massless quark, *Phys. Rev.* **D95**, 5, 054012, arXiv:1702.00426 [hep-ph].

Mertig, R. and van Neerven, W. L. (1996). The calculation of the two loop spin splitting functions P(ij)(1)(x), *Z. Phys.* **C70**, 637–654, arXiv:hep-ph/9506451 [hep-ph].

Metz, A. and Vossen, A. (2016). Parton fragmentation functions, *Prog. Part. Nucl. Phys.* **91**, 136–202, arXiv:1607.02521 [hep-ex].

Michel, L. and Radicati, L. A. (1970). Breaking of the SU_3xSU_3 symmetry in hadronic physics, in M. Conversi (ed.), *Evolution of particle physics: A volume dedicated to Edoardo Amaldi in his sixtieth birthday* (Academic Press (New York)), 191–203.

Moch, S., Ruijl, B., Ueda, T., Vermaseren, J. A. M., and Vogt, A. (2017). Four-loop non-singlet splitting functions in the planar limit and beyond, *JHEP* **10**, 041, arXiv:1707.08315 [hep-ph].

Moch, S., Vermaseren, J. A. M., and Vogt, A. (2004). The three loop splitting functions in QCD: The nonsinglet case, *Nucl. Phys.* **B688**, 101–134, arXiv:hep-ph/0403192 [hep-ph].

Moch, S. *et al.* (eds.) (2014). *"High precision fundamental constants at the TeV scale", Proceedings of the Mainz Institute for Theoretical Physics program, March 10–21, 2014*, arXiv:1405.4781 [hep-ph].

Mours, B. *et al.* (1996). The design, construction and performance of the ALEPH silicon vertex detector, *Nucl. Instrum. Meth.* **A379**, 101–115.

Mueller, A. H. (2001). Parton saturation: An overview, in *QCD perspectives on hot and dense matter. Proceedings, NATO Advanced Study Institute, Summer School, Cargese, France, August 6–18, 2001*, 45–72, arXiv:hep-ph/0111244 [hep-ph].

Muggli, P., Marsh, K. A., Wang, S., Clayton, C. E., Lee, S., Katsouleas, T. C., and Joshi, C. (1999). Photo-ionized lithium source for plasma accelerator applications, *IEEE Trans. Plasma Sci.* **27**, 3, 791–799.

Muselli, C. (2017). Double differential high energy resummation, arXiv:1710.09376 [hep-ph].

Muselli, C., Bonvini, M., Forte, S., Marzani, S., and Ridolfi, G. (2015). Top quark pair production beyond NNLO, *JHEP* **08**, 076, arXiv:1505.02006 [hep-ph].

Nakayama, K., Fahy, B., and Hashimoto, S. (2016). Short-distance charmonium correlator on the lattice with Møbius domain-wall fermion and a determination of charm quark mass, *Phys. Rev.* **D94**, 5, 054507, arXiv:1606.01002 [hep-lat].

Nason, P. (2016). Theory summary, in *Proceedings, 8th International Workshop on Top Quark Physics (TOP2015): Ischia, Italy, September 14–18, 2015*, Vol. TOP2015, 056, arXiv:1602.00443 [hep-ph].

Neumann, T. and Williams, C. (2017). The Higgs boson at high p_T, *Phys. Rev.* **D95**, 1, 014004, arXiv:1609.00367 [hep-ph].

Newman, P. and Stasto, A. (2013). Dig deeper, *Nature Phys.* **9**, 448.

Newman, P. and Wing, M. (2014). The hadronic final state at HERA, *Rev. Mod. Phys.* **86**, 3, 1037, arXiv:1308.3368 [hep-ex].

Nobel Committee (2004). https://www.nobelprize.org/nobel_prizes/physics/laureates/2004/, Nobel Prize in Physics.

Nocera, E. R. (2016). Unbiased polarized PDFs upgraded with new inclusive DIS data, in *Proceedings, 16th Workshop on High Energy Spin Physics (DSPIN-15): Dubna, Moscow region, Russia, September 8–12, 2015*, Vol. 678, 1, 012030, arXiv:1510.04248 [hep-ph].

Nocera, E. R. (2017). Impact of recent RHIC data on helicity-dependent parton distribution functions, in *22nd International Symposium on Spin Physics (SPIN 2016) Urbana, IL, USA, September 25–30, 2016*, arXiv:1702.05077 [hep-ph], http://inspirehep.net/record/1513801/files/arXiv:1702.05077.pdf.

Nocera, E. R. (2018). Fragmentation functions of charged hadrons, in *Proceedings, 25th International Workshop on Deep-Inelastic Scattering and Related Topics (DIS 2017): Birmingham, UK, April 3–7, 2017*, Vol. DIS2017, 231, arXiv:1709.03400 [hep-ph].

Nocera, E. R., Ball, R. D., Forte, S., Ridolfi, G., and Rojo, J. (2014). A first unbiased global determination of polarized PDFs and their uncertainties, *Nucl. Phys.* **B887**, 276–308, arXiv:1406.5539 [hep-ph].

Olive, K. A. *et al.* (2014). Review of particle physics, *Chin. Phys.* **C38**, 090001.

Ooguri, H. and Vafa, C. (2007). On the geometry of the string landscape and the swampland, *Nucl. Phys.* **B766**, 21–33, arXiv:hep-th/0605264 [hep-th].

Ossola, G., Papadopoulos, C. G., and Pittau, R. (2007). Reducing full one-loop amplitudes to scalar integrals at the integrand level, *Nucl. Phys.* **B763**, 147–169, arXiv:hep-ph/0609007 [hep-ph].

Öz, E. and Muggli, P. (2014). A novel Rb vapor plasma source for plasma wakefield accelerators, *Nucl. Inst. Meth.* **740**, 197.

Pais, A. (1986). *Inward bound. Of matter and forces in the physical world* (Oxford University Press (Oxford)).

Pais, A. (1996). https://www.nobelprize.org/nobel_prizes/themes/physics/pais/, "Physics in Denmark: The first four hundred years", Address for the 75th Anniversary of the Niels Bohr Institute, Copenhagen, 6 March 1996".

Pak, A., Rogal, M., and Steinhauser, M. (2010). Finite top quark mass effects in NNLO Higgs boson production at LHC, *JHEP* **02**, 025, arXiv:0911.4662 [hep-ph].

Pak, A., Rogal, M., and Steinhauser, M. (2011). Production of scalar and pseudo-scalar Higgs bosons to next-to-next-to-leading order at hadron colliders, *JHEP* **09**, 088, arXiv:1107.3391 [hep-ph].

Pakvasa, S. and Sugawara, H. (1978). Discrete symmetry and Cabibbo Angle, *Phys. Lett.* **73B**, 61–64.

Parisi, G. (1980). Summing large perturbative corrections in QCD, *Phys. Lett.* **90B**, 295–296.

Parisi, G. (2017). Historical and personal recollections of Guido Altarelli, in *Proceedings, 5th International Conference on New Frontiers in Physics: Kolymbari, Crete, Greece, July 6–14, 2016*, Vol. 164, 02001, arXiv:1712.10030 [physics.hist-ph].

Parisi, G. and Petronzio, R. (1979). Small transverse momentum distributions in hard processes, *Nucl. Phys.* **B154**, 427–440.

Parker, B. (2017). "SC magnet concepts for *eh* colliders IRs", Talk at LHeC Workshop, CERN.

Passarino, G. and Veltman, M. J. G. (1979). One loop corrections for e^+e^- annihilation into mu+ mu- in the Weinberg model, *Nucl. Phys.* **B160**, 151–207.

Patrignani, C. *et al.* (2016). Review of particle physics, *Chin. Phys.* **C40**, 10, 100001.

PBCWG (2016–2017). https://indico.cern.ch/category/7885/, *Physics Beyond Colliders*, CERN Working Group.

Peccei, R. D. and Quinn, H. R. (1977). CP conservation in the presence of instantons, *Phys. Rev. Lett.* **38**, 1440–1443.

Perez, E. and Rizvi, E. (2013). The quark and gluon structure of the proton, *Rept. Prog. Phys.* **76**, 046201, arXiv:1208.1178 [hep-ex].

Perimeter Institute for Theoretical Physics (2015). http://www.perimeterinstitute.ca/research/research-initiatives/tensor-networks-initiative/, Tensor Networks Initiative.

Pich, A. (2013). Review of α_s determinations, in *Proceedings, 10th Conference on Quark Confinement and the Hadron Spectrum (Confinement X): Munich, Germany, October 8–12, 2012*, arXiv:1303.2262 [hep-ph], [PoSConfinementX,022(2012)].

Pich, A. and Rodríguez-Sánchez, A. (2016). Determination of the QCD coupling from ALEPH τ decay data, *Phys. Rev.* **D94**, 3, 034027, arXiv:1605.06830 [hep-ph].

Pierog, T., Karpenko, I., Katzy, J. M., Yatsenko, E., and Werner, K. (2015). EPOS LHC: Test of collective hadronization with data measured at the CERN Large Hadron Collider, *Phys. Rev.* **C92**, 3, 034906, arXiv:1306.0121 [hep-ph].

Pineda, A. (2001). Determination of the bottom quark mass from the Upsilon(1S) system, *JHEP* **06**, 022, arXiv:hep-ph/0105008 [hep-ph].

Polini, A. *et al.* (2007). The design and performance of the ZEUS micro vertex detector, *Nucl. Instrum. Meth.* **A581**, 656–686, arXiv:0708.3011 [physics.ins-det].

Politzer, H. D. (1973). Reliable perturbative results for strong interactions? *Phys. Rev. Lett.* **30**, 1346–1349.

Politzer, H. D. (1974). Asymptotic freedom: An approach to strong interactions, *Phys. Rept.* **14**, 129–180.

Politzer, H. D. (1977). Gluon corrections to Drell–Yan processes, *Nucl. Phys.* **B129**, 301–318.

Pukhov, A. (1999). Three-dimensional electromagnetic relativistic particle-in-cell code VLPL (Virtual Laser Plasma Lab), *J. Plasma Phys.* **61**, 425.

Pukhov, A., Kumar, N., Tuckmantel, T., Upadhyay, A., Lotov, K., Muggli, P., Khudik, V., Siemon, C., and Shvets, G. (2011). Phase velocity and particle injection in a self-modulated proton-driven plasma wakefield accelerator, *Phys. Rev. Lett.* **107**, 145003, arXiv:1108.0071 [physics.plasm-ph].

Pumplin, J., Stump, D., Brock, R., Casey, D., Huston, J., Kalk, J., Lai, H. L., and Tung, W. K. (2001). Uncertainties of predictions from parton distribution functions. 2. The Hessian method, *Phys. Rev.* **D65**, 014013, arXiv:hep-ph/0101032 [hep-ph].

Quigg, C. (2013). "DIS and beyond", Summary talk at the DIS workshop at Marseille.

Ramond, P. (2010). *Group theory: A physicist's survey* (Cambridge University Press (Cambridge)).

Ratcliffe, P. G. (2004). Hyperon beta decay and the CKM matrix, *Czech. J. Phys.* **54**, B11–B20, arXiv:hep-ph/0402063 [hep-ph].

Ravasio, S. F., Jezo, T., Nason, P., and Oleari, C. (2018). A theoretical study of top-mass measurements at the LHC using NLO+PS generators of increasing accuracy, arXiv:1801.03944 [hep-ph].

Rijken, P. J. and van Neerven, W. L. (1995). Order α_s^2 contributions to the Drell–Yan cross-section at fixed target energies, *Phys. Rev.* **D51**, 44–63, arXiv:hep-ph/9408366 [hep-ph].

Rolandi, G. (2017). https://indico.cern.ch/event/550509/contributions/2413232/, "Detector concepts for FCC-ee", talk at the First FCC Physics Workshop, CERN, 16-20 January 2017.

Ross, G. and Serna, M. (2008). Unification and fermion mass structure, *Phys. Lett.* **B664**, 97–102, arXiv:0704.1248 [hep-ph].

Rückl, R. (1983). Effects of compositeness in deep inelastic scattering, *Phys. Lett.* **129B**, 363–369.

Rückl, R. (1984). Probing lepton and quark substructure in polarized $e^{\mp}N$ scattering, *Nucl. Phys.* **B234**, 91–105.

Sachdev, S. (2010). Condensed matter and AdS/CFT, arXiv:1002.2947 [hep-th], [*Lect. Notes Phys.* **828**, 273 (2011)].

Salam, G. P. (1998). A resummation of large subleading corrections at small x, *JHEP* **07**, 019, arXiv:hep-ph/9806482 [hep-ph].

Salam, G. P. (2010). Towards jetography, *Eur. Phys. J.* **C67**, 637–686, arXiv:0906.1833 [hep-ph].

Salam, G. P. (2016). http://lhcp2016.hep.lu.se/, "Theory overview of QCD physics at hadron colliders", Talk at the Fourth Annual Large Hadron Collider Physics Conference, Lund, Sweden, 13–18 June, 2016.

Sasaki, M., Suyama, T., Tanaka, T., and Yokoyama, S. (2016). Primordial black hole scenario for the gravitational-wave event GW150914, *Phys. Rev. Lett.* **117**, 6, 061101, arXiv:1603.08338 [astro-ph.CO].

Sasaki, S. and Yamazaki, T. (2009). Lattice study of flavor SU(3) breaking in hyperon beta decay, *Phys. Rev.* **D79**, 074508, arXiv:0811.1406 [hep-ph].

Sato, N., Ethier, J. J., Melnitchouk, W., Hirai, M., Kumano, S., and Accardi, A. (2016a). First Monte Carlo analysis of fragmentation functions from single-inclusive e^+e^- annihilation, *Phys. Rev.* **D94**, 11, 114004, arXiv:1609.00899 [hep-ph].

Sato, N., Melnitchouk, W., Kuhn, S. E., Ethier, J. J., and Accardi, A. (2016b). Iterative Monte Carlo analysis of spin-dependent parton distributions, *Phys. Rev.* **D93**, 7, 074005, arXiv:1601.07782 [hep-ph].

Sauli, F. (2004). From bubble chambers to electronic systems: 25 years of evolution in particle detectors at CERN (1979-2004), *Phys. Rept.* **403-404**, 471–504.

Savage, M. J. and Walden, J. (1997). SU(3) breaking in neutral current axial matrix elements and the spin content of the nucleon, *Phys. Rev.* **D55**, 5376–5384, arXiv:hep-ph/9611210 [hep-ph].

Schael, S. *et al.* (2005). Branching ratios and spectral functions of tau decays: Final ALEPH measurements and physics implications, *Phys. Rept.* **421**, 191–284, arXiv:hep-ex/0506072 [hep-ex].

Schael, S. *et al.* (2006). Precision electroweak measurements on the Z resonance, *Phys. Rept.* **427**, 257–454, arXiv:hep-ex/0509008 [hep-ex].

Schael, S. *et al.* (2013). Electroweak measurements in electron-positron collisions at W-boson-pair energies at LEP, *Phys. Rept.* **532**, 119–244, arXiv:1302.3415 [hep-ex].

Schellekens, A. N. (2013). Life at the interface of particle physics and string theory, *Rev. Mod. Phys.* **85**, 4, 1491–1540, arXiv:1306.5083 [hep-ph].

Schieck, J., Bethke, S., Kluth, S., Pahl, C., and Trocsanyi, Z. (2013). Measurement of the strong coupling α_s from the three-jet rate in e+e- — annihilation using JADE data, *Eur. Phys. J.* **C73**, 3, 2332, arXiv:1205.3714 [hep-ex].

Schopper, H. (2009). *LEP: The lord of the collider rings at CERN 1980-2000: The making, operation and legacy of the world's largest scientific instrument* (Springer (Berlin)), ISBN 9783540893004.

Schroder, Y. and Steinhauser, M. (2006). Four-loop decoupling relations for the strong coupling, *JHEP* **01**, 051, arXiv:hep-ph/0512058 [hep-ph].

Schroeder, C. B., Benedetti, C., Esarey, E., Grüner, F., and Leemans, W. (2011). Growth and phase velocity of self-modulated beam-driven plasma waves, *Phys. Rev. Lett.* **107**, 145002.

Schwanenberger, C. (2017). "Top physics at the LHeC", Talk at the LHeC Workshop at CERN.

Simons Foundation, Mathematical and Physical Sciences (1996). https://www.simonsfoundation.org/mathematics-physical-sciences/it-from-qubit/, It from Qubit: Simons Collaboration on Quantum Fields, Gravity and Information.

Sirunyan, A. M. *et al.* (2017a). Evidence for the Higgs boson decay to a bottom quark-antiquark pair, arXiv:1709.07497 [hep-ex], cMS-HIG-16-044, CERN-EP-2017-233.

Sirunyan, A. M. *et al.* (2017b). Measurement of the triple-differential dijet cross section in proton-proton collisions at $\sqrt{s} = 8$ TeV and constraints on parton distribution functions, *Eur. Phys. J.* **C77**, 11, 746, arXiv:1705.02628 [hep-ex].

Sirunyan, A. M. *et al.* (2017c). Measurements of the pp $\rightarrow W\gamma\gamma$ and pp $\rightarrow Z\gamma\gamma$ cross sections and limits on anomalous quartic gauge couplings at $\sqrt{s} = 8$ TeV, *JHEP* **10**, 072, arXiv:1704.00366 [hep-ex].

Sirunyan, A. M. *et al.* (2017d). Measurements of the pp$\rightarrow$ZZ production cross section and the Z$\rightarrow 4\ell$ branching fraction, and constraints on anomalous triple gauge couplings at $\sqrt{s} = 13$ TeV, arXiv:1709.08601 [hep-ex], cMS-SMP-16-017, CERN-EP-2017-219.

Sirunyan, A. M. *et al.* (2017e). Observation of electroweak production of same-sign W boson pairs in the two jet and two same-sign lepton final state in proton-proton collisions at $\sqrt{s} = 13$ TeV, arXiv:1709.05822 [hep-ex], cMS-SMP-17-004, CERN-EP-2017-232.

Sjöstrand, T. (2009). Monte Carlo tools, in *Proceedings, 65th Scottish Universities Summer School in Physics: LHC Physics (SUSSP65): St. Andrews, UK, August 16–29,*

2009, 309–339, arXiv:0911.5286 [hep-ph], http://inspirehep.net/record/838269/files/arXiv:0911.5286.pdf.

Sjöstrand, T., Ask, S., Christiansen, J. R., Corke, R., Desai, N., Ilten, P., Mrenna, S., Prestel, S., Rasmussen, C. O., and Skands, P. Z. (2015). An introduction to PYTHIA 8.2, *Comput. Phys. Commun.* **191**, 159–177, arXiv:1410.3012 [hep-ph].

Sjöstrand, T., Mrenna, S., and Skands, P. Z. (2006). PYTHIA 6.4 physics and manual, *JHEP* **05**, 026, arXiv:hep-ph/0603175 [hep-ph].

Sjöstrand, T., Mrenna, S., and Skands, P. Z. (2008). A brief introduction to PYTHIA 8.1, *Comput. Phys. Commun.* **178**, 852, arXiv:0710.3820 [hep-ph].

Skands, P., Carrazza, S., and Rojo, J. (2014). Tuning PYTHIA 8.1: The Monash 2013 Tune, *Eur. Phys. J.* **C74**, 8, 3024, arXiv:1404.5630 [hep-ph].

Somogyi, G., Trocsanyi, Z., and Del Duca, V. (2007). A subtraction scheme for computing QCD jet cross sections at NNLO: Regularization of doubly-real emissions, *JHEP* **01**, 070, arXiv:hep-ph/0609042 [hep-ph].

South, D. M. and Turcato, M. (2016). Review of searches for rare processes and physics beyond the standard model at HERA, *Eur. Phys. J.* **C76**, 6, 336, arXiv:1605.03459 [hep-ex].

Stewart, I. W., Tackmann, F. J., and Waalewijn, W. J. (2010). N-jettiness: An inclusive event shape to veto jets, *Phys. Rev. Lett.* **105**, 092002, arXiv:1004.2489 [hep-ph].

Strikman, M. (2015). "Probing BFKL dynamics in *eh* scattering in large t semi-inclusive rapidity gap processes", Talk at the LHeC workshop at Chavannes-de-Bogis.

Stump, D., Pumplin, J., Brock, R., Casey, D., Huston, J., Kalk, J., Lai, H. L., and Tung, W. K. (2001). Uncertainties of predictions from parton distribution functions. 1. The Lagrange multiplier method, *Phys. Rev.* **D65**, 014012, arXiv:hep-ph/0101051 [hep-ph].

Sun, H. (2017). "Top Physics at the LHeC", Talk at the DIS workshop, Birmingham.

Sun, H. and Wang, X. (2016). Searches for the anomalous FCNC Top-Higgs couplings at the LHeC, arXiv:1602.04670 [hep-ph].

Susskind, L. (2003). The anthropic landscape of string theory, arXiv:hep-th/0302219 [hep-th].

't Hooft, G. and Veltman, M. J. G. (1979). Scalar one loop integrals, *Nucl. Phys.* **B153**, 365–401.

Tajima, T. and Dawson, J. M. (1979). Laser electron accelerator, *Phys. Rev. Lett.* **43**, 267–270.

Talamini, V. (2006). Affine-P-matrices in orbit spaces and invariant theory, in *8th International School on Theoretical Physics: Symmetry and Structural Properties of Condensed Matter (SSPCM 2005): Dedicated to Memory of Prof. Brian G. Wybourne Myczkowce, Poland, August 31–September 7, 2005*, arXiv:hep-th/0607165 [hep-th], [*J. Phys. Conf.* Ser.30, 30 (2006)].

Thorne, R., Harland–Lang, L. A., Martin, A. D., and Motylinski, P. (2015). MMHT 2014 parton distribution functions, in *Proceedings, 23rd International Workshop on Deep-Inelastic Scattering and Related Subjects (DIS 2015): Dallas, Texas, USA, April 27–May 01, 2015*, Vol. DIS2015, 056.

Thorne, R. S. (2006). A variable-flavor number scheme for NNLO, *Phys. Rev.* **D73**, 054019, arXiv:hep-ph/0601245 [hep-ph].

Thorne, R. S. (2012). Effect of changes of variable flavor number scheme on parton distribution functions and predicted cross sections, *Phys. Rev.* **D86**, 074017, arXiv:1201.6180 [hep-ph].

Thorne, R. S. (2014). The effect on PDFs and $\alpha_S(M_Z^2)$ due to changes in flavour scheme and higher twist contributions, *Eur. Phys. J.* **C74**, 7, 2958, arXiv:1402.3536 [hep-ph].

Thorne, R. S. and Roberts, R. G. (1998). An ordered analysis of heavy flavor production in deep inelastic scattering, *Phys. Rev.* **D57**, 6871–6898, arXiv:hep-ph/9709442 [hep-ph].

Trott, M. (2014). "Physics of the Higgs boson", Talk at the LHeC workshop at Chavannes-de-Bogis.

Tückmantel, T., Pukhov, A., Liljo, J., and Hochbruk, M. (2010). Three-dimensional relativistic particle-in-cell hybrid code based on an exponential integrator, *IEEE Trans. Plasma Science* **38**, 2383.

Vafa, C. (2005). The string landscape and the swampland, arXiv:hep-th/0509212 [hep-th].

van Hameren, A. (2011). OneLOop: For the evaluation of one-loop scalar functions, *Comput. Phys. Commun.* **182**, 2427–2438, arXiv:1007.4716 [hep-ph].

van Neerven, W. L. and Zijlstra, E. B. (1992). The $O(\alpha_s^2)$ corrected Drell–Yan K factor in the DIS and MS scheme, *Nucl. Phys.* **B382**, 11–62, [Erratum: *Nucl. Phys.* **B680**, 513 (2004)].

Veltman, M. J. G. (1977). Second threshold in weak interactions, *Acta Phys. Polon.* **B8**, 475.

Veneziano, G. (1968). Construction of a crossing-symmetric, Regge behaved amplitude for linearly rising trajectories, *Nuovo Cim.* **A57**, 190–197.

Vieira, J., Fang, Y., Mori, W. B., Silva, L. O., and Muggli, P. (2012a). Transverse self-modulation of ultra-relativistic lepton beams in the plasma wakefield accelerator, *Phys. Plasmas* **19**, 063105, arXiv:1206.0792 [physics.plasm-ph].

Vieira, J., Fonseca, R. A., Mori, W. B., and Silva, L. O. (2012b). The ion motion in self-modulated plasma wakefield accelerators, *Phys. Rev. Lett.* **109**, 145005, arXiv:1208.3242 [physics.plasm-ph].

Vilenkin, A. (1983). The birth of inflationary universes, *Phys. Rev.* **D27**, 2848.

Virasoro, M. A. (1969). Alternative constructions of crossing-symmetric amplitudes with regge behavior, *Phys. Rev.* **177**, 2309–2311.

Vogt, A., Moch, S., and Vermaseren, J. A. M. (2004). The three-loop splitting functions in QCD: The singlet case, *Nucl. Phys.* **B691**, 129–181, arXiv:hep-ph/0404111 [hep-ph].

von Weizsäcker, C. F. (1934). Radiation emitted in collisions of very fast electrons, *Z. Phys.* **88**, 612–625.

Voss, G. A. and Wiik, B. H. (1994). The electron proton collider HERA, *Ann. Rev. Nucl. Part. Sci.* **44**, 413–452.

Wallny, R. (2001). *A measurement of the gluon distribution in the proton and of the strong coupling constant α_s from inclusive deep-inelastic scattering*, Ph.D. thesis, Zurich U., https://inspirehep.net/record/572390/files/desy-thesis-01-058.ps.gz.

Wandelt, B. D., Dave, R., Farrar, G. R., McGuire, P. C., Spergel, D. N., and Steinhardt, P. J. (2000). Selfinteracting dark matter, in *Sources and detection of dark matter and dark energy in the universe. Proceedings, 4th International Symposium, DM 2000, Marina del Rey, USA, February 23–25, 2000*, 263–274, arXiv:astro-ph/0006344 [astro-ph], http://www.slac.stanford.edu/spires/find/books/www?cl=QB461:I57:2000.

Wang, X. *et al.* (2013). Quasi-monoenergetic laser-plasma acceleration of electrons to 2GeV, *Nature Comm.* **4**, 1988.

Weinberg, D. H., Bullock, J. S., Governato, F., Kuzio de Naray, R., and Peter, A. H. G. (2014). Cold dark matter: Controversies on small scales, *Proc. Nat. Acad. Sci.* **112**, 12249–12255, arXiv:1306.0913 [astro-ph.CO].

Weinberg, S. (1987). Anthropic bound on the cosmological constant, *Phys. Rev. Lett.* **59**, 2607.

Weinzierl, S. (2009). Event shapes and jet rates in electron-positron annihilation at NNLO, *JHEP* **06**, 041, `arXiv:0904.1077 [hep-ph]`.

Weinzierl, S. (2011). Jet algorithms in electron-positron annihilation: Perturbative higher order predictions, *Eur. Phys. J.* **C71**, 1565, `arXiv:1011.6247 [hep-ph]`, [Erratum: *Eur. Phys. J.* **C71**, 1717 (2011)].

Wilczek, F. and Zee, A. (1977). Discrete flavor symmetries and a formula for the Cabibbo angle, *Phys. Lett.* **70B**, 418, [Erratum: *Phys. Lett.* **72B**, 504 (1978)].

Williams, E. J. (1934). Nature of the high-energy particles of penetrating radiation and status of ionization and radiation formulae, *Phys. Rev.* **45**, 729–730.

Wing, M. and Caldwell, A. (2016). A proposed very high energy electron-proton collider, VHEeP, in *Proceedings, 24th International Workshop on Deep-Inelastic Scattering and Related Subjects (DIS 2016): Hamburg, Germany, April 11–15, 2016*, Vol. DIS2016, 248, `arXiv:1609.03907 [physics.acc-ph]`.

Witten, E. (2017). Symmetry and emergence, `arXiv:1710.01791 [hep-th]`.

Wolf, G. (2010). Review of high energy diffraction in real and virtual photon proton scattering at HERA, *Rept. Prog. Phys.* **73**, 116202, `arXiv:0907.1217 [hep-ex]`.

Yao, W. M. *et al.* (2006). Review of particle physics, *J. Phys.* **G33**, 1–1232.

Zel'dovich, Y. B., and Novikov, I. D. (1967). The hypothesis of cores retarded during expansion and the hot cosmological model, *Sov. Astr.* **10**, 602.

Zenaiev, O. *et al.* (2015). Impact of heavy-flavour production cross sections measured by the LHCb experiment on parton distribution functions at low x, *Eur. Phys. J.* **C75**, 8, 396, `arXiv:1503.04581 [hep-ph]`.

Zhao, Y. X., Deshpande, A., Huang, J., Kumar, K. S., and Riordan, S. (2017). Neutral-current weak interactions at an EIC, *Eur. Phys. J.* **A53**, 3, 55, `arXiv:1612.06927 [nucl-ex]`.

Zheng, X. *et al.* (2004). Precision measurement of the neutron spin asymmetries and spin-dependent structure functions in the valence quark region, *Phys. Rev.* **C70**, 065207, `arXiv:nucl-ex/0405006 [nucl-ex]`.

Zhu, S.-L., Sacco, G., and Ramsey-Musolf, M. J. (2002). Recoil order chiral corrections to baryon octet axial currents and large N_c QCD, *Phys. Rev.* **D66**, 034021, `arXiv:hep-ph/0201179 [hep-ph]`.

Zimmer, B. (1959). Language Log, 27 March 2007, "Crisis = danger + opportunity: The plot thickens".

Zimmermann, F. (2017). https://indico.cern.ch/event/556692/contributions/2483407/, "HE-LHC and FCC-eh — CDR plan and status", talk at the FCC week 2017, Berlin, 29 May–June 2 2017.

Zimmermann, F., Brüning, O., and Klein, M. (2013). The LHeC as a Higgs boson factory, in *Proceedings, 4th International Particle Accelerator Conference (IPAC 2013): Shanghai, China, May 12–17, 2013*, MOPWO054, `http://JACoW.org/IPAC2013/papers/mopwo054.pdf`.